Foreign Object Debris and Damage in Aviation

Foreign Object Debris and Damage in Aviation

Ahmed F. El-Sayed

CRC Press is an imprint of the
Taylor & Francis Group, an **informa** business

Cover image: © Remo Guidi/Stocktrek Images

First edition published 2022
by CRC Press
6000 Broken Sound Parkway NW, Suite 300, Boca Raton, FL 33487-2742

and by CRC Press
2 Park Square, Milton Park, Abingdon, Oxon, OX14 4RN

CRC Press is an imprint of Taylor & Francis Group, LLC

Library of Congress Cataloging-in-Publication Data
Names: El-Sayed, Ahmed F., author.
Title: Foreign object debris and damage in aviation / Ahmed F. El-Sayed.
Description: First edition. | Boca Raton, FL: CRC Press, 2022. |
Includes bibliographical references and index. | Summary: "The book discusses biological and non-biological Foreign Object Debris (FOD) and associated Foreign Object Damage (FOD) in aviation. Written for aviation industry personnel, aircraft transport and ground operators, and aircraft pilots, the reader will learn to manage FOD to guarantee air traffic safety with minimum costs to airlines and airports. Management control for the debris begins at the aircraft design phase, and the book includes numerical analyses for estimating damage caused by strikes. The book explores aircraft operation in adverse weather conditions and inanimate FOD management programs for airports, airlines, airframe, and engine manufacturers"– Provided by publisher.
Identifiers: LCCN 2021041151 (print) | LCCN 2021041152 (ebook) |
ISBN 9780367678418 (hbk) | ISBN 9780367678432 (pbk) |
ISBN 9781003133087 (ebk)
Subjects: LCSH: Foreign object debris (Aeronautics) |
Aircraft accidents–Prevention. | Aircraft bird striikes. |
Windborne debris. | Space debris. | Drones.
Classification: LCC TL553.545 .E43 2022 (print) |
LCC TL553.545 (ebook) | DDC 629.13–dc23/eng/20211103
LC record available at https://lccn.loc.gov/2021041151
LC ebook record available at https://lccn.loc.gov/2021041152

ISBN: 978-0-367-67841-8 (hbk)
ISBN: 978-0-367-67843-2 (pbk)
ISBN: 978-1-003-13308-7 (ebk)

DOI: 10.1201/9781003133087

Typeset in Times
by Newgen Publishing UK

Dedication

To the flowers in my life: my beloved mother and wife:
Fatma and Amany
To my sons: Mohamed, Abdallah, and Khalid
and
My daughter-in-laws: Hoda and Zahra, and my sweet
granddaughter Hana

Contents

Preface

The acronym "FOD" has two interrelated meanings: namely, Foreign Object Debris (FO Debris) and Foreign Object Damage (FO Damage). The first known FOD incident occurred in Ohio, United States, on September 7, 1905, when Orville Wright (one of Wrights Brothers) struck a bird near Dayton. However, the first victim of FOD was Callbraith Rodgers on April 3, 1912, when he flew into a flock of birds during an exhibition flight in Long Beach, California, United States. A gull got caught in his aircraft control cable, and the plane crashed into the ocean, where he died. Since then, hundreds of thousands of incidents/accidents and some hundreds of catastrophic accidents have occurred due to FOD. At present, the annual direct and indirect costs of FOD are nearly $14.0 billion. However, little has been done to understand and thwart the FOD dangers, other than investigating damage after each incident. Though thousands of reports and circulars dealing with FOD were published, no single comprehensive book was published about FOD and its management. Moreover, many aviation texts include only a brief review of FOD, which lack the depth necessary to understand the detection and mitigation of FOD at airports fully. Many other aviation texts do not even include superficial coverage of FOD, which places the aviation student at a disadvantage.

For these reasons, I concluded that it is time to write this book to fill a gap in the current library of information on FOD. It aims to provide a thorough understanding and analysis of FOD, its causes, and management. It will interest engineers, pilots, managers in airports and airlines, and commanders in airbases. I hope it will be attractive in different parts of the world since aviation is a global business and has universal challenges.

The book is a standard text for programs/courses desiring an in-depth review of FOD for industry and academia. For industry, it may be a text for flying training, airline and airport management, and air basis commanders. It could be a reference book for professionals in aircraft or helicopters dealing with failure analysis and design. It may be appropriate for any FOD courses organized by ICAO, FAA, EASA, IATA, and other aviation authorities. For academia, it will have an academic appeal and maybe a text for separate engineering graduate or undergraduate courses or separate senior design projects. It could be an accompaniment or supplement to graduate-level courses on aircraft design, aerodynamics, aircraft propulsion, helicopters, material analysis, numerical methods, airport management, pavement methods, etc.

My first involvement in FOD topics dates back to the 1970s of the last century during my ten years of work for EgyptAir Airline, Cairo airport, Egypt. Repair and maintenance of several aircraft and engine modules due to the FODs illuminated the impact of FOD on airlines' operation and revenue. This was coincident with my Ph.D. research in the Aerospace department, Cairo University, which handled solid particulate flows and the erosion of centrifugal compressors. Later on, I extended my erosion research to include axial compressors, internal combustion engines, pipe bends, axial fans of high bypass ratio (HBPR) turbofan engines, and propellers of turboprop engines. Next, I examined icing problems for aircraft frames and engines.

I kept a close eye on the performance deterioration of aircraft and engines due to sandstorms and snowstorms during writing my books: *Aircraft Propulsion and Gas Turbine Engines*, Taylor & Francis/CRC Press, 2008 and 2017, and *Fundamentals of Aircraft and Rocket Propulsion*, Springer, 2016. More recently, I added bird strikes and their damage to aircraft and engines in my book: *Bird Strike in Aviation*, Wiley 2019. I carried out most of my experimental and numerical work on FOD topics in Egypt, the United States, the UK, and Austria. I was honored to exchange expertise with many distinguished specialists in FOD areas in several institutions and organizations in the United States, Canada, Brazil, the UK, Belgium, Austria, Germany, France, Italy, Japan, China, Russia, India, Malaysia, Algeria, and Australia.

This book includes ten chapters that account for almost every facet of the subject. Chapter 1 is an introductory description of the topic. It starts by listing definitions for FOD given by several aviation authorities, including FAA, EASA, Australasian Aviation Ground Safety Council (AAGSC), Vancouver International Airport Authority, and military air forces. FO Debris is any substance or live animal close to the operation area of the aircraft. FO Damage is any damage to an aircraft, a missile, a drone, systems, and stores, or malfunction attributed to FO Debris. FO Damage can be expressed in physical (human injury or death) or economic (monetary) terms, downgrade the product's safety, stability, efficiency, or performance characteristics.

I proposed three different classifications for FO Debris. The first classifies them as animate (birds, mammals) or inanimate (fasteners, tire fragments). The second categorizes them as ground (reptiles, freight pallets) or airborne debris (bats, insects). The third divides debris into internal (within aircraft-like flight crew items) or external debris (mechanic's tools). Finally, debris may be classified based on the matter (material, colors, and size). Adverse weather like sand storms, volcanic ash, rainfalls, and snowstorms may generate FO debris. Helicopters rotor downwash also generates sandstorms or snowstorms depending on operation in deserts or snowy terrains. Unmanned aircraft vehicles (UAVs), or drones, and parts of space vehicles in an uncontrolled reentry also resemble a rather new FO debris.

FO Damage is the damage to aircraft and equipment and may cause human injury or death. FO Damage may cause trouble to airports, airlines, and military bases. The damage of flying vehicles may be categorized into four types: namely, acceptable with no maintenance requirements, acceptable with some minor repair, unacceptable with either in-site repair or module removal for shop repair, and finally catastrophic.

Chapter 2 lists many aviation accidents and incidents caused by FO Debris. The aviation industry is more determined to improve its safety record each successive year via daily training and research to keep air transportation the safest way to travel.

After listing several definitions for incidents and accidents by different aviation authorities and industries, statistics and a review of the details of some aviation accidents/incidents and their causes for the civilian, military aircraft, and helicopters are discussed. The catastrophic accidents of fixed-wing civil aircraft are categorized based on the different types of FO Debris. Fallen debris from a preceding aircraft led to the crash of Concorde in 2000. Birds led to aircraft ditching in the Hudson River. Mammals (especially deers) caused numerous catastrophic accidents. Rare accidents were encountered by other animates, including reptiles, insects, and humans. Wildlife, particularly birds, were responsible for more than 14,600 collisions with civil fixed-wing aircraft in 2018, resulting in an average of 40 strikes per day.

Adverse weather, including sandstorms, rainstorms, snow, hail, and volcanic ash, caused several accidents for civilian and military aircraft. Other debris related to military aircraft only is bomb fragments, fragments of bombed runways, and air refueling baskets. Helicopters suffered from debris, including birds, rainstorms, volcanic ash, maintenance tools, rags, and towels. Moreover, helicopters hovering or landing in desert and icy terrains may cause their crash. Spacecraft like Space Shuttle encountered bird strikes.

Studying the causes of air disasters helps engineers to design safer planes and prompts airlines and airports officials to initiate better air travel guidelines to improve aircraft safety and survivability.

Chapter 3 focuses on inanimate FO Debris generated by the ground operation of civil and military aircraft. Ground debris in airports and military bases depends on the type of runways. Runways are categorized as paved and unpaved or unimproved ones. Paved runways are either made of asphalt or concrete. Unpaved runways include gravel, coral, turf, sand, clay, hard-packed soil mixtures, grass, turf, dirt, and ice. FAA classifies runway conditions based on its contamination. Military bases are vulnerable to airstrikes and thus may have bomb debris or damaged concrete/asphalt of large sizes on their runways.

Ground operation of aircraft may generate debris in several cases. Tires rolling may threaten aircraft as it may loft debris that strikes aircraft frames and may be ingested into the engines. Also, tires may be cut and blown out into aircraft frames, as Concorde encountered in its last flight in 2000.

Ground run of engines develops dangerous areas upstream its intakes and downstream its exhausts as identified in their operation manual. A ground vortex may be generated, which extends from the ground up to their intakes. Such vortices are strong enough to pick up debris and push them into the engine. The exhaust gases resemble another hazard source that endangers both personnel and nearby aircraft. Thrust reversal may also push ground debris into the engines' intakes. Aircraft operation from wet runways will reduce the visibility of flight crews and influences the thrust force.

Chapter 4 treats inanimate debris caused by adverse weather and includes heavy rain, drizzle, snow, hail, fog, mist, haze, ice crystals, sand, dust, and volcanic ash. For example, the eruption of Eyjafjallajökull volcano on April 14, 2010 led to the cancellation of 108,000 flights and revenue loss of US$ 1.7 billion in the first six days. Concerning military operations, snowstorms, rainfalls, and sandstorms influenced the combat zones in Russia, Ukraine, Korea, Vietnam, the Persian Gulf, and Afghanistan. Statistics for weather-aviation incidents are first reviewed. The necessary inspection procedure if an aircraft encounters hazardous weather is described in detail. De-icing and anti-icing systems are reviewed. Circulars for aircraft protection against weather hazards like title 14 of the Code of Federal Regulations (14 CFR) issued by FAA are reviewed. Corresponding circulars issued by EASA and other aviation authorities are also discussed.

Chapter 5 defines a rather new category of FO Debris. These are UAVs or drones and space debris. They have brought the attention of aviation authorities in recent decades.

Drones have combined advantages and disadvantages for civil aviation and military operation. In civil applications, UAVs are used extensively in photography, mapping, surveying, wildlife management, an inspection of wind farms, agriculture, parcel delivery, and communications. They proved a great help during the COVID-19 pandemic. According to the latest industry studies, the unmanned aerial system (UAS) market volume reached 4.7 million units by 2020. For military applications, UAS has a noticeable contribution in communications and working as a team with military manned aircraft.

However, drones can pose hazards to aircraft and airports through the collision between them and manned aircraft and disrupt air transport in busy airports and hubs like Dubai and Gatwick airports. Numerical and experimental studies for collision between a UAV and a manned aircraft were studied by few universities, research institutes, and companies. FAA and EASA issued several circulars limiting the operation of drones close to airports.

Space debris is defined as anything in orbit that is man-made and is no longer in use. Based on NASA research, more than 500,000 space debris are orbiting the Earth. Typically, 200 to 400 big debris reenter Earth each year. Fortunately, until now, no human was struck by such reentered debris.

Chapter 6 describes animate (or biological debris) that includes humans, wildlife, insects, and grass. Strike of this animate debris with aircraft may damage the aircraft and result in the death and wound of living creatures. Most strikes are caused by wildlife, but in rare cases, humans and grass are sucked into aircraft engines. Statistics for wildlife strikes with civil and military aircraft are discussed. Birds are responsible for more than 90% of strikes, while mammals, reptiles, and rodents share the remaining 10%. The annual, monthly, daily, and hourly strikes are reviewed. Both critical flight phase and frequently impacted parts of aircraft are also discussed. A review of wildlife strikes with military aircraft is next described. Deer resembles the most dangerous mammals for both civil and military aircraft. Insects, grass, and humans rarely strike aircraft.

Chapter 7 describes how to control, prevent, and eliminate FOD in manufacturing processes (FOE). Different phases of the manufacturing process include design, machining, assembly, testing, inspection, packaging, shipping, receiving, repair, and all facilities and services operations. The design of aircraft frames, engines, components, systems, and instruments must ensure their protection from various FO Debris. Such threats necessitate the selection of costly material and high technological production methods which in turn reduce the product profitability. A review for design features like selection of composite materials, engine installation, de-icing and anti-icing systems of different civilian aircraft including Boeing 787, 737, Airbus A350, Antonov An225, Embraer 145, and military aircraft including B-2, F-16, Su-27, Mig-29 is discussed. Additional design and production features for helicopters like Chinook, Apache, including inlet particle separators and anti-icing systems, are employed to protect both rotor, windshield, and engine from icing. Next, during manufacturing processes, FOD prevention program based on FAA and EASA is employed. Also, during the transportation of subsystems to the final assembly destination, all precautions are made.

Chapter 8 identifies methods for management of inanimate (non-biological) FO Debris. Based on the National Aerospace Standards (NAS 412), most FO Damage is attributable to four factors: poor housekeeping, deterioration of facilities and equipment, improper maintenance, and inadequate operational practices. A review of the FOD management program by the Federal Aviation Administration (FAA), the FOD Control Corporation, the US National Research Council, and Boeing Company is given. Proactive-Reactive Measures for FOD are described. Detailed explanation for the management procedure proposed by FAA (being used by most airlines in worldwide airports) includes FOD prevention, detection, removal and evaluation is discussed in detail.

Chapter 9 discusses the management of animate or biological FO Debris by establishing the Wildlife Hazards Working Group (WHWG). WHWG should be ready to implement the FAA 14 CFR 139.337(a)–(c) to minimize any risks to aviation safety, airport structures or equipment, humans, and populations hazardous wildlife on and around the airport. Management of animate or biological FOD is performed via two control methods: namely, active and passive. Since airports attract animates due to the availabilities of life necessities such as water, food, and shelter, then passive control is achieved via habitat modification and exclusion. Active control is performed through repellent and harassment techniques and wildlife removal. Repellent includes chemical, audio, visual, falconry, trained dogs, radio-controlled aircraft, and nonlethal projectiles. Removal of wildlife includes capturing using chemicals, traps, and killing. Control methods for the most hazardous birds and mammals are identified. Training for airport personnel includes birds and mammals' identification, control, and reporting methods. Finally, the impact of COVID-19 on airports and airlines' operation and storage of aircraft is discussed.

Chapter 10 provides a review of numerical methods applied to particulate flow problems around aircraft. They may cause erosion, corrosion, fouling, built-up dirt, FO Damage, and icing problems, which will influence the aerodynamics, weight of aircraft, thrust force or power, and aircraft safety. Modeling of these particulate flows uses two numerical methods: namely, Eulerian and Lagrangian ones. For the Eulerian approach, the commercial codes FENSAP-ICE, CANICE, CIRAAMIL FLUENT, and CFX are used. For the Lagrangian approach, the codes LEWICE 2D/3D, ONICE 2D/3D, ONERA2D, and TRAJICE2 are used.

The first part of this chapter handles solid debris, including sand, dust, and volcanic ash. These may cause erosion and fouling of aeroengines. Erosion and fouling methods employ three steps: namely, calculation of air/gas flow, tracing of solid particle trajectories, and estimation of erosion or deposition rates. For a fixed-wing aircraft, the following case studies are discussed: the intake and fan of high bypass ratio turbofan engines like GE CF-6, axial and centrifugal compressors, axial and radial turbines, the propeller of turboprop engines like that installed to C-130 aircraft, and piston engines powering small aircraft. The role of secondary flow in modifying the erosion pattern is investigated. The erosion of the rotor of helicopters is reviewed in detail. Numerical modeling of fouling due to volcanic ash of compressors and turbines is presented.

The second part of the chapter focuses on liquid particles including rain and ice. Ice accretion is encountered when an aircraft flies through clouds below 26,000 ft at subsonic speeds. Ice accretion results from small (5 to 50 μm) supercooled droplets which can freeze upon impact with the aircraft surface. Rainfall on a wing having a NACA64-210 airfoil section is reviewed. Several case studies for ice accretion, including wing having airfoil section of NACA 23012, intake of turbofan engines like CF6-50, and Fan of GE-NASA energy efficient engine, are thoroughly examined. The icing of several propellers is also analyzed. An anti-icing analysis for NACA 23012 aircraft wing section involving hot air jets from a piccolo tube is analyzed.

I deeply acknowledge the support of distinguished professors and experts who exchanged their expertise through their valuable criticism and suggestions: Darrell Pepper, UNLV, Nevada, United States; C. Daniel Prather, California Baptist University, California, United States; and Philip Owen, To70 Aviation Australia, Australia.

Several universities, companies, and organizations supported my writing via their valuable publications and illustrations. These include Moscow Institute of Physics and Technology, Airbus SAS, atr-aircraft.com, The FOD Control Corporation, The World Bird Strike Association, The FOD*BOSS, BlueStreak Equipment Magnetics.

I was given great help by Sebastien Barthé, BEA Aerospace Inc., France, Gary Cooke, President of the World Birdstrike Association, United States, Saeeda Mohamed Ahmed, Air Accident Investigation, General Civil Aviation Authority, UAE, David Bragi, The FOD Control Corporation, United States, Axel Rossmann, Technical University of Graz (Tu-Graz), Austria, Barbara Murphy, National Academies Press, United States, Mary Ann Muller, Taylor & Francis, United States, Steven Calvillo, TYMCO, Inc, United States, Igor Bosnyakov, and Joseph Gourgy, Moscow Institute of Physics and Technology (State University) – MIPT, Phystech, Russia, Phil Shaw, Avisure: aviation, wildlife, safety, Australia.

I deeply acknowledge the support of Ahmed Hamed Elakel, Senior Technical Service Engineer, SANAD Aerotech, Abu Dhabi (UAE), who devoted plenty of his time reviewing the manuscript. Mohamed El-Sayed, CEO of EA Energy Solutions (United States), Islam Mohamed Asoliman Zaid, Khalifa University, Abu Dhabi, (UAE), Abdallah El-Sayed, Prevedere Inc (United States), and Mohamed Ahmed Aziz, Suez University (Egypt) sacrificed their valuable time for preparing the manuscript.

The continuous support and elegant management of Nora Konopka; Editorial Director of Engineering; Taylor & Francis is deeply acknowledged and appreciated. My editor Kyra Lindholm has been a great help since day one.

1 Introduction

Foreign Object Damage (FO Damage) is defined as any damage to an aircraft, a missile, a drone, their systems and stores, or malfunction attributed to a Foreign Object Debris (FO Debris). FO Damage can be expressed in physical (human injury or death) or economic (monetary) terms, which may downgrade the product's safety, stability, efficiency, or performance characteristics.

Different classifications for FO Debris are defined. FO Debris may be classified as animate (biological) or inanimate (non-biological) objects. It may be also defined as ground or airborne debris. Moreover, it may be defined as internal (to aircraft) or external debris. Finally, it may be classified based on matter (material, colors, and size).

Inanimate (inorganic or non-biological) foreign object debris include personal items like coins, pens, mobiles, watches, eyeglasses, hats, gloves, soda cans, paperwork, and others. FO debris may also include aircraft and engine parts like fasteners, fuel caps, oil sticks, metal sheets, and technician tools like wrench, socket, hammer, etc. Other debris related to the airport may include lights, signs, rocks, sand, pebbles, and loose vegetation. Moreover, FO Debris may include ground support equipment (GSE) like maintenance equipment, fueling trucks, tow tractors, and catering supplies like spoons, food wrappers, freight pallets, and baggage tags and pieces. In addition, adverse weather may add additional debris like hail, snow, rain, sand, and volcanic ash. Concerning military airbases debris may contain bomb-damaged runways, bomb fragments, parts of a refueling basket during refueling process. Recently, FO Debris included drones, Unmanned Aircraft Vehicles (UAVs), and parts of space vehicles that reenter the Earth orbits like the core stage of the Chinese Long March 5B rocket that crashed in the Indian Ocean north of the Maldives on May 8, 2021. Finally, helicopters flying close to the ground may generate sandstorms if flying in deserts or snow and icy particles when flying near snowy terrains.

Animate foreign object debris (living organisms) may be airborne like birds and bats, insects, or ground ones, including mammals (deer, moose, etc.) and reptiles (crocodiles, alligators, and snakes), grass, and plant fragments. The most precious debris is humans who are in rare cases sucked into aeroengines.

Foreign Object Damage is the damage caused by the foreign object debris to aircraft parts and equipment and may cause human injury or death. Damage to aircraft

DOI: 10.1201/9781003133087-1

parts could be in the form of punctured airframes, torn or punctured tires, worn compressors and fans, nicked turbine, or propeller blades, and, in rare instances, complete damage to aircraft or engine failure. The injury to airport employees may be caused by debris propelled by the jet blast, propeller, or rotor wash. FO Damage may cause trouble to airports, airlines, and military bases. In brief, the damage of flying vehicles may be categorized into four types namely: acceptable with no maintenance requirements, acceptable with some minor repair, unacceptable with either in-site repair or module removal for shop repair, and catastrophic.

1.1 INTRODUCTION

Since the first man-powered flight in 1903, the aviation industry recorded the fastest-growing transportation field. With the increasing world population and high-flying rate, the aviation industry faces a great financial loss due to aircraft crashes. Detailed investigation of these air crashes reveals many reasons. One of them is FOD.

In the aviation industry, the acronym "FOD" has two interrelated meanings, namely:

- Foreign Object Debris
- Foreign Object Damage

Foreign Object Debris – is any object ranging from small fasteners, rocks to luggage racks and wild animals. Adverse weather may add additional debris like hail, ice, snow, sand, and volcano ash, which may cause damage to aircraft, equipment, or people. Foreign object debris at airports includes any object found in an inappropriate location. It is available at terminal gates, aprons, taxiways, runways, and run-up pads. Foreign object debris is sometimes abbreviated as FO Debris.

Foreign Object Damage – is the damage caused by the foreign object debris to equipment and aircraft parts and human injury or death. Damage to aircraft parts may be in the form of torn or punctured tires, punctured airframes, nicked turbine, or propeller blades, and, in rare instances, complete damage to aircraft or engine failure. The injury to airport employees may be caused by debris propelled by the jet blast, propeller, or helicopter's rotor wash. Similarly, foreign object damage is sometimes abbreviated as FO Damage.

The impact of foreign object debris (FOD) remains a major concern in the aviation, defense, and space industries. It resembles a major potential source of risk for regulators and customers. Damage may influence both airframe and engines (or powerplants) of aircraft.

FOD may influence both fixed-wing aircraft and rotary-wing vehicles (helicopters). Moreover, both seaplanes and amphibious aircraft are vulnerable to FOD. Also, FOD may affect satellite, satellite launch vehicles, missiles, and spacecraft.

- Examples for civilian fixed-wing aircraft are: Boeing series (B-787, 777, 737), Airbus series (A320, 330, 380), Embraer series (Embraer-170, 190, 195), Tupolev series (Tu-124, 34, 154)

- Examples for military fixed-wing aircrafts: F-15, 16, 18, 22, 35, C-130, Airbus A400, Rafale A, B, C, Mig-21, 29, Sukhoi Su-25, 27, 30, 33, 37, 47, Shenyang J-31, Comac C919
- Examples for seaplanes are the Float boats Grumman G-111 Albatross and De Havilland Canada DHC-6 Twin Otter, and amphibious aircrafts ASW A-40 and Be-200
- Examples for civilian helicopters Airbus Super Puma, H125, H130, H135
- Examples for military helicopters MD Helicopters MH-6 Little Bird, Boeing-Sikorsky RAH-66 Comanche
- Examples for aero-engines are Rolls Royce Trent Family, General Electric GE90, Pratt & Whitney PWA4000, Russian PD-14, Kuznetsov NK-32.02
- Examples for space crafts are Space shuttle Atlantis and Discovery

Lack of FOD prevention plans often leads to significant problems with the quality or cost of the delivered product and ultimately threatens people's lives.

The annual direct cost of FOD to the US aviation industry is $474 million and $1.26 billion to the global aviation industry. Direct plus indirect costs, such as flight delays, damaged equipment, reduced efficiency litigation, and others, cost the US aviation industry $5.2 billion annually and the global aviation industry $13.9 billion annually [1–3]. Owing to such extensive and costly damage of FOD, detecting and removing FOD from an Air Operation Area (AOA) is an extremely critical task for ensuring safety.

FOD also influences aviation safety. Several catastrophic accidents encountered by aircraft are due to foreign object debris. An example is the Air France Concorde flight # 4590 on July 25, 2000, departing Charles de Gaulle Airport, Paris to JFK International in New York (Figure 1.1). The right front tire of the Concorde aircraft

FIGURE 1.1 Concorde aircraft in the fire before the crash in Paris [4].

FIGURE 1.2 Titanium alloy strip fallen from the engine of DC-10-30 aircraft [4].

ran over a titanium alloy strip, as shown in Figure 1.2. That strip fell from the engine of a Continental Airlines McDonnel Douglas DC-10-30 aircraft, which took off from the same runway 5 min earlier [4]. The debris cut the tire and sent a large chunk of the tire (4.5 kg) into the underside of the aircraft's left wing at an estimated speed of 140 m/s, which punctured a fuel tank. The two engines (#1 and #2) caught fire and stalled. Finally, the aircraft crashed into a hotel in nearby Gonesse 2 min after takeoff, killing all 109 people aboard and four people in the hotel, with another person in the hotel critically injured.

1.2 DEFINITIONS FOR FOREIGN OBJECT DEBRIS

The term Foreign Object Debris (FOD) has several definitions.

- Based on the Federal Aviation Administration (FAA) [5]

  "Any object, live or not, located in an inappropriate location in the airport environment that can injure airport or air carrier personnel and damage aircraft."
- Based on the European Aviation Safety Agency (EASA) [6]

  "An inanimate object within the movement area of an aerodrome which has no operational or aeronautical function, and which has the potential to be a hazard to aircraft operation."
- Based on the Australasian Aviation Ground Safety Council (AAGSC) [7], p. 21

  "Any object that is left in an area where it could cause damage. Such debris includes, but is not restricted to, metal (e.g., tools, nuts, bolts, and lock wire), wood, stones, pavement fragments, sand, plastic wrapping, and paper."
- Based on Vancouver International Airport Authority [8]

  "Any substance, part, component, natural element or live animal that, because of its proximity to the area of the aircraft in motion, has the potential to encounter an aircraft accidentally and threaten its safe operation and or require repair."

  Other definitions frequently found in the open literature for foreign object debris are:

- A substance, debris, or article alien to a vehicle or system which would potentially cause damage to aircraft, equipment, or people [9].
- Any object, particle, substance, debris, or agent is not where it is supposed to be [2].
- A substance, debris, or article alien to a vehicle or system that could potentially cause damage. It's *anything – large or small* – inside or around aircraft and Flight line operations *that do not belong there* – which could create a hazard to equipment or personnel [10].
- Anything that is where it does not belong on the runway, taxiway, or ramp area [11].
- Any alien substance or particle that invades any component of an aircraft that causes or has the potential to cause damage to aircraft, persons, or otherwise diminishes safety [1].

Foreign Object Debris is abbreviated as (FOD) by Aerospace Industries Association (ASA) [12].

In airports or similar aviation environments, Foreign Object Debris could create a hazard to aircraft, equipment, cargo, personnel, or anything else of value. While in a manufacturing or similar environment, FOD could contaminate the product or otherwise undermine quality control standards or injure personnel.

Foreign Object Debris (FOD) is available in many different forms and comes from many different sources. These belong to the following categories:

- Apron items (debris from catering including paper and plastic items, freight pallets and luggage parts, as well as debris from ramp equipment);
- Runway and taxiway materials (asphalt and concrete chunks, rubber materials, and paint chips);
- Construction debris (pieces of wood, stones, fasteners, and miscellaneous metal objects);
- Plastic or polyethylene materials; and
- Contaminants from weather conditions (snow, ice, rain, hail, sand, volcanic ash).

1.3 DEFINITIONS FOR FOREIGN OBJECT DAMAGE

The term Foreign Object Damage (FOD) also has several definitions.

- Based on the Federal Aviation Administration [5]

  "Any damage attributed to a foreign object that can be expressed in physical or economic (monetary) terms which may or may not downgrade the product's safety or performance characteristics."
- Based on the Air Force definition [13, 14]

FOD is damage to, or malfunctioning, an aircraft, a missile, a drone, their systems, or stores. Damage may be caused by an alien object to an area or system, being ingested by, or lodged in a mechanism. FOD may cause material damage, or it may make the system or equipment unusable, unsafe, or less efficient.

Other definitions are listed hereafter:

- *Damage caused by the foreign object (that 99% of the time was missed or ignored by you or me).*
- Any damage is done to aircraft, helicopters, drones, launch vehicles, or other aviation equipment by foreign object debris entering the engines, flight controls, or other operating systems [10].
- *Damage caused by Foreign Object Debris that compromises the quality, functionality, or economic value of a manufactured item* [2].

At best, Foreign Object Damage will cost only money, however, at worst, it results in injury or death.

1.4 CLASSIFICATION OF FOREIGN OBJECT DEBRIS AS ANIMATE AND INANIMATE (INORGANIC) CATEGORIES

Foreign Object Debris is either animate (biological) or inanimate (non-biological) objects. Figure 1.3 illustrates such a classification.

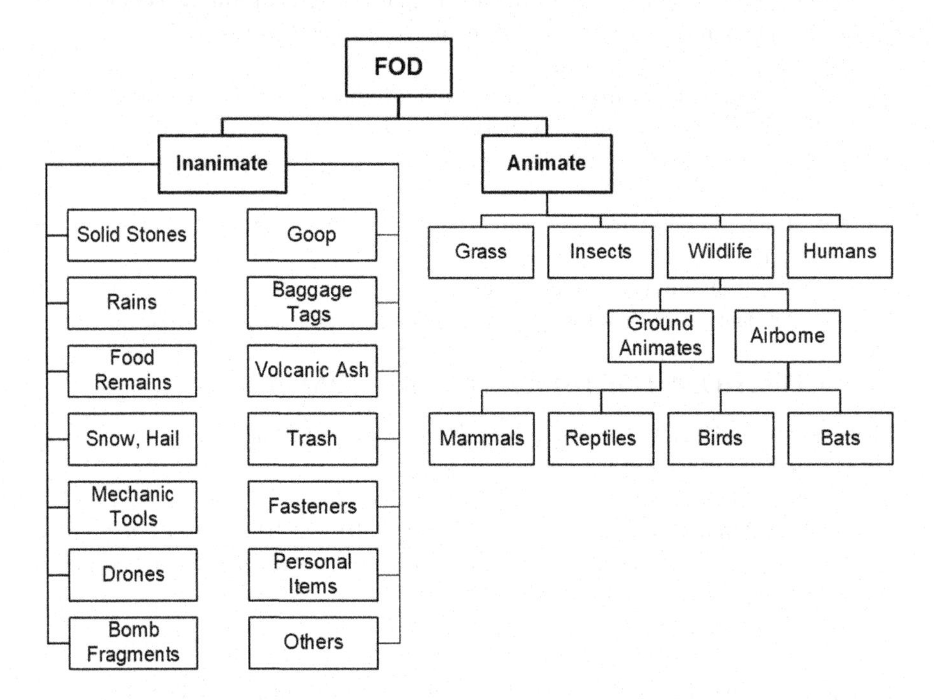

FIGURE 1.3 Classification for Foreign Object Debris based on animate and inanimate categories.

1.4.1 Examples for Inanimate (Inorganic or Non-Biological) Foreign Object Debris

As listed in [5, 13, 14], they include:

- Personal items (jewelry, sunglasses, flashlights, metal shavings, coins, paperclips, staples, pens and pencils, ID cards, mobiles, watches, eyeglasses, identification badges, tape, cameras, hairpins, keys, clipboards, hats, rags, gloves, soda cans, water bottles, caps, cups, paperwork, and any other object that airport or airline personnel can carry
- Mechanics' tools/shop aids (wrench, socket, hammer)
- Aircraft and engine parts (fasteners, washers, safety wire, fuel caps, landing gear fragments, oil sticks, metal sheets, trapdoors, and tire fragments)
- Airport infrastructure (lights and signs)
- Fragments of a broken pavement of runways, taxiways, and apron
- Ground Support Equipment (GSE) operating on the airfield (aircraft, airport operations vehicles, maintenance equipment, fueling trucks, tow tractors, and diesel power units, other aircraft servicing equipment, and construction equipment).
- Rubber deposits
- Goop – grease, paint
- Spilled or splashed corrosive liquids
- Catering supplies (paper and plastic debris, spools and forks, food wrappers, and beverage containers)
- Trash (plastic wrap, used tape)
- Freight pallets, baggage tags, and pieces of luggage
- Parts dropping off the plane in front of the next flying plane on the taxiway
- The environment (rain, hail, snow, and ice)
- Rocks, mud, dirt, dust, sand, pebbles, and loose vegetation
- Volcanic ash
- Bomb damaged runways
- Bomb fragments
- Parts of a refueling basket during refueling of military aircraft
- Drones and Unmanned Aircraft Vehicles (UAV)

FODs can be relocated by the airliners' operation in the following cases [10]:

- The jet blast of aircraft engines during taxiing can blow FOD into personnel and aircraft.
- Transitions from a relatively large-width runway onto a smaller-width taxiway. In this case, the outboard engines may blow any loose sand and materials from the shoulder and infield areas onto the runway.
- The outboard engines of four-engine aircraft can move debris from the runway edge and shoulder areas and accumulate it toward the center of the runway or taxiway.

FIGURE 1.4 The helicopter's rotor downwash [52].

Operation of helicopters may also generate additional FODs in the following cases:

- **Maneuver** over freshly mowed or loose-dirt areas can move FOD onto runways, taxiways, and ramps [15]. The helicopter's rotor downwash (Figure 1.4) can
  - blow any loose sand and materials
  - splash seawater (Figure 1.5)
  - propel lightweight ground support equipment (GSE), or materials staged nearby.

The effect of downwash from a hovering Sikorsky Seahawk is visible on the surface of the water below (Figure 1.5).

Volcanoes may threaten both helicopters and fixed-wing aircraft flying close to them.

Drones or UAVs operating close to airports may collide with aircraft during takeoff and landing.

1.4.2 Examples for Animate Foreign Object Debris (Living Organisms)

- Birds and bats
- Insects
- Mammals (deer, moose) and reptiles (crocodiles, alligators, and snakes)
- Grass and plant fragment
- Humans

FIGURE 1.5 The splashing seawater due to downwash from a hovering Sikorsky Seahawk close to the surface [50].

FIGURE 1.6 A flock of birds threatening an Airbus A330 aircraft.

In the past decade, the FAA has recorded a total of 85,998 wildlife strikes across the United States, while another reasonable number of wildlife strikes go unrecorded. More than 95% are due to birds' and bats' strikes, as reviewed by the author [16].

Figure 1.6 illustrates a flock of birds surrounding an Airbus A330 of China Eastern at London Heathrow.

Numerous mammals, including deer [17], coyotes, raccoons, skunks, opossums, foxes, etc., may also result in aircraft accidents if living close to airports.

Reptiles like turtles, alligators, and crocodiles are another group of wild animals that may collide with aircraft.

As documented by the FAA, some 25 people were killed due to airplane wildlife strikes between 1990 and 2013.

Figure 1.7 illustrates Cessna C550-Citation II aircraft [17] in fire after hitting a white-tailed deer during landing at an airport in South Carolina in 2012. The collision ruptured a fuel tank, sparking a fire that destroyed the plane. The flight crew escaped unharmed.

Figure 1.8 illustrates a bald eagle grasping a fish and flying back to his nest on January 18, 2017. However, this fish dropped and collided with a Gulfstream G-IV aircraft during its takeoff roll [18].

Insects, especially bugs hitting the leading edge of the wing and tail during takeoff or landing, could decrease aircraft performance and increase fuel consumption. These little blasts of bug guts disrupt the smooth laminar airflow over the wings/tails, creating more drag on the airplane and contributing to increased fuel consumption.

FIGURE 1.7 Cessna C550-Citation II aircraft after hitting a deer during landing [17].

Courtesy: DoD, FAA.

FIGURE 1.8 Bald eagles that fly in the vicinity of airliners and losing their grip.

Permission from Brian E. Kushner.

An extensive study at Schiphol Airport, Amsterdam, was carried out by a special team who performed a weekly inspection within one year on eight aircraft to quantify the number of insect strikes on the thorough leading edges (Figure 1.9).

Another group of researchers at NASA's Langley Research Center – the "bug team" – recently ran several flight tests of coatings that may reduce the amount of bug contamination on the wings of commercial aircraft [19].

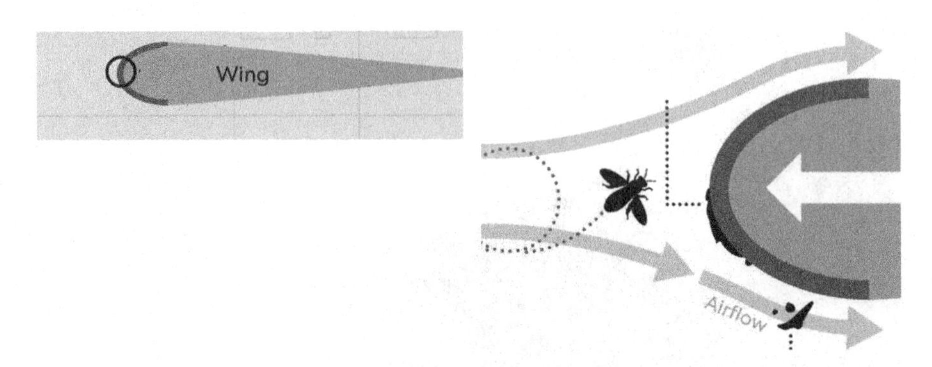

FIGURE 1.9 Insect strike with the leading edge of aircraft; Schiphol Airport.

FIGURE 1.10 Female keyhole wasp (*Pachadynerus nasiden*) on 3-D printed DHC8 probe, Brisbane Airport, May 2016 [20].

With permission of "Brisbane Airport Corporation and Ecosure."

For floatplanes operating in the buggy environment, it is often necessary to clean the wing's leading edge before liftoff.

A new study was performed in Australia concerning the mud-nesting keyhole wasp (Pachodynerus nasidens), which views aircraft pitot probes as an attractive nesting opportunity at Brisbane airport [20]. Since pitot probes measure aircraft speed and altitude, then its obstruction by foreign objects such as insects may cause inaccurate airspeed-sensing and altitude-sensing signals, which may lead to loss of safe flight (Figure 1.10).

1.5 CLASSIFICATION OF FOREIGN OBJECT DEBRIS AS AIRBORNE AND GROUND DEBRIS

Another categorization for FOD (debris) is illustrated in Figure 1.11 as ground, airborne, and space objects.

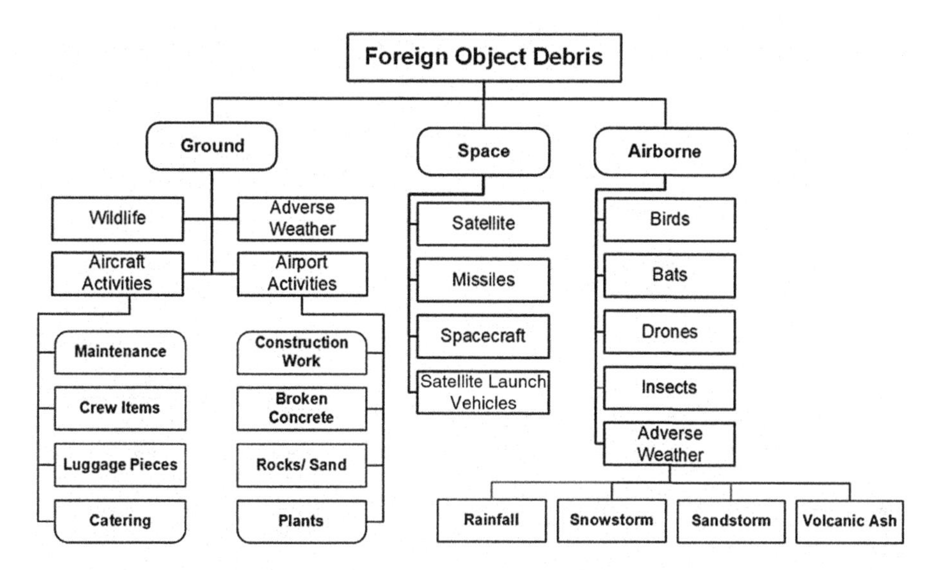

FIGURE 1.11 Classification for Foreign Object Debris based on ground, airborne, and space categories.

1.5.1 Ground Debris

Ground objects include humans and wildlife. Other ground objects are generated from aircraft activities like maintenance and overhaul (tools fasteners), crew items (pens, keys, hats, eyeglasses), luggage pieces like tags, portable electronic devices (PEDs), paint chips, and metal shavings.

Missing parts and missing tools are maintenance foreign object debris, which normally causes accidents [21, 22].

Ground objects may also be generated from airport activities like construction work, as well as any broken runways and taxiways, rocks/sand, and plants.

Catering and remains of flight meals, food wrappers, and plastic forks and spoons which during loading and unloading could be ground foreign object debris.

1.5.2 Airborne Debris

Airborne debris may include living creatures or inanimates including drones, UAVs, and space debris. Flying creatures may include birds, bats, and insects. Drones or small unmanned aerial vehicles (UAV) are new airborne objects threatening aircraft flying close to airports (in climb and descent flight phases).

Several airborne debris is associated with adverse weather conditions, including solid particles like sand, dust, and volcanic ashes. Others include liquid particles like rainfalls and snowstorms.

When airplanes encounter sandstorms, or while taking off/landing on unpaved runways, sand particles may collide with them. Some aircraft like C130 and Boeing 737-200 can land on unpaved runways. Flight in sandstorms is very challenging for pilots as they result in low visibility ($\pm 800m$) like that in very thick fog. Moreover,

FIGURE 1.12 A helicopter operating in a sandstorm [23].

Courtesy: US Army.

sand particles will be ingested into the core of the aircraft engine and abrade the fan or the compressor blades, which results in power reduction and poor efficiency.

Helicopters may be subjected to more severe conditions when flying close to the ground in deserts. Rotor downwash generates a severe sandstorm that influences the helicopter itself and any nearby vehicles, objects, or people. Figure 1.12 illustrates the threat of sandstorms to helicopters [23].

When a helicopter is hovering or landing on snowy terrain, snow and icy particles blow everywhere.

1.5.3 Space Debris

Since man first ventured into space (in the 1950s), space has been riddled with debris. Thousands of satellites and space vehicles are orbiting space in the 21st Century. Moreover, since hundreds of satellites are uncontrollable or leftover, there will be an increased probability of collision at an enormous velocity between a satellite and another or space vehicle. Each collision results in hundreds or even thousands of pieces of debris that do not immediately fall to Earth but continue to orbit around it.

In 2016, NASA managed to track more than 500,000 pieces of debris (or space junk) orbiting the Earth at speeds of more than 28,000 km/h; refer to Figure 1.13. They range from the size of the tennis ball to a tiny marble [24–26].

An example of such a strike was encountered in April 2016, where a small piece of orbiting junk hit the International Space Station (ISS), carrying six astronauts at a hypersonic speed. The small object impacted one of the observation windows of the

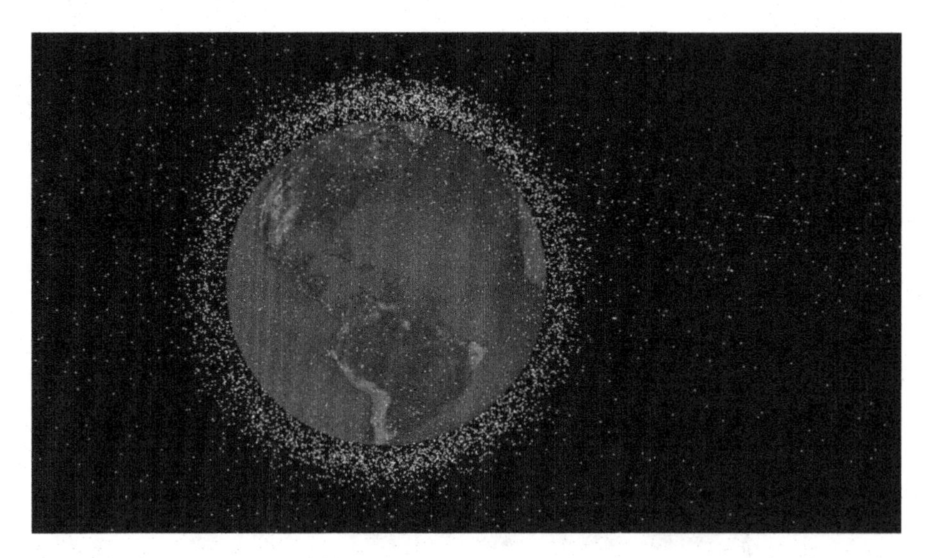

FIGURE 1.13 Space debris [26].

Courtesy: NASA.

Cupola module added to the ISS in 2010 for monitoring Earth and celestial objects. Such a glass window was chipped by the collision resulting in 7 mm in diameter [25].

Also, on August 23, 2106, a solar panel on the Sentinel-1A satellite, which was in a routine environmental monitoring mission, had been hit by too small debris to be tracked [24]. After spacecraft (or parent body) breakup, individual components, or fragments, will continue toward the Earth and heated due to aerodynamic friction until they either demise or impact the Earth.

Through a controlled entry, and due to aerodynamic forces, spacecraft will break up at altitudes between 84 and 72 km. Large, dense satellites generally break up at lower altitudes. Solar arrays will break off their parent spacecraft around 90–95 km.

According to NASA and ESA (European Space Agency), an average of uncontrolled re-entry debris (one to two) cataloged pieces of debris has fallen back to Earth each day since 1959. Some have a cross-section of one square meter or more.

Well-known examples of large-mass uncontrolled re-entries:

- Skylab (74 tons, re-entry July 1979)
- Salyut-7/Kosmos-1686 (40 tons, re-entry February 1991)

Up to date, there have been no known injuries resulting from re-entering space debris. However, the risk of re-entry debris is from mechanical impact, chemical or radiological contamination to the environment [27–32].

Figure 1.14 displays the main propellant tank of the second stage of a Delta II launch vehicle, which landed near Georgetown, TX, on January 22, 1997. It is an approximately 250-kg tank primarily a stainless-steel structure and survived reentry relatively intact [31].

Examples of recovered satellite components.

FIGURE 1.14 Recovered satellite components [31].

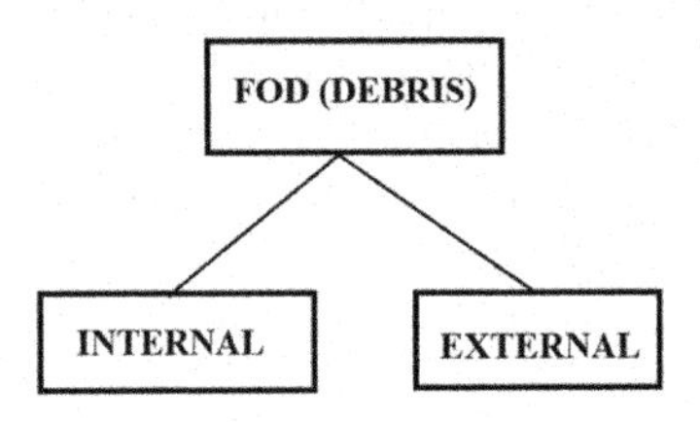

FIGURE 1.15 Classification for Foreign Object Debris based on external and internal sources.

1.6 CLASSIFICATION OF FOREIGN OBJECT DEBRIS AS EXTERNAL AND INTERNAL SOURCES

Figure 1.15 displays another categorization for foreign object debris based on existence and impact threat to the aircraft. External sources include animate and inanimate (inorganic) sources, while internal sources are mostly inanimate (non-biological).

1.6.1 External FOD

They include items outside the aircraft, such as birds, bats, mammals, snakes, alligators, hail, ice, snowstorms, sandstorms, volcanic ash, or objects left on the runway.

Figure 1.16 illustrates a bird strike with a helicopter. The windshield is the most critical part vulnerable to bird strikes [33].

FIGURE 1.16 Bird strike with a helicopter [51].

Courtesy: USDA-APHIS.

FIGURE 1.17 A mammal is approaching an F-16 aircraft.

Courtesy: USDA [35].

Aircrew must carefully search any openings, cavities in the airplane, as well as jet engines in their pre-flight checks. Virgin Australia engineers discovered an owl hidden in the aircraft's engine [34] on April 17, 2019.

Mammal strike contributes to a reasonable percentage of FOD damage. For example, at GA airports, between 1990 and 2007, mammal strikes accounted for 14% of all reported strikes occurring at GA airports (Figure 1.17) [35].

FIGURE 1.18 Hail damage to T-1A Jayhawk [49].

Courtesy: USAF.

Hail is another external source for foreign objects. There are two types of hails, namely, small and large size ones. The dangerous problem of hail is that it is invisible until the pilot runs into it.

Airplanes traveling at high altitudes (30,000 ft+) and speeds close to 500 miles/h if struck by chunks of ice will be in great danger. Windows will be cracked, but structurally intact is a testament to their engineering design. A complete windscreen failure would mean a high probability of loss of the aircraft.

Figure 1.18 illustrates the hail damage to T-1A Jayhawk military version of the Beech 400A twin-engine jet trainer. It was grounded after being severely damaged in a Texan hailstorm. The average repair cost was $3.4 million.

1.6.2 Internal FOD

During assembly, maintenance, and repair processes, FOD can drop into and remain suspended in the radial crevasses of an airframe remaining secure and undetected during a pre-flight inspection. Such internal FOD include:

- Items inside the aircraft, such as those which get loose in the cockpit. It can jam or restrict the operation of the control cables. Moreover, it may short-out electrical connections, overheat avionics and navigation systems. More dangerously, it may cause a puncture in the aircraft's skin and block airflow to critical components.
- Tools mistakenly left inside the aircraft after manufacturing or servicing.

Both groups can influence aircraft flight safety. Military aircraft are especially prone to these critical coincidences due to their wider flight envelope. Also, negative (G) forces in flight may dislodge debris, causing it to interact with the aircraft's systems.

Any unsecured items in the cockpit, like pencils, pens, and the relief jar, which during sharp maneuvers may be thrown away [36]. Both fighter airplanes and aerobatic ones may encounter such an inside FOD.

1.7 CLASSIFICATION OF FOREIGN OBJECT DEBRIS BASED ON MATTER

Many types of FOD vary in material, colors, and size. In general, there are four classes of FOD, namely,

- Metal
- Stone
- Birds/bats
- Miscellaneous

Moreover, debris can be categorized into two groups only, namely, soft, and hard bodies. Soft debris is like birds, bats, ice slabs, and plastics, while hard debris is like stones, fasteners, tools, etc.

A one-year airport study for the collected known FOD items revealed that [37]

- Based on Material: over 60% of the items were made of metal, followed by 18% of the items being made of rubber
- Based on color: dark-colored items made up nearly 50% of the FOD collected
- Based on size: a common FOD was 1 in by 1 in (2.50 cm x 2.50 cm) or smaller

1.8 FOD CAUSES

The causes of Foreign Object Damage are plenty. Hereafter, some of them:

- Poor awareness of the actual hazard of FOD [38, 39]
- Poor maintenance of aircraft, building, and equipment [40]
- Deterioration of aircraft maintenance facilities [41]
- Fueling and construction equipment [42]
- Improper tools
- Leaving tools on the ground after working
- Inadequate staff training
- Additional unscheduled workloads [43]
- Insufficient housekeeping, training, or controls
- Fatigue of staff not to delay aircraft schedule for inspection [44]
- Failure to do rollover FOD checks at entry control points onto the airfield
- Personnel and material hazards [43]
- Breaking and chipping of pavements

- Signboards and lights
- Mowers on the infield
- Weather hazardous conditions icing, snow, hail, sand
- Winds blowing debris around the airfield
- Lightning strikes the pavement
- Birds/bats collision
- Wildlife collision
- Debris Lofting by Tires
- Inlet vortex formed between a jet engine inlet and ground surface at certain conditions [45]
- Jet engine blasts
- Propeller slipstreams
- Thrust reverser action especially for aft-mounted engines, which can blow any loose material on the runway forward into the engine inlet
- Presence of uncontrolled, for example, contractors' vehicles on the airfield [46]
- Hovering of helicopters close to ground or sea
- Bomb fragments

1.9 FOD DAMAGE CATEGORIES

Damage of flying vehicles may be categorized into the following four types:

1. ***Acceptable*** with **no** maintenance requirements
2. ***Acceptable*** with some **minor repair**
3. ***Unacceptable*** with either in-site repair or module removal for shop repair
4. ***Catastrophic***, where aircraft/engine failure leading to an aircraft accident

1.10 IMPACTS OF FOD

Foreign Object Debris greatly influences airports, airlines, military bases, and their staff and passengers.

Damages influence both aircraft, airport, and military bases as well as in contact personnel.

1.10.1 Aircraft and Engine

The aircraft problems caused by debris [43] may be summarized as follows:

- Catastrophic damage and loss of aircraft like the Air France Concorde, flight # 4590 on July 25, 2000 (see Chapter 2 for detailed accidents and incidents) [47, 48].
- Damaging aircraft systems
- Damaging aircraft jet engines or their modules (intake, fan, compressor, etc.) if debris is ingested
- Nicking or damaging propellers of piston or turboprop engines
- Cutting, tearing, puncturing, or blowout of aircraft tires

- Jamming flight control mechanisms when lodged next to levers and handles
- Damaging delicate components when trapped inside of an equipment housing
- Puncturing or damaging airframe parts (landing gear, wing, fuselage, tail, etc.)
- Failure to retract the landing gear
- Premature wear of seals causing leaks
- Premature wear of internal components

1.10.2 Airport, Airline, and Military Bases

FOD can lead to the following troubles:

- Unplanned landings
- Delayed and canceled flights as well as schedule disruptions
- Damage to airport buildings or nearby premises
- A significant threat to the safety of air travel if encountered in the air operation area (AOA)
- The financial impact to the organization due to a massive amount of direct and indirect costs
- Reducing air defense and other mission capabilities which can affect national security

1.10.3 Personnel

FOD have severe influences on personnel, including:

- Catastrophic loss of life if encountered during critical phases of flight
- Injuries of employees and passengers after being propelled by a jet blast or prop wash
- Injuries of employees or workers in factories associated with aerospace industries
- Injuries of the pilots if penetrated the windshield
- Injuries, killing, maiming, producing orphans and puts folks out of work
- Increased maintenance and repair time
- Individuals and corporations may take their business elsewhere
- Passengers wait and fume

REFERENCES

[1] Foreign Object Debris and Damage Prevention, www.boeing.com/commercial/aeromagazine/aero_01/textonly/s01txt.html

[2] What is FOD? www.fodcontrol.com/what-is-fod/

[3] Prather, C.D. Current Airport Inspection Practices Regarding FOD (Foreign Object Debris/Damage), A Synthesis of Airport Practice, ACRP Synthesis 26, 2011 National Academy of Sciences.

[4] Bureau Enquêtes-Accidents. Preliminary Report – Translation, Accident on July 25, 2000 at "La Patte d'Oie" in Gonesse (95) to the Concorde registered F-BTSC operated by Air France, http://212.155.144.30/docs/anglais/htm/f-sc000725pa.html

[5] Airport Foreign Object Debris (FOD) Management, FAA Advisory Circular 150/5210-24, September 30, 2010.

[6] Runway safety, RMT.0703, Notice of Proposed Amendment 2018–14, European Union Aviation Safety Agency. [www.easa.europa.eu/sites/default/files/dfu/NPA%202018-14.pdf]

[7] Australasian Aviation Ground Safety Council, AAGSC Recommended Industry Practice FOD Removal Equipment, August 12, 2003, www.aagsc.org/ members/aagsc_adm/UploadFiles/RIP21_FODRemoval Equipment2003.pdf.

[8] Mason F.A., Kraus D.C., Johnson W.B., and Watson J., Reducing Foreign Object Damage Through Improved Human Performance: Best Practices, FAA 1989–2002, January 31, 2005, www.faa.gov/about/initiatives/maintenance_hf/library/documents/media/human_factors_maintenance/reducing_foreign_object_damage_through_improved_human_performance.best_practices.pdf

[9] Foreign Object Damage Prevention, Lockheed Martin Aeronautics Company, www.lockheedmartin.com/content/dam/lockheed-martin/aero/documents/scm/tandc/FOD/fod.pdf

[10] FOD Prevention–It's Up to You!, FOD Control Program Issues, Ideas & Actions, January 2013, www.fodcontrol.com

[11] FOD Prevention Guideline, National Aerospace FOD Prevention, Inc., http://as9100store.com/downloads/NAFPI-FOD-Prevention-Guideline.pdf

[12] Best In Class – FOD Prevention Guidance: NAS 412, A Holistic Approach to Risk Mitigation for the Aerospace and Defense Industry, 2017, https://global.ihs.com/images/SUPPLEMENTAL_DOCUMENTS/92/nas412_white_paper_2017-06.pdf

[13] Air Force Regulation (AFR) 66-33, Prevention of Foreign Object Damage (FOD) to Aircraft. Missiles or Drones.

[14] Beatty, D.N, Gearhart, J.J., Readdy, F., Duchatellier, R., The Study of Foreign Object Damage Caused by Aircraft Operations on Unconventional and Bomb-Damaged Airfield Surfaces, ESL-TR-81–39, June 1981.

[15] U.S. Dept. of Transportation, FAA, Flight Standards Service – Helicopter Flying Handbook, FAA-H-8083-21A, www.faa.gov/regulations_policies/handbooks_manuals/aviation/helicopter_flying_handbook/media/hfh_ch11.pdf

[16] El-Sayed A. F., Bird Strike In Aviation: Statistics, Analysis and Management, Wiley, July 2019, www.amazon.com/Bird-Strike-Aviation-Statistics-Management/dp/1119529735

[17] DeFusco, R.P. and Unangst, E.T. (2013). Airport Wildlife Population Management, A Synthesis of Airport Practice, ACRP Synthesis Report 39.

[18] Kushner, B.E. (2017). Flying Fish and Other Oddball Midair Collisions, Air & Space Magazine, January 18, 2017, www.airspacemag.com/as-next/flying-fish-and-other-oddball-midair-collisions-180961857/

[19] "NASA researchers to flying insects: 'Bug off!'." ScienceDaily. ScienceDaily, November 5, 2013, www.sciencedaily.com/releases/2013/11/131105122725.htm

[20] House, A.P.N., Ring, J.G., Hill, M.J., and Shaw, P.P., Insects and aviation safety: The case of the keyhole wasp Pachodynerus nasidens (Hymenoptera: Vespidae) in Australia, Transportation Research Interdisciplinary Perspectives, February 2020, http://dx.doi.org/10.1016/j.trip.2020.100096

[21] Christopher, A.H., NTSB Update: Maintenance Related Accidents, www.ntsb.gov/news/speeches/CHart/Documents/Hart_140407.pdf

[22] FOD (Foreign Object Damage) – Aviation Maintenance Tool Management Transport Canada, www.youtube.com/watch?v=dd0Z2PKUA-g

[23] McNally, D., Army researchers, tackle a tiny enemy: sand, US Army, November 4, 2016, www.army.mil/article/177698
[24] Rathi, A., Photos: This is the damage that tiny space debris traveling at incredible speeds can do, QUARTZ, September 3, 2016, https://qz.com/773511/photos-this-is-the-damage-that-tiny-space-debris-traveling-at-incredible-speeds-can-do/
[25] Van Zijl, J., Space Debris Hit the International Space Station Causing Small Crack in Window, The Science Explorer; Technology, May 13, 2016, http://thescienceexplorer.com/technology/space-debris-hit-international-space-station-causing-small-crack-window
[26] Orbital Debris Program Office: Debris Protection, NASA, Astromaterials Research & Exploration Science, November 2019, https://orbitaldebris.jsc.nasa.gov/protection/
[27] Guidelines For FOD Prevention And Control Engineering Essay, www.ukessays.com/essays/engineering/guidelines-for-fod-prevention-and-control-engineering-essay.php
[28] Standard for Foreign Object Damage/Foreign Object Debris (FOD) Prevention, MSFC-STD-3598; Revision: B, NASA George C. Marshall Space Flight Center, May 21, 2019.
[29] Report on Space Debris, United Nations, New York, 1999, www.orbitaldebris.jsc.nasa.gov/library/un_report_on_space_debris99.pdf
[30] United Space in Europe; ESA: Reentry and collision avoidance, www.esa.int/Safety_Security/Space_Debris/Reentry_and_collision_avoidance
[31] NASA, Orbital Debris Management & Risk Mitigation, www.nasa.gov/pdf/692076main_Orbital_Debris_Management_and_Risk_Mitigation.pdf
[32] Space Debris Page Covering 2004–Present, Space Debris Page Covering 1981–2003, https://eclipsetours.com/paul-maley/space-debris/space-debris-1960-1980/
[33] Keirn, G., Helicopters and Bird Strikes; Results from First Analysis Available Online, U.S. Department of Agriculture, February 21, 2017, www.usda.gov/media/blog/2013/06/06/helicopters-and-bird-strikes-results-first-analysis-available-online
[34] Dodgson L., Engineers found an owl napping inside the engine of a Virgin Australia plane, Insider, April 23, 2019, 6:11 AM, www.insider.com/owl-found-napping-inside-virgin-australia-plane-engine-2019-4
[35] Cleary, E. C. and Dolbeer, R. A. (2005). Wildlife Hazard Management at Airports: A Manual for Airport Personnel, *USDA National Wildlife Research Center – Staff Publications*. 133, http://digitalcommons.unl.edu/icwdm_usdanwrc/133
[36] Udris A., "3 Examples Of How FOD In The Cockpit Is Distracting And Dangerous +Video", Boldmethod, February 14, 2014, www.boldmethod.com/blog/2014/02/fod-cockpit/
[37] Aftab H. and Minhas R.S.M., Still Image-based foreign object debris (FOD) detection system, FEIIC 5th World Engineering Congress, September 2013, www.researchgate.net/publication/257934906
[38] Ringer, G., FOD Awareness, ASA Annual Conference 2015 Phoenix, AZ, www.faa.gov/about/initiatives/maintenance_hf/library/documents/media/human_factors_maintenance/maint_product781.pdf
[39] Foreign Object Elimination (FOE), Elements of Basic Awareness, National Center for Aerospace & Transportation Technologies, May 2009.
[40] Kraus, D. C., and Watson, J. Guidelines for the Prevention and Elimination of Foreign Object Damage/Debris (FOD) in the Aviation Maintenance Environment through Improved Human Performance, December 21, 2001, www.faa.gov/about/initiatives/maintenance_hf/library/documents/media/human_factors_maintenance/maint_product781.pdf

[41] Foreign Object Damage/Debris, (FOD) International Aerospace Quality Group (IAQG), SCMH 7.3, April 13, 2010.
[42] Procedure for the Control and Prevention of Foreign Object Debris, VITRON MF7P210, January 9, 2012, http://theatlasgroup.biz/wp-content/uploads/2013/05/MF7P210-NC.pdf
[43] Foreign Object Debris (FOD), Skybrary, August 4, 2017, www.skybrary.aero/index.php/Foreign_Object_Debris_(FOD)
[44] Reason, J. and Maddox, M. (1998). FAA Research 1989 – 2000/Human Factors in Aviation Maintenance and Inspection/Human Factors Guide for Aviation Maintenance: Chapter 14 Human Error. Washington, DC, FAA.
[45] El-Sayed, A.F., OD in Intakes – A Case Study for Ice Accretion in the Intake of a High Bypass Turbofan Engine, RTO-EN-AVT-195, Von Karman Institute lecture series on “Engine Intake Aerothermal Design: Subsonic to High-Speed Applications,” November 14–17, 2011.
[46] Foreign Object Debris and Foreign Object Damage (FOD) Prevention For Aviation Maintenance & Manufacturing, www.hestories.info/foreign-object-debris-and-foreign-object-damage-fod-prevention.html
[47] International Civil Aviation Organization (ICAO), Accident prevention manual (1st Edition). ICAO Document Number 9422-AN/923. Montreal, CN (1984).
[48] Current Mishap Definitions and Reporting Criteria, Naval Safety Center, January 29, 2018, https://navalsafetycenter.navy.mil/Resources/Current-Mishap-Definitions/, (accessed November 18, 2021).
[49] Newdick, T., The Air Force Spent $134M To Repair 39 Hail-Damaged T-1 Jet Trainers It’s About To Retire, The WarZone, March 25, 2021, www.thedrive.com/the-war-zone/39945/the-air-force-spent-134m-to-repair-39-hail-damaged-t-1-jet-trainers-its-about-to-retire
[50] https://en.wikipedia.org/wiki/Downwash#/media/File:US_Navy_050419-N-5313A-414_Search_and_Rescue_(SAR)_swimmers_attached_to_the_Kearsarge_Expeditionary_Strike_Group_conduct_search_and_rescue_training_during_routine_helicopter_operations.jpg
[51] Keirn, G. (2013). Helicopters and Bird Strikes; Results from First Analysis Available Online, www.usda.gov/media/blog/2013/06/06/helicopters-and-bird-strikesresults-first-analysis-available-online (accessed April 29, 2021).
[52] Vortex ring state. https://en.wikipedia.org/wiki/Vortex_ring_state#/media/File:Vortex_ring_state.png

2 Accidents and Incidents Due to FOD and Its Costs

2.1 INTRODUCTION

Foreign object debris threatens the aviation industry. The impact of potential accidents and incidents is the damage of aircraft, potential loss of lives, and the drastic direct and indirect costs, which accumulate to nearly $14 billion/year. Consequently, air safety concerns airlines, airports, governments, authorities like ICAO, IATA, etc. Every accident, of course, is a tragedy. That makes the aviation industry more determined to improve its safety record each successive year. A huge sum is daily spent on maintaining air safety on training, technology, installation and maintenance of equipment and devices, research, etc. That may be one of the important reasons rare incidents and accidents in aviation either in the air or on the ground, occur. Statistically speaking, flying is the safest way to travel. According to statistics for passenger fatalities in the US in 2000–2009, for every 1 billion passenger miles traveled by car, 7.28 people die, whereas, by plane, it is only 0.07 people [1].

In this chapter, statistics and a review of the details of some aviation accidents/incidents and their causes for the civilian, military aircraft, and helicopters will be discussed. Studying the cause of air disasters helps engineers design safer planes and prompts officials to initiate better air travel guidelines to improve aircraft safety and survivability. There were more than 14,600 reported collisions with wildlife in 2018. That equates to an average of 40 bird strikes per day.

2.1.1 Definitions of Aviation Accidents/Incidents Based on Annex 13 (ICAO)

The following definitions for aircraft accidents and incidents [2] are identified as.

2.1.1.1 Accident

An accident is an occurrence anytime between the boarding and disembarking of an aircraft, in which:

a. A person is fatally or seriously injured because of:
   - being in the aircraft, or in contact with any part of the aircraft; or
   - exposed to the jet blast.

DOI: 10.1201/9781003133087-2

These injuries do not include natural causes or infliction by other persons, or if the injuries are stowaways hiding outside the aircraft cockpit.

2.1.1.2 Serious Injury

As also defined in Annex 13 [2], it is an injury which is sustained by a person in an accident and which is described by one of the following:

a. requires hospitalization for more than 48 h, commencing within seven days from the date the injury was received
b. results in a fracture of any bone (except simple fractures of fingers, toes, or nose)
c. involve lacerations that cause severe hemorrhage, nerve, muscle, or tendon damage
d. involves injury to any internal organ
e. involves second or third-degree burns or any burns affecting more than 5% of the body surface
f. involves verified exposure to infectious substances or injurious radiation.

2.1.1.3 Incident

An incident is defined as an occurrence, other than an accident, associated with the operation of an aircraft which affects or could affect the safety of the operation.

The types of aircraft incidents described below are of main interest to the ICAO for its accident prevention studies [3].

a. Engine failure. Failures of more than one engine on the same aircraft and failures which are not confined to the engine, excluding compressor blade and turbine bucket failures.
b. Fires. Fires that occur in flight, including those engine fires which are not contained in the engine.
c. Terrain and obstacle clearance incidents. Occurrences result in the danger of collision or actual collision with terrain or obstacles.
d. Flight control and stability problems. Occurrences that have caused difficulties in controlling the aircraft, for example, aircraft system failures, weather phenomena, operation outside the approved flight envelope.
e. Takeoff and landing incidents. Incidents such as undershooting, overrunning, running off the side of runways, wheels-up landing.
f. Flight crew incapacitation. The inability of any required flight crew member to perform prescribed flight duties resulted from reduced medical fitness. Decompression is resulting in an emergency descent.
g. Near collisions and other air traffic incidents. Near collisions and other hazardous air traffic incidents, including faulty procedures or equipment failures.

2.1.2 Definitions of Aviation Accidents according to Airbus Industries [4]

Airbus gives the following definitions for accidents, which are slightly different from ICAO definitions [4]:

2.1.2.1 Operational Accident

An accident occurs between the time any person boards the aircraft with the intention of flight until all such persons have disembarked, excluding sabotage, military actions, terrorism, suicide, and the like.

2.1.2.2 Fatal Accident

An event in which at least one person is fatally or seriously injured as a result of:

- being in the aircraft, or
- direct contact with any part of the aircraft, including parts that have become detached from the aircraft, or
- direct exposure to jet blast, except when the injuries are from natural causes, self-inflicted or inflicted by other persons, or when the injuries are to stowaways hiding outside the areas normally available to the passengers and crew.

2.1.2.3 Hull Loss Accident

An event in which an aircraft is destroyed or damaged beyond economical repair. The threshold of economical repair is decreasing with the residual value of the aircraft. Therefore, as an aircraft is aging, an event leading to damage economically repairable years before may be considered a hull loss.

2.2 STATISTICS FOR FATAL ACCIDENTS

2.2.1 Civilian Aircraft

The air transport industry plays an essential role in global economic activity and development. The primary causes of accidents can be broken down into five categories, namely: pilot error, mechanical, weather, sabotage, and others [5].

1. The pilot error includes improper procedure, descending below minima, excessive landing speed, missed runway, fuel starvation, navigation error, etc.
2. Mechanical includes engine failure, equipment failure, structural failure, design flaw, maintenance error
3. Weather, which includes severe turbulence, wind shear, poor visibility, heavy rain, severe winds, icing, lightning strike, etc.
4. Sabotage, which includes hijacking, being shot down, terrorism, suicide, and the like
5. Other, which include bird strike, ATC error, ground crew error, overloading, improperly loaded cargo, mid-air collision with other aircraft, pilot incapacitation, fuel contamination, obstruction on the runway, fire/smoke in flight, etc.

Re-examining the above groups, we can figure FOD sources listed either explicitly or implicitly. For example, (pilot error) may include running the engines on the ground in the presence of personnel or luggage carts in dangerous areas of aircraft, as they will be vulnerable to ingestion into engines – also, pilot error if taking off without de-icing any iced surface. Moreover, an erroneous decision to fly through cumulus clouds full of supercooled droplets instead of flying above/below or around it. The (mechanical) group includes maintenance errors like lost tools, unfixed or dropped fasteners, and missed gloves, rags, and towels. The (weather) group includes the following FOD sources: icing, heavy rain, and other threats like a thunderstorm and severe wind. The final group (others) includes bird strikes explicitly. Also, implicitly it includes obstruction on runways [5], which could be broken asphalt of the runway, presence of dropped personal items, cargo and luggage carts, construction material and trucks, etc.

Table 2.1 presents accident records for civilian aircraft carrying at least 19 passengers from January 1, 1950 to June 30, 2019. Military helicopters and private aircraft are not included. This list includes only accidents with at least two fatalities.

Table 2.2 presents the percentages of fatal accidents by flight phase as outlined in the studies by Airbus industries through nearly 20 years (1999–2018); [4, 6] and also includes the percentages of fatal accidents by flight phase as assembled by Boeing company through 50 years (1959–2008); [7]. Both statistics assured that the approach and landing are when the most fatal accidents occur. The least accidents occur during cruises.

The most comprehensive survey for fatal accidents and fatalities per year for airliners (14+ passengers) since 1942 and up to 2019 is given in [8]. The maximum number of airliner accidents was in 1948 (99 accidents), while the minimum number was in 2015 (15 accidents). The maximum number of fatalities was in 1972 (2389 fatalities), while the minimum number was in 2017 (59).

From the worldwide (561) fatalities in 2018, the European share was (189) persons who died in aviation accidents on EU territory as outlined by the Aviation Safety Agency (EASA) [9]. A breakdown of these 189 fatalities is as follows:

- 84% (159 persons) belong to the category "General aviation," which a maximum takeoff mass (MTOM) of less than 2250 kg. This sub-category comprises small airplanes, gliders, and "microlights."
- 4.2% (8 persons) of "Commercial air transport" (8 fatalities)
- 6.3% (12 persons) "General aviation" of large aircraft over 2250 kg MTOM
- 5.5% (10 persons) "Aerial work" aircraft (aircraft used for specialized services such as agriculture, construction, photography, surveying, observation and patrol, search and rescue, aerial advertisement, etc.)

TABLE 2.1
Causes of Accidents, 1950–2019 [5]

Causes of Fatal Accidents by Decade

Decade	1950s	1960s	1970s	1980s	1990s	2000s	2010s	All
Pilot Error	50%	53%	49%	42%	49%	50%	57%	49%
Mechanical	26%	27%	19%	22%	22%	23%	21%	23%
Weather	15%	7%	10%	14%	7%	8%	10%	10%
Sabotage	4%	4%	9%	12%	8%	9%	8%	8%
Other	5%	9%	13%	10%	14%	10%	4%	10%

TABLE 2.2
Statistics for Commercial Aircraft Accidents by Airbus and Boeing Companies

Flight Phase	Parking/ Taxi	Takeoff/ Run/ Aborted	Climb	Cruise	Descent	Approach	landing	Other Go-around
Airbus 1999-2018 [4, 6]								
Airbus Fatal Accidents%	0.2	2.5	8.5	5.5	1.4	11	4.	2.8%
Airbus Non-Fatal Accident%	4.5	7.7	3.7	1.7	0.3	5.5	40	0.2%
Airbus (Total)%	4.7	10.2	12.2	6.2	1.7	16.5	44.5	3.0
Boeing Fatal Accidents of Commercial Jet Airplane (1959–2008) [7]								
Boeing Fatal Accident%	12.0	12.0	18.0	8.0	4.0	21.0	25.0	–

Fatalities in the accidents of the commercial air transport category from 1990–2019 on EU territory include fatalities of crashes of non-EU airplanes, like the Ukrainian-registered passenger aircraft close to Thessaloniki/Greece in 1997 (70 fatalities) and the mid-air collision between a Russian passenger aircraft and a Bahraini-registered cargo aircraft over south Germany in 2002 (71 victims) [9].

Statistics for accidents due to icing of aircraft are given here as an example for FOD accidents. NASA surveyed and analyzed icing accidents [10] for subsonic aircraft for 16 years (1988–2003). A summary of this study is given in Table 2.3.

2.2.2 Military Aircraft

Military aircraft have a vital role in any national defense. Normally accidents in military aircraft are identified as mishaps [11]. Three types of mishaps are presented in Table 2.4.

Damage class "H" was divided into two based on the observation that some wildlife strikes were non-damaging but were being reported with an associated monetary cost. The cost reported ($55) is associated with the labor required to collect the wildlife strike sample. Therefore, if a strike was considered an H class but resulted in a cost greater than $55, that meant that some level of damage occurred. Relative Hazard Scores are dependent upon comparing the number of damaging strikes to the number of non-damaging strikes. Splitting up Class H allowed us to identify non-damaging strikes.

The US Naval aircraft has slightly different definitions [12] for mishaps, as displayed in Table 2.5.

Mishaps for military aircraft in different branches of the Department of Defense (DoD) are displayed in Figures 2.1–2.3 for the seven years from 2012 to 2018.

TABLE 2.3
Types of Icing Events by Flight Operation (Without Ramp and Security Events) [10]

	Operation			
Type	**Part 121**	**Schedule Part 135**	**Nonscheduled Part 135**	**Part 91**
Total flight hours	232,868,640	25,050,928	46,350,000	416,319,000
Total fatal accidents	42	48	275	4511
Fatal accidents with icing conditions	4 (9.5%)	4 (8.3%)	22 (8.0%)	165 (3.7%)
Total fatalities	1603	327	660	8628
Fatalities with icing conditions	99 (6.2%)	53 (16.2%)	27 (4.1%)	350 (4.1%)
Total accidents	534	203	1057	23055
Accidents with icing conditions	6 (1.1%)	17 (8.4%)	58 (5.5%)	321 (1.4%)

FAA identifies aircraft categories by part as listed above, where
FAR Part 91 resembles small non-commercial aircraft
FAR Part 121 handles scheduled air carrier (airliners); Domestic, Flag, and Supplemental Operations
FAR Part 135 includes small aircraft, under 30 seats.

TABLE 2.4
Mishap Definitions of United States Navy (USN) and United States Air Force (USAF) Damage Classes in US Dollars [11]

Class	Cost	Injuries
A	>US $ 2.0 million Or aircraft destroyed	Fatality or permanent disability
B	US$ 500,000–US$ 2,000,000	Permanent partial disability Or three or more people hospitalized
C (Non-fatal)	US$ 50,000–US$ 500,000	Nonfatal injury resulting in lost work days
D	US$ 20,000–US$ 50,000	
E	<US$ 50,000	
H (Damaging)	>US$ 55	
H (Non-damaging)	<US$ 55.0	

Mishaps for both Air Force aircraft; Figure 2.1, and Navy aircraft; Figure 2.3, has continuously increased, while the Army aircraft, Figure 2.2 has decreased during the same period. However, the total mishaps of all the aircraft of the Department of Defense (DoD), Figure 2.4, increased in these seven years.

Luckily, non-fatal "Class C" mishaps which cause modest damage and injuries, continue to increase while "Class A" mishaps that result in deaths and cost millions in damages are ticking downward; Figure 2.4.

TABLE 2.5
Mishap Definitions for Naval Aircraft [12]

Class	Cost	Injuries
A	US$ 2.5 million and aircraft destroyed	Fatality or permanent disability
B	Greater than US$ 600,000 But Less than US$ 2.5 million	Permanent partial disability Or three or more people hospitalized as inpatients
C (Non-fatal)	Greater than US$ 60,000 But less than US$ 600,000	Nonfatal injury resulting in loss of time from work beyond day/shift when the injury occurred
D*	Greater than US$ 25,000 But less than US$ 60,000	Recordable injury or illness not otherwise classified as Class A, B, or C

* Class D was promulgated by DoDI 6055.07 on June 6, 2011.

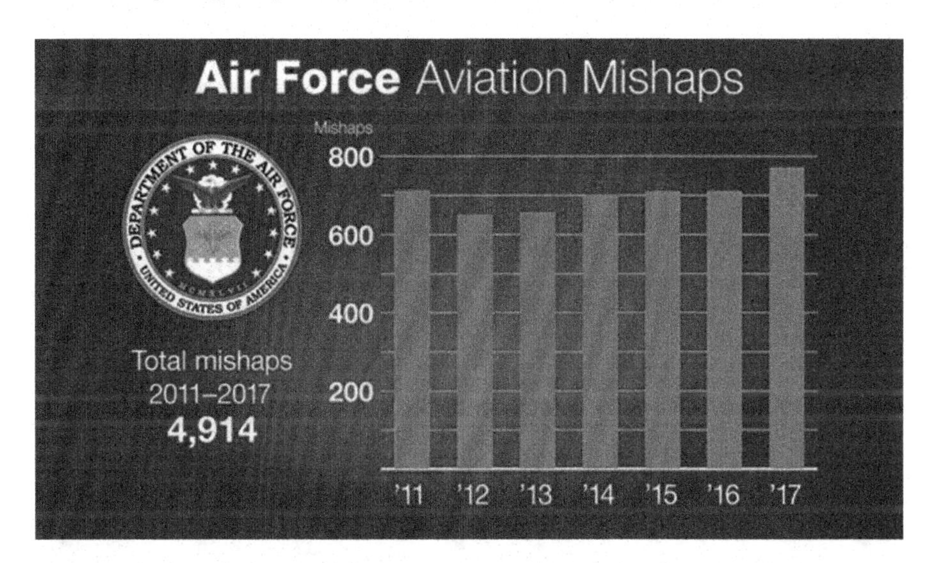

FIGURE 2.1 Air Force Aviation mishaps in the period (2011–2017).

Courtesy: US Air Force [13].

It is worth mentioning that the total mishaps in 2011–2017 (7,590) include the mishaps of Engineering Corps Helicopters [13].

Foreign object debris is responsible for many mishaps to military aircraft. It puts important operations and very expensive aircraft at risk. Hereafter, some details for mishaps due to different types of FOD will be discussed. The US Air Force Safety Center [14] assembled details of mishaps for all aircraft belonging to different military categories, including bombers, cargo, attack/fighters, trainers, and a helicopter. It includes statistics for strikes by fiscal year, month, hour, phase of the operation from 1985 to 2016. Also, it lists all data for strikes by aircraft impact point and altitude in

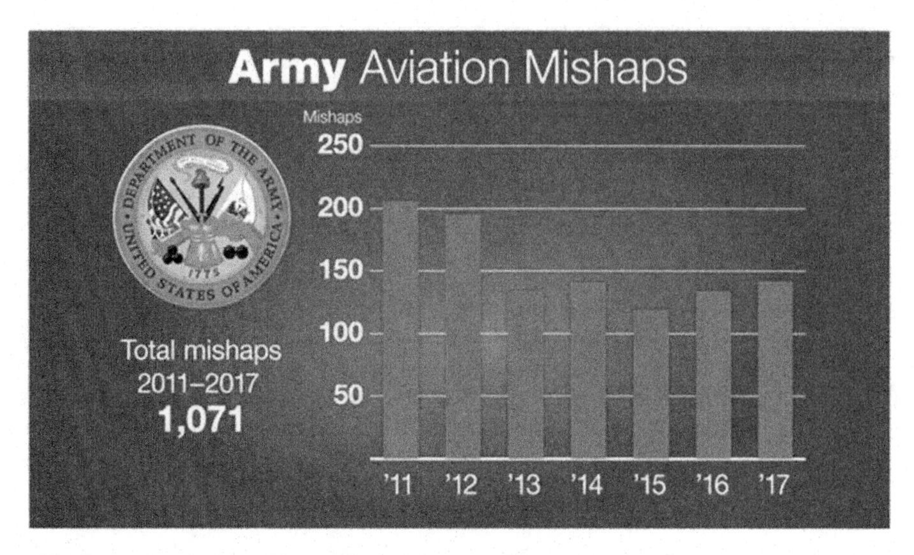

FIGURE 2.2 Army aviation mishaps in the period (2011–2017).

Courtesy: US Army [13].

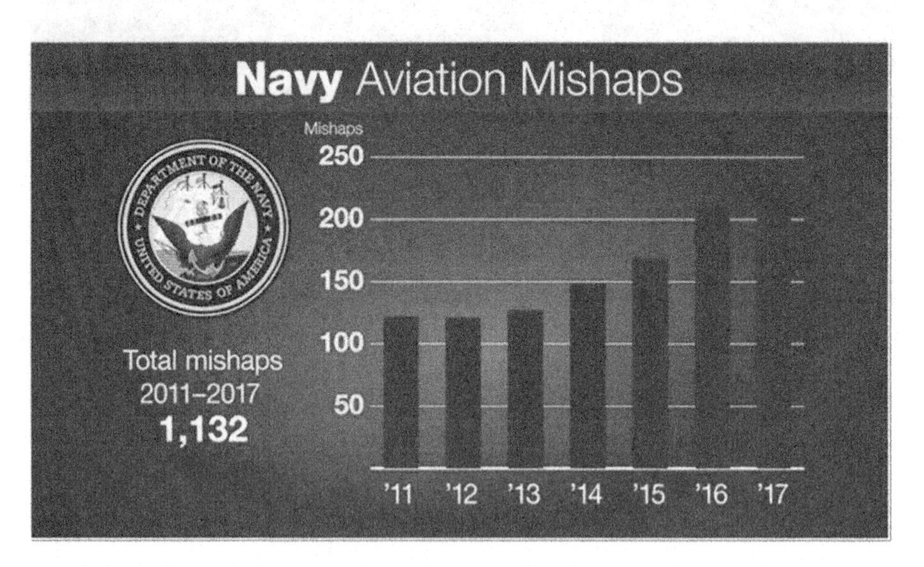

FIGURE 2.3 Navy aviation Mishaps in the period (2011–2017).

Courtesy: US Navy [13].

the same period. As a sample for these data; wildlife strikes (birds, bats, mammals, and reptiles) with the aircraft of the US Air Force in the period FY1985–FY2016 are given in Table 2.6. These statistics [14] also clarify the top 50 species of wildlife that struck its aircraft from FY1995–FY2016.

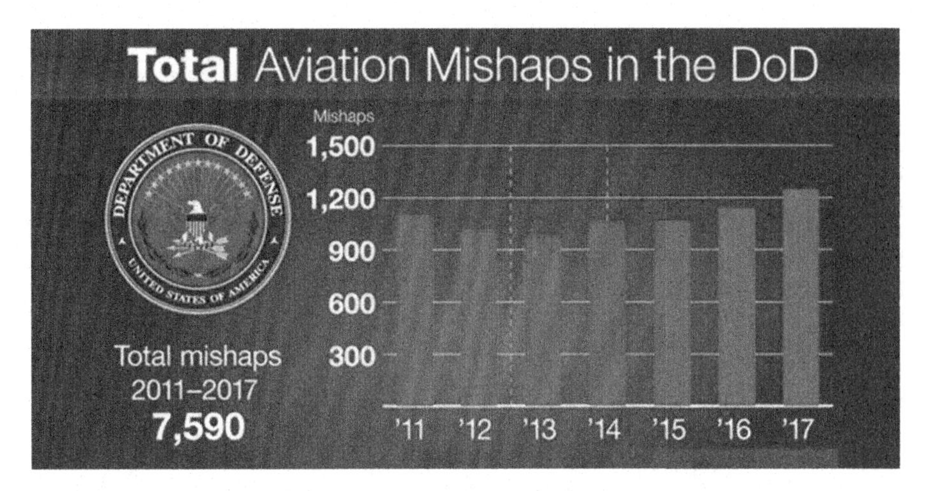

FIGURE 2.4 Total aviation mishaps in the period 2011–2017.

Courtesy: US Department of Defense [13].

TABLE 2.6
USAF Wildlife Strike by Fiscal Year 1985–2017

Year	Cost	Count	Cost per 100K flying hours
FY85	$4,427,829	1960	$127,304
FY86	$3,584,216	2820	$103,533
FY87	$251,444,565	2789	$7,257,707
FY88	$6,227,297	2652	$186,146
FY89	$24,321,211	2999	$714,009
FY90	$6,775,932	3056	$201,298
FY91	$18,595,994	2692	$504,676
FY92	$17,813,130	2539	$638,952
FY93	$18,422,719	2410	$729,343
FY94	$15,904,499	2281	$704,908
FY95	$91,042,795	2693	$4,109,709
FY96	$5,389,154	2960	$248,383
FY97	$10,401,928	2886	$490,731
FY98	$25,225,890	3116	$1,194,886
FY99	$29,066,667	3690	$1,364,019
FY00	$34,894,624	3275	$1,713,244
FY01	$13,498,877	3969	$653,033
FY02	$7,933,233	3660	$333,443
FY03	$54,037,497	4409	$2,255,223
FY04	$53,602,894	4679	$2,334,669
FY05	$21,770,621	5107	$1,015,988
FY06	$16,236,524	5019	$768,173

(continued)

TABLE 2.6 (continued)

Year	Cost	Count	Cost per 100K flying hours
FY07	$28,390,000	5002	$1,384,494
FY08	$11,042,000	4819	$562,720
FY09	$15,193,136	4707	$716,974
FY10	$22,341,664	4723	$1,023,694
FY11	$13,061,140	4471	$564,903
FY12	$10,178,368	4768	$527,650
FY13	$46,424,978	4230	$2,197,112
FY14	$55,062,223	4289	$3,445,696
FY15	$29,227,360	4342	$1,623,271
FY16	$20,633,787	4089	$1,170,139
Total	$982,172,752	117,101	N/A

TABLE 2.7
The Number and Rate of FOD Mishaps per 100,000 Flight Hours by Aircraft Model (2004–2009) [15]

Aircraft Model	FOD Event	Bird Events	Bird Strike Rate	Other FOD	Rate of Other FOD	Flight Hours	Rate
F-16	3	3	123.46	–	–	2,430	123.46
T-45	14	9	2.5	5	1.39	359,914	3.89
F/A-18	32	16	1.555	16	1.555	1,028,719	3.11
E-6	2	1	1.43	1	1.43	69,994	2.86
EA-6B	4	3	1.98	1	0.66	151,886	2.64
P-3/EP-3	9	6	1.72	3	0.86	348,632	2.58
E-2	2	1	1.025	1	1.025	97,602	2.05
T-34	0	0	0.00	0	0.00	584,463	0.00
T-44	0	0	0.00	0	0.00	126,004	0.00
T-39	0	0	0.00	0	0.00	47,881	0.00

Next, mishaps of Naval Air Forces due to FOD are considered [15]. Only class C or more severe events were included during the five years (FY2004–FY2009). Table 2.7 displays the number and rate of FOD mishaps per 100,000 flight hours by aircraft model.

The aircraft F-16 has the maximum rate, and all its strikes are with birds. Other aircraft are struck with different kinds of FOD, as described in Table 2.8.

2.2.3 Helicopter

Helicopters manage to achieve numerous missions both in peace and war. Helicopter accidents are categorized into six groups as per the extent of injuries and the amount of damage.

TABLE 2.8
Aircraft, Foreign Object Debris, and Damaged Part [15]

Aircraft Model	Object	Damage	Total
F/A-18	Bird	Engine	12
		Forward-Looking Infra-Red (FLIR) Pod	1
		Multiple	1
		Port Aileron	1
		Vertical Stabilizer	1
	Refueling Basket	Engine	8
	Fastener/Bolt	Engine	2
	Hail	Airframe	1
	Antenna	Engine	1
	Engine Intake	Engine	1
	Ground Refueling Receptacle	Engine	1
	Metal Box	Canopy	1
	Unsecured Object	Flap	1
P-3	Bird	Engine	4
		Radome	1
		Vertical stabilizer	1
	Clamp	Engine	1
	Coin	Engine	1
	Unknown Object	Engine	1
T-45	Bird	Engine	6
		Canopy	2
		Wing	1
	Canopy Cover	Engine	1
	Cell Phone	Engine	1
	Cranial	Engine	1
	Gear Pin	Engine	1
	Turn Screen Strap	Engine	1
E-2	Bird	Engine	1
	Tire	Propeller	1

Table 2.9 identifies these six groups [16].

The fatal accidents for Western-built turbine helicopters in 1990–2018 for both single-engine and multiple-engine categories are given in [17]. The number of fatal accidents experienced by single-engine western helicopters is nearly double that of multi-engine ones. The maximum number of accidents for a single-engine group was in 1992 and 2000. For a multi-engine helicopter, the worst year was 1998, where the number of accidents reached 25.

Next, considering the total fatal accidents, the least was in 2014, only 38, while the worst year was 2011, with 69 fatal accidents [17].

The annual total fatalities in single- and multi-engine western helicopters from 1990 to 2018 are listed in [17]. The worst year was 2011 when 183 passengers and crew were killed, while the best year was 2014, with 38 fatalities.

TABLE 2.9
Helicopter Accident Classifications [16]

Class	Cost	Injuries
A	US$ 1.0 million or more	Fatality or total disability
B	US$ 200,000–US$ 1,000,000	Permanent partial disability Or five or more personnel hospitalized
C	US$ 10,000–US$ 200,000	Injury or illness that cause loss of time from work or later disability
D	US$ 2,000–US$ 10,000	Nonfatal injuries/illness in conjunction with property damage
E	Less than US$ 2,000	No injuries or fatalities. Mission (either operational or maintenance) is interrupted or not completed
F	Any amount due to foreign objects	No injuries or fatalities

TABLE 2.10
US Helicopter Accidents in the Period 2013–2019

Year	Total accidents	The accident rate per 100,000 flight hours	Fatal accidents	The fatal accident rate per 100,000 flight hours	Number of fatalities
2013	146	4.95	30	1.02	62
2014	138	4.26	21	0.65	37
2015	121	3.67	17	0.52	28
2016	108	3.48	17	0.54	29
2017	123	3.70	20	0.60	34
2018	121	3.62	24	0.72	55

In 2011, for example, the number of fatal accidents by the US registered helicopter is 18, while all the western-built turbine helicopters encountered 68 accidents. As identified by the FAA [18–21], the number of accidents and fatal accidents for US Helicopters in the five years (2013–2018) is mixed. A reasonable drop in its numbers up to 216, then an increase up to 2018; Table 2.10.

Icing is responsible for a reasonable number of US Helicopters. Table 2.11 illustrates a breakdown of the total and icing-related accidents/incidents according to class is shown in the period from 1985 to 1999, and the share of icing from these overall accidents [16]. During the FY85–FY99 period, there was a total of 54,081 aircraft accidents/incidents according to Army Safety Office data. Its majority (90%) was in the Class E category. During the same period, there were 255 recorded icing accidents/incidents, representing only 0.5% of the total. Most of the icing accidents/incidents were also in the Class E category.

TABLE 2.11
Comparison of Icing Accidents/Incidents to Total Helicopter Accidents/Incidents according to Classification [16]

Class	Total in class	Total in class due to icing	Percent due to icing (%)
A	399	1	0.3
B	188	1	0.5
C	1,294	27	2.1
D	3,047	25	0.8
E	48,956	184	0.4
F	197	17	8.6
Total	54,081	255	0.6

2.3 SOME CATASTROPHIC ACCIDENTS DUE TO FOD

- Sandstorm
- Wildlife
- Birds
- Human
- A fallen part from a preceding aircraft
- Icing
- Fasteners
- Hail

2.3.1 Civilian Fixed-Wing Aircraft

2.3.1.1 Sandstorm

On March 18, 1966, the United Arab Airlines (UAA) Antonov An-24 Flight 749 was a scheduled international passenger flight from Nicosia Airport to Cairo International Airport.

En route, the aircraft encountered bad weather conditions in Cairo due to sandstorms [22]. The flight crew also reported they were flying through thunderstorms with icing conditions, that two of the aircraft's altimeters were giving different readings. The aircraft descended below the safe flight altitude in the final approach, and the port wing impacted the dunes lying to the northeast of the aerodrome. As a result, the pilot lost control of his aircraft and hit the ground. All thirty passengers and crew on board were killed.

2.3.1.2 Deer (Wildlife)

On November 17, 2012, a Cessna Citation-550 jet (N6763L) operated by Stevens Aviation, Inc. was substantially damaged during a collision with a deer after landing on Runway 9 at Greenwood County Airport Greenwood, South Carolina [23, 24].

The deer appeared from the wood line and ran into the path of the aircraft, struck the jet at the leading edge of the left-wing above the left main landing gear, and ruptured an adjacent fuel cell. The pilot maintained directional control, and the jet was stopped on the runway, spilling fuel and on fire. The crew performed an emergency shutdown and escaped without injury, but a post-crash fire subsequently consumed the jet.

2.3.1.3 Birds

2.3.1.3.1 Miracle on the Hudson

On January 15, 2009, Airbus A320-214; US Airways Flight #1549, during the climb phase after takeoff from LaGuardia Airport, New York City, encountered a flock of Canada geese. As reported by the National Transportation Safety Board report [25]:

> Both engines were operating normally until they each ingested at least two large birds (weighing about 8 pounds each), one of which was ingested into each engine core, causing mechanical damage that prevented the engines from being able to provide sufficient thrust to sustain flight.

Unable to reach any airport, the pilots Chesley Sullenberger and Jeffrey Skiles decided to glide the plane and ditch on the Hudson River, which provided the highest probability that the accident would be survivable. The captain started the auxiliary power unit (APU) to provide the airplane with a primary electrical power source and maintain the airplane flight envelope protections against a stall. All 155 people aboard were rescued by nearby boats, and there were few serious injuries. According to the NTSB report, this was the most successful ditching in aviation history, and the accident came to be known as the "Miracle on the Hudson."

Figure 2.5 displays the flight path followed by the pilots for ditching in the Hudson River.

Figure 2.6 illustrates the rescue process for the passengers following the successful ditching in the Hudson River [27].

2.3.1.3.2 Miracle over Ramensk

On Thursday, August 15, 2019, a Russian Airbus A321 Ural Airlines Flight #U6178 operating from Moscow to Simferopol and carrying 226 passengers and 7 crew (233 in total) made an emergency landing in a cornfield outside Moscow's Zhukovsky International Airport (ZIA) after hitting a flock of seagulls during takeoff [28]. The bird strike occurred shortly after takeoff, causing stalls of both engines. Thus the crew decided to land immediately. The plane landed in the cornfield about a kilometer from the runway, on the outskirts of Moscow, with both its engines off and landing gear retracted. Passengers were evacuated on inflatable ramps. A total of 23 people was hospitalized, including nine children, but there were neither fatalities nor serious injuries. Ural Airlines pilot Damir Yusupov was called a "hero" by Russian media, and the accident is identified as "Miracle over Ramensk."

2.3.1.3.3 DC-10

On November 12, 1975, McDonnell Douglas aircraft DC-10-30CF registration (N1032F) and operated by Overseas National Airways (ONA), struck many birds during departure roll from JFK International Airport (New York) [29]. During acceleration, but before reaching the V1 speed, a flock of birds suddenly rose from the runway and were ingested into engine #3 (GE CF6-50). The pilot aborted takeoff. The

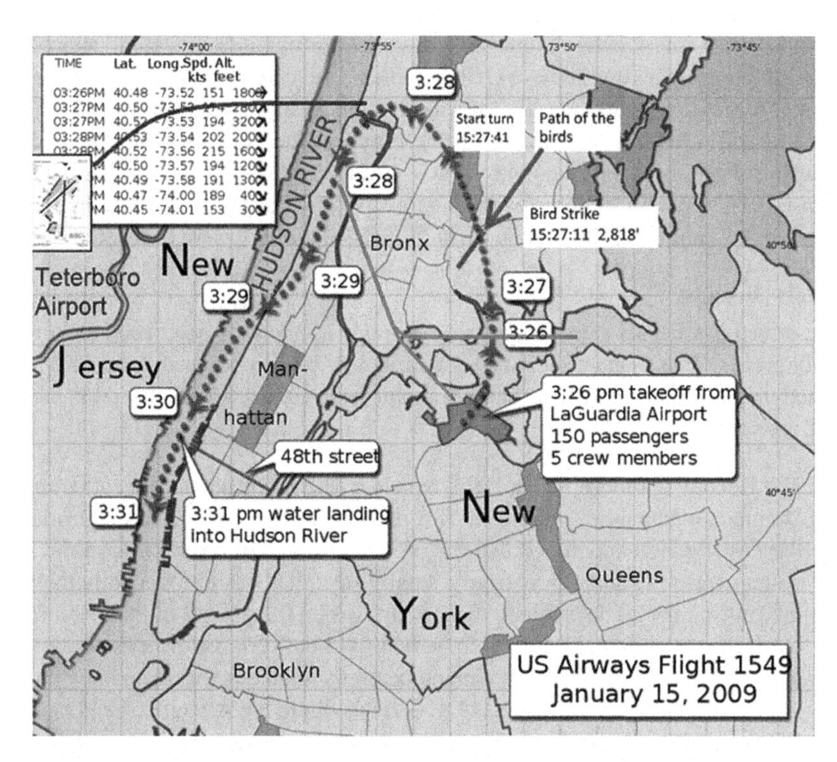

FIGURE 2.5 Flight path flown (red) and the alternative trajectories to Teterboro (dark blue) and back toward La Guardia (magenta) were simulated for the investigation [26].

Permission (GFDL).

FIGURE 2.6 US Airways Flight 1549 in the Hudson River, New York, United States on January 15, 2009 [27].

Permission (CC BY 2.0).

fan blades of engine #3 were damaged, causing rotor imbalance. Debris entered the engine's high-pressure compressor (HPC), ignited, and caused the compressor case to separate. A fire erupted in the right-wing and pylon of engine #3 [29].

The aircraft couldn't be stopped on the runway. The pilot elegantly steered the aircraft off the runway onto taxiway Z at a 40-knot speed. The main undercarriage collapsed, and the aircraft came to rest against the shoulder of the taxiway. Luckily, there were no fatalities. All 138 people on board and aircrew were evacuated safely.

2.3.1.4 Insects

As described in Chapter 1, insects may cause tragic accidents if obstructed the airflow in pitot probes. Two tragic accidents were attributed to the mud-nesting keyhole wasp (Pachodynerus nasidens), which caused anomalous readings from the pitot probes. These are:

- The Boeing 757 in February 1996, which crashed shortly after takeoff from the Dominican Republic, killing all 189 passengers and crew. The authorities of the investigation concluded that the probable source of obstruction in the pitot system was mud and debris from a small insect that was introduced in the pitot tube during the 20 days where the aircraft was on the ground in Plata before the crash. The aircraft was returned for service without verification of the static pitot system as recommended by the manufacturer's maintenance procedures [30].
- In an earlier incident, a Douglas DC-3 crashed into the Atlantic Ocean near the Bahamas after takeoff from West Palm Beach International Airport, Florida, in 1980 (National Transport Safety Bureau, 1981), with 34 lives lost [31].

In both cases, mud-dauber wasp nest material was found in the pitot heads and considered the most likely cause of the pitot system malfunction.

2.3.1.5 A Fallen Part from a Preceding Aircraft

In rare situations, an aircraft taking off from a certain runway may be subject to a FOD fallen from a preceding aircraft that took off from the same runway. An example of such a case is the Concorde aircraft (registered F-BTSC) described in Chapter 1.

On July 25, 2000, Continental Airlines McDonnel Douglas DC-10-30 aircraft took off from runway 26 right at Roissy Charles de Gaulle Airport. A titanium alloy strip has been fallen from one of its engines. Concorde aircraft (registered F-BTSC) in its charter Flight #AFR 4590 to JFK International airport, New York, took off from the same runway only 5 min after. The front right tire (tire No 2) of the left landing gear ran over the fallen titanium strip.

The debris cut the tire and sent a large chunk of the tire at high speed to puncture fuel tank #5 in the left-wing. Engines #1 and #2 caught fire and stalled. The landing gear did not retract. The aircraft flew for around a minute at a speed of 200 knots and an altitude of 200 ft.

Concorde was unable to gain height or speed. Engine #1 lost thrust at first, and next, engines #3 and #4 also lost thrust. The aircraft crashed onto a hotel killing all 109 people aboard. Four people in the hotel at "La Patte d'Oie" in Gonesse and another person in the hotel were critically injured [32].

2.3.1.6 Icing

2.3.1.6.1 The Icing of Pitot Tube

On June 1, 2009, an Air France Airbus A330-203 aircraft Flight #447 from Rio de Janeiro to Paris stalled and crashed into the Atlantic Ocean at 02:14 UTC, killing all 228 persons on board (216 passengers and 12 crew) [33].

The final report of the French Bureau d'Enquêtes et d'Analyses pour la Sécurité de l'Aviation Civile (BEA) [34], released on July 5, 2012, concluded that the aircraft crashed due to the following reasons:

- Encountering icing conditions and the accumulation of ice crystals in the pitot tubes caused temporary inconsistencies between the airspeed measurements and the disconnection of the autopilot.
- Inappropriate control inputs by the crew caused the aircraft to enter an aerodynamic stall.
- The crew's failure to diagnose the stall situation and, consequently, a lack of inputs would have made it possible to recover from it.

This accident was the deadliest in the history of both Air France and the deadliest aviation accident involving the Airbus A330; Figure 2.7.

2.3.1.6.2 The Icing of Wing, Tail, and Fuselage

On February 4, 2012, at 13:35 UTC (07:35 local time in Tyumen city Eastern Russia), an ATR72-201 aircraft operated by UTAir Aviation, JSC, took off from Roschino International Airport to Surgut. Onboard there were 4 crew members and 39 passengers, all RF citizens (Figure 2.8).

As described in the final report [35, 36], immediately after the landing gear and the flaps retraction, the aircraft started descending with a significant left bank and then collided with terrain. The ground collision first led to the structural damage of the left-wing, followed by the fuel spillage and fire and further to the destruction of aircraft with the right-wing, cockpit, and rear section with empennage separation. Out of the 43 persons on board, 4 crew members and 29 passengers were killed. Others received serious injuries.

The final report for the ATR 72-201 VP-BYZ aircraft accident stated that the accident causes were:

1. The wrong decision of the Pilot in Charge (PIC) to take off without de/anti-icing treatment, regardless of the presence of the snow and ice deposits on the aircraft surface (wing, fuselage, and tail plane). It resulted in the degradation of aircraft aerodynamic performance and stall during climbing after takeoff as well as the inability of the crew to recognize stall and, consequently, failure to undertake recovery procedure.
2. The erroneous evaluation of aircraft conditions by the PIC and aircraft mechanic after the aircraft has been on the ground in icing conditions for a long time and in the release of the aircraft to fly without de/anti-icing treatment.

FIGURE 2.7 The aircraft's vertical stabilizer after its recovery from the ocean [33].

2.3.1.6.3 Icing of Wing

On January 13, 1982, Air Florida Boeing 737-222 aircraft registered an N62AF during flight #90 smashed into the 14th Street Bridge, collapsing into the icy Potomac River shortly after takeoff, and 78 people perished.

The National Transportation Safety Board (NTSB) report stated that the probable cause of the accident was pilot error. The pilots failed to use anti-icing during ground operation and takeoff, the wrong decision to take off with snow/ice on the wing and horizontal tail surface, and failure to abort the takeoff even after detecting a power problem while taxiing and seeing ice and snow buildup on the wings [37].

FIGURE 2.8 Horizontal stabilizer right side with traces of snow and ice deposits [36].

2.3.1.7 Human

2.3.1.7.1 Ground Run

Two danger areas are identified when an aircraft operates its engine on the ground, namely in front of the engines (upstream) and aft (downstream). These critical areas depend on the engine power as well as its rotational speed. Intake and Exhaust danger areas of the A/C A320neo, when powered by engine PW 1100 G at maximum takeoff power conditions, are defined in [38]. Anything (animate or inanimate) in the area denoted by inlet suction will be sucked into the engine. Unfortunately, these accidents happen very often.

Also, anything in the exhaust wake area will be subjected to high-speed and high-temperature exhaust.

2.3.1.7.2 Victims

- A human working for Air India was sucked in the engine of aircraft India Airline Flight #Al 619 from Mumbai to the southern city of Hyderabad on Thursday, December 17, 2015 [39]. The accident happened as the plane pushed back for takeoff, and the co-pilot mistakenly switched on the engine, while a maintenance man was standing close by.
- On May 13, 2010, a Boeing 737-500 Continental Airlines Flight #1515 taking off from El Paso International Airport in Texas to Bush Intercontinental Airport in Houston, a mechanic was sucked into one of the engines and killed [40]. The incident occurred during a maintenance check in preparation for the plane's departure.

- On November 4, 2004, a Boeing 737 plane owned by Kazakhstan airline Air Astana was taking off from Moscow to London when its jet engine sucked an engineer while he was examining them [41]. The pilot started up the plane's two engines, not realizing the engineer was there. The airport staff noticed blood spattered on the ground behind one engine. The aircraft was taken out of service while the engine was cleaned.

2.3.1.8 Fasteners and Personal Items

2.3.1.8.1 Kitfox Series 5

On April 28, 2012, a Kitfox Series 5 aircraft crashed into the Atlantic Ocean and sank as the elevator control "bound up." The pilot could not regain elevator control [42]. Examination revealed A 6-inch long, plastic, spring-loaded clamp and a leather work glove beneath the boot and between the tube seat structure and the control column bearing (Figure 2.9). This FOD impinged the elevator control, leading to the loss of control.

FIGURE 2.9 Kitfox Series 5 aircraft and typical gloves and clamp.

Courtesy: NTSB.

2.3.1.8.2 Grumman American AA-5

On January 5, 2020, the Grumman American AA-5 Cheetah aircraft left the airport in Henderson County, NC, United States. But the engine lost power because paint chips obstructed the fuel filter, which led to fuel starvation. The pilot subsequently had a forced landing on wooded terrain in Rutherford County, North Carolina. The airplane sustained substantial damage, and the four occupants were hospitalized (1 adult, three children). The pilot reported to the NTSB that the aircraft had recently been painted [43].

2.3.1.8.3 China Airlines 737-800 registration B18616

On August 20, 2007, a Boeing 737-800, registered B18616, operated by China Airlines, took off from Taiwan Taoyuan International Airport on a regularly scheduled Flight #120 and landed at Naha Airport.

At about 10:33, immediately after the aircraft stopped, the fuel was leaking from the fuel tank on the right-wing caught fire, and the aircraft was badly damaged and destroyed by fire, leaving only part of the airframe intact. The reason for fuel leakage was a bolt, which had come loose from the main track of #5 slat, and punctured the right-wing fuel tank, creating a hole 2–3 cm in diameter (Figure 2.10).

There were 165 people on board (the Captain, 7 other crew members, and 157 passengers, including two infants). Everyone on board was evacuated from the aircraft, and there were no dead and wounded, although one ground crew was injured [44].

2.3.1.9 Hail

On April 4, 1977, a Southern Airways DC-9-31 aircraft, registration N1335U during Flight #242 from Northwest Alabama Regional Airport to Atlanta Municipal Airport, United States, crashed in New Hope, Paulding County, Georgia.

NTSB outlined that the probable cause of the accident was the wrong decision of the pilot to penetrate rather than avoid the area of a severe thunderstorm. The penetration resulted in losing thrust on both engines due to the ingestion of massive amounts of water and hail, which resulted in a severe stall and major damage to the engine compressors. The engines could not be restarted, and the pilots had to make an emergency landing. The aircraft was destroyed.

FIGURE 2.10 Burned remains of China Airlines 737-800 at Naha Airport on August 20, 2007 [45].

FIGURE 2.11 The remnants of the tail section. The separated right-wing is in the foreground [47].

Of the 85 persons aboard, 62 people on the aircraft (including both pilots) were killed; 22 passengers were seriously injured, and 1 was slightly injured. One passenger died on June 5, 1977. Additionally, eight persons, one ground, were killed, and one person was seriously injured [46]. Figure 2.11 illustrates the remnants of the tail section and the separated right-wing on the ground [47].

2.3.1.10 Rainstorm

On August 6, 1997, Boeing 747 aircraft of Korean Air, Flight #801 flying from Seoul, Korea to Guam with 2 pilots, 1 flight engineer, 14 flight attendants, and 237 passengers on board. It encountered heavy rainfall, and high winds were determined to have contributed to the crash of this Korean Air flight from Seoul, Korea to Guam in 1997. The Boeing 747 was making a landing approach at the airport in Guam when it slammed into Nimitz Hill just three miles short of the runway at 660 ft. The NTSB report outlined that the heavy rain shower and the pilot error when he attempted to make an instrument landing even though the glideslope landing system (ILS) was not in operation at the airport at the time. Of the 254 persons on board, 229 were killed, 22 passengers and 3 flight attendants survived the accident with serious injuries [48].

2.3.1.11 Volcanoes

On June 24, 1982, British Airways Boeing 747-200, called (The City of Edinburgh), BA Flight #009, from Heathrow airport; London to Auckland; Captain Eric Moody. The aircraft flew into a cloud of volcanic ash thrown up from Mount Galunggung

Volcano in West Java (180 km from Jakarta), Indonesia, which failed all four engines [49].

Accident description:

1. The aircraft had 248 passengers, and 15 crew was traveling at its cruising altitude of 36,000 ft (11,000 m) above the Indian Ocean south of Java when one by one, all four engines surge and quit.
2. The giant airliner was converted into a large and not very efficient glider with a glide ratio of 15:1 (15 m of forwarding travel for every meter drop in altitude)
3. The pilot headed the glider back to Jakarta, the nearest airport.
4. Oxygen masks deployed from the ceiling as the cabin pressure dropped since the engines were not running.
5. The Captain had to put the aircraft into a shallow dive as some masks were not working to prevent those passengers without masks from asphyxiating to death. Thus aircraft dropped at 1,800 m per minute.
6. At this point, none of the crew knew why all four engines had quit.
7. At the altitude of 4,100 m, the crew managed to get engine #4 started and then #3, which allowed the Captain to put the 747 into a gentle climb.
8. Shortly afterward, Senior Engineer Officer Barry Townley-Freeman managed to start the other two engines.
9. The aircraft City of Edinburgh had been gliding for 13 min–about 80 nautical miles (longest glide in a non-purpose-built aircraft).
10. The windshield had been sandblasted by volcanic grit, and the pilots had almost zero forward visibility.
11. The Captain elegantly flew an improvised instrument approach, and he put the 747 down perfectly on Jakarta's runway.
12. The hands and Clothes of the Pilot and first officer were covered in fine black dust.
13. No injuries
14. No fatalities

The aircraft and engines experienced the following damages:

1. The windshields were sandblasted.
2. Ash (mostly Silicate with a mean diameter of 0.075 mm) was found in pitot tube, thus giving wrong airspeed readings.
3. Engine nacelle and nose cone stripped of paint as if the aircraft and engine sandblasted.
4. Erosion of compressor blades
5. Engines suck in the debris, turning it into glass in the hot section. Turbine blades were fouled and were the worst damaged parts; Figure 2.12.
6. The four engines, windshield, and fuel tanks were replaced.

FIGURE 2.12 Some damaged engine parts from BA 9 on display at Auckland Museum, Parts (from top): Fan blade – Compressor blade – Turbine stator [50].

2.3.1.12 Debris from Cracked Turbine Disc

On September 5, 2000, a Boeing B747-300 of Japan Airlines (JAL) registration JA8178 was operating on a scheduled international flight, JL726 from Soekarno-Hatta International Airport Jakarta, experienced a serious incident shortly after takeoff at 16:39 UTC.

The disk of the fifth stage low-pressure turbine of engine #1 failed, ejecting debris [51]. Fragments of the failed engine separated from the engine and were ejected through the exhaust nozzle, damaging the aircraft lower L/H wing, the L/H flaps, and the L/H side of the airframe skin at several places. Moreover, fragments fell on the ground damaging 21 houses. Pilots shut down the engine and, in one hour, dumped 163,000 lbs. of fuel over the open sea, and the airplane returned to Soekarno-Hatta Airport. The landing at 17:36 UTC was uneventful, and no problems were encountered. No injuries were reported among the passengers and crew, and persons on the ground.

2.3.2 Military Fixed-Wing Aircraft

2.3.2.1 Air-Refueling basket

On August 22, 2018, during the mid-air refueling operation of the F-35C jet, by a Super Hornet F/A 18F (Figure 2.13), a part of the refueling basket broke off, and that debris was sucked and ingested into the engine of the F-35 [52]. Its engine needed to be replaced, and the repair of the damage cost roughly $12.7 million.

2.3.2.2 Fasteners

On May 6, 2016, a Marine Corps AV-8B Harrier jet was in a low-altitude training mission off the coast of Wrightsville Beach. The pilot heard three distinct “thumps.” Moments later, his engine had failed catastrophically. The pilot ejected from his

FIGURE 2.13 The next-generation F-35C fighter refueled by a Super Hornet F/A 18F.

Courtesy: US Navy.

aircraft, which then crashed into the Atlantic Ocean about two miles offshore [53]. US Marine Corps investigated the incident and classified it as a "Class A Mishap." The aircraft was fully recovered on June 17, 2016. A post-mishap assessment showed significant damage to the entire airframe and engine, attributed to the high-speed water entry impact. The engineering investigation report determined the catastrophic engine failure stemmed from foreign object damage to the seventh stage, high-pressure compressor blades. The object was a 28 threads-per-inch fastener with a 5/16th-inch double hexagon bolt. The damage was severe enough to degrade all available stall margins rendering the engine inoperable, There were two options where the fastener came from. The first option was that the aircraft "ingested" the fastener on the runway at ILM or Marine Corps Air Station Cherry Point. But if this were true, the aircraft would not have generated sufficient power for takeoff. The second option was that the fastener likely "dislodged from inside a boundary layer door." However, it was unable to determine the exact origin of the fastener. The total estimated damage costs for the aircraft are $62.8 million. Additionally, estimated damage costs for an advanced targeting device recovered from the jet are $1.646 million.

2.3.2.3 Rainstorm

On March 3, 2001, a short C-23 Sherpa military transport aircraft belonging to the Florida Army National Guard was en route from Hurlburt Field Florida, to NAS Oceana, Virginia. Aircraft crashed during heavy rainstorms around 1,100 h in Unadilla, Georgia, in the United States. All 21 people on board are killed [54].

2.3.2.4 Sandstorm

On April 4, 2001, a Sudanese Air Force Antonov An-24 crashed during a sandstorm in Adar Year, Sudan. Of the 30 people on board, 14 were killed; among them, Sudan's deputy defense minister and 13-high-ranking officers [55].

FIGURE 2.14 Remains of Boeing E-3A Sentry (707-300B) [57].

2.3.2.5 Birds

On September 22, 1995, A Boeing E-3B Sentry 77-0354 with callsign Yukla 27 (a military Boeing 707 derivative equipped with airborne warning and control system (AWACS)) operated by the USAF was on a training mission was waiting for takeoff. Another aircraft, Lockheed Hercules, departed and disturbed a flock of Canada geese. The Yukla 27 crew were not warned about this by the tower controller and thus as the plane rotated for lift-off, numerous Canada geese were ingested in the #1 and #2 engines, resulting in a catastrophic failure of the #2 engine and a stalling of the #1 engine [56]. The pilots tried to take off, slowly climbed, and began to dump fuel. Losing two engines, the plane could not get enough power and plunged into a heavily wooded area less than a mile from the runway, broke up, exploded, and burned. All 4 crew and 20 passengers were killed (Figure 2.14) [57].

2.3.2.6 Icing

On June 7, 2017, a Shaanxi Y-8-200W 5820 aircraft of the Myanmar Air Force during a trip from Myeik to Yangon crashed, and all the 122 people on board were killed in one of the deadliest aviation accidents in Myanmar [58]. There were no reports of bad weather at the time, so the pilot flew into the storm. Investigators found the pilot had lost control after entering thick storm clouds. Ice had formed on the wings, causing a stall and eventually an unrecoverable spin.

2.3.3 Rotary-Wing Aircraft (Helicopter)

2.3.3.1 Rainstorm

On Monday, June 10, 2019, the helicopter Agusta A109E flown by Timothy McCormack had a horrifying crash on the roof of a Midtown Manhattan skyscraper (54-story AXA Equitable Center on Seventh Ave. at W. 51st St.) just minutes after taking off in heavy rain and dense clouds [59]. Crash obliterated the chopper as it tore a gigantic hole in the roof and sent vibrations through the building. Aviation fuel leaked down the sides of the building in the aftermath of the crash, but the fire was extinguished in 30 min.

2.3.3.2 Volcanoes

2.3.3.2.1 Volcanic Ash

As identified by the National Transportation Safety Board (NTSB), the probable causes of the accidents associated with volcanoes are the flight in and near a volcanic gas cloud, which induces a partial loss of engine power due to a lack of combustible oxygen in the atmosphere.

On Monday, December 9, 2019, a pilot and four passengers were touring White Island when their chopper was blown off its launch pad when the volcano erupted at 2:11 pm [60]. The group, led by pilot Brian De Pauw, went into the water and fled by boat. The group was forced to flee by boat as their chopper's rotors were destroyed, blasting off the launch pad.

2.3.3.2.2 Lava

On June 15, 2003, at 0935 Hawaiian standard times, a McDonnell Douglas Helicopter, Inc. (MDHI) 369D, N4493M, impacted a lava field on the Pulama Pali in the Volcanoes National Park, Volcano, Hawaii. K & S Helicopters, d.b.a . This tour was scheduled as a 45-min flight that would have entailed flying over the Pu'u O'o Vents, down towards the shoreline to see molten lava flowing into the ocean, and then back to ITO.

The helicopter was fully engulfed in flames. A post-impact fire destroyed the helicopter. The fuselage of the helicopter, the main wreckage, was in a small hole (made by the lava). The majority of the helicopter was within a 10-ft radius of the main wreckage. Other portions, the landing gear skids, cross tubes, pieces of Plexiglass windshield, the passenger side door, and the tail boom, had separated into two pieces and was laying 5 to 10 ft from the main wreckage. One passenger was found outside of the helicopter, 10 ft from the main wreckage, while the pilot and the other two passengers were found within the main wreckage. The cause of death for four occupants was listed as thermal injuries that ranged from 90 to 100% of the body due to a helicopter accident.

2.3.3.3 Birds

On January 7, 2014, a USAF Sikorsky HH-60 Pave Hawk crashed into marshes near Cley-next-the-Sea, Norfolk, United Kingdom. The helicopter had experienced multiple strikes of geese (2.7–5.4 kg), which are substantially above any rotorcraft certification requirements. The pilot and co-pilot were incapacitated, which resulted in a loss of control. The helicopter rolled left and impacted the ground approximately 3 sec after being struck by the birds. The four-crew member was killed, and the aircraft was destroyed, which costs were $40.3 million [62].

2.3.3.4 Sandstorm

2.3.3.4.1 Case (1)

On January 26, 2005, a United States Marine Corps CH-53 Sikorsky Super Stallion helicopter ferrying troops crashed during a storm near Al Rutba, Iraqi killing all 31 onboard [61].

2.3.3.4.2 Case (2)

On September 19, 2009, a United States Army Sikorsky UH-60 Black Hawk crashed at Joint Base Balad (formerly Al-Bakr Air Base), Balad, Iraq. The accident occurred during a storm including high winds and a sandstorm resulting in 12 crew being injured and one fatality [61].

2.3.3.4.3 Case (3)

On July 28, 2010, an Iraqi Air Force Mil Mi-17 crashed in a sandstorm southwest of Baghdad, killing five [63].

2.3.3.5 Icing

On January 17, 2012, Royal Canadian Mounted Police (RCMP) Eurocopter AS350B3 C-FMPG was destroyed, and its pilot was killed at Cultus Lake, near Chilliwack, British Columbia. The temperature was approximately -10°C, and there was snow on the blades and the fuselage and tail boom [64]. After lifting off into a hover, the helicopter climbed to approximately 50 ft above ground level (AGL) and slowly traveled up and forward about 260 ft in a straight line before hovering for about 30 sec at the height of 80 ft (AGL). Soon afterward, there was a bang and a puff of grey/white vapor from the exhaust area, and the rotor revolutions per minute (rpm) decayed immediately. The helicopter began to descend, pitched nose-down briefly, and then descended almost vertically. It collided with the terrain in a nose-down, right-side-down attitude. The helicopter was destroyed, and the pilot, the sole person on board, was fatally injured. The investigation determined that the cause of the loss of engine power was a sudden change in the critical air/fuel ratio due to soft ice ingestion.

The lack of auto-ignition systems for single-engine helicopters like the AS350 places pilots and passengers at increased risk of an accident following an engine power loss at low altitude due to snow, water, and ice ingestion.

2.3.3.6 Towel

On May 07, 2015, the pilot of a Rotorcraft Development Corporation (formerly Garlick) UH-1B helicopter registration N46969 reported that she heard a loud noise, followed by a "bang"; the low rotor rpm horn sounded, the pilot entered an autorotation, and the helicopter crashed on a hillside and rolled. The helicopter

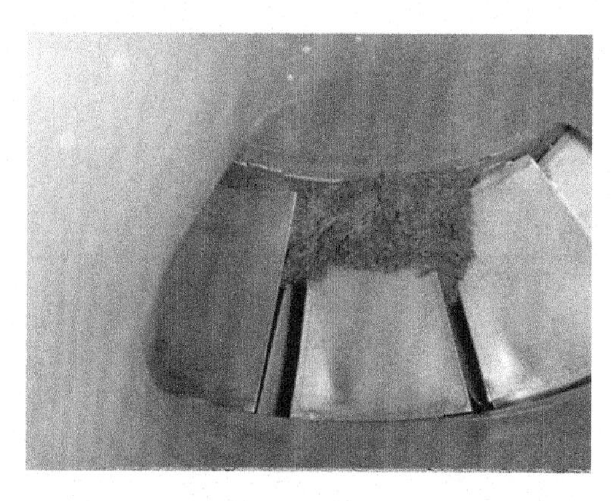

FIGURE 2.15 IGVs with the material.

Courtesy: NTSB.

records indicated that a newly-overhauled engine was installed on the day of the accident.

Hereafter, the summary for the post-accident examination was conducted of the engine (Lycoming T53-L-13B shaft turbine) and airframe [65]:

- The helicopter sustained substantial damage.
- The inlet guide vanes (IGV) had material consistent with a towel (Figure 2.15).
- The trailing edges of the IGVs appeared to be displaced forward and distorted and displayed signs consistent with foreign object damage.
- The compressor section contains material consistent with a towel.
- The interior of the engine exhaust tailpipe showed signs consistent with internal overheating.
- The power turbine section could not be rotated manually, and various power turbine blades were damaged.

2.3.4 Spacecrafts

On January 16, 2003, Space Shuttle Columbia, Columbia's 28th mission (STS-107) [66], was launched from Kennedy Space Center (Figure 2.16). About 81.7 sec after launch, a foam piece, Figure 2.17, having a size of a suitcase, broke off from the external tank (ET) and struck Columbia's left-wing reinforced carbon-carbon (RCC) panels [67]. Such FOD (a piece of foam) damaged the heat shield tiles on the leading edge of the left-wing. At the foam strike, the orbiter was traveling at Mach 2.46 (1,870 mph). Foam shedding had been observed on four previous flights: STS-7 (1983), STS-32 (1990), STS-50 (1992), and STS-112. All caused minor damages, but the shuttle missions were completed successfully.

On February 1, 2003, during re-entry to the atmosphere, the aircraft structure (no longer protected by the tiles) began to melt, causing the wing to fail and the Columbia to break apart, killing all 7 crew members.

FIGURE 2.16 Columbia is lifting off on its final mission.

The outcome of this accident is the following instructions for debris prevention [68]:

- The Space Shuttle System, including the ground systems, shall be designed to preclude the shedding of ice and other debris from the Shuttle elements during prelaunch and flight operations that would jeopardize the flight crew, vehicle, mission success or would adversely impact turnaround operations.
- No debris shall emanate from the critical zone of the External Tank on the launch pad or during ascent except for such material, which may result from a normal thermal protection system recession due to ascent heating.

2.4 ACCIDENTS/INCIDENTS ON-GROUND OPERATION

These may be due to:

- Humans
- Lost tools
- Personal items of maintenance staff (Clamp, leather work glove, rags, etc.)
- Snow
- Mammals (dogs, deer, rabbits, alligators, Giraffe, etc.)

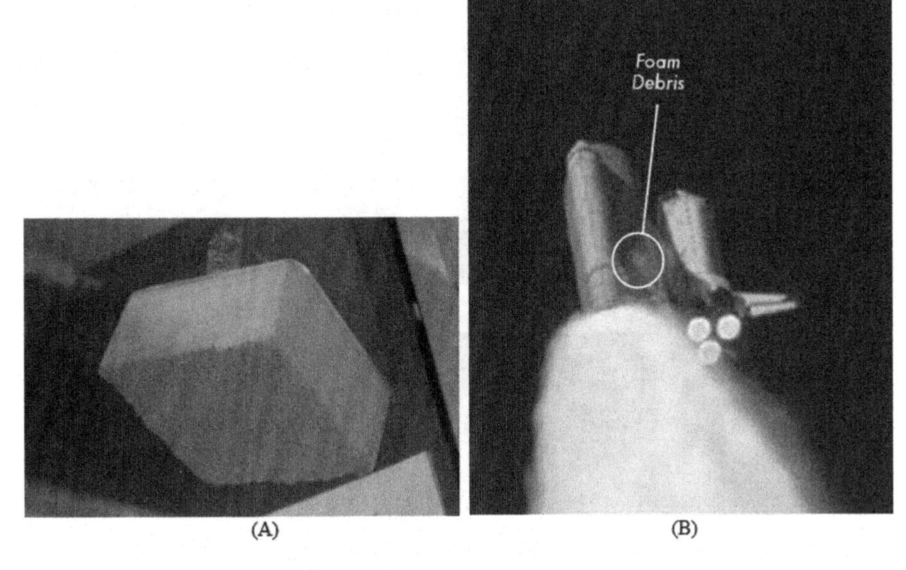

FIGURE 2.17 Space shuttle foam debris.
(a) External tank foam block [68].

Permission (CC BY-SA 3.0).

(b) A shower of foam debris after the impact on Columbia's left-wing (It was not observed in real-time) [67].

- Eastern Grey Kangaroo (Australia)
- Corrosion of aircraft modules

2.4.1 Lost Tool

On March 20, 2017, a Swearingen SA226-TC Metro II aircraft operated by Western Air Express, during takeoff/climb from Boise Air Terminal/Gowen Field (BOI), Boise, Idaho, United States, encountered a strike with foreign object debris, which resulted in substantial damage of the airplane. As the pilot was rotating from the runway, he experienced vibration, and so he decided to return to BOI and landed safely.

Examination of the aircraft identified the following damages (Figure 2.18):

- A 4-inch (100 mm) of the tip of a left propeller blade was missing.
- A piece of the propeller blade found in the cabin.
- A puncture hole of 4-inch by 4-inch on the left forward side of the fuselage just aft of the main air stair door.

As per the NTSB report [69], the probable cause is the missed screwdriver left lodged in the windshield wiper area of the forward fuselage during maintenance and out of sight of the pilot (Figure 2.19). Subsequently, it became dislodged on takeoff/initial climb and collided with a left propeller blade.

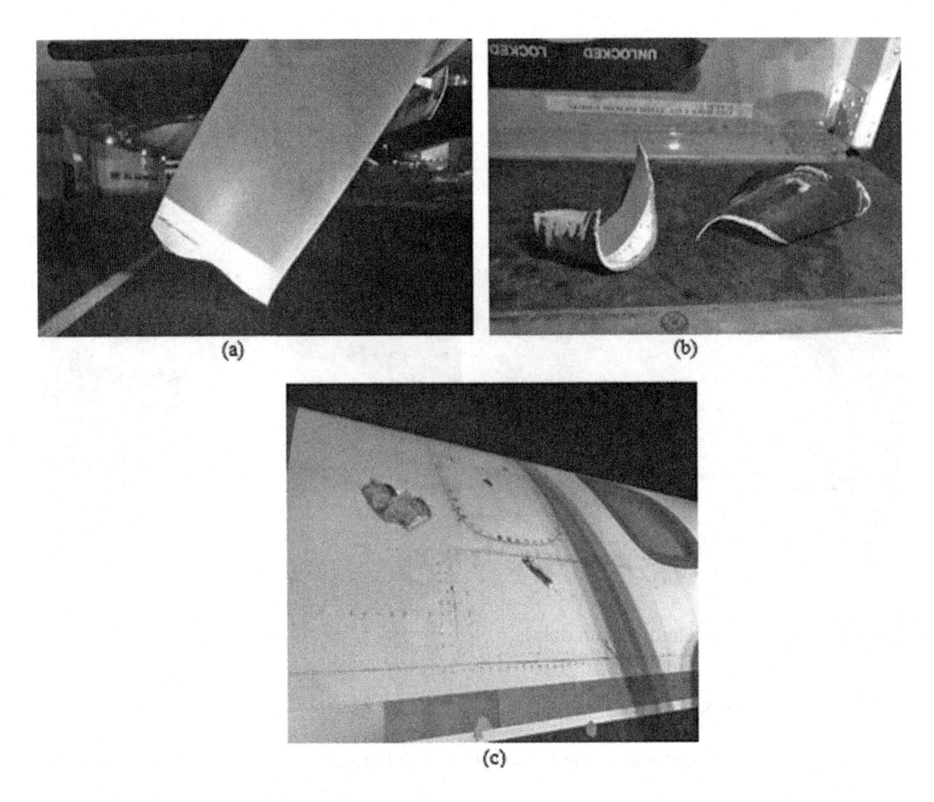

(a) (b) (c)

FIGURE 2.18 Damage of the SA226-TC Metro II aircraft.

Credit: NTSB.

(a) Left Propeller with a missing 4-inch segment of its tip.

(b) Propeller debris that penetrated the fuselage.

(c) Fuselage damage.

2.4.2 Wildlife

2.4.2.1 Gnus (Wildebeests)

On August 16, 2019, the De Havilland DHC-8-200 Dash 8 aircraft registration 5Y-SLM was scheduled from Nairobi Wilson to Maasai Mara Kichwa Tembo (Kenya). During landing at Kichwa Tembo Airstrip in a westerly direction at about 11:00 L (08:00), it collided with some gnus (also known as Wildebeests) [70].

The aircraft suffered substantial damage:

- The left main gear collapsed and separated from the aircraft.
- The left propeller impacted the ground, ejecting one blade.
- The aircraft veered left off the runway.

The passengers and crew were safe and uninjured.

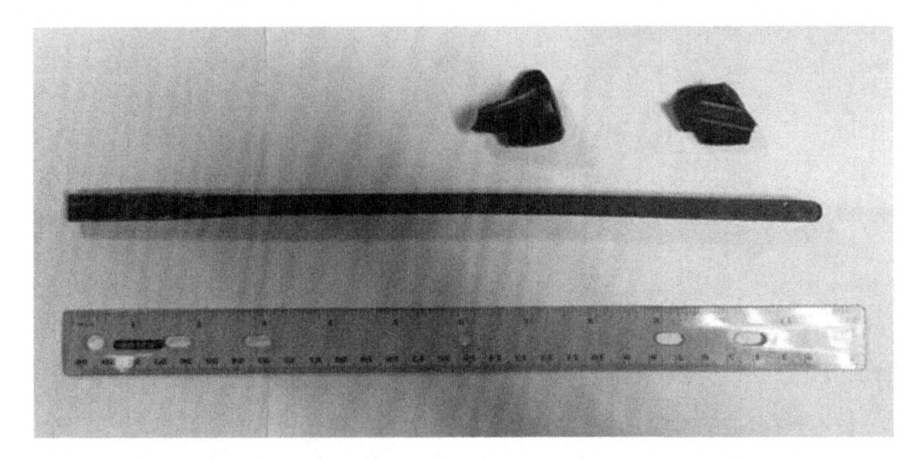

FIGURE 2.19 Lost tool: Screwdriver Foreign Object Debris (FOD) recovered from the runway at Boise after damaging SA226 N158WA.

Credit: NTSB.

2.4.2.2 Eastern Grey Kangaroo

On February 25, 2014, a Mooney M20TN aircraft impact an Eastern Grey Kangaroo [71]. It landed at Kempsey Aerodrome under VFR Aircraft encountered minor damage to the wing and propeller.

2.4.2.3 Alligators

Alligators are extensively found in Florida and other areas, marshy lands near airports. The number of alligator impacts has been increasing from 1998 to February 2018. Examples for these incidents are given hereafter:

In June 2017, a pilot was crossing the runway at the Orlando Executive Airport near downtown Orlando in a small prop plane, and when an alligator leaped into the air and hit the wing [72]. The alligator was killed upon impact.

In 2013, an alligator was involved in a wildlife FOD incident at the Southwest Florida International Airport in Fort Meyers.

On September 12, 2012, at the Orlando International Airport, an alligator was hit by a plane.

Figure 2.20 illustrates an alligator in MacDill Air Force Base in Florida.

2.4.3 Birds

On January 27, 2020, during the takeoff of a Wizzair Airbus A320-200 aircraft registration HA-LPU (flight W6-3351) from Cluj (Romania) to Brussels Charleroi (Belgium) with 173 persons on board, the aircraft received a bird strike [73]. The crew continued their flight to Brussels since there were not any abnormal indications. However, en route about 10 nm southeast of Brno (Czech Republic), the crew decided to divert to Katowice (Poland), where the aircraft landed safely about 75 min after departure.

FIGURE 2.20 An alligator in MacDill Air Force Base in Florida.

Courtesy: US Air Force.

The aircraft was replaced by another Airbus A320-200 aircraft, which landed in Brussels Charleroi Airport with a delay of about 2 h. The HA-LPU aircraft remained on the ground for about 9.5 h after landing.

2.4.4 Runway Edge Lights Fragments

On October 23, 2010, Finnair Embraer ERJ-190, registration OH-LKL was performing flight AY-658 from Oslo (Norway) to Helsinki (Finland) with 31 passengers. The flight crew mistook the left runway edge lights for runway centerline lights while lining up. During takeoff, the tires on the left (LH) main gear and nose gear collided with three runway edge lights [74]. The tires sustained minor damage, and the runway edge light fragments were flung into the right (RH) engine.

The pilot then corrected onto the centerline, and the aircraft and its engines normally operated at takeoff. The flight was continued to Helsinki-Vantaa airport, the destination. The strike with the edge lights resulted in the damage of:

- The nose gear tire
- The left-hand main gear tires
- The right-hand engine;

The damage was so substantial that the engine and both tires had to be replaced.

2.4.5 Snow

On December 12, 2019, a Russian Helicopters (Mil) Mi-8 (reg. RA-22720) operated by KrasAvia, during takeoff with 24 people on board fell while taking off from an airfield in the remote settlement of Baykit due to a snow tornado that damaged the tail boom [75].

Six people were injured. Doctors are examining four of them, and two others are undergoing surgery. The condition of all those injured is assessed as moderate, and they mostly have fractures.

The aircraft captain was sent to a hospital, and the number of those wounded in the crash later rose to 15.

The copter experienced substantial damage due to its rough landing.

2.5 ACCIDENT/INCIDENT DURING THE FLIGHT

2.5.1 Fixed-Wing Aircraft

2.5.1.1 Possible Causes

These may be due to:

- Birds and bats
- Drones
- Fasteners; Harrier Aircraft
- Fog
- Snowstorm
- The icing of aircraft parts and instruments
- Hail
- Mechanic's personal items
- Rainfall
- Wind shear
- Sandstorm
- Volcanoes
- Parts of a refueling basket (military aircraft)

2.5.1.2 Icing

On January 29, 2018, the two-seater EA-18G was cruising at 25,000 ft, about 60 miles south of Seattle, on a flight from Washington state's Naval Air Station Whidbey Island to Naval Weapons Station China Lake [76]. The system that controls the cockpit air temperature and cabin pressure, known as the environmental control system, was icing.

The temperature inside the cockpit suddenly dropped to -30 degrees, and a mist pumped into the cockpit, covering the instruments and windows in a layer of ice, rendering the pilots almost completely blind.

The fog inside the aircraft iced over the instrument panel, forcing the pilot and electronic warfare officer (EWO) to use a Garmin watch to track their heading and altitude. At the same time, air controllers began relaying instructions to the crew. The pilot and EWO were forced to use the emergency oxygen supply, which was completely depleted by the end of the flight.

A heroic effort by the two-person crew and the ground-based controllers managed to guide the aircraft back to Whidbey Island. The aircrew suffered from "severe blistering and burns on hands" which was treated upon landing.

2.5.1.3 Hail

Two problems are associated with hail storms:

1. Hail is more or less invisible until the aircraft runs into it
2. Hail does not just come out of the bottom of clouds but can be ejected from the top of some larger storms. Large thunderstorms are maelstroms of updrafts and downdrafts. When hail is caught in an updraft, it can be thrown out the top of the storm for dozens of miles.

Hereafter, examples for two case studies of hailstorms striking aircraft.

2.5.1.3.1 Case (1)

On Friday August 7, 2015, Delta Airlines Airbus A320, Flight #1889, was flying from Boston to Salt Lake, with 125 passengers and 5 crew. En route, it ran into an area of heavy hail intensity near Denver [77]. Hail forced the pilots to divert the flight making an emergency landing at Denver International Airport.

Hail resulted in the following damages [78]:

- The nose cone, housing the plane's radar, had almost entirely collapsed
- The paint had been shredded off
- The windshield has been shattered, causing visibility issues for the pilots

There were no injuries other than some passenger anxiety.

2.5.1.3.2 Case (2)

On July 27, 2017, Airbus A320 passenger aircraft registration UR-AJC operated by Atlasglobal Ukraine during initial climb suffered hail strike resulting in substantial damage [79].

2.5.1.4 Snow

On Monday, February 27, 2017, a single-engine 6-seat airplane SOCATA TBM 700 crashed in a snowstorm at Bellingham International Airport (KBLI), Whatcom County, Washington, leaving a pilot with minor injuries.

As per the final report of the National Transportation Safety Board (NTSB), during the preflight, it was snowing, and the pilot wiped the snow. Snow continued to fall and accumulate on the wing during taxiing and takeoff roll [80]. During the takeoff roll, the snow was sloughing off the wings. As the airspeed increased, the airplane yawed to the left, then stalled. Finally, the airplane impacted the ramp in a left-wing-down attitude and slid 500 to 600 ft.

The airplane sustained substantial damage to the fuselage and left-wing, but luckily the pilot had minor injuries.

2.5.1.5 Birds

A bird strike becomes a daily event, but only two cases will be discussed hereafter.

2.5.1.5.1 Case (1)

On June 3, 1995, during the landing of Concorde aircraft in JFK airport (New York, United States) and at 10 ft AGL, one or two Canadian geese were ingested into the #3 engine. Shrapnel from the #3 engine destroyed the #4 engine and cut several hydraulic lines and control cables [81]. The pilot landed safely, but the runway was closed for several hours. Damage to the Concorde was estimated at over US $7 million, as well as compensation of US $5.3 million for the Authority of New York and New Jersey and eventually settled out of court (total cost US $12.3 million).

2.5.1.5.2 Case (2)

On July 3, 2017, AirAsia X1 flight D7207, an Airbus 330 aircraft (powered by two Rolls Royce Trent 700 engines), registered 9M-XXT, took off at Gold Coast Airport, Queensland, for a scheduled passenger transport flight to Kuala Lumpur, Malaysia [82].

As the flight crew commenced the takeoff roll, the flight data recorder data showed an increase in the vibrations of Engine #2. Next, at nearly 2,300 ft, Engine #2 Stalled with loud banging accompanied by fire. Accident analysis identified that two masked lapwings (commonly known as a plover) struck engine #2. Their remains included one complete carcass, as well as additional debris from another bird.

Examination of the engine at Brisbane Airport showed that:

- A single fan blade-tip section, approximately 140 mm x 125 mm, had fractured from one fan blade.
- The fan rear seal was found broken into pieces and scattered throughout the bypass areas of the engine.
- Fragments of this seal entered the core of the engine leading to significant damage to the compressors and resulting in compressor stalls
- Evidence of fire within the engine

2.5.1.6 Volcanic Ash

Jet aircraft are damaged when they fly through clouds containing finely fragmented rock debris and acid gases produced by explosive volcanic eruptions [83]. Clouds of volcanic ash and corrosive gases cannot be detected by weather radar currently carried aboard airplanes, and such clouds are difficult to distinguish visually from meteorological clouds.

The explosive eruptions of volcanos inject enormous clouds of volcanic ash and gases into the stratosphere to very high altitudes exceeding 100,000 ft. The eruption may continue for several days damaging commercial jet airplanes during their in-flight encounters with the drifting ash clouds.

FIGURE 2.21 DC-10-30 resting on its tail due to volcanic ash [84].

For nearby airports, the falling ash may damage several aircraft on the ground. Volcanic ash deposits on a parked McDonnell-Douglas DC-10-30 during the 1991 eruption of Mount Pinatubo caused the aircraft to rest on its tail [84]. The falling ash deposits were able to cause the 120-ton airliner's center of gravity to shift (Figure 2.21).

2.5.1.7 Missing Pin

On December 18, 2017, Jetstar Airways Airbus A320 aircraft registered VH-VQG Flight #JQ452 from Adelaide, South Australia to Gold Coast Airport, Queensland, was on the final approach; after a normal descent and touchdown, the captain selected both engine thrust reversers. The left engine thrust reverser did not activate [85]. The aircraft decelerated using normal braking and taxied to the gate without further incident. During overnight maintenance, staff in Adelaide installed the thrust reverser for the left engine but did not remove it.

The next engineering inspection found the left engine thrust reverser lockout pin installed, effectively deactivating the reverser. The lockout pin was removed, the thrust reverser confirmed to be operating normally, and the aircraft returned to service. There was no damage to the aircraft or injuries as a result of the incident.

2.5.1.8 UAV/Drones

On Thursday, October 15, 2017, a Skyjet plane was struck by a drone approaching the Jean Lesage Airport, Quebec City, Canada. This was the first time in Canada that a drone has collided with a commercial aircraft [86]. The Skyjet plane carrying eight passengers sustained minor damage but landed safely. The strike of any drone with aircraft might be catastrophic if it collided with the cockpit or the engine.

Transport Canada stated that in 2017, 1,596 drone incidents had been reported, 131 of which were deemed aviation safety concerns. This Transport Canada has issued a series of interim safety measures for drone operators, including:

- It is illegal to fly a recreational drone within 5.5 km from an airport and 1.8 km from a heliport without special permission
- Anyone who endangers the safety of an aircraft could face a $25,000 fine or prison time.
- Drone operators must have:
  1. a minimum age
  2. a mandatory written test
  3. Registered their names and addresses on the drone itself

2.5.2 Military Aircraft

2.5.2.1 Birds

On October 2, 2019, an E-6B Mercury airborne communications aircraft was forced to land after a bird was sucked into one of its four engines, destroying it.

No one on the aircraft was injured. It's unclear what species of birds were involved.

The incident was classified as a "Class A" incident since the cost of engine repair and replacement exceeded US$ 2 million [87].

2.5.3 Rotary-Wing Aircraft (Helicopter)

2.5.3.1 Rags

On June 25, 2013, when the pilot of MD (formerly McDonnell Douglas) Helicopters MD-369E was conducting a post-maintenance check flight, the engine lost power, and the pilot entered an autorotation. He managed to land hard, its right skid collapsed, and the helicopter rolled on its right side [88, 89]. A post-accident examination of the helicopter's engine revealed that:

- Cloth material had been ingested into the engine air intake (Figure 2.22).
- Cloth material was found in the engine in several locations, which blocked the airflow through the engine and caused it to flame out (Figure 2.23)
- The cloth material found in the engine was consistent with maintenance rags found in a box at the operator's hangar facility.

The events leading to the accident started when maintenance personnel removed the engine from the helicopter to exchange the helicopter's air inlet barrier filter system. During which maintenance personnel covered vulnerable areas of the engine with shop rags to prevent contamination. But, during the reinstallation of the engine, they did not remove all of the shop rags. Due to the installation of the engine air inlet barrier system, the shop rags would not have been visible during the preflight inspection.

FIGURE 2.22 Engine intake partially blocked with rags [88].

Courtesy: NTSB.

2.5.3.2 Missed Wrench

On March 16, 2010, the Airbus AS350 B3 helicopter experienced moderate damage during a post-maintenance check flight. As the main rotor speed reached 100%, the pilot and the two mechanics heard a "bang." After the helicopter landed, the mechanics discovered that the adjustable wrench was missing [90].

Post-accident investigation revealed the damage of:

- One main rotor blade
- Tail boom
- Lower vertical stabilizer

2.5.3.3 UAV (Drones)

On September 21, 2017, a civilian UAV collided with a Black Hawk helicopter in the evening over the eastern shore of Staten Island, New York City, United States. The drone slammed into the side of the Helicopter. Nobody was hurt, but a part of the drone was found on the transmission deck right at the bottom of the main rotor system.

The NTSB report issued in December 2017 about the collision found the UAV pilot at fault. He deliberately flew the UAV 2.5 miles away from himself and was unaware of the helicopters' presence. Under FAA regulations, UAVs are prohibited from flying beyond the line of sight [91].

FIGURE 2.23 Mechanics rags inside the compressor [88]: (top) compressor case halves removed; (bottom) successive stages of the compressor.

Courtesy: NTSB.

2.5.4 Space Vehicles

2.5.4.1 Birds

On April 8, 2002, and from Kennedy Space Center, FLA, the Space Shuttle Atlantis [92] seems surrounded by birds – most likely pelicans – as it roars into the clear blue sky on mission STS-110 (Figure 2.24). No damage to the space shuttle, though many birds were killed.

FIGURE 2.24 Space shuttle.

Courtesy: NASA.

2.5.4.2 Generation of Foreign Object Debris in Space

Space debris result from collisions, defunct satellites, and decades of ill-regulated activities in space [93]. By 2016, there were nearly 500,000 pieces of human-generated space debris in orbit, of which approximately 23,000 are pieces of junk larger than 10 cm, causing an increasingly high-risk space domain. Moreover, collisions between satellites generate extra hundred debris. Since the objects in orbit are moving up 17,000 miles per hour in space, then if one of these pieces hits another satellite, it can cause damage that might make a satellite inoperable. Moreover, some of these pieces are potentially quite small, making them hard to track.

To avoid possible collisions, satellite operators are forced to perform almost routine maneuvers to change their satellite's trajectory to avoid a collision with debris.

In recent decades, ground-air launched anti-satellite (ASAT) weapons are used in military and civilian activities. It is the capability of destroying satellites in orbit by shooting them with a missile launched from Earth.

Experts are concerned with such action as destroying satellites with missiles can potentially create hundreds of thousands of pieces of debris that remain in space for many years. Pieces created from an ASAT test can stretch out in a stream over long distances, covering many miles in space along the satellite's original orbit, even spreading to slightly higher and lower altitudes. Within weeks and months, that stream will get broader and wider and more diffused.

Shooting satellites using intercontinental ballistic missiles must not break any international laws. The Outer Space Treaty, which entered into force in 1967, governs how nations behave in space. That treaty bans the use of weapons of mass destruction in orbit, but it does not explicitly ban the use of missile technologies used for ASAT.

2.5.4.3 Foreign Object Damage in Space

Since 2016, NASA is tracking more than 500,000 pieces of debris, or space junk, orbiting the Earth at speeds greater than 28,000 km/h. Their size ranges from that of a tennis ball to a tiny marble, though most of them are millimeter-sized particles, which are too small to track. The greatest risk to space missions comes from such non-trackable debris.

The problem of space debris gets worse, even if no more objects are sent in space around the Earth. The reason is that the objects already there will slowly disintegrate into smaller particles, which may collide with one another and create more particles and so on (Figure 2.25).

2.5.4.3.1 Case (1)

In March 2017, India mastered what is known as anti-satellite, or ASAT, and claimed its ability to destroy one of its satellites in orbit by shooting it with a missile launched from Earth [94].

2.5.4.3.2 Case (2): Sentinel-1A satellite

On August 23, 2016, a solar panel on the Sentinel-1A satellite, which does routine environmental monitoring, had been hit by space debris of a millimeter-sized

FIGURE 2.25 Debris in space.

Courtesy: NASA.

FIGURE 2.26 Damage of a solar panel on the Sentinel-1A satellite [95].

Courtesy: NASA.

particle (Figure 2.26). However, the damage caused was about 100 times its diameter [95].

2.5.4.3.3 Case (3)

On February 10, 2009, a Russian satellite collided with a US satellite 800 km above Siberia, which left a cloud of debris consisting of approximately 500–600 pieces [96].

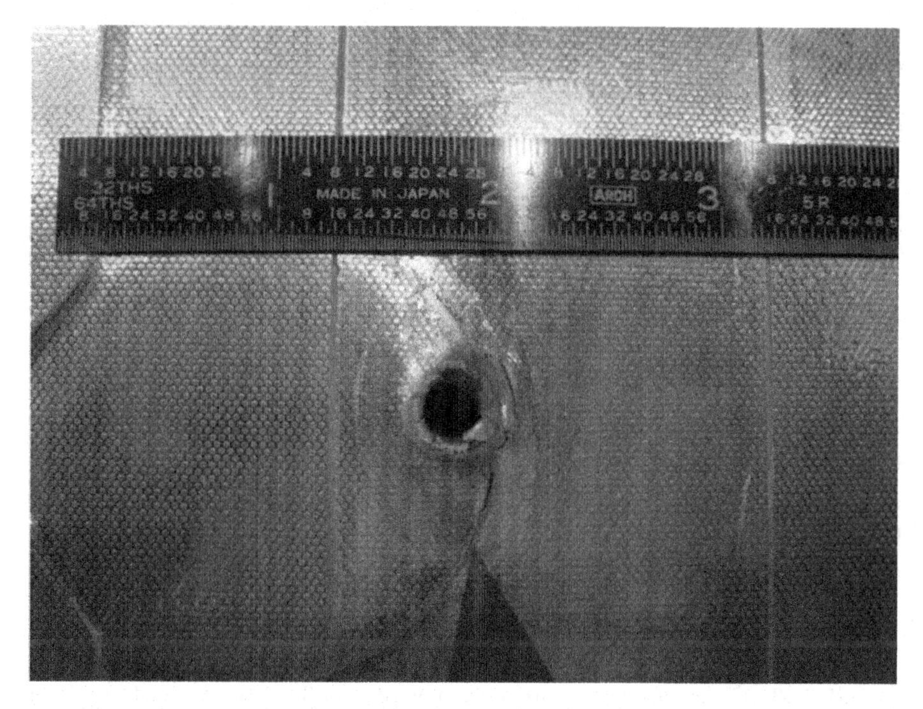

FIGURE 2.27 Entry-hole damage (5.5 mm diameter hole) to Endeavour's left-side aft-most radiator panel observed during post-flight inspection [95].

Courtesy: NASA.

2.5.4.3.4 Case (4): The Radiator of the Endeavour Space Shuttle Mission STS-118

In August 2007, during its mission to the International Space Station, either a micrometeoroid or orbital debris (MMOD) particle impacted and completely penetrated one of the shuttle Endeavour's radiator panels and the underlying thermal control system (TCS) blanket, leaving deposits on (but no damage to) the payload bay door [95]. The damage from this impact by small MMOD particles is larger than any previously seen on the shuttle radiator panels. A close-up photograph of the radiator impact entry hole is shown in Figure 2.27.

2.5.4.3.5 Case (5)

In 2007, China destroyed its weather satellite during an ASAT test. Amateur satellite trackers estimated that that test created more than 3,000 objects, many of which have remained in orbit for years since the incident [94].

2.5.4.3.6 Case (6): Space Shuttle Challenger's Front Window on Mission STS-7

There are also risks from meteorites coming from space in addition to the human-made threats from debris. For example, a micrometeoroid left the crater shown in

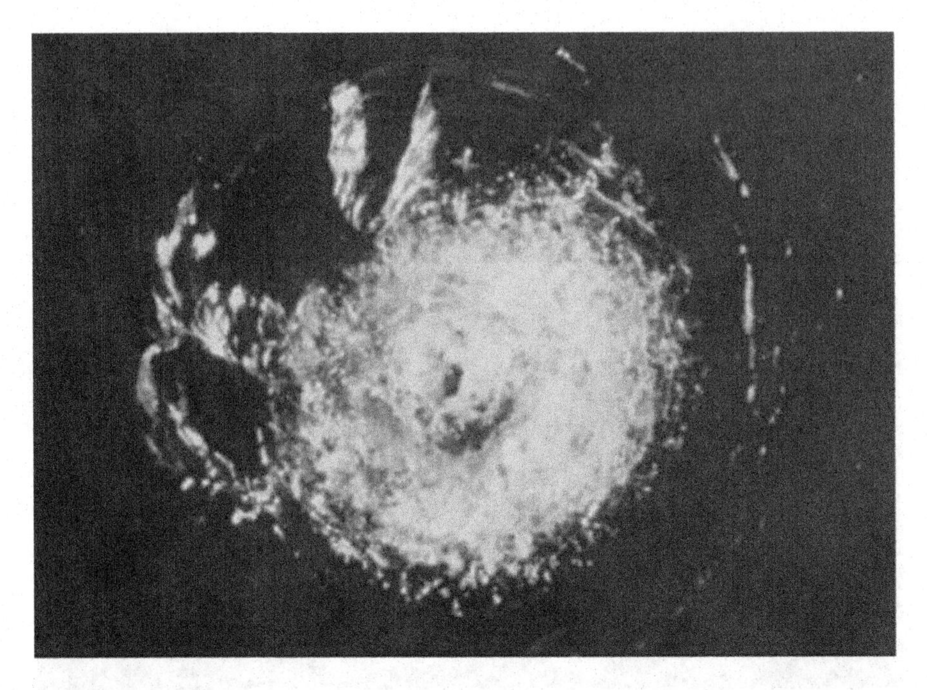

FIGURE 2.28 Crater on the surface of the front window of Space Shuttle Challenger's on Mission STS-7 [95].

Figure 2.28 on the surface of the Space Shuttle Challenger's front window on STS-7 in 1983 [95].

It is important to minimize the space debris in orbit around Earth, as this will, in turn, reduce the risk of causing severe damage to valuable space assets that help humankind in his activities; like:

- Using Google Maps
- Space missions, such as sending rovers to Mars
- Supply missions to the International Space Station

2.6 SAFETY ALERT FOR MINIMIZING FOD

The NTSB highlights the problem as follows [4]:

- Mechanics, or others who help with aircraft maintenance, might leave items or residual debris behind after performing maintenance tasks that could become foreign object debris (FOD). Examples of FOD include tools, hardware, eyeglasses, keys, portable electronic devices (PEDs), paint chips, and metal shavings.

- If mechanics and others do not account for every item that they use in or around an aircraft and clean as they go, this FOD can be ingested into the engine or interfere with critical flight systems, leading to an accident

Next, it states the following Safety Alert actions that workers can take to minimize the chances of FOD being left behind to do damage [90]:

- Perform an inventory of tools, personal items, and personal protective equipment before working on an aircraft. Take only what is necessary for the specific maintenance task. Consider placing nonessential personal items such as jewelry, coins, keys, and PEDs in a secure location instead of keeping them with you during maintenance tasks.
- Prepare the workspace on the aircraft by covering engines, static pitot ports, air inlets, and other areas with protective materials to reduce the likelihood of FOD migration (including residual debris, such as paint chips or metal shavings) to critical flight systems.
- While working in low-visibility areas (ramp/hangar), ensure proper lighting is used to check for FOD left behind during maintenance.
- Keep hardware and consumables in appropriate containers to prevent them from becoming FOD. Store tools in toolboxes and bags, then organize them so that you can easily recognize if one or more is missing.
- Distractions can cause you to forget things during maintenance tasks. Always follow the maintenance manual/task card and use a checklist. If you get distracted, go back three steps when restarting your work.
- As you perform the maintenance task, clean as you go to reduce the likelihood of leaving any items. Keep a FOD container next to you during the maintenance task for easy FOD disposal.
- Perform a second inventory of tools, any essential personal items, and personal protective equipment (such as safety glasses, gloves, and hearing protection) after you have completed the maintenance task to ensure that items have not been left behind. Remove any aircraft protective materials so that they do not become FOD.
- Ask another mechanic to visually inspect your work area for any items that may become FOD. The second set of eyes might see something that you missed.
- Recognize that human factors such as complacency, fatigue, pressure, stress, and a lack of situation awareness can contribute to FOD.
- Consider conducting daily FOD walks in areas such as hangars, ramps, and runways to identify and remove FOD.

For safety's sake, it is always a good idea to be on the lookout for FOD. Check your hangar, your vehicle, your toolbox, and even your passengers for FOD. Take a look up and down the ramp now and then to make sure areas you and fellow aviators use stay clear of rock chips, concrete pieces, screws, and fasteners of all kinds. If any construction has been performed on the hangar, the taxiway, or nearby buildings, the potential for damaging FOD goes way up. Wave has all seen odd things on airplane taxiways and in ramp areas over the years, and in both military and commercial environments, FOD demands attention. The same is true for general aviation.

Help each other do these things and encourage everyone on the flight line to become FOD-aware. If you think doing a FOD check takes too much time, it's nothing compared to waiting for investigators and attorneys to let you get on with your life.

REFERENCES

[1] Ian Savage, Comparing the Fatality Risks in United States Transportation Across Modes and Over Time, *Research in Transportation Economics: The Economics of Transportation Safety*, 43(1), 9–22, 2013.

[2] Aircraft Accident and Incident Investigation. Annex 13, International Civil Aviation Organization. www.pilot18.com/wp-content/uploads/2017/10/Pilot18.com-ICAO-Annex-13-Aircraft-Accident-and-Incident-Investigation.pdf (accessed January 11, 2020).

[3] Accident/Incident Reporting Manual (ADREP Manual), 2e. International Civil Aviation Organization (Doc 9156-AN/900), 1987.

[4] A Statistical Analysis of Commercial Aviation Accidents 1958–2018, Airbus, https://aviation-safety.net/airlinesafety/industry/reports/Airbus-Commercial-Aviation-Accidents-1958-2018.pdf

[5] Causes of Fatal Accidents by Decade, June 2013, http://planecrashinfo.com/cause.htm

[6] Accidents by Flight Phase, http://accidentstats.airbus.com/statistics/accident-by-flight-phase

[7] Statistics, Causes of Fatal Accident by Decades, 2021, http://planecrashinfo.com/cause.htm

[8] Fatal Accidents for Airliners by Period, ASN, 2021, https://aviation-safety.net/statistics/period/stats.php?cat=A1

[9] Air safety statistics in the EU, Eurostat, Statistics Explained, ISSN 2443-8219, November 2021, https://ec.europa.eu/eurostat/statistics-explained/index.php/Air_safety_statistics_in_the_EU#Overview_of_fatalities_in_air_transport_in_the_EU

[10] Sharon Monica Jones, Mary S. Reveley, Joni K. Evans, and Francesca A. Barrientos, Subsonic Aircraft Safety Icing Study, NASA/TM—2008–215107, 2008.

[11] Morgan B. Pfeiffer, Bradley F. Blackwell, and Travis L. DeVault. "Quantification of avian hazards to military aircraft and implications for wildlife management," *PLOS ONE*, November 1, 2018, USDA, https://doi.org/10.1371/journal.pone.0206599

[12] The Current Mishap Definitions and Reporting Criteria, file:///C:/short%20courses%20KANSAS%20University/ACCIDENTS/MILITARY/NAVAL%20SAFETY%20CENTER.html

[13] Stephen Losey, Air Force aviation accidents reach seven-year high as low-level mishaps soar, April 8, 2018, www.airforcetimes.com/news/your-air-force/2018/04/08/air-force-aviation-accidents-reach-seven-year-high-as-low-level-mishaps-soar/?utm_expid=.jFR93cgdTFyMrWXdYEtvgA.1&utm_referrer=#jwvideo

[14] Aviation Statistics, Air Force Safety Center, www.safety.af.mil/Divisions/Aviation-Safety-Division/Aviation-Statistics/

[15] Edward Hobbs and Krystyna Eaker, Foreign Object Damage Mishap Analysis, Commander Naval Air Forces, 2020, www.public.navy.mil/NAVSAFECEN/Documents/statistics/ops_research/PDF/10-026.pdf

[16] Lindamae Peck, Charles C. Ryerson, and C. James Martel, Army Aircraft Icing, US Army Corps of Engineers, Engineer Research and Development Center, Technical Report, ERDC/CRREL TR-02-13, 2002.
[17] Helicopter Safety and Losses Annual Review 2018, file:///C:/Users/dr_ah/Downloads/Helicopter%20Safety%20and%20Losses%202018.pdf
[18] Lee Roskop, U.S. Rotorcraft Accident Administration Data and Statistics, 2012 FAA/Industry Safety Forum, January 2012, http://aea.net/events/rotorcraft/files/US_Rotorcraft_Accident_Data_And_Statistics.pdf
[19] Aviation safety network, https://aviation-safety.net/database/record.php?id=19950922-0
[20] U.S. Helicopter Accidents Decrease, FAA, February 10, 2017, www.faa.gov/news/updates/?newsId=87406
[21] Vertical: Mixed results for U.S. helicopter accident rates during 2018, March 11, 2019; www.verticalmag.com/press-releases/mixed-results-for-u-s-helicopter-accident-rates-during-2018/
[22] Aviation Safety Network, ASN, Friday 18 March 1966, https://aviation-safety.net/database/record.php?id=19660318-0
[23] Crash of a Cessna 550 Citation II in Greenwood, Bureau of Aircraft Accident Archives, www.baaa-acro.com/crash/crash-cessna-550-citation-ii-greenwood
[24] National Transportation Safety Board, FACTUAL REPORT – AVIATION, ID: ERA13LA061, Occurrence Date: 11/17/2012, www.baaa acro.com/sites/default/files/import/uploads/2013/12/N6763L.pdf
[25] Accident Report: Loss of Thrust in Both Engines After Encountering a Flock of Birds and Subsequent Ditching on the Hudson River US Airways Flight 1549 Airbus A320-214, N106US Weehawken, New Jersey January 15, 2009, NTSB/AAR-10/03 PB2010-910403, file:///C:/short%20courses%20KANSAS%20University/ACCIDENTS/HUDSON%20MIRACLE.pdf
[26] Flight path flown (red). Alternative trajectories to Teterboro (dark blue) and back toward La Guardia (magenta) were simulated for the investigation, March 12, 2010. https://en.wikipedia.org/wiki/US_Airways_Flight_1549#/media/File:Flight_1549-OptionsNotTaken.PNG
[27] US Airways Flight 1549, 2021. https://en.wikipedia.org/wiki/US_Airways_Flight_1549#/media/File:US_Airways_Flight_1549_(N106US)_after_crashing_into_the_Hudson_River_(crop_2).jpg,
[28] ABC NEWS, Jon Haworth, 23 injured when Russian jet makes emergency landing in corn field after striking birds, August 15, 2019, https://abcnews.go.com/International/23-injured-russian-jet-make-emergency-landing-corn/story?id=64987115
[29] Aviation Safety Network, ASN, Wednesday 12 November 1975, https://aviation-safety.net/database/record.php?id=19751112-1
[30] Junta lnvestigadora de Accidentes Aereos (1996) Final aviation accident report Birgenair Flight ALW-301 Puerto Plata, Dominican Republic February 6, 1996. Dominican Republic.
[31] National Transport Safety Bureau, Florida Commuter Airlines Inc. Douglas DC-3, N75KW Grand Bahama Island, Bahamas September 12, 1980: Accident Report, Washington DC, 19.
[32] BEA Report: Accident on 25 July 2000 at La Patte d'Oie in Gonesse (95) to the Concorde registered F-BTSC operated by Air France, translation f-sc000725a, F-BTSC – July 25, 2000, https://reports.aviation-safety.net/2000/20000725-0_CONC_F-BTSC.pdf
[33] The aircraft's vertical stabilizer after its recovery from the Ocean https://en.wikipedia.org/wiki/Air_France_Flight_447#/media/File:Voo_Air_France_447-2006-06-14.jpg

[34] Final report: On the accident on 1st June 2009 to the Airbus A330-203 registered F-GZCP, Bureau d'Enquêtes et d'Analyses pour la Sécurité de l'Aviation Civile, Final Report, July 27, 2012, www.bea.aero/docspa/2009/f-cp090601.en/pdf/f-cp090601.en.pdf

[35] Final Report on Results of Investigation of Accident, ATR72-201 VP-BYZ, Interstate Aviation Committee Air Accident Investigation Commission, www.aviation-accidents.net/report-download.php?id=94

[36] The wreckage of UTair 120. https://en.wikipedia.org/wiki/UTair_Flight_120#/media/File:UTair_120_wreckage.jpg

[37] Aircraft Accident Report, Air Florida Inc. Boeing 737-222, N62AF, Collins with 14th Street Bridge, Near Washington National Airport, Washington DC, January 13, 1982, NTSB-AAR-82-8, www.ntsb.gov/investigations/AccidentReports/Reports/AAR8208.pdf

[38] Aircraft Characteristics Airport and Maintenance Planning, Airbus A320, November 01/2019, file:///C:/Users/dr_ah/Downloads/Airbus-Commercial-Aircraft-AC-A320.pdf

[39] Man Dies After Being Sucked into Jet Engine, Sky News, December 17, 2015. https://news.sky.com/story/man-dies-after-being-sucked-into-jet- engine-10335701

[40] Mechanic Sucked into Boeing 737-500 Engine, Key.Aero, May 17, 2010, www.key.aero/forum/commercial-aviation/102843-mechanic-sucked-into-b737-500-engine

[41] Engineer Sucked Through 737 Engine, PPRuNe, Professional Pilots, November 4, 2004, www.pprune.org/engineers-technicians/150865-engineer-sucked-through-737-engine.html

[42] Aviation Accident Summary ERA12LA307, 2012, www.accidents.app/summaries/accident/20120428X53650

[43] Aviation Safety Network, ASN Wikibase Occurrence # 231966, https://aviation-safety.net/wikibase/231966

[44] Aircraft Accident Investigation Report, China Airlines B 1 8 6 1 6, AA2009-7, August 28, 2009, Japan Transport Safety Board, www.mlit.go.jp/jtsb/eng-air_report/B18616.pdf

[45] Burned remains of China Airlines 737-800 registration number B-18616. https://en.wikipedia.org/wiki/China_Airlines_Flight_120#/media/File:China_Airlines_B-18616_fire.jpg

[46] Aircraft Accident Report, Southern Airways DC-9-31 aircraft, registration N1335U, New Hope, Georgia, April 4, 1977, NTSB REPORT AAR 78-3, www.ntsb.gov/investigations/AccidentReports/Reports/AAR7803.pdf

[47] The remnants of the tail section. The separated right wing is in the foreground, https://en.wikipedia.org/wiki/Southern_Airways_Flight_242#/media/File:Flight_242_tail_and_wing.jpg

[48] Aircraft Accident Report Controlled Flight Into Terrain Korean Air Flight 801 BOEING 747-300, HL7468 Nimitz Hill, Guam, August 6, 1997, NTSB/AAR-00/01 DCA97MA058, www.ntsb.gov/investigations/AccidentReports/Reports/AAR0001.pdf

[49] British Airways Flight 9, https://en.wikipedia.org/wiki/British_Airways_Flight_9

[50] Damaged engine parts from BA009 on display at Auckland Museum. https://en.wikipedia.org/wiki/British_Airways_Flight_9#/media/File:British-Airways-Flight-9_turbine_and_compressor_blades.JPG

[51] EDITOR, Aviation accidents, Japan Airlines – BoeingB747-300 (JA8178) flight JL726, May 5, 2019, www.aviation-accidents.net/japan-airlines-boeing-b747-300-ja8178-flight-jl726/

[52] Geoff Ziezulewicz, Navy F-35C suffers first major airborne mishap, September 7, 2018, www.navytimes.com/news/your-navy/2018/09/07/f-35-suffers-first-major-airborne-mishap/
[53] Brandon Wissbaum, WECT NEWS 6, Bottom of FormA Marine Corps Harrier jet pilot heard three 'thumps.' Then his engine failed catastrophically, December 6, 2017, www.wect.com/story/37005907/a-marine-corps-harrier-jet-pilot-heard-three-thumps-then-his-engine-failed-catastrophically/
[54] Deadliest American Disasters and Large Loss-of-Life Events, 2001-March 3, USAF C-23 Plane Crash (VA-NG(18) FL ANG (3)), Unadilla, GA-21, March 3, 2001, www.usdeadlyevents.com/2001-march-3-usaf-c-23-plane-crash-va-ng-18-fl-ang-3-unadilla-ga-21/
[55] "Chronological Listing of Sudanese Air Force." Ejection-history. Retrieved August 19, 2011.
[56] ASN Aircraft accident Boeing E-3A Sentry (707-300B) 77-0354,1995
[57] Aviation safety network, Photo of Boeing E-3A Sentry (707-300B) 77-0354, https://aviation-safety.net/photo/5830/Boeing-E-3A-Sentry-77-0354
[58] Skynews, Military aircraft with 120 people on board goes missing off Myanmar, June 7, 2017, https://news.sky.com/story/military-aircraft-with-120-people-on-board-goes-missing-off-myanmar-10907332
[59] Ennica Jacob, Kerry Bruke, Catherina Gioino, and Cathy Bruke, New York Daily News, Helicopter crash-lands on the roof of Midtown NYC skyscraper in heavy rain, thick clouds, killing veteran pilot, June 10, 2019, www.nydailynews.com/new-york/ny-helicopter-crash-midtown-20190610-yrgqgkqmvveibjhzqsqz3unezm-story.html
[60] Zoe Zaczek, Haunting photo of helicopter blasted off its launch pad and covered in ash reveals an incredible story of survival in the wake of the NZ volcano disaster, Daily Mail Australia, 12 December 2019. www.dailymail.co.uk/news/article-7784653/Photo-helicopter-tells-story-survival-White-Island-volcano-disaster.html
[61] National Transportation Safety Board, NTSB Identification: LAX03FA200, www.ntsb.gov/_layouts/ntsb.aviation/brief2.aspx?ev_id=20030630X00970&ntsbno=LAX03FA200&akey=1
[62] USAF HH-60G Downed by Geese in Norfolk, 7 January 7, 2014, Aerosurrance, Accidents, and Incidents Helicopter, Military/Defence Safety Management, Special Mission Aircraft, January 3, 2019, http://aerossurance.com/helicopters/usaf-hh60g-birdstrike-uk/
[63] List of accidents and incidents involving military aircraft (2000-present) https://military.wikia.org/wiki/List_of_accidents_and_incidents_involving_military_aircraft_(2000%E2%80%93present)
[64] RCMP AS350B3 Left Uncovered During Snowfall Fatally Loses Power on Take Off Aerosurrance, Accidents and Incidents Helicopter, Human Factors/Performance, Safety Management, Special Mission Aircraft, January 16, 2019. http://aerossurance.com/helicopters/rcmp-as350b3-ice-ingestion/
[65] Aviation Safety Network, ASN Wikibase Occurrence # 175980. https://aviation-safety.net/wikibase/175980
[66] Columbia lifting off on its final mission https://en.wikipedia.org/wiki/Space_Shuttle_Columbia_disaster#/media/File:Close-up_STS-107_Launch_-_GPN-2003-00080.jpg
[67] Columbia Accident Investigation Board (August 2003). Volume 1 (PDF). NASA Distribution. https://web.archive.org/web/20060630042334/http://caib.nasa.gov/news/report/pdf/vol1/full/caib_report_volume1.pdf

[68] Space Shuttle external tank foam Block https://en.wikipedia.org/wiki/Space_Shuttle_Columbia_disaster#/media/File:Space_Shuttle_external_tank_foam_block.JPG
[69] National Transportation Safety Board Aviation Accident Final Report, WPR17LA078, March 20, 2017, https://reports.aviation-safety.net/2017/20170320-1_SW4_N158WA.pdf
[70] AEROINSIDE, Safarilink De Havilland DHC-8-200 Dash 8 at Maasai Mara on August 16, 2019, gnu strike on landing, 2019-08-16 15:20:13 GMT www.aeroinside.com/item/13578/safarilink-dh8b-at-maasai-mara-on-aug-16th-2019-gnu-strike-on-landing
[71] Phil Shaw and Richard Robinson, Wildlife Strikes: What your airport needs to know to keep it out of the Courts, Avisure, Australian Airports Association annual conference, November 16, 2018. https://avisure.com/uploads/doc/AAA_Presentation_PS_16nov18.pdf
[72] Eric Grundhauser, An Alligator's Deadly Encounter With a Plane in Florida, "One of the craziest things I've ever seen in all my years in aviation." Atlas Obscura, June 7, 2017, www.atlasobscura.com/articles/alligator-airplane-florida
[73] AeroInside, Wizz Airbus A-320 at Cluj on Jan 27th, 2020, bird strike, Jan 27, 2020, www.aeroinside.com/item/14316/wizz-a320-at-cluj-on-jan-27th-2020-bird-strike?utm_source=newsletter&utm_campaign=20200128
[74] Simon Hradecky, Incident: Finnair E190 at Oslo on Oct 23rd, 2010, struck runway edge lights during takeoff, The Aviation Herald, November 4, 2010, http://avherald.com/h?article=43301 3ef
[75] ASN Wikibase Occurrence # 231682, https://aviation-safety.net/wikibase/231682
[76] Flying blind and freezing: Navy investigating terrifying EA-18G Growler flight, February 23, 2018, David B. Larter, www.defensenews.com/breaking-news/2018/02/23/flying-blind-and-freezing-navy-investigating-terrifying-ea-18g-growler-flight/
[77] ASN Wikibase Occurrence # 178444 https://aviation-safety.net/wikibase/178444
[78] Jason Abbruzzese and Andrew Freedman, Hailstorm dents nose, shatters windshield on Delta flight, 2015, https://mashable.com/2015/08/09/hail-storm-delta-flight/
[79] IATA: Safety Report 2017, 54th edition, 2018. https://cdn.aviation-safety.net/airlinesafety/industry/reports/IATA-safety-report-2017.pdf
[80] Caleb Hutton, One person hurt as plane crashes at Bellingham airport; flights grounded, The Bellingham Herald, February 27, 2017, www.bellinghamherald.com/news/local/article135280419.html
[81] El-Sayed, A.F., Bird Strike in Aviation, Statistics, Analysis and Management, Wiley, July 2019.
[82] Australian Government, Australian Transport Safety Bureau, Birdstrike and engine failure involving Airbus A330, 9M-XXT Gold Coast Airport, Queensland, 3 July 2017, ATSB Transport Safety Report Aviation Occurrence Investigation AO-2017-070 Final – 2 May 2018. www.skybrary.aero/bookshelf/books/4441.pdf
[83] Casadevall, T.J., 1992, Volcanic hazards and aviation safety. *Federal Aviation Administration Aviation Safety Journal*, 2(3), pp. 9–17.
[84] Volcanic Ash Deposits on DC-10 aircraft, https://en.wikipedia.org/wiki/Volcanic_ash#/media/File:DC-10-30_resting_on_its_tail_due_to_Pinatubo_ashfall.jpg
[85] Australian Government, Australian Transport Safety Bureau, Undetected engine thrust reverser deactivation involving Airbus A320, VH-VQG Gold Coast Airport, Queensland on 18 December 2017, ATSB Transport Safety Report Aviation Occurrence Investigation AO-2017-117 Final – September 5, 2019, www.aviation-accidents.net/report-download.php?id=684

[86] CBC News, A first in Canada: Drone collides with passenger plane above Quebec City airport Social Sharing, October 15, 2017. www.cbc.ca/news/canada/montreal/garneau-airport-drone-quebec-1.4355792
[87] Courtney Mabus, A Single Bird Takes Down the U.S. Navy's 'Doomsday' Plane, Your Navy, October 17, 2019, www.navytimes.com/news/your-navy/2019/10/17/bird-strike-grounds-navy-doomsday-aircraft/
[88] FOD, Account for all items after performing maintenance tasks! June 14, 2019, https://fodprevention.com/ntsb-safety-alert-control-foreign-object-debris/
[89] NTSB, NTSB Identification: WPR13LA290, https://ntsb.gov/_layouts/ntsb.aviation/brief.aspx?ev_id=20130625X05535&key=1&queryId=58d16bea-2e1f-419d-bbe3-8a97504e9187&pgno=19&pgsize=50
[90] Goglia, J., Torqued: Foreign Object Debris, Small Pieces Matter, Ainonline September 2, 2016, www.ainonline.com/aviation-news/blogs/torqued-foreign-object-debris-small-pieces-matter
[91] Goglia, J., "NTSB Finds Drone Pilot At Fault For Midair Collision With Army Helicopter". Forbes. Retrieved December 15, 2017. www.forbes.com/sites/johngoglia/2017/12/14/ntsb-finds-drone-pilot-at-fault-for-midair-collision-with-army-helicopter/#44b85aff7b3f
[92] NASA–http://spaceflight.nasa.gov/gallery/images/shuttle/sts-110/html/ksc02pd0456.html
[93] Grush, L., India shows it can destroy satellites in space, worrying experts about space debris, The Verge, March 27, 2019. www.theverge.com/2019/3/27/18283730/india-anti-satellite-demonstration-asat-test-microsat-r-space-debris
[94] Akshat Rathi, Photos: This is the damage that tiny space debris traveling at incredible speeds can do, Quartz, September 3, 2016, https://qz.com/773511/photos-this-is-the-damage-that-tiny-space-debris-traveling-at-incredible-speeds-can-do/
[95] Lear, D., Hyde, J., Christiansen, E., Herrin, J., Lyons, F., STS-118 Radiator Impact Damage, https://ntrs.nasa.gov/archive/nasa/casi.ntrs.nasa.gov/20080010742.pdf
[96] CNN, "Russian, U.S. satellites collide in space." February 12, 2009. http://edition.cnn.com/2009/TECH/02/12/us.russia.satellite.crash/index.html?_s=PM:TECH

3 Inanimate Debris Generated by Aircraft Ground Operation

3.1 RUNWAY TYPES AND CONDITIONS

3.1.1 Introduction

Aircraft takeoff and landing performance are affected by many factors, among them the following [1]:

- Manufacturer's or operator's limitations (maximum permitted takeoff or landing weight, maximum crosswind, or tailwind component).
- Airfield: elevation, ambient temperature, surface wind, obstacle clearance data.
- Runway characteristics: length, width (a wide runway is especially welcome in a strong crosswind), slope (uphill is always preferred), degree of contamination.

According to the International Civil Aviation Organization (ICAO), a runway is defined as a rectangular area on a land of aerodrome prepared for the takeoff and landing of aircraft. A runway surface may be an artificial (man-made) normally identified as a paved or improved runway or a natural surface identified as an unpaved or unimproved runway.

Runway lengths are given in meters all over the world apart from North America where lengths are given in feet. Typical runway dimensions are as follows:

- The smallest is 245 m long and 8 m wide (804 ft × 26 ft) in smaller general aviation airports.
- The longest is 5,500 m long and 80 m wide (18,045 ft × 262 ft) at large for international airports to match the largest jets.
- A special runway is arranged to land the Space Shuttle (Edwards Air Force Base) in California. Its dimensions are 11,917 m × 274 m (39,098 ft × 899 ft).

Runways are laid out according to the numbers on a compass where 0° and 360° are North, 90° is East, 180° is South, and 270° is West. At the end of each runway, its compass direction is indicated by a painted large number.

DOI: 10.1201/9781003133087-3

There are four types of runways in airports: single, parallel, open V, and intersecting [2].

- Single runway

A single runway position is selected based on the prevailing wind, noise, land use, and other factors. It should accommodate 42–53 operations per hour. It is used for both takeoff and landing.

- Parallel runway

Parallel runways should accommodate 64–128 operations per hour. There are four types of parallel runways:

- Close runways where the distance between them is less than 2,500 ft
- Intermediate parallel runways which are 2,500–4,300 ft apart
- The far parallel runways which are 4,300 ft or greater apart
- The dual-line parallel runways which are 4,300 ft or greater between each pair
- Open V

Two runways that diverge in different directions but do not intersect thus look like an "open-V." They are used when there is little to no wind and thus allows for both runways to be used at the same time. In case of strong winds, only one runway is used. There are two layouts:

- Open V with dependent operation toward the intersection (converging)
- Open V with dependent operation away from the intersection (diverging)
- Intersecting runways

Two or more runways cross paths and share ground. This layout is used when there are relatively strong prevailing winds in more than one direction during the year. In case of strength in one direction, operations will be only from one runway. There are three configurations:

- Two intersecting near the threshold
- Two intersecting at the middle of each
- Two intersecting at the far threshold

Runways are subject to four safety risks: foreign object debris (FOD), birds, incursions, and excursions. FOD is the costliest of them. Aircraft are exposed to strike risk along the entire operational length of the runway. During landing, aircraft will be subject to the highest risk when thrust reversers are applied [3].

Data indicate that debris is evenly distributed along the runway. Numerous small FOD items exist on the runway, but the greatest risk and highest costs come from larger debris. Surveys have found that runway debris falls into the distinct size and weight ranges. The three most common groups are: 2–4 g, 2–20 cm (10% of debris), 150–200 g, 6–38 cm (9% of debris), and 400–500 g, 14–60 cm (9% of debris).

3.1.2 Definitions

1. Paved or improved runways
   They are runways constructed from asphalt, concrete, or a mixture of both.
2. Unpaved runways (or unimproved runways) [4]
   a. Unpaved runway: A runway pavement constructed with an unpaved surface.
   b. Unpaved surface: A runway surface composed of unbound or natural materials. It may include gravel, coral, turf, sand, clay, hard-packed soil mixtures, grass, turf, dirt, and ice.
   c. Unprepared runway: Any natural surface used as a runway that man has not altered.
3. Pavement Classification Number (PCN) [5]
   A number expressing the bearing strength of pavement for unrestricted operations (ICAO pavement strength reporting format).
4. Aircraft Classification Number (ACN)
   A number expressing the relative structural loading effect of an aircraft on a pavement for a specified pavement type and a specified standard subgrade category (ICAO pavement strength reporting format) [5].
5. Pavement Load Rating (PLR)
   A number expressing the bearing strength of pavement for unrestricted operations.
6. Surface treated runway
   A gravel pavement structure covered with a thin layer of asphalt stabilized material to prevent water penetration and facilitate drainage.
7. MU friction
   Designate a friction value representing runway surface conditions [6]. MU (friction) values range from 0 to 100, where 0 is the lowest friction value and 100 is the maximum friction value obtainable. The MU value for frozen contaminants on runway surfaces is 40 or less. The lower the MU value, the less effective braking performance becomes, and the more difficult directional control becomes.
8. California bearing ratio (CBR)
   A measure of the load-bearing capacity of a given sample of soil expressed as a ratio relative to the load-bearing capacity of crushed limestone. The load-bearing capacity of crushed limestone is expressed as a CBR of 100 [5].
9. Segregation
   Segregation is the accumulation of loose, non-cohesive aggregates on the runway surface. This is due to jet or propeller blast, tire action, and weathering [5].
10. Rutting
    Rutting is a longitudinal deformation in the wheel path [5].
11. Surface drainage
    Poor surface drainage is indicated by damp surface areas persisting after rainfall or snowmelt without rutting [5].

12. Frost action
    Frost action is indicated by differential heaving of the surface or depressions which is found in a certain place every year during the frozen season [5].
13. Roughness
    Roughness is an unevenness in the longitudinal profile. It may be caused by loss of material, frost action, or settlement.
14. Vegetation
    Uncontrolled vegetation growth may occur on the graveled operational surface due to poor drainage or accumulation of organic soils (earth) in the surface. Low traffic volume may increase the vegetation growth.
15. Wake turbulence
    Phenomena are resulting from the passage of an aircraft in flight and when operating on the airport movement area [7]. The term includes vortices, thrust stream turbulence, jet blast, jet wash, propeller wash, and rotor wash both on the ground and in the air.

3.1.3 Unpaved Runways

Unpaved runways form a huge part of North American aviation infrastructure, particularly in remote areas. According to Index Mundi [8], there are in the United States:

- 9,885 airports have UNPAVED (gravel) runways
- 5,045 airports have paved runways

An unpaved or gravel runway is essentially a flexible pavement with a surface course of unbound granular material. Its surface strength depends on the surface composition, moisture content, gradation, compaction, and aggregate interlock. It is susceptible to weakening from moisture penetration and frost action. Loose material associated with gravel runway surfaces needs some methods to protect the aircraft from debris. The rough texture of gravel surfaces increases tire wear. Runway conditions based on the pilot braking action range from good to medium, medium, medium to poor, poor, and nil.

Operation of aircraft from unpaved runways has the following effects on aircraft performance:

- Increased takeoff distance due to the increased rolling resistance.
- Increased stopping distance due to reduced braking performance.
- Increased accelerated stop distances in case of aborted takeoff.
- Degraded ground handling during takeoff, landing, and maneuvering.

3.1.4 FAA Runway Conditions and Contamination

3.1.4.1 Runway Categories

FAA classifies the runway condition into dry, wet, and contaminant.

a. Dry runway
   - Neither wet runway nor contaminated.

    - No more than 25% of the runway surface area is covered by visible moisture or dampness, frost, slush, snow (any type), or ice.

b. Wet runway
    - A runway is wet when it is neither dry nor contaminated.
    - More than 25% of the runway surface area is covered by any visible dampness or water that is 1/8 inch (3 mm) or less in depth.

c. Contaminant runways

It is a runway with more than 25% of its surface covered by frost, ice, and depth of snow, slush, or water having a depth greater than 1/8 inch in depth on an airport pavement where the effects could be detrimental to the friction characteristic of the pavement surface. The contaminant may be:

1. Water
   Water which depth exceeds 1/8 inch. If lesser, it is considered a wet runway.
2. Snow
   Snow has four types: dry snow, wet snow, slush, and compacted snow.
3. Other contaminant runways

They include mud, ash, sand, or oil.

- Mud
  It is a wet, sticky, and soft earth material which measurable depth will be reported.
- Ash
  It is a grayish white to black solid particles ejected by a volcanic eruption.
- Sand
  It is a sedimentary material, finer than a granule and coarser than a silt.
- Oil
  It is a viscous fluid derived from petroleum or synthetic material used as fuel or lubricant.

4. Multiple contaminants

It is a combination of contaminant types on the runway. When this happens, the two types most prevalent for each third of the runway are reported. The runway condition is based on the most hazardous type in each third of the runway.

3.1.4.2 Runway Conditions Based on Airport Operator and Pilots

Effective from October 1, 2016, FAA implemented significant changes to the reporting of runway conditions and contamination. The new system referred to as:

- Takeoff and Landing Performance Assessment (TALPA) [9]
- Runway Condition Assessment Matrix (RCAM) that airport operators use to assign [10]

- Runway Condition Codes (RwyCC) of between zero and six for each third of the runway [11]
- Field Condition (FICON).

Figure 3.1 illustrates RCAM. As outlined in [11], RCAM is read from left to right. The unshaded portion of the RCAM is associated with how an airport operator conducts a runway condition assessment. At first, the contaminant on the runway is defined. Next, determine the appropriate code. These codes (expressed in numbers varying from six to zero) are based on defined terms and increments which harmonize with ICAO Annex 14, Aerodrome Design and Operations [12]. Higher numbers represent more favorable conditions based on objective measurements of the type and amount of surface contamination.

The following column identifies the coefficient of friction Mu. The correlation of the Mu (μ) values with runway conditions and condition codes in the RCAM is only

Assessment Criteria		Control/Braking Assessment Criteria			
Runway Condition Description	RwyCC	Mu		Deceleration or Directional Control Observation	Pilot Reported Braking Action
• Dry	6		40 or HIGHER	—	—
• Frost • Wet (Includes damp and 1/8 inch depth or less of water) 1/8 Inch (3 mm) Depth or Less of: • Slush • Dry Snow • Wet Snow	5			Braking deceleration is normal for the wheel braking effort applied AND directional control is normal.	Good
–15 °C and Colder Outside Air Temperature: • Compacted Snow	4	39		Braking deceleration OR directional control is between Good and Medium.	Good to Medium
• Slippery When Wet (wet runway) • Dry Snow or Wet Snow (any depth) over Compacted Snow Greater Than 1/8 Inch (3 mm) Depth of: • Dry Snow • Wet Snow Warmer Than –15 °C Outside Air Temperature: • Compacted Snow	3	to 30	29	Braking deceleration is noticeably reduced for the wheel braking effort applied OR directional control is noticeably reduced	Medium
Greater Than 1/8 Inch (3 mm) Depth of: • Water • Slush	2		to	Braking deceleration OR directional control is between Medium and Poor.	Medium to Poor
• Ice	1	20 or	21	Braking deceleration is significantly reduced for the wheel braking effort applied OR directional control is significantly reduced.	Poor
• Wet Ice • Slush over Ice • Water over Compacted Snow • Dry Snow or Wet Snow over Ice	0	LOWER		Braking deceleration is minimal to nonexistent for the wheel braking effort applied OR directional control is uncertain.	Nil

FIGURE 3.1 Runway Condition Assessment Matrix (RCAM) [11].

Courtesy: U.S. Department of Transportation, FAA.

approximate ranges for generic friction measuring device and is intended to be used only to upgrade or downgrade a runway condition code.

The shaded portion of the RCAM is associated with the experience of the pilot with braking action. The pilot/aircraft operator, operational RCAM illustration, is different from the RCAM illustration used by airport operators. The pilot reporting braking conditions categorizes the runway conditions into good, good to medium, medium, medium to poor, poor, nil.

The RwyCC divides the runway into three segments, and the contamination condition is assessed for each third. Hereafter explanation for some examples for the runway contamination conditions:

(RWY 04L FICON 5/5/5 50 PRCT 1/8IN DRY SN).
The runway is uniform coverage in each third having 50% Coverage of 1/8 Dry Snow
(RWY 36 FICON 4/3/3 60 PRCT COMPACTED SN, 75 PRCT IN WET SN OVER COMPACTED SN, 80 PRCT 1 IN WET SN OVER COMPACTED SN)

This is an example of having a different contaminant in each third:

- The first (touchdown) third is "4" and is described as having a 60% coverage of compacted snow.
- The second "3" is the RCC for the middle third of the runway and described as 75% covered with wet snow over the top of compacted snow.
- The last "3" is the rollout portion of the runway, and it has 80% coverage of one inch of wet snow over compacted snow.

The National Business Aviation Association (NBAA) Access Committee has developed these comprehensive educational resources to assist aircraft operators becoming familiar with all aspects of TALPA [13].

3.1.5 Paved (Asphalt and Concrete) Runway

Asphalt is called a flexible surface; concrete is called a rigid surface. PCN is normally used with the ACN to indicate the strength of a runway, taxiway, or apron (ramp).

3.1.5.1 Asphalt Runways

Asphalt runways are cheaper, easier, and quicker to build. They are used for a small airport that handles mostly small, light aircraft and is not very busy. Perhaps cost is controlling at the smaller fields that use asphalt. According to the Asphalt Pavement Alliance, more than 85% of general aviation airports use asphalt.

Generally, concrete certainly is not preferred for road construction in the UK. Many major non-US airports have asphalt runways. Heathrow, Gatwick, Schiphol, Charles de Gaulle CDG, Barcelona, Dubai, Narita, Kansai, and Bangkok Suvarnabhumi all have only asphalt runways. Frankfurt has two asphalt and two concrete runways. These airports handle plenty of Superjumbos like the Airbus A380. For larger commercial airports, an asphalt surface can be laid over a worn concrete base much more cheaply (and quickly) than redoing concrete from scratch. All runways in big airports, even those covered in asphalt, have reinforced concrete foundations that vary in depth.

The advantages of Asphalt runways are:

- More cost-effective
- Gives better friction
- It can also drain when raining due to its porosity
- Faster to repair, whereas a concrete runway may be out of order for a while if a slab crack
- Takes less time to lay down than concrete. Newly laid asphalt may be allowed to aircraft operations as it is compacted and cool.

However, asphalt is susceptible to fuel degradation, as it could infiltrate it and dissolve the actual asphalt holding together the aggregate.

3.1.5.2 Concrete Runways

Concrete runways are much more durable and last a lot longer. Nearly all major aerodrome runway pavements are originally constructed using concrete (PCCP).

Reinforced concrete runways are a much better option for busy airports with a lot of heavy jet traffic. They are far more capable of handling that traffic. Larger airports such as Charlotte, Louisville, Sea-Tac, Munich, Zurich, and so on use concrete.

For all stationary aircraft (in parking spots and holding areas) and in runway exits, concrete is used as it handles such stationary loads well.

Concrete is more useful in hotter climates as on a warm day; the asphalt would deform.

Exposed concrete is notorious for cracking when exposed to extremely high and low temperatures.

However, concrete is very much harder to repair when it develops holes, and those holes tend to be deeper and sharper-edged. Another disadvantage concrete cures after few days of pavement.

3.1.6 Operation from Unpaved or Contaminated Runways

As defined before, runways classified as "unpaved or "unimproved" include gravel runways that are defined as a runway normally constructed from a mixture of compacted soils and stones, with a surface that is not bound by any additives (neither asphaltic nor cementitious). A grass runway usually does not qualify as a gravel runway [14]. Operation from unpaved runways have several concerns:

1. Whether the landing strip can handle the weight of the aircraft and landing impact(s).
2. Whether the aircraft landing gear itself survives repeated abuse.
3. How much debris does get sucked into the engines.

For these difficulties, not all aircraft can operate from unpaved runways. Examples of aircraft capable of operating from unimproved runways are Boeing 737-200, Boeing 727, Pilatus PC-24 business jet, ATR 72-600, BAe 146, L-410 Turbolet, TU-154B, and Yak-40.

Boeing 727 was used on dirt airfields in Africa quite often. Installation of its engines in the rear fuselage helps to protect them from FOD ingestion.

The British aircraft BAe 146-100, -200, and -300 models, which equivalents are Avro RJ70, RJ85, and RJ100, can be fitted with a gravel kit to enable operations rough, unprepared airstrips. Also, its high wing provides reasonable protection against FOD.

The aircraft Yakovlev Yak-40 can operate from poorly equipped airports with short unpaved runways (less than 700 m or 2,300 ft) in poor weather.

Finally, it is worth mentioning that the Honda HA-420 HondaJet has over-wing mounted twin turbofans. The aircraft is well suited for unprepared runways.

T-tail aircraft like a DC-9, B727, MD83, and so on reduces FOD damage since their wings screen FOD debris.

3.1.6.1 Boeing 737

An optional Unpaved Strip Kit was made available for the 737-100/200 from February 1969. It allowed aircraft to make more than 2,000 movements a year on gravel, dirt, or grass strips (Figure 3.2). These runways were mostly in Alaska and Canada, serving remote locations. The kit includes [15–17]

- *Nose-gear gravel deflector to keep gravel off the underbelly when the gear was lowered.*
- *Smaller deflectors on the oversized main gear to protect the flaps.*
- *Protective metal shields over hydraulic tubing and brake cables on the main gear strut.*

FIGURE 3.2 Gravel kit on a landing Air North 737-200 [17].

- *Protective metal shields over speed brake cables.*
- *Glass fiber reinforced underside of the inboard flaps.*
- *Metal edge band on elephant ear faring.*
- *Abrasion-resistant Teflon-based paint on wing and fuselage undersurfaces.*
- *Strengthened under-fuselage aerials.*
- *Retractable anti-collision light.*
- *Vortex dissipators fitted below the engines.*
- *Screens in the wheel well to protect components against damage.*

The nose gear gravel deflector is made of corrosion-resistant steel. It has a sheet metal leading edge that acts as an airfoil to ensure aerodynamic stability. When the gear retracts, the deflector is hydraulically rotated around the underneath of the nosewheel before seating into its faring at the front of the nosewheel bay. The rotation is programmed to maintain the deflector in a nose-up attitude during transit. The maximum speed for gear operation is reduced to 180 knots while the max speed with the gear extended is only 200 knots.

As of early 2020, few airlines in Canada, including Nolioner Aviation, Air Inuit, Canadian North, and Chrono Aviation, are still flying 737s with Unpaved Strip Kits.

Figure 3.3 illustrates the vortex dissipator [17]. It prevents vortices from forming, thus preventing gravel from ingestion into the engine. Air bled from the engines

FIGURE 3.3 Canadian North 737-200 engine vortex dissipator [17].

is blown to break up vortices. The vortex dissipator operates during take-off and landing.

Boeing reported that tire wear for 737 aircraft operating on gravel runways is generally increased and can be as much as four times that of 737 aircraft operating on conventional runways [18].

3.1.6.2 Pilatus PC-24 Business Jet

Pilatus PC-24 has earned approval from Europe and the United States to operate on unpaved runways including dirt and gravel. Runway length is as short as only 856 m (2,810 ft) (Figure 3.4). It has the following design features: dual 73 psi main tires, long-travel trailing link landing gear, triple-disc steel brake, clever flap systems, and special wing design to operate from rough fields [19]. Thus, PC-24 can operate from over 20,000 landing sites around the globe.

PC-24 is powered by two engines mounted on the rear fuselage, which also protect the engines from ingesting FOD.

3.1.6.3 ATR 72-600

Russian Interstate Aviation Committee (IAC) approved in June 2019 the operation of ATR 72-600 from a new unpaved runway to be a total of four unpaved runways (Mys Kamenny, Lensk, Bodaibo, and Igrim) in the remote parts of Russia. The certification allows ATR72-600 aircraft to operate from gravel, dirt, and grass strips (Figure 3.5).

3.1.6.4 Tu-154B

Tu-154 is capable of operating from unpaved and gravel airfields with its basic facilities. it was extensively used in the extreme Arctic conditions of Russia's northern/

FIGURE 3.4 Pilatus PC-24 business jet operating from the unpaved runway.

FIGURE 3.5 ATR72-600 aircraft operating from unpaved runways.

eastern regions where other airliners were unable to operate. The Tu-154 has this engine placement in the rear fuselage, which enables it to operate from unpaved fields (Figure 3.6).

3.1.7 Military Bases

Military aircraft may operate from unimproved runways (grass, dirt, and gravel strips) and from bomb-damaged runways. For better FOD resistance, aircraft designers modified military aircraft for this endeavor by strengthening the undercarriage and upgrading the tires for more stress, and sometimes strengthening the fuselage frame.

1. With regard to landing gear special designs for operation from unpaved runways:
   a. Use longer landing gear to be able to land on the roads, refuel and rearm and fly again as Swedish SAAB fighters
   b. Have multiple wheels to distribute the landing load so that the plane does not sink into the runway as in the C-130
   c. Larger and higher number of tires as in C17. For example, both C17 and Boeing 777 have nearly same maximum takeoff weight (MTOW), but C17's tires are significantly wider than the 777's (estimated a 2:1 height: width ratio vs. 3:1)

FIGURE 3.6 TAROM Tu-154B-1 [20].

Photo permission GFDL 1.2.

2. Concerning wings
   A high wing design is employed; thus, the aircraft engines are mounted further above the ground which reduces the possibilities of FOD ingestion (e.g., C-130, C-17, IL-76)
3. Concerning tires use
   a. A larger number of tires
   b. Greater dimensions

Examples for these military aircraft are Airbus A400, Hercules C130, Galaxy C-5, C-17, *P-3,* Harrier, CN-235, A-10 Warthog, Mig 29, and Sukhoi Su-25.

The Harrier V/STOL aircraft can retain a functional air force even if the airports are destroyed. Thus, it can take off from, and land in, any space large enough for it to fit into, which includes unpaved dirt pads. However, if operating from dirt fields it ingests considerable quantities of dirt and dust since it does not have any mechanism for preventing engine dirt ingestion.

3.1.7.1 Airbus A400M

Airbus Defense and Space A400M has successfully demonstrated the capability to operate on gravel runways [21]. Airbus Military's A400M has completed an important set of tests for operation from gravel runways in Ablitas, Spain (Figure 3.7). Tests included ground maneuvering, rejected takeoffs, and use of propeller reverse thrust at speeds as low as 70 knots [21].

FIGURE 3.7 Airbus A400 during unpaved runway tests [21].

With Permission: © Airbus – EXM Company – photo by Alexandre Doumenjou.

3.1.7.2 C130

The C-130 aircraft can operate from unimproved runways [22] because it is:

- A high lift and high-wing aircraft which is suited for slow-speed landings and short takeoffs.
- Can also operate from beach lands. Moreover, it has multiple wheels that distribute the landing load, so the plane does not sink into the runway [22, 23].

Figure 3.8 illustrates C-130 in one of its missions operating from a gravel runway.

3.1.7.3 C17

C17 aircraft has the capability of following features that enable it for operation from unpaved runways (Figure 3.9):

- High wing
- Wider tires
- The engine's thrust reversers of the engines (four Pratt & Whitney F117-PW-100 turbofan) direct engine exhaust air upward and forward, reducing the chances of foreign object damage by ingestion of runway debris.

FIGURE 3.8 C-130 in OPERATION from a gravel runway [22].

Courtesy: Department of Defense.

FIGURE 3.9 C-17 aircraft.

3.1.7.4 MiG-29UN

Most aircrafts designed in the former Soviet Union or presently Russia are capable of operating from unpaved runways because they needed to service remote locations in

Siberia that never had paved runways.

The fighters MiG-25 or the MiG-29 have the following design features which enable them to operate from unpaved runways [24]:

1. Relatively low-pressure tires (10 bar/150 psi).
2. Beefy main tires allow MiG-29UN to roll easily over minor imperfections in the runway/taxiway surface, including dirt, gravel, or grass strips.
3. The MiG-29 can even close its air intakes to prevent foreign object damage, sucking air through louvers on the upper wing (Figure 3.10). When operating from unpaved runways, during takeoff and landing, the main air inlet "C" is closed. Air is drawn through a secondary inlet "D" on top of the wing "A," which is opened through a shutter system "C." Such inlet resembles slits over the top of the wing strakes like shark's gills. This prevents the engine from ingesting dust, rocks, birds, rodents, etc. on takeoff and landing from unpaved runways.

During flight at subsonic speeds (bottom diagram), the main air inlets are opened and the secondary inlet "D" is closed by the shutter system "C."

3.1.7.5 Su-27

Su-27 fighter aircraft intended to operate from unpaved runway outfit its nose gear by a mud guard/FOD screen with downward facing deflectors louvers guiding all debris away from the large air intakes. It pivots around the wheel axis and is connected to the nose gear leg by two struts [24, 25] (Figure 3.11).

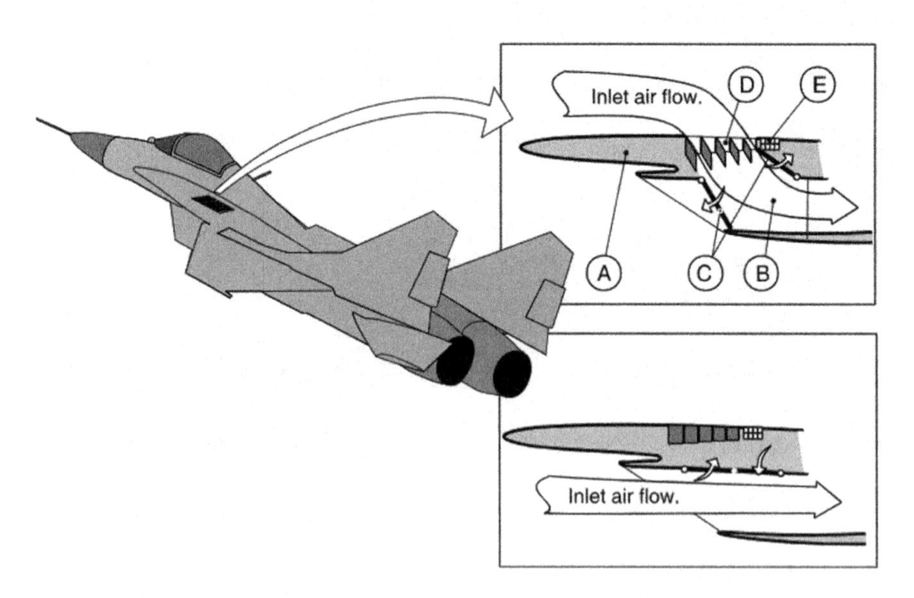

FIGURE 3.10 MiG-29 intake design for FOD debris protection [24].

With author permission.

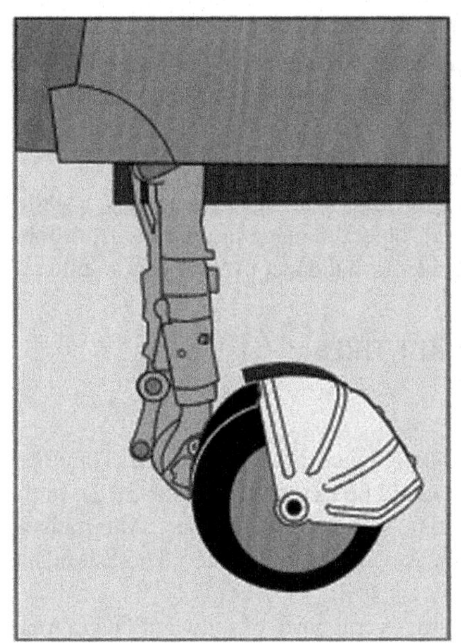
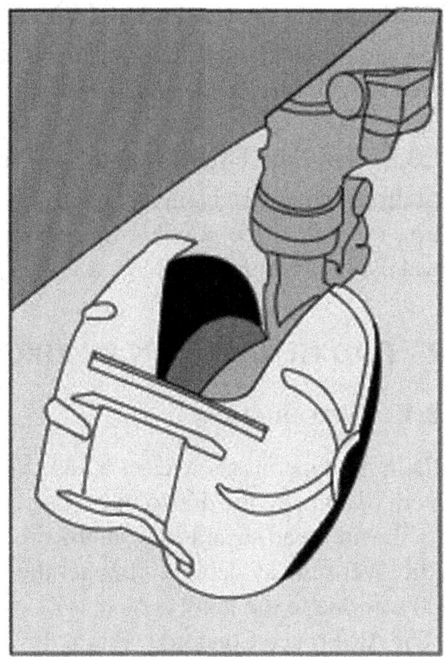

FIGURE 3.11 Nosewheel of Su-27 fighter [24].

With author permission.

3.1.8 Maintenance of an Unpaved Runway

A guide that presents the best practices on runway maintenance according to ICAO's SARP's (Standards and Recommended Practices) and other related documentation is given in [26]. This guide provides references from IATA, Airport Council International (ACI), the Flight Safety Foundation, and FAA Advisory Circulars. It focuses on runway maintenance that may prevent runway excursions responsible for many incidents.

The maintenance of pavement is most important for FOD reduction. Airports are encouraged to follow ICAO's guidelines to ensure adequate texture depth, rapid drainage of rainfall runoff water, and adequate friction characteristics. ACI recommends regular surface drainage, removal of rubber, and contaminants from the runway surface. The frequency of measuring the surface friction depends on the daily number of landings of aircraft powered by turbine-based engines. It is once a year if the number of aircraft is less than 15, and every week if the number of aircraft exceeds 250. Rubber removal is performed via different methods including water blasting, shotblasting, chemical, and grinders.

In brief, here are the three main steps for maintenance of unpaved runways [27].

3.1.8.1 Treat Every Day Like Inspection Day

All runways must be assessed daily and must pass certain specifications. Eye test must assure that no standing water, deep ruts, or deep loose gravel on the runway.

3.1.8.2 Apply Proven Soil Stabilizers for Pavement-Like Strength

Using some type of stabilization agent is necessary. Cold weather loosens the surface and creates irregularities in the strip, along with the typical wear and tear from plane traffic.

3.1.8.3 For Safe Flights, Keep Dust to a Minimum

Dust from inadequately maintained gravel runways can cause hazardous visibility issues, as well as the possible risk of foreign object damage to aircrafts from loose gravel. Dust can be removed by running a grader or bulldozer over it every so often.

3.2 FOD GENERATION BY AIRCRAFT TIRES

3.2.1 Introduction

FOD in Air Operations Areas (AOA) can damage both tires and aircraft. The debris can cut or damage the tire so that a takeoff would be hazardous (Figure 3.12). In this case, the tire tread separates from the tire and becomes a foreign object. Alternatively, the tire loft runway debris to impact the surfaces of aircraft. Concorde crash in July 2000 was due to the first case.

The Air France Concorde (Flight 4590) ran over a piece of runway FOD (a metal piece) which initiated a tire failure that resulted in large pieces of rubber being thrown against the underside of the left wing [28]. The tire ruptured the fuel tank,

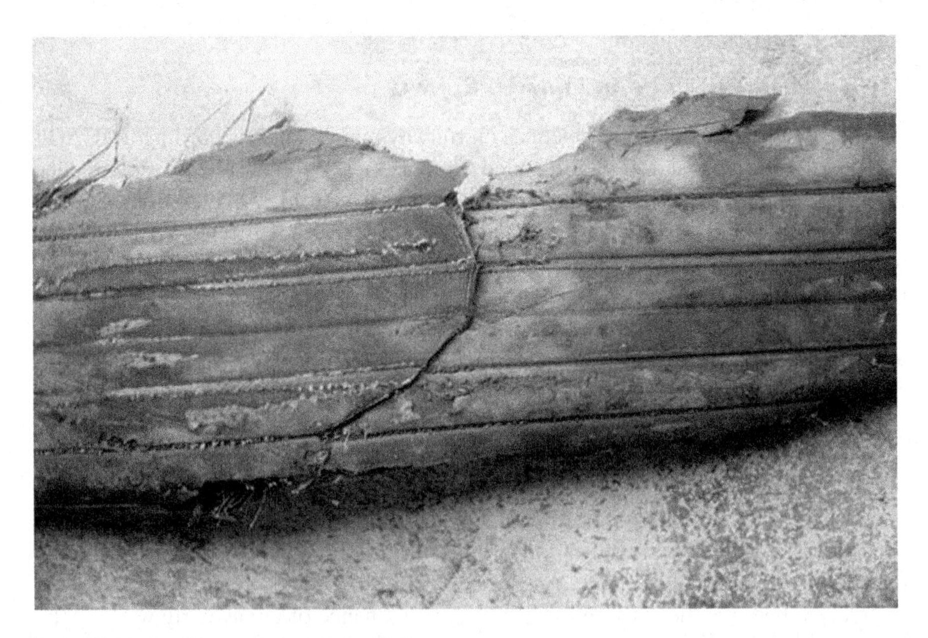

FIGURE 3.12 Piece of aircraft tire [75].

(c) BEA / bea.aero.

induced a fire, and the separation of a section of the tank lower surface (which was later found on the runway). The tire was cut and burst releasing debris which size was 100 × 33 cm and weighing 4.5 kg. This resulted in a loss of aircraft control that ended with the Concorde crashing into a nearby hotel. Such a piece of metal resulted in the death of 113 people, destroyed a 46-US million-dollar aircraft, and destroyed a hotel. However, the Concorde crash may have been prevented if the routine full runway inspection was not cancelled just hours before the crash. Thus, regular inspections of AOA's are required to remove debris.

A similar recent accident but had a happy end was on September 27, 2016 in Abu Dhabi International Airport (OMAA) [29]. An Etihad Airways Boeing 777-3FXER, registration A6-ETL, flight EY450 from OMAA to Kingsford Smith Airport (YSSY), Sydney, at 1130 LT. During takeoff, shortly after rotation, the flight crew heard a loud bang associated with No.1 engine high exhaust gas temperature (EGT), followed by warning messages regarding No.1 engine which was next auto shut down. The takeoff continued and the flight crew decided to return to the airport. The runway safety team inspected the runway and discovered tire debris on it. The flight crew managed to land safely with no injuries at 1203:44 LT. Investigation revealed that:

- More than 90% of the outer layer of the No.1 nosewheel tire had shed its tread and the nose gear steering cable was damaged (Figure 3.13).
- The lower fuselage, aft of the NLG bay was damaged (Figure 3.14).

FIGURE 3.13 No.1 nosewheel tire and steering cable damage [29].

With permission of General Civil Aviation Authority, Abu Dhabi, United Arab Emirates.

FIGURE 3.14 Damage to No.1 engine fan blades [29].

With permission of General Civil Aviation Authority, Abu Dhabi, United Arab Emirates.

- The No.1 engine fan blades and engine inlet sustained damage (Figure 3.14).
- Tire debris impacted on the inboard fan cowling of the No.2 engine.

They concluded that: "*the shedding of the No.1 nosewheel tire tread occurred because of the tire contacting foreign object debris. Subsequently, the damaged tire debris was ingested by the No.1 engine causing engine failure*" [29].

> Runway debris lofted by aircraft tires can cause considerable damage to aircraft structures.
>
> There are four mechanisms for lofting both dust and macro-debris: "tread envelopment," "pinch lofting," "tread gripping," and "drag acceleration by lofted water".

Generally, aircraft tires are highly stressed. Aircraft tires compared to automobile tires are: 3 times the speed, 3 times the deflection (30% vs. 10%), 6 times the pressure, and 13 times the load.

Lofted macroparticles due to debonding and breakup of tire treads is a type of damage that occurs in nearly all aircraft tires. Tire tread breakup is caused by the growth of incipient flaws in the tire tread. The size of the fragments projected varies from chunks of the tread thickness to large segments of the entire tread. Tire tread is generally confined to the plane of the wheel.

3.2.2 Field Studies

3.2.2.1 Civil Aircraft

Field studies were carried out in [30–32] to examine:

a. Projection of gravel from landing gear tires, and
b. Ingestion of gravel resulting from engine inlet vortex.

Experimental studies conducted by the Boeing Corporation [30, 32] supporting the operation of Boeing 727 and 737 commercial aircraft from unpaved runways indicated the following findings:

- Macro-debris (chunks with characteristic dimensions greater than 10 mm) is projected by aircraft tires in relatively intense fields extending in sectors 30° horizontally and vertically from the tire (30 degrees up from runway surface and 30°on each side of the tire plane). Velocity of the objects in the spray was comparable to the aircraft velocity. Less intense concentration of debris is projected at angles between 30° and 60° (Figures 3.15 and 3.16).
- These results indicate that for all types of aircraft, debris lofted by their nosewheel can impact most of the underside of the fuselage and wings. Engine air intake ducts will ingest debris if they are located well to the rear of the nosewheel. Also, debris sprayed by the nose landing gear could be projected against external stores.
- Ingestion resulting from engine inlet vortex formation is also a significant contributor to engine FOD while operating on unprepared surfaces. Surface debris is disturbed by the rotational vortex flow and projected upward where it is entrained in the main inlet airflow and then ingested. In addition to the Boeing 737, vortex removal systems or dissipators have been installed in DC-8, F-18,

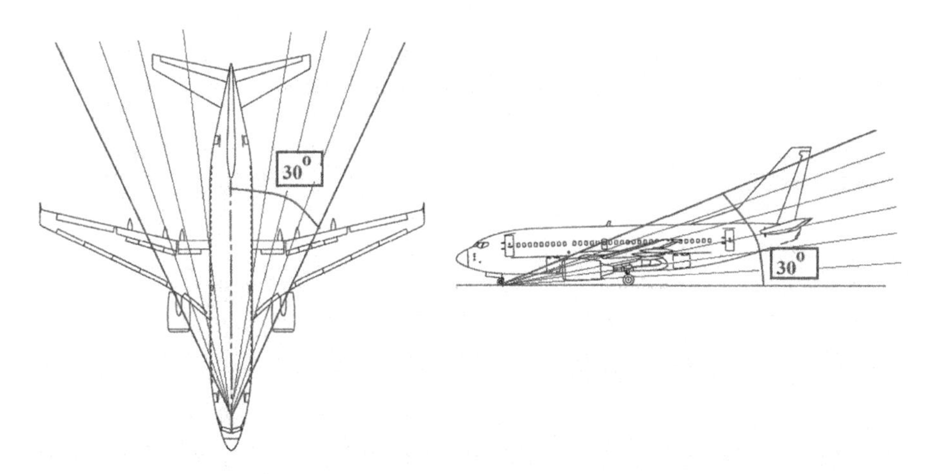

FIGURE 3.15 Sprayed rocks from nose landing gear [30].

Courtesy: US Air Force.

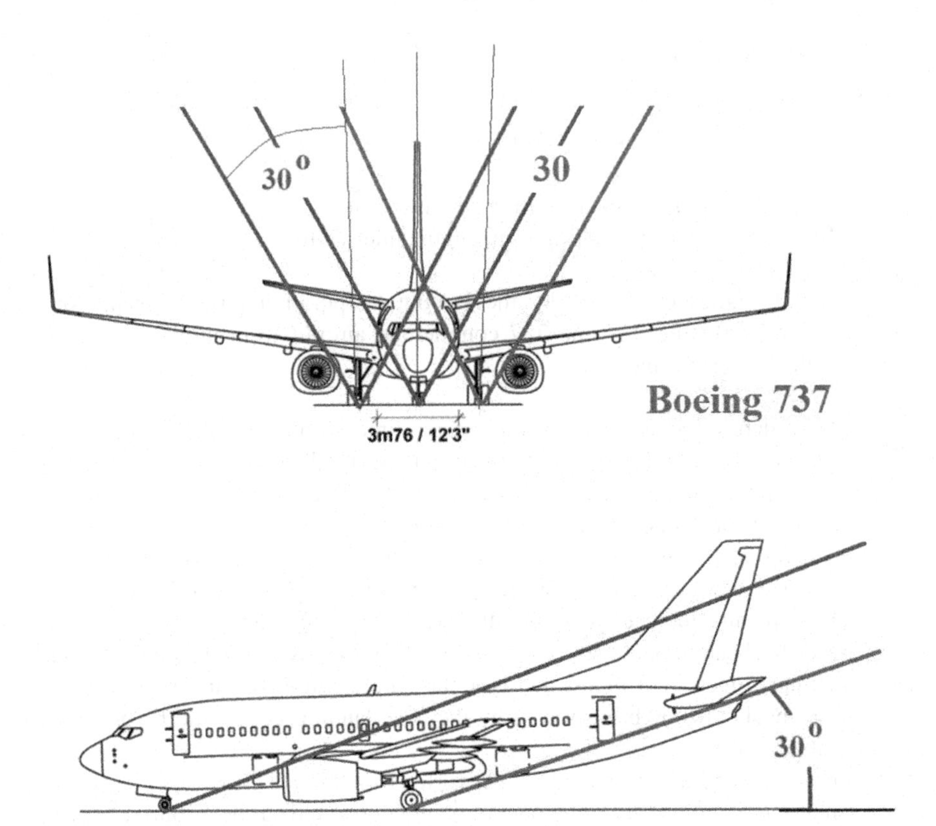

FIGURE 3.16 Sprayed rocks (greater than 10 mm diameter) from nose and main gear tires [30].

Courtesy: US Air Force.

> F-1ll, and A-6 aircraft, but their use has been discontinued in all but the A-6 aircraft because of inconclusive evidence of their effectiveness.

Photographs of aircraft traveling through standing water show similar behavior of the spray produced.

3.2.2.2 Military Aircraft

The detailed study for the FOD caused by aircraft operation on unconventional and bomb-damaged airfield surfaces had the following conclusions [30]:

3.2.2.2.1 Airframe

Debris will impinge on the airframe and cause surface abrasion and not penetration as the spray velocity is not sufficient to cause penetration. For military aircraft, it may cause a mission abort and damage of antennas to the extent that communications or navigation systems would be degraded.

3.2.2.2.2 Landing gear damage

Spray from the tires will impinge on the several items attached to the landing gear or located in the wheel wells. They will cause moderate abrasion to the hydraulic lines, and electrical cables, but may beak the taxi lights. The landing gear, itself, would probably not be damaged.

3.2.2.2.3 External stores damage

Spray from the tires will impinge on external stores and cause surface abrasion. These debris will not penetrate external fuel tanks but may cause sufficient damage to radar or infrared domes on external stores to a level of degrading their operation.

3.2.2.2.4 F15 Aircraft

As shown in Figure 3.17, the debris projected from the nosewheel tire influenced the aircraft modules as follows:

- Most of the debris were sprayed on the lower surfaces of the fuselage, wings, and tail.
- Few debris sprayed into the engine inlets.
- Other debris damaged the radomes of fuselage-mounted SPARROW missiles.

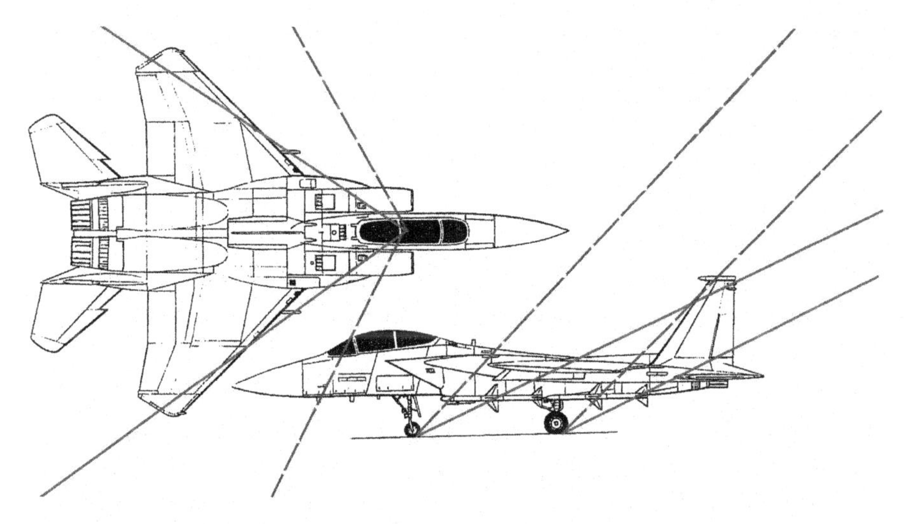

FIGURE 3.17 F15-tire spray pattern [30].

Courtesy: US Air Force.

United States Air Force (USAF) data indicated a moderate rate of F-15 FOD incidents with a very small percentage specifically attributed to asphalt, rocks, and concrete.

3.2.2.2.5 F16 Aircraft

For F16, its engine inlet is in front of the nosewheel. Thus, debris projected from nosewheel will not be sprayed into the engine. These debris impacted the lower surfaces of the fuselage, wings, and tail. However, the low location of the engine inlet helped the formation of an inlet vortex which helped in the ingestion of moisture that led to inlet icing as well as solid particle ingestion.

3.2.3 Experimental Studies

3.2.3.1 Test Facility

A detailed experimental and analytical study to investigate tire lofting mechanics is described in detail in [34]. This work was among the research programs of the United States Air Force Engineering and Services Center (AFESC), Rapid Runway Repair Branch (AFESC/RDCR). Since airfield runways are prime targets in military conflicts, then postattack there will be bomb-damaged runways with debris of different material and sizes. The objectives of that study were:

1. Experimental study for the mechanisms responsible for debris lofting by tires.
2. Analytical modeling for debris lofting.

Lofting action of tires was experimentally studied on a reduced-scale test track using high-speed photography.

The experimental facility was the indoor Air Force Mobility Development Laboratory (MDL) located at Wright-Patterson Air Force Base. The MDL has a reduced-scale test track having 27 m diameter, where the speed of test vehicles was up to 20 m/s relative to the track and the loads on tire was up to 1,000 pounds.

The following modifications were performed to the test facility:

- One test section from the MDL track was modified to include 6-inch-thick concrete slab, or a 6-inch deep.
- Bed of graded aggregate connected to a water tank.
- A tire carriage employing a goodyear 13-inch diameter tire and 14 plies tires used in Falcon jet.
- A screening to protect the facility from lofted stones and water.

For photographing, the following equipment were installed: an orthogonal camera, a computer program for reducing the data, a TV monitor, and recording units.

A parametric study was carried out including the following variables:

a. Debris: stone shape (angular and rounded), pebble size (1 inch or ¼ inch).
b. Tire: speed (11–18 m/s), pressure (maximum 200 psi), load (maximum 1,000 pound), single or dual tire.
c. Presence or absence of standing water (maximum depth 0.5 inch).

3.2.3.2 Conclusions

1. Lofting depends on runway coverage with stones of different sizes. Maximum lofting occurs at coverage of about 15%.
2. Saturation occurs when there is more than one stone per tire footprint area. Above saturation, the launching debris decreases.
3. Approximately 90% of lofted stones had speeds of less than 2.5 m/s.
4. Most stones are launched within 40° to the normal tire plane. The launch angles relative to runway surface exceeds 8° in the rearward direction.
5. Standing water suppresses lofting of 1-inch stones.
6. Angular stones are at least three times more likely to be lofted than smooth stones.
7. Most stones are launched with forward spins.
8. Large stones are more likely to be lofted than small stones.
9. Tire speed increases the launch probability, but does not affect the launch velocity.
10. Tire pressure or load has minor effect on lofting.
11. Large marbles are launched at very high speed in the forward direction by a different mechanism than stones.

3.2.4 Analytical Modeling

A closed-form analytical model to predict critical runway stone lofting parameters was developed in [35–37]. The parameters examined influencing the vertical loft of debris were the debris type and size collected from different airfield in civil airports and RAF bases in UK (stone circularity and aspect ratio as well as other debris represented by aircraft fasteners like hexagonal nut), the friction between the stone and ground, the density of the stone, aircraft speed, the tire shape, diameter, footprint and stiffness and the presence of grooves in the tire tread. The threat to aircraft from runway debris depends on the distribution of objects on the runway and to the speeds at which such objects are lofted.

These mechanisms were studied by developing numerical models using a dynamic explicit finite element software package, LS-DYNA [38].

A finite element model of a tire rolling over a stone was developed to predict vertical loft speeds of stones under typical takeoff conditions [38].

There was an agreement between the numerical and experimental techniques that for an inflated aircraft tire rolling over stones during takeoff conditions a modest vertical loft velocity of less than 5 m/s is encountered. The study proved that horizontally oriented structures could receive highly oblique impacts during only takeoff rotation and landing touchdown. The assumption that the impact velocity equivalent to the instantaneous aircraft speed was the real reason for aircraft structure damage was verified.

Another study examined the effect of stone overlap defined as the distance between the outer edge of the tire and the edge of stone closest to the tire to the stone diameter expressed as a percentage of the stone diameter [35, 36].

Large stone diameter was considered (10–30 mm), the tire diameters varied from 0.4 to 1.4 m which represent the nose and main wheel tire sizes respectively [39].

The tire speed was varied from 0 to 100 *m/s* to simulate the takeoff and landing speeds of military and civil aircraft [39]. The baseline values considered were 10 mm diameter spherical stone, 0.4 m for tire diameter and 70 m/s for tire speed. The overlap was varied from 0% to 50% of the stone diameter. LS-DYNA software package [38] was used. The original model assumed that the stone is a sphere, the tire is cylindrical, and the ground is flat surface. The tire was modeled as a rigid wheel rim having a solid elastic tire tread rather than inflated tire. Solid elements were chosen to capture the 3D deformation of the tire tread. Near the contact area between the tire and stone, the tire tread element size was $1.7 \times 1.3 \times 1.0\ mm^3$, while the stone element size was $0.70 \times 0.55 \times 0.33\ mm^3$. The stone had 7,000 constant stress elements and 7,350 nodes, while the wheel had 8,940 fully integrated 8-node elements and 11,532 nodes. At the contact point, a finer mesh on the tire tread moves was assumed to move with the tire motion using LS-DYNA software. The coefficient of friction between the stone and tire was greater than that between the stone and ground. The tire tread at regions of interaction with the runway and stone was modelled as a rubber rich area.

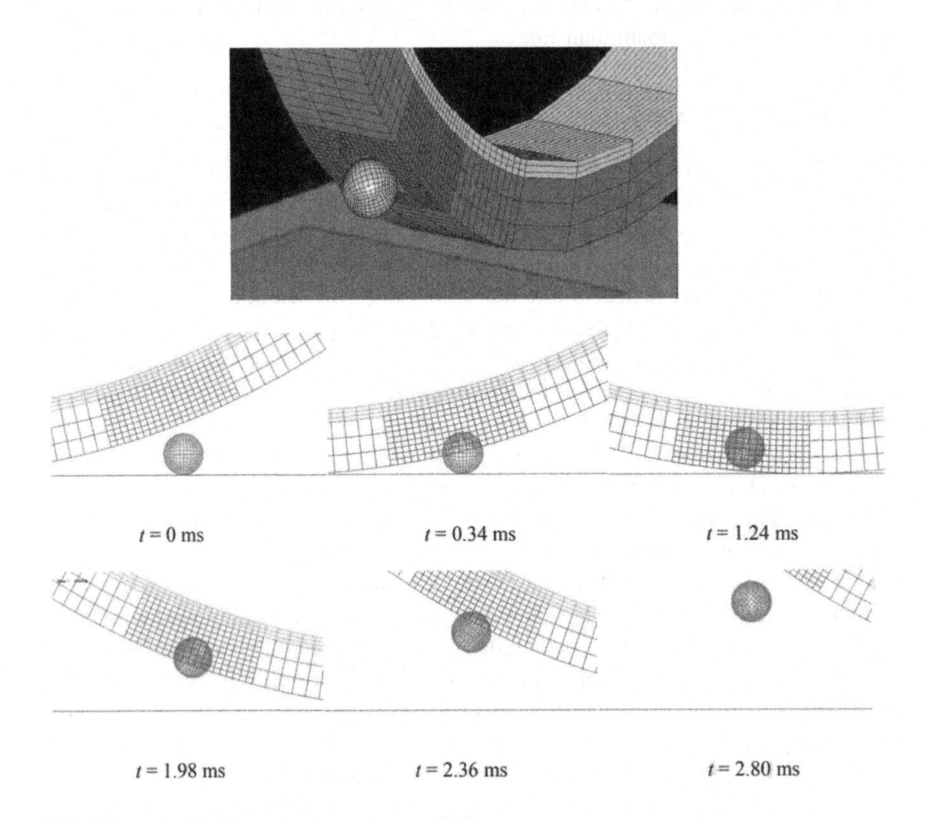

FIGURE 3.18 Lofting of spherical stone [35].

A case study was examined for a stone having diameter equal to the width of groove in the tire tread (Figure 3.18). If the stone was gripped inside the groove, then the rubber around the groove pressed the stone so that it was lifted when this point on the tire was raised from the ground. Thus, after the tire had rolled forward a short distance of 0.15 m, the stone came out of the groove and moved with a high velocity of magnitude 17.1 m/s, and possibly damage the undersurface of the aircraft.

A second case study for lofting a hexagonal nut having a 7.9 mm thread diameter which represents typical parts that may have fallen from an aircraft was modelled. The nut was only lofted to very modest vertical speeds of less than 10 m/s (Figure 3.19). The flat bottom of the nut prevented any tilting or spinning to occur so that the nut could not escape before the wheel axle passed over it.

Other results of this analytical lofting model outlined that:

- The vertical loft speed was directly proportional to the overlap up to the value 20% and approached a maximum at 50% overlap.
- The vertical loft speed was directly proportional to the tire speed.
- At any given overlap, larger stones were lofted at higher speeds.
- For a given stone size, increasing the overlap also led to greater vertical loft speeds.
- Larger tires resulted in lower vertical loft speeds.

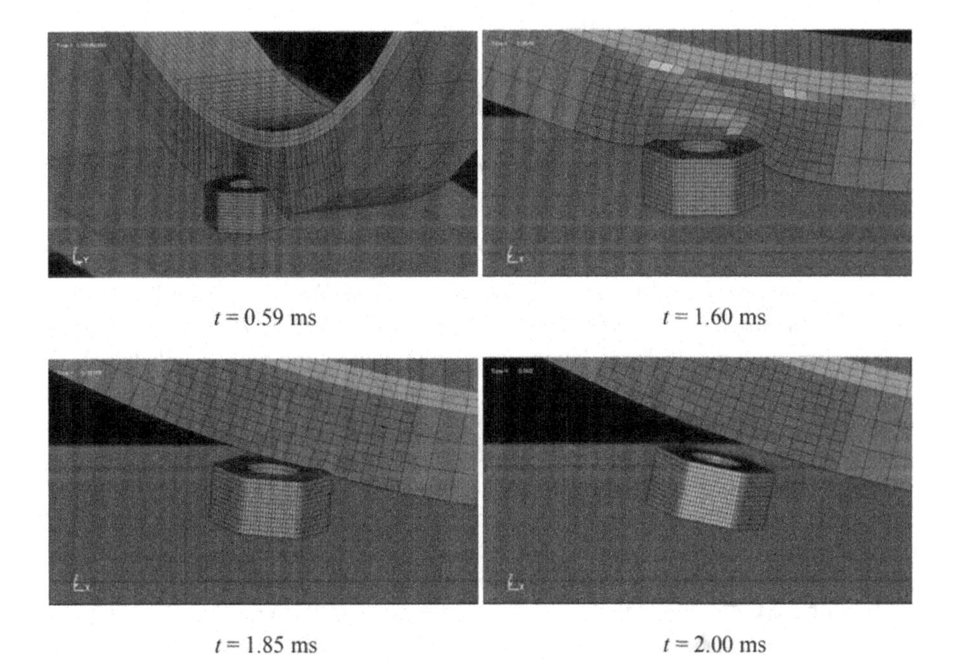

FIGURE 3.19 Lofting of hexagonal nut [35].

3.2.5 Aquaplaning or Hydroplaning

As defined in [33, 40], Aquaplaning is "*a condition that can exist when an aircraft is landed on a runway surface contaminated with standing water, slush, and/or wet snow.*"

> A runway is contaminated when more than 25% of its surface area is covered by standing water or slush more than 3mm (1/8 inch) deep [41]. Aquaplaning may have serious adverse effects on aircraft control and braking on landing roll. It may also reduce the effectiveness of wheel braking during takeoff abortion causing the aircraft to run off the end of the runway.

Aircraft traveling through flooded runways encounter tire-generated water spray like tire stone lofting discussed earlier. Such water spray may cause one or more of these drawbacks:

- Impingement on any airframe component which creates additional drag and increase the takeoff ground run, and may damage some airframe components.
- Ingestion into the engines which may result in engine surge, stall, or even flameout.
- Reduction of braking and steering forces which influences aircraft controllability during ground operation.

Therefore, the airworthiness requirements state that every commercial aircraft must pass the water spray tests before operating. Airbus A350 XWB performed spray test in May 2014 [42], while the Boeing 737 MAX passed the water spray test in January 2017 [43].

Water spray tests have been conducted on many other airplanes as Airbus A400M, Boeing 747-8 Freighter, Sukhoi Superjet 100, and Dassault Falcon 7X.

Experimental testing for water aquaplaning was carried out at NASA Langley Research Center to measure the flow rate and trajectory of water spray generated by an aircraft tire operating on a flooded runway [44]. The effects of forward speed, tire load, and water depth on the spray pattern were evaluated.

Very limited numerical simulations were employed to provide a detailed analysis for the spray location, water flow rates, and engine water ingestion levels. A numerical simulation for the water spray of the Embraer E170 with a deployed thrust reverser was performed using the discrete phase model (DPM) [45].

Another numerical simulation for water spray generated by the rolling motion of airplane tire using the smoothed particle hydrodynamics (SPH) method was carried out in [46]. This work provided the detailed results of the whole spray process and the effects of tire forward speed and water film thickness on water spray.

A third work developed the "contaminated runway (CR)-spray," method [47] for calculating the water spray generated by the tires and possible engine ingestion. The CR-spray method is based on the Lagrangian droplet trajectory calculation approach.

A detailed numerical treatment is given in [48] for tracing the trajectories of water droplets around a full aircraft. The treatment of such case was divided into two phases: the first one dealt with the formation of water spray due to rolling tires and the second phase handled the droplet motion in the airflow field around the aircraft.

The trajectories of droplets are dependent on the initial velocity, gravity, and aerodynamic forces. Though the aerodynamic forces acting on a particle include many types as will be thoroughly described in Chapter 10, only the drag force was considered using Morsi and Alexander empirical formulae [49].

The numerical method followed the typical procedure for particulate flows employing coupled two-phase flow simulation (will also described in detail in Chapter 10). The air flow around the aircraft resembles the continuous phase which is governed by Reynolds-averaged Navier–Stokes equations with the realizable k–ε turbulence model. The coefficient of viscosity is the sum of molecular viscosity and turbulent eddy viscosity. The effects of the discrete phase (water droplet) were included in the momentum equation as a source (or a force) term. For the discrete phase (water droplets), a numerical simulation platform was achieved by integrating the SPH method and the DPM method considering the different physical characteristics of the two phases. The SPH method was employed in the first phase using the commercial software LS-DYNA [50], while the DPM method was used in the second phase. DPM simulation is conducted using the commercial software ANSYS FLUENT [51].

A case study was used to predict the tire generated water spray of a regional jet with rear mounted engines. In the SPH simulation, the tire is considered as a rigid shell body with 39,036 nodes, and the minimum size of cells is 3 mm.

In the DPM simulation, the landing gear geometry was simplified and the winglets were ignored. The sea-level standard atmospheric pressure and temperature conditions are used.

A cuboid computational domain was employed, where velocity and pressure boundary conditions were assumed for inlet and outlet boundaries, respectively, while moving boundary condition was assumed for ground. The aircraft surfaces were treated as stationary wall condition, and the engine intakes are treated as the mass flow outlet conditions. A medium grid having 11.6 million cells was adopted in the simulations.

The diameter of the DPM particles is considered uniformly as of 1 mm in the DPM method.

The accuracy of the simulation platform was validated by comparing the simulation results with the NASA tire-generated water spray experiment of a regional jet with rear-mounted engines.

3.2.6 Effects of Engine Installation

3.2.6.1 Civil Aircraft

FOD on the tarmac, taxiways, and runways may be thrown up by the landing gear or even parts of the landing gear itself, and next ingested into the aircraft engines. The number of debris injected into engines increases with the increase of engine inlet diameter. There two cases to discuss for engine installation: namely wing and fuselage.

3.2.6.1.1 Wing Installation

Most airliners and heavy transport aircraft are powered by wing installed engines. When nosewheels roll over any debris it may be lofted and sucked into the engines.

As described in Section 3.1.6.1, the engines of Boeing 737 series are the closest engines to ground in all wing installed engines. For this reason, there is a high probability for ingestion of FOD debris into its engines. Some series like 737-200 is fitted with nose-gear gravel deflector as discussed earlier. However, engines in recent series like 737-600, 737-800 have ovalities in its lower part of intake to increase the clearance spacing between engine and ground.

3.2.6.1.2 Fuselage Installation

Many Regional aircraft are powered by engines mounted to the rear of fuselage (MD-80, Embraer E175 series as examples). Debris on runways may be thrown by the main landing gear. Other debris including ice breaking off from the fuselage or broken parts of the fuselage (fragments of the radome, windshield wipers, etc.) may be also sucked up by the engine [24].

3.2.6.2 Military Aircraft

In fighter aircraft, the nosewheel is responsible for the ingestion of foreign objects into the engine. The tendency to ingest foreign objects depends on:

- The inlet size (diameter (D), diagonal, cross section)
- The height of the inlet above the ground (H)
- The location of inlet relative to the nosewheel.

For engine inlet upstream of the nose gear (like F16), no debris lofted by nose tire will be ingested into the engine.

3.3 INGESTION OF FOD INTO AIRCRAFT ENGINES DUE TO INLET VORTEX

During ground operation of aircraft, airflow drawn into its engines can be affected by air currents. Such currents could be

- Surface winds
- Jet efflux from another aircraft in close proximity
- Inlet or ground vortices.

This section is devoted to the inlet (ground) vortex. When a jet engine operates near the ground at low aircraft speeds and high thrust, a vortex can be formed that starts at the ground and bends around into the inlet and known as the inlet vortex (more details are given in Chapter 10). Such a vortex consists of rapidly whirling air that creates a tube-like under pressure zone. The strongest vortex is encountered when the airplane is stationary or taxiing and fortunately disappears at higher runway speeds [58]. These small vortices create suction forces strong enough to pick up foreign objects from the runway and ingest them into the internal machinery of the engine causing an aircraft safety hazard. Such FOD may lead to compressor blade erosion and stall, deteriorating engine performance and reducing service life. This risk increases with higher bypass ratios and/or inlets that are located to the ground [52]. For example, an engine inlet with a 2 m diameter, and its center of about 2 m off the ground, can suck up a concrete sphere with a diameter of up to 30 cm! [24].

This unusual type of vortical flow can occur with all classes of military and commercial airplanes.

Ground vortices can also occur for propeller airplanes as for a C-130 transport. Fighters are especially susceptible to ground vortices as their jet engines are close to the runway or air carrier deck. The F-15 airplane powered by FI00 engines can affect an area up to 25 ft in front of the inlet with the engines at idle. At air bases, ground crews use a variety of maintenance techniques to minimize possibilities of ingesting foreign objects. This include regularly walking up to ten abreast to clear runways and taxiways.

Vortex formation requires specific conditions such as sufficient air flow speed and prevailing side wind (strength and direction). Details of vortex formation are given by the author [53–55].

Figure (3.20) illustrates the maximum ingestion size. The top right diagram shows a test assembly described in [56] with a H/D = 1.2, and inlet diameter D= 30 cm

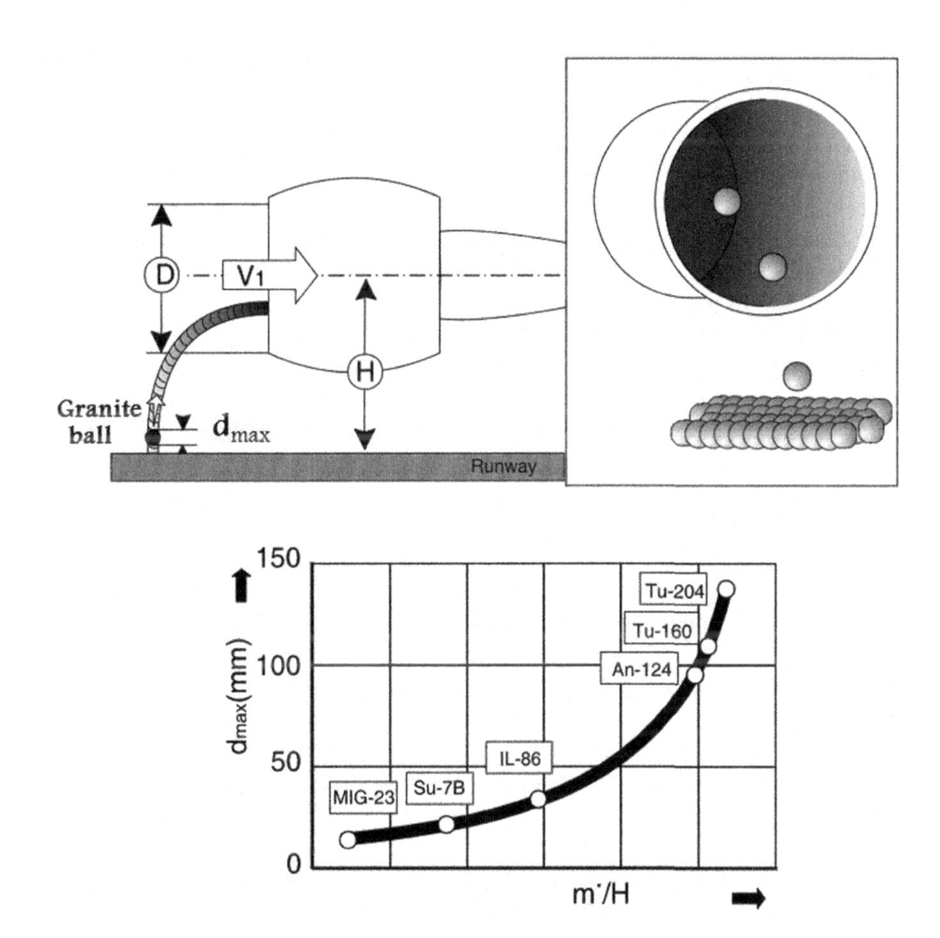

FIGURE 3.20 The maximum ingestion size of debris into jet engines [24].

With author permission.

sucking up 25 mm diameter glass spheres. The lower diagram is based on [57] and giving the maximum ingestible foreign object sizes of granite spheres (vertical axis) against the ratio of mass flow rate to engine height ($\dot{m} / H$) expressed as

$$\dot{m} / H = k \times V_1 \times D / H \tag{3.1}$$

for several Russian aircraft engine types. As the figure shows, for heavy transport aircraft rocks with diameters of considerably more than 100 mm can be sucked up, depending on the air flow rate. For fighters, debris with a maximum diameter of 15 mm can be sucked.

3.4 ENGINE THRUST HAZARD

3.4.1 Introduction

Commercial airplanes are now powered by powerful engines which thrust ranges from 18,000 to 100,000 pound [59]. Apart from the pressure thrust, most of the thrust is a velocity generated thrust which equals the product of the velocity difference between exhaust and intake the air mass flow rate [60]. However, the exhaust wake from these high-power engines may pose hazards in commercial airport environments and is known as jet blast hazard. A jet blast hazard is defined as the phenomenon of rapid exhaust air/gas movement produced by a running jet engine of aircraft. The exhaust speed increases at high power settings during taxiing, before and during takeoff, as well as during engine maintenance [61].

Airlines and airport authorities should consider these hazards to personnel, or unsecured objects behind the aircraft like baggage carts, service vehicles, airport infrastructure, and other nearby airplanes. Pilots of light aircraft frequently stay off to the side of the runway, rather than follow in the center, to negate the effect of the blast of other aircraft. For recent jet engines operating at rated thrust level, the speed of the exhaust wake can exceed 325 knots (603 km/h) immediately aft of the engine exhaust nozzle. This exhaust flow field extends aft in an expanding cone and reach the empennage with a speed of 260 knots (483 km/h) at maximum rated power. The operation manual of every commercial jet has a section titled "Jet Blast Data." On this section there is a diagram of the aircraft's jet blast "damage profile," as measured from the tail and with engines at low RPM settings (usually 35%–40% of N1). At 200 ft behind the outboard wing-mounted engines the exhaust speeds approach 100 knots [62].

Another hazard is where the jet efflux dislodges sections of taxiway or stop way paving, deflecting it rearward and upward causing it to hit and damage the stabilizer and/or elevators. This could lead to loss of control during rotation and initial climb [63, 64].

As will be described in detail in Chapter 10, mathematical modeling of debris trajectories is composed of main three steps: air/gas flow calculation, particle trajectories, and impact with the surface of candidate object. Commercial codes are now available for such endeavor including FLUENT, CFDRX, NASTRAN, and so on.

3.4.2 Jet Blast Statistics

Based on the incident reports due to jet blast [62], a breakdown of the associated damage incidents is as follows:

- 85% were to the wings, props, flaps, and rudders of other aircraft, especially to light aircraft weighing 5,000 pounds or less.
- 11% involved building structures, objects, or vehicles.
- 4% caused injuries to people.

The source of jet blast incidents was as follows:

- 45% caused by Large Transport (LGT) aircraft weighing between 150,001 and 300,000 pounds, examples were the B-727, B-757, and A320.
- 25% are due to medium-size transports weighing 60,001-150,000 pounds, including aircraft such as the DC-9, BA-146, MD-80, and B-737.
- 24% by widebody aircraft weighing over 300,000 pounds, such as the DC-10, L-1011, B-747, and B-767, were the source of jet blast.

There are three hazardous areas associated with jet blast (efflux): namely,

1. Ramp
   a. Damage other aircraft
   b. Uproot trees
   c. Blow over ground equipment (vehicles, baggage carts, aircraft steps, etc.)
   d. Cause structural damage to buildings, shatter windows
   e. Injure or kill passengers, crew, and ground personnel who may be in the vicinity.
2. Taxiways or runways
   a. Damage other aircraft – and especially jet engines – by FOD
   b. Blow over smaller or light aircraft or result of loss of directional control.
3. Maneuvering areas
   a. Damage other aircraft due to the high-speed efflux of engines running at high power.

In a total 50% of these incidents occurred on taxiways, in run-up areas, and adjacent to or on runways for relatively uncongested airport areas. The other 50% occurred on ramps, because of close aircraft parking and tight maneuvering conditions. Incidents of jet blast damage on ramps were associated with sharp turns of the aircraft during pushback, power back, taxi-out, or taxi into a gate.

3.4.3 Minimizing Jet Blast Hazard

Jet blast hazard can be minimized through two different methods:

- Installation of jet blast deflector (or fence)
- Planning new airports or runways with unharmful jet blast contours to other aircraft.

3.4.3.1 Jet Blast Deflector (JBD) or Blast Fence

It is a safety device that redirects the jet blast to prevent its hazards. Its structure should be strong enough to withstand hot, high-speed exhausts, and debris carried by the turbulent air/gas.

Jet blast deflectors are either fixed or movable which can be raised and lowered by hydraulic arms and actively cooled. Blast deflectors can be also used as protection from helicopter and propeller wash. Jet blast deflectors can be combined with sound silencers.

3.4.3.2 Airport Planning

The design of an airport movement area must consider the hazard of "jet blast." Planners use special software to determine the jet blast contours for different throttle lever settings (say idle, takeoff, or others) of different aircraft expected to use such an airport. Based on such studies, types and positions of jet blast deflectors will be defined [65]. It is a tedious work for two reasons. First of all, simulation must be done for all aircraft even if they have close configuration (dimensions, mass, and engines). For example, Boeing 737-800 and Airbus A320-200 are nearly identical; however, they have different hazard areas. The same is for Boeing 747-400 and Airbus A380 aircraft. The second reason is that specific maneuvers on the apron (e.g., turns, engine-out operations) should be considered for each aircraft, and not only a "one-fit-all" value for all aircraft.

However, these computational codes provide more accurate design and more efficient space utilization, less operational restrictions, and at the same time maintaining or increasing the safety level.

3.4.4 Jet Blast in Airports

Jet blast deflectors were first erected at airports in the 1950s. In the 1960s jet blast deflectors had a height of 1.8–2.4 m, while in the 1990s deflectors' height was greatly increased to reach 11 m high to match airliners fitted with engines installed to the rear of fuselage like McDonnell Douglas DC-10 and MD-11 (Figure 3.21). Deflectors are installed at the beginnings of runways, especially when roadways or structures are adjacent.

Airports that are in dense urban areas often have deflectors between taxiways and airport borders. Jet blast deflectors usually direct exhaust gases upward to avoid ground debris.

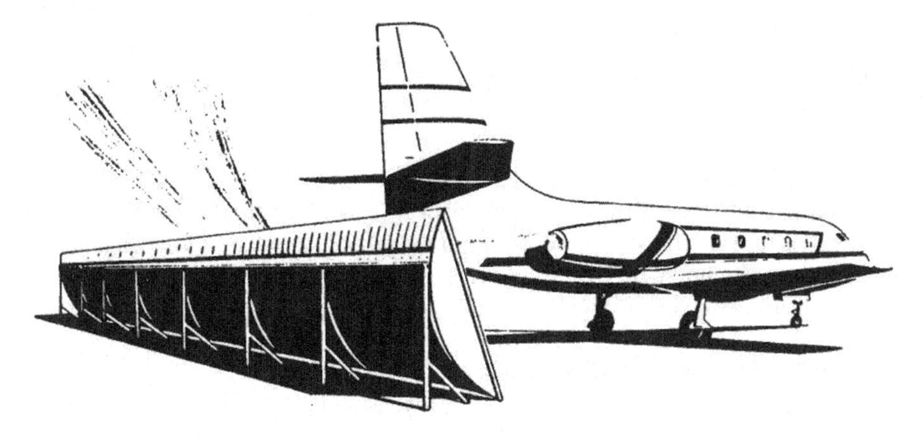

FIGURE 3.21 A typical blast fence at an airport [74].

3.4.5 Jet Blast in Air Carrier

Aircraft carriers use jet blast deflectors at the rear of aircraft catapults since the late 1940s and early 1950s to protect other aircraft from exhaust blast damage. Such deflectors could be raised and lowered by hydraulic actuators. The jet blast deflector lies flush with and serves as a portion of the flight deck. When the aircraft on its way to the catapult, the jet blast is raised into position to redirect the hot jet blast. As soon as the deflector is raised, another aircraft can be brought into position behind it [66].

Since fighter airplanes use afterburners during takeoff, jet blast deflectors are placed in very close proximity to the 1,300 °C temperatures. To mitigate such a heat problem, active cooling systems using seawater circulating through water lines within the deflector panel were employed. Recently, the US Navy in 2008 used heavy-duty metal panels covered in heat-dissipating ceramic tiles in the USS George H. W. Bush air carrier like that used in Space Shuttle.

3.5 THRUST REVERSERS

3.5.1 Function

Thrust reversal (also called reverse thrust) is a form of diverter that, when activated, reverses the thrust, and thus provides a powerful stopping force used for ground roll reduction [67]. The aircraft decelerating forces consist of wheel brakes, aerodynamic braking (flaps and speed brakes as examples), and thrust reversers [68]. The contribution of thrust reversers is significant on contaminated (wet, slushy, and icy) runways for and for events of refused takeoffs, when wheel braking effectiveness is greatly diminished. When the thrust reverse is applied, the flow does not reverse through 180°; however, the final path of the exhaust gases is about 45° from straight ahead. To achieve the greatest effect on deceleration, thrust reversers should be used at high speeds.

3.5.2 Types

Normally, a jet engine has one of three types of thrust reversers: a cascade reverser, a clamshell reverser, or a target (or bucket) reverser [69, 70].

3.5.2.1 Bucket Target Type (Hot Stream)

It uses a pair of hydraulically actuated "bucket" type doors to reverse the hot gas stream. For forward thrust, these doors form the propelling nozzle of the engine. For thrust reverse condition, two reverser buckets are deployed to block the rearward flow of the exhaust and redirect it with a forward component [60].

3.5.2.2 Clam-Shell Type (Hot Stream)

The clam-shell door system is a pneumatically operated system. When activated, the doors rotate to uncover the ducts and close the normal hot gas stream. Cascade vanes then direct the gas stream in a forward direction causing the thrust to be directed forward [71–73].

3.5.2.3 Cascade Type (Cold Stream)

The cascade thrust reverser is commonly used on turbofan engines and are often designed to reverse only the fan air portion [75, 76]. During normal operation, the bypass cold stream nozzle is open. When activating the reverse thrust system, the actuation system moves a translating cowl rearward and at the same time folds the blocker doors to close the cold stream final nozzle, thus diverting the airflow through the cascade vanes. Cascades are generally less effective than target reversers, especially used in unmixed turbofan engines that reverse only fan air, as they do not affect the engine core, which continues to produce forward thrust. However, it is more effective with mixed turbofan engines [60].

3.5.3 FOD Ingestion into Engines due to Thrust Reverse

Activation of the thrust reverser can blow debris from the runway into the engines [24] as shown in Figure 3.22. This is also dependent on the direction of the deflection system. If possible, the gas or air jet should be directed away from the engine inlet. The location of the engines on the side of the fuselage at the rear fuselage reduces the possibilities of FOD ingestion into the engines compared to wing installation. Due to such FOD reingestion concerns, thrust reversers are typically not used at speeds below about 60 knots.

3.5.4 Contaminated Runways

Thrust reversers when activated on contaminated runways it reduces visibility. In addition, the power of the engines lifts water droplets from the tarmac and ingests them into the engines.

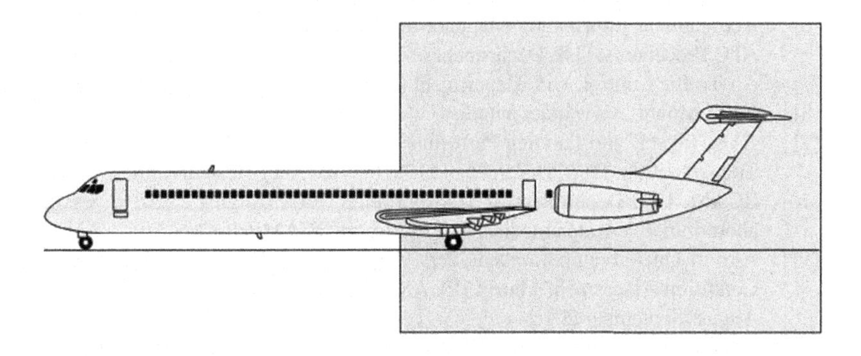

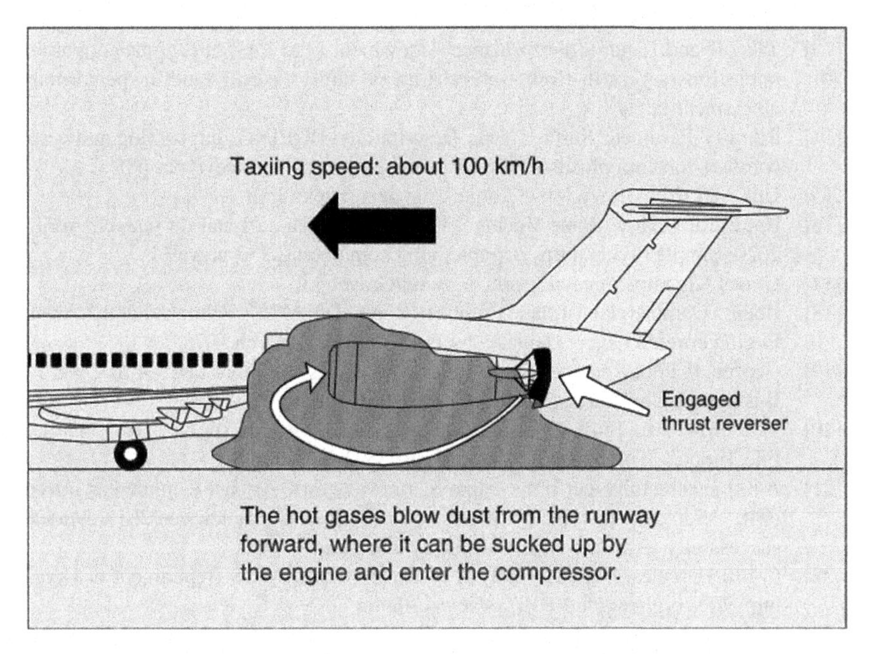

FIGURE 3.22 Thrust reverser blows foreign object debris into the engine [24].

With author permission.

REFERENCES

[1] The Briefing Room – Learning from Experience, Getting the Wind-up, July 2008, Hindsight No.7, 20–22

[2] Kennedy, R.J. Four Runway Configuration Types and Their Relation to Arrival Delays, M.Sc. Thesis, Purdue University, May 2015

[3] McCreary, I. Runway Safety: FOD, Birds, and the Case of Automated Scanning, Insight SRI, 2010

[4] www.skybrary.aero/index.php/Work_in_progress:Operations_from_Unpaved_Runways

[5] Transport Canada, Advisory Circular: Unpaved Runway Surfaces, Civil Aviation, Standards, AC 300-004, 2017-12-05

[6] Aeronautical Information Manual: Official Guide to Basic Flight Information and ATC Procedures, U.S. Department of Transportation, FAA, December 10, 2015
[7] Air Traffic Control, U.S. Department of Transportation, FAA, April 3, 2014
[8] Index Mundi, www.indexmundi.com/map/?v=123
[9] FAA, Takeoff and Landing Performance Assessment (TALPA), U.S. Department of Transportation, FAA, December 19, 2018, www.faa.gov/about/initiatives/talpa/
[10] RCAM, U.S. Department of Transportation, FAA, August 2, 2016, www.faa.gov/about/initiatives/talpa/media/TALPA-Airport-RCAM.pdf
[11] Airport Operators and Stakeholders, Airport Condition Reporting and the Runway Condition Assessment Matrix (RCAM), U.S. Department of Transportation, FAA, August/September 2016
[12] Annex 14 – Aerodromes – Volume I – Aerodromes Design and Operations, 8th Edition, July 2018, https://store.icao.int/en/annex-14-aerodromes
[13] Takeoff and Landing Performance Assessment (TALPA), https://nbaa.org/aircraft-operations/safety/in-flight-safety/runway-safety/takeoff-landing-performance-assessment-talpa/
[14] Runway Pavement Surface Type Descriptions, BOEING, https://blog.midwestind.com/wp-content/uploads/2017/11/boeing_pavement_surface_types.pdf
[15] Unpaved strip kit, www.b737.org.uk/unpavedstripkit.htm
[16] Hayward, J. How Some Boeing 737s are Equipped to Land on Gravel, April 20, 2020, Simple Flying, https://simpleflying.com/boeing-737-gravel/
[17] Gravel Kit, https://en.wikipedia.org/wiki/Gravel_kit
[18] Boeing Commercial Airplane Company Report D6-44767, Volumes 1 and 2, Airplane Engine Foreign Object Damage, by J. A. McEntire, March 1978.
[19] George, F. Pilot Report – Pilatus PC-24: A business Jet in a Class All Its Own, BCA Business & Commercial Aviation, From August 2018
[20] TAROM Tu-154B-1, https://en.wikipedia.org/wiki/Tupolev_Tu-154#/media/File:Tupolev_Tu-154B-1,_Tarom_AN0193531.jpg
[21] A400 successfully performs unpaved runway tests, Airbus, Commercial Aircraft, (04), September 2013, www.airbus.com/newsroom/news/en/2013/09/a400m-successfully-performs-unpaved-runway-tests.html
[22] C-130 Hercules, Military Analysis Network, Courtesy US Department of Defense, https://fas.org/man/dod-101/sys/ac/c-130.htm
[23] https://commons.wikimedia.org/wiki/File:C-130_Mk3_Hercules_Transport_Aircraft_landing_at_Saunton_Sands_air_strip._MOD_45151000.jpg
[24] Aeroengine safety: 5.2.1.2 Intake of Foreign Objects, https://aeroenginesafety.tugraz.at/doku.php?id=5:52:521:5212:5212#prettyPhoto
[25] Su-27 Flanker, Maybach, November 12, 2012, http://maybach300c.blogspot.com/2012/11/su-27-flanker.html
[26] Guide for maintaining runways in accordance to Annex 14 of ICAO, Airport Council International, 28/6/2012, https://applications.icao.int/tools/RSP_ikit/story_content/external_files/Guide-for-Maintaining-Runways-in-Accordance-to-Annex-14-of%20ICAO.pdf
[27] Burnett, J. 3 Vital Maintenance Considerations for Gravel Runways, EK35®, Fines Preservation®, Gravel Runway February 10, 2016, https://blog.midwestind.com/unpaved-airports-safety/
[28] 'Protection from debris impacts', Notice of Proposed Amendment (NPA), European Aviation Safety Agency (EASA), NPA22013-02, 18 January 2013, www.easa.europa.eu/sites/default/files/dfu/NPA%202013-02.pdf

[29] Nose Gear Tire Tread Separation and Number 1 Engine Failure, Air Accident Investigation Sector, Serious Incident Final Report: AAIS Case No: AIFN/0011/2016, 2016
[30] Beatty, D.N., Gearhart, J.J., Readdy, F., and Duchatellier, R. The Study of Foreign Object Damage Caused by Aircraft Operations on Unconventional and Bomb-Damaged Airfield Surfaces, ESL-TR-81-39, June 1981
[31] Boeing Commercial Airplane Co., Airplane Requirements for Operations on Gravel Runways, D6-45222-1, March 10, 1980
[32] Boeing Commercial Airplane Divisions, Substantiation for 727 Gravel Runway Operation, D6-18498, October 11, 1966
[33] A diagram of an aquaplaning tire, https://en.wikipedia.org/wiki/Aquaplaning#/media/File:Hydroplaning.svg
[34] Bless, S.J., Cross, L., Piekutowski, A.J., and Swift Fod, H.F. Generation by Aircraft Tires, ESL-TR-82-47, August 1983, US Air Force Engineering and Services Laboratory
[35] Nguyen S.N., Greenhalgh E.S., Olsson R., Iannucci L., and Curtis P.T. Parametric Analysis of Runway Stone Lofting Mechanisms, International Journal of Impact Engineering (2009), doi: 10.1016/j.ijimpeng.2009.11.006, https://hal.archives-ouvertes.fr/hal-00665448
[36] Nguyen, S.N., Greenhalgh, E.S., and Olsson, R. Analytical Modeling of Runway Stone Lofting, Journal of Aircraft, Vol. 48, No. 4, July–August 2011
[37] Nguyen, S.N., Greenhalgh, E.S., Iannucci, L., Olsson, R., and Curtis, P.T. Improved models for runway debris lofting simulations, The Aeronautical Journal, November 2009, Vol. 113, No. 1148
[38] Hallquist J.O. LS-Dyna Theoretical Manual. Version 970 ed. Livermore Software Technology Corporation. Livermore, California, 1998.
[39] Jackson, P. (ed.), Jane's All the World's Aircraft 2007–2008, Jane's Information Group, London, 2007.
[40] Aquaplaning-Hydroplaning, Skybrary, www.skybrary.aero/index.php/Aquaplaning (accessed May 1, 2021)
[41] "Supplementary Performance Information for Take-Off from Wet Runways and for Operations on Runways Contaminated by Standing Water, Slush, Loose Snow, Compacted Snow or Ice," Joint Aviation Authorities, JAR-25, AMJ 25X1591, Hoofddorp, The Netherlands, 1993.
[42] "A350 XWB MSN004 Successfully Undertakes Water Ingestion Tests at Istres," Airbus S.A.S Press Release, Toulouse, France, May 2014, www.airbus.com/newsroom/press-releases/en/2014/05/a350-xwb-msn004-successfully-undertakes-water-ingestion-tests-at-istres.html [retrieved May 2021].
[43] "Boeing 737 MAX Makes a Splash," The Boeing Company, Chicago, IL, Jan. 2017, www.boeing.com/features/2017/01/max-water-spray-testing-01-17.page [retrieved May 2021].
[44] Daugherty, R.H. and Stubbs, S.M., "Measurements of Flow Rate and Trajectory of Aircraft Tire-Generated Water Spray," NASA Rept. TP-2718, 1987.
[45] Trapp, L.G. and Oliveira, G.L., "Aircraft Thrust Reverser Cascade Configuration Evaluation Through CFD," 41st Aerospace Sciences Meeting and Exhibit, AIAA Paper 2003-0723, Jan. 2003.
[46] Qu, Q., Zhang, F., Liu, P., and Agarwal, R.K., "Numerical Simulation of Water Spray Caused by a Rolling Airplane Tire," Journal of Aircraft, Vol. 53, No. 1, 2016, pp. 182–188. doi:10.2514/1.C033276

[47] Gooden, J.H.M., "Engine Ingestion as a Result of Crosswind During Take-Offs from Water Contaminated Runways," National Aerospace Laboratory/NLR Rept. TP-2013-201, Amsterdam, June 2013

[48] Zhao, K., Liu, P., Qu, Q., Lin, L., Lv J., Ding, W., and Agarwal, R.K. Numerical Simulation of Aircraft Tire-Generated Spray and Engine Ingestion on Flooded Runways, Journal of Aircraft, Vol. 54, No. 5, 1840–1849, September–October 2017

[49] Morsi, S. and Alexander, A., "An Investigation of Particle Trajectories in Two-Phase Flow Systems," Journal of Fluid Mechanics, Vol. 55, No. 2, 1972, pp. 193–208. doi:10.1017/S0022112072001806

[50] LS-DYNA Keyword User's Manual, Vol. 1, Livermore Software Technology Corp., Troy, MI, 2015, pp. 1097–1100, 2189–2191.

[51] ANSYS FLUENT 14.0 Documentation, ANSYS, Inc., Ann Arbor, MI, 2012,

[52] Bore, C.L. "Scaling Laws For Vortex Induced Debris Ingestion Into Air Intakes", Research Note BAe-KRS-N-GEN-274, March 1983.

[53] El-Sayed, A.F. and Emeara, M.S. "Aero-Engines Intake: A Review and Case Study", *Journal of Robotics and Mechanical Engineering Research*, Verizon Publisher, Vol: 1, Issue: 3, 10 March 2016.

[54] El-Sayed, A.F. and Emeara, M.S. "Intake of Aero-Engines: A Case Study", *The International Conference of Engineering Sciences and Applications*, Aswan, Egypt, January, 29 – 31, 2016

[55] El-Sayed, A.F. and Emeara, M.S. Aerodynamics of intakes of high bypass ratio (HBPR) turbofan engines, International Robotics & Automation Journal, Volume 6 Issue 2, pages 88–97, 2020

[56] Glenny, D.E. "Ingestion of debris into intakes by vortex action", U.D.C. No. 621-757: 533.697.2:532.527, N.G.T.E. Peyestock, G.P. 1114, December 1968.

[57] Ivanyushkib, A.K. and Pavlyukov, E.V. "Aerodynamic Problems of Propulsion System Operation Safety", Proceedings of the "Aircraft Flight Safety Conference", Zhukovsky, Russia, August 31–September 5, 1993, pages 148–162.

[58] Campbell, J.F. and Chamber, J.R. Patterns in the Sky: Natural Visualization of Aircraft Flow Fields, NASA. Langley Research Center, NASA-SP-514, 1994

[59] Engine Thrust Hazards in the Airport Environment, www.boeing.com/commercial/aeromagazine/aero_06/textonly/s02txt.html

[60] El-Sayed, A.F. Aircraft Propulsion and Gas Turbine Engines, Taylor & Francis, 2nd edition, 2017

[61] Jet Efflux Hazard, www.skybrary.aero/index.php/Jet_Efflux_Hazard

[62] ASRS Directline, to the Aviation Safety Reporting System (ASRS), Issue Number 6: August 1993, https://asrs.arc.nasa.gov/publications/directline/dl6_blast.htm

[63] Jet blast warning sign, https://en.wikipedia.org/wiki/Jet_blast#/media/File:SXMDanger.jpg

[64] De Courville, B. and Thisselin, J.-J. Applying takeoff thrust on unsuitable pavement surface may have hidden dangers, Aircraft Operation, N0. 3, 2004, pp 7–8

[65] A safety-based approach to assess jet blast at aerodromes, Airsight, 2016, www.airsight.de/projects/item/a-safety-based-approach-to-assess-jet-blast-at-aerodromes/

[66] Jet blast deflector hydraulically raised to protect F/A-18 Hornet from the exhaust of another aboard of USS Abraham Lincoln (CVN-72) air carrier in 2003 https://en.wikipedia.org/wiki/Jet_blast#/media/File:US_Navy_030405-N-9951B-021_Two_F-A-18_Hornets_prepare_to_launch.jpg

[67] NASA, Part II: THE JET AGE: Chapter 10: Technology of the Jet Airplane https://history.nasa.gov/SP-468/ch10-3.htm

[68] Yetter, J.A. Why Do Airlines Want and Use Thrust Reversers? A Compilation of Airline Industry Responses to a Survey Regarding the Use of Thrust Reversers on Commercial Transport Airplanes, NASA Technical Memorandum 109158, Jan 1995
[69] The Jet Engine, Rolls-Royce plc 1986, Fifth edition, Reprinted 1996 with revisions
[70] Siddiqui1, M.A. and Haq, M.S. Review of Thrust Reverser Mechanism used in Turbofan Jet Engine Aircraft, International Journal of Engineering Research and Technology. ISSN 0974-3154 Volume 6, Number 5 (2013), pp. 717–726 © International Research Publication House www.irphouse.com
[71] Target 'bucket' thrust reverser deployed on the Tay engines of a Fokker 100 https://en.wikipedia.org/wiki/Thrust_reversal#/media/File:Klm_f100_ph-kle_arp.jpg
[72] Cold-stream type thrust reverser being deployed on a Boeing 777-300. https://en.wikipedia.org/wiki/Thrust_reversal#/media/File:PH-BVC_KLM_(3701878334).jpg
[73] A Mahan Air Airbus 310 using reverse thrust in rainy weather at Dusseldorf Airport https://en.wikipedia.org/wiki/Runway#/media/File:Mahan_Air_A310_EP-MNO.jpg
[74] A typical blast fence at an airport https://en.wikipedia.org/wiki/Jet_blast_deflector#/media/File:Blast_fence_FAA.jpg
[75] Bureau Enquêtes-Accidents. Preliminary Report – Translation, Accident on 25 July 2000 at "La Patte d'Oie" in Gonesse (95) to the Concorde registered F-BTSC operated by Air France http://212.155.144.30/docs/anglais/htm/f-sc000725pa.html
[75] El-Sayed, A.F. Fundamentals of aircraft and rocket propulsion, Springer, 2016
[76] "A U.S. Air Force Airman inspects the landing gear of a C-17 Globemaster III aircraft, July 1, 2014, on the flight line at Joint Base Charleston, S.C. The maintainers performed checks and maintenance around 140701-F-EV310-009" by A1C Clayton Cupit – www.defenseimagery.mil/imageRetrieve.action?guid=18df79c5d609dbaa0679ab32d9e6aea6e76200c6&t=2.

4 Inanimate Debris Generated by Adverse Weather Conditions

4.1 INTRODUCTION

Weather shocks frequently impair the smooth functioning of transportation systems and disrupt aviation safety. Weather proves hazardous to aviation (on ground and in-flight). It may result in aviation accidents/incidents and flight disruption (which is defined as the scheduled flight is canceled, or delayed for 2 h or more within 48 hours of the originally scheduled departure time [1]). Among the greatest aviation threats are heavy rain, drizzle, snow, hail, fog, mist, haze, visibility, icing, engine icing, microburst, cumulonimbus and thunderstorm, wind, turbulence and wind shear, low cloud and poor visibility, lightning, ceiling, squalls/line squalls, hot and high, sand and dust storms (SDS), and volcanic ash [2–4].

Weather severely impacts both civilian and defense aviation operations [5–7]. Aircraft may be either fixed- or rotary-wing vehicles. Concerning civil aviation, weather influences its two communities: namely, the airlines (pilots, flight crew, and passengers) and the airports (air traffic controllers, dispatchers, and maintenance staff). Civilian aircraft may encounter tremendous changes in environmental conditions during the same trip. For example, a commercial aircraft may travel from one climatic condition at the start of its trip to a different environment at the end of the trip.

Military activities are also impacted by bad weather conditions [8, 9]. Such activities are very sensitive to weather which affects navigation, safety of flight, and tactics.

Heavy rainfalls, snowstorms, and thunderstorms led to the defeat of Napoleon in 1812 (the French Russian Campaign) and in 1815 (Waterloo) during ground battles. During World War II, commanders checked whether the airfields would experience bad weather. Air Force weather personnel were among the first US forces deployed to combat zones in Korea, Vietnam, the Persian Gulf, and Afghanistan. The US Air Force in the Vietnam war experienced much worse flying weather specially during the cool season (cloudiness, thunderstorms, precipitation, visibility, etc.). During Operation Desert Storm in 1991 and Iraq freedom in 2003, clouds, sandstorms, rainfalls, and solar critically influenced the battles. Fighters and military helicopters in the Russian–Ukraine war of February 2022 suffered a lot from the bad weather and snow storms.

Helicopters also suffer from bad weather conditions. Fog, rainfall, snowfall, dust, and sand storms reduce visibility and may lead to accidents.

DOI: 10.1201/9781003133087-4

From the above description, constant monitoring of the day-to-day weather situation forms a key part of handling any flight or mission. In addition, the environmental conditions must be carefully considered in the design of airframe, aeroengines, systems, and equipment of aircraft of both fixed- and rotary-wing vehicles.

Aviation authorities, including Federal Aviation Administration (FAA) and European Union Aviation Safety Agency (EASA), issued several circulars for aircraft protection against weather hazards. Also, in compliance with Title 14 of the Code of Federal Regulations (14 CFR), all pilots should get a complete weather briefing before each flight [10]. The pilot is responsible for ensuring that they have all information needed to make a safe flight.

4.2 DEFINITIONS

A brief description of such adverse weather that disrupts aviation is given hereafter [11].

4.2.1 Thunderstorms

Thunderstorms are responsible for most weather-related aircraft accidents/incidents and many delays [12]. Thunderstorms are born from cumulonimbus clouds (CB) and are so-known as convective weather [11]. They are accompanied by severe turbulence, lightning, hail, heavy precipitation, and severe icing. There are as many as 40,000 thunderstorm occurrences each day worldwide. Thunderstorms are barriers to air traffic because they are usually too tall to fly over, too dangerous to fly through or under, and can be difficult to circumnavigate [11]. There are three principal thunderstorm types: single-cell, multicell (cluster and line), and supercell.

A thunderstorm can combine several aviation weather hazards like lightning, adverse winds, downbursts, turbulence, icing, hail, rapid altimeter changes, static electricity, and tornadoes.

4.2.2 Cumulonimbus (CB) Clouds

Cumulonimbus cloud [13] is a heavy and dense cloud of considerable vertical extent in the form of a mountain or huge tower. It is often associated with heavy precipitation, lightning, and thunder.

4.2.3 Precipitation

Precipitation is any form of liquid or solid particle that falls from the atmosphere and reaches the ground. The different types of precipitation are rain, drizzle, ice pellets (sleet), hail, small hail (snow pellets), snow, snow grains, and ice crystals [14]. The most effective tool to detect precipitation is radar (radio detection and ranging), which was utilized since the 1940s. The ground used radar by the National Weather Service is called the Weather Surveillance Radar-1988 Doppler (WSR-88D) built in 1988. Pilots use their aircraft radar in addition to weather detection [15].

4.2.3.1 Drizzle

Fairly uniform precipitation is composed of fine drops very close together. It falls from stratiform clouds. Drizzle appears to float while following air currents, but unlike fog droplets, it falls to the ground. Quite often, fog and drizzle occur together [14]. Drizzle usually restricts visibility to a greater degree than rain [15]. When drizzle changes to light rain, visibility improves as the droplet size increases, which implies that there are fewer droplets per unit area.

4.2.3.2 Rain

Drops larger than drizzle (0.02 inch/0.5 mm or more) are considered rain. However, smaller drops are also considered raindrops if, in contrast to drizzle, they are widely separated [14].

4.2.3.3 Heavy Rainfall

High rainfall rates will:

- Increase the aircraft's weight
- Reduce visibility to less than 400 m, and windshield wipers become ineffective
- Increase surface roughness of the wings and tails which reduces lift and increase drag
- Reduce aircraft safety particularly during takeoff, approach, and landing
- Reduce the tire friction coefficient due to the standing water on runway
- Lead to performance deterioration of aircraft engines
- Lead to the ingestion of water droplets into the aircraft engines if thrust reverse is operated during landing
- Lead to engine flameout of helicopters.

Until now aviation specialists are still unable to overcome the threat of heavy rain to aircraft. They just advise to stop flying in heavily rainy days.

4.2.3.4 Icing and Snow

Ice may have different forms including pellets and crystals. Pellets (sleet) are round or irregular hard grains of ice consisting of frozen raindrops. Ice crystals have the form of needles, columns, or plates. It is also called "diamond dust." Snow crystals are mostly branched and have the form of six-pointed stars [14]. Icing endangers both ground and flight operation of aircraft. It may disrupt flight schedules with frequent delays and cancellation. Accidents caused by aircraft icing represent 12% of the total accidents due to weather. On ground, a standing water near-freezing conditions will lead to ice accumulation on aircraft during its parking, taxiing, or takeoff. During flight, ice accumulates on every exposed frontal surface of the airplane, wings, propeller, flight deck windshield, the control surfaces, antennas, instrument offices, air intake, and cowlings [16].

For fixed-wing aircraft, ice on a wing surface will increase the weight and drag forces and decrease the lift and thrust forces. Thus, the airplane may stall at higher speeds and lower angles of attack than normal. Ice accretion may have rime, clear, or mixed patterns.

For helicopters ice accretion may occur on its rotors and front parts (windshield and fuselage). Both de-icing and anti-icing systems are employed in fixed- and rotary-wing aircraft. They should be de-iced before takeoff to ensure its smooth lifting surfaces. Next, anti-icing fluids are applied to delay the reformation of ice for a certain period or prevent the adhesion of ice to ease its removal. Boeing and Airbus have their guidance to pilots and flight planners to reduce the risk of flights at such adverse weather.

4.2.3.5 Hail

Hail is a precipitation in the form of small balls or other pieces of ice falling separately or frozen together in irregular lumps. An individual unit of hail is called a hailstone. Small hail stones are equal or greater than 0.25 inch or 5 mm in diameter which is either round or conical [14]. Large hail stones may be equal or greater than 4.5 inch or 11.25 cm which indicate severe thunderstorms. Hailstones that are equal or greater than 0.75 inch in diameter can cause significant damage to aircraft and make it difficult to control. A severe case of hailstone was encountered at Vivian, South Dakota, on July 23, 2010, which diameter was 8 inches. Hailstones may influence aircraft during takeoff and cruise operation. Its strike with the windshields may yield zero visibility and influence its radome, propeller, and engines [15]. Hailstones of a sufficient size can cause damage too.

4.2.4 Obscuration Types

Obscuration types are phenomena in the atmosphere (other than precipitation) that reduce the horizontal visibility [14]. They include mist, fog, smoke, volcanic ash, sand, dust, and haze.

4.2.4.1 Mist

Mist is a visible fine aggregate of water particles or ice crystals suspended in the atmosphere. It is identified as mist when the difference between the air temperature and dew point is less or equal to 3 °F (1.7 °C), which reduces visibility to the range (1–11 km).

4.2.4.2 Haze

Haze is a suspension of extremely small dry particles in the air. It is identified as "haze" when the difference between the air temperature and dew point is greater than 3 °F (1.7 °C). Mist is invisible to the naked eye and sufficiently numerous to give the air an opalescent appearance. Certain haze particles increase in size with increasing relative humidity, thus greatly decreasing visibility [17].

4.2.4.3 Fog and Ice Fog

Fog is a visible small water droplets at the Earth's surface that reduce horizontal visibility to less than 1 km. Ice fog has the following names: radiation fog, advection fog, upslope fog, frontal fog, and steam fog [17]. Ice fog is a type of fog formed by direct freezing of supercooled water droplets. Ice fog is rare at temperatures warmer than

−30 °C. At temperatures warmer than −30 °C, a steam fog of liquid water droplets is formed, which may turn into ice fog when cooled.

4.2.4.4 Smoke

Small particles suspended in the air and produced by combustion. A transition to haze may occur when smoke particles have traveled great distances, 25–100 miles (40–160 km) or more.

4.2.4.5 Dust Storm

Dust storm is composed of dust particles having fine diameters exposed to strong winds (15 knots or more) and lofted airborne. Intense dust storms may reduce visibility to near zero close to source regions in few seconds. The average height of a dust storm is 3,000–15,000 ft (1–4.6 km). Without turbulence, dust generally settles at a rate of 1,000 ft (300 m) per hour, which needs several (or days) for the dust to settle [17]. Operation of aircraft in a dust storm can be very hazardous. Dust can clog the air intake of engines, damage electro-optical systems, and result in human health problems.

4.2.4.6 Sandstorm

Sandstorm is sand particles raised by the wind to a height sufficient to reduce horizontal visibility. Such sand particles are mostly confined to the range 3.5–15 m (10–50 ft) above the ground [17]. Sandstorms are similar to dust storms, but larger and heavier than dust particles. Sandstorms are generated in desert regions where there is loose sand and dunes.

4.2.4.7 Volcanic Ash

Volcanic ash is fine particles of rock powder that issued from an erupting volcano. It may remain suspended in the atmosphere for long periods. These particle may abrade the aircraft surface. It may be ingested into the avionics, navigation systems, and aircraft engine [18].

The ash plume may not be visible especially at night. However, it is difficult to distinguish visually between an ash cloud and an ordinary cloud. Though volcanic ash cannot be detected by air traffic control radar, it may be detected by weather radar especially during the early stages of a volcanic eruption when the ash is more concentrated [17].

4.3 STATISTICS FOR WEATHER-AVIATION INFLUENCES

Meteorological parameters of weather are observed [19] using

1. New remote sensing platforms, namely wind LIDAR, sodars, radars, and geostationary satellites
2. In-situ instruments at the surface and in the atmosphere
3. Aircraft sensors
4. Unmanned aerial vehicles mounted sensors.

4.3.1 Accidents and Incidents

4.3.1.1 Survey

Most aviation accidents occur during the takeoff, climb, descent, approach, and landing phases of the flight [11]. Weather accounts for 10%–15% of all aviation accidents [20].

Most accidents due to hazardous weather occur during landing (43%), approach (18%), and takeoff (12%) [21]. Timing (hour and day) of snow&icing, fog, sandstorm, and volcanic ash can be predicted. Ground transport will be disrupted by snow&icing, thunderstorm, and sandstorm. Runways and taxiways will be disrupted by snow&icing and sandstorm. Influence of snow&icing, fog, and sandstorm may be propagated from one airport to multiple other ones for several hours or even days, while the effect of volcanic ash may be propagated for several others for days or even weeks.

4.3.1.2 Accidents in the Period from 1994 to 2000

Among the 8,657 aviation-related accidents, weather caused 1,784 event of these accidents [22]. Visibility, ceiling height, and precipitation-related conditions occurred 485 times, wind and turbulence 1,149 times, and airframe icing and engine icing 150 times. Weather-related accidents are caused by wind and visibility, which had the percentages 48% and 20%, respectively, of the total accidents [23, 24, 25].

4.3.1.3 Accidents in the Period from 2000 to 2011

From the statistics of National Transportation Safety Board, 29% out of the 19,441 were related to weather conditions [26]. For small, noncommercial aircraft (Part 91 class), the primary cause of weather-related accidents was adverse winds (52%), followed by low ceilings (7%). For commercial jet aircraft (Part 121 class), more than 70% of weather-related accidents were due to turbulence followed by adverse wind 14% [27].

A short list for contribution of weather to aviation accidents is given hereafter for few countries:

United States: 8% of commercial incidents in the United States is due to weather [28].

Indonesia: 58% of the incidents in Indonesia is due to weather [29]. Adverse weather contributed to loss of visual reference for flights in the mountainous areas, which contributed to the majority of Indonesian fatal accidents

Taiwan: Aviation Safety Council (ASC) [30] found that accidents/incidents due to weather were 16.3%, while others noted that the global average ranged from 21% to 26% [31].

4.3.2 Delays

Delays may be observed in arrivals and departures of flights. Generally, aircraft delays emerge when there are interactions between air transport partners (i.e., airlines, airports, and air traffic control units) and external parameters (i.e., adverse weather conditions, strikes, political reasons, and others) which lead to airport congestion [32]. Adverse weather conditions at one airport will affect flights downstream [33, 34]. Delays add extra costs including additional airport charges, maintenance, and crew costs and passenger compensation [35]. Delays can be divided into avoidable and unavoidable. Unavoidable delays are directly related to the severity of the weather [35]. The National Center of Atmospheric Research clarifies that nearly 60% of daily delays and cancellations due to weather are avoidable [35]. Effective delay management and improved weather forecasting will reduce the impact and duration of delays. Several studies examining weather effects are listed in [36–38] for the United States, [39] for Europe, [40] for Asia, [41] for China, [42] for Indonesia, [10] for Africa, and [43] for Australia.

Few studies have been done for departure delays. A comprehensive analysis for the departure of 2.14 million flight from ten large US airports between January 2012 and September 2017 was performed [44]. The departure delays for 90 connections from these ten airports were analyzed.

A summary for departure delays in the six-year period (2012–2017) for the selected ten airports is as follows. A significant increase in departure delay up to 23 min was concluded depending on the weather type and intensity of the disturbance. The mean delay is equal to 28 min with a standard deviation of 40.2 min. Based on the hours over the day, departure delay peaks at around 8 p.m. (20 min) and recedes thereafter, having a minimal at 8 a.m. (8 min). Considering the day of the week, flights departing on Mondays, Thursdays, and Fridays have slightly higher delays (12 min). A minimum value is observed on Saturday (8 min). Based on the month of a year, the highest delays are during summer months (June, July, August) and December. Due to the holiday season, the average departure delay in July and August amounts to around 15 min, whereas departure delays in September, October, and November on average are around 9 min. Moreover, convective weather during summer, snow storms and cold temperatures in December increase the delays to around 15 min. For the ten airports, the mean departure delay ranges between 9.5 min (SeattleTacoma International) and 16 min (Chicago O'Hare International). The maximum delays are in Chicago and Denver and the minimum delays are in Seattle, Charlotte, and NYC's JFK Airport.

The following weather events impact departure delay [44]:

- Rainfall exceeding 0.1 inches/h results in increases departure delays by on average 13 min. For stronger intensities, departure delay increases by around 23 min.
- Snowfall (temperature below zero) causes on average an additional delay of around 11 min.

4.3.3 Costs

Reducing aviation delays due to adverse weather is a part of cost cutting solutions. Some statistics are given here.

A 65% of the delays experienced by US domestic airlines are attributable to adverse weather, with estimated costs of US$ 3 billion per year [45]. The massive flight delays and cancellations due to the winter storms and cold weather in January 2014 cost the airlines and passengers US$ 1.4 billion according to masFlight, an airline consulting firm [46]. A three-days freezing fog event in the UK resulted in about 50 M US$ in financial losses for businesses [47]. A US$ 250,000 loss per fog hour experienced in Frankfurt airport [48]. Based on National Oceanic and Atmospheric Administration, US airports can save US$ 600 million per year from improved winter weather forecasting and icing diagnostics [49]. Based on NavCanada, a 100% accurate terminal aerodrome forecasts at Canadian airports would save US$ 12.5 million annually [49]. Flight delays due to weather at Atlanta Hartsfield International Airport, United States costs one airline US$ 6 million annually [50]. A study on 35 commercial US airports in 2008 revealed that 81,429 hours of arrival delays could be avoided which will save over US$ 258 million [49]. The percentage of avoidable delays due to inaccuracy of terminal weather forecast amounted to 12.2% [49].

Based on the US Bureau of Transportation Statistics, more than 720 million passengers boarded a domestic flight in the United States in 2017. With 6.2% delays due to weather, 44,640,000 passengers experience some kind of weather-related delay. The cost of flight delay for freight transport is estimated as US$ 0.77 per package for a 15-min delay and US$ 3.92 per package for a 60-min delay [51]. These costs accumulate between US$ 12,000 and US$ 25,000 per hour of delayed freight per aircraft. If all the weather-related departure delays in 2017 were due to precipitation (which range between 10 and 23 min), this corresponds to costs between US$ 670 million and US$ 1.54 billion [52]. Passengers value on-time performance and are willing to pay to avoid delays [53]. Moreover, the flight delays lead to an average reduction in ticket prices by US$ 1.42 per additional minute of flight delay [54]. The above listed numbers reflect only the value of time from passengers' perspective. However, for airlines, delays cause additional costs due to expenses for crews, fuel, and maintenance.

Finally, it is worth mentioning that reducing aviation delays due to adverse weather could be considered as part of cost cutting solutions [55].

4.4 SAND AND DUST STORMS (SDS)

4.4.1 Introduction

SDS have been understood as a natural phenomenon of wind carrying dust from desert areas. In desert areas, SDS are caused by either thunderstorm outflows or by strong pressure gradients. Such strong turbulent winds erode sand and silt small particles from arid and semiarid landscapes and then lift and launch them into the air [56]. The main terrestrial sources of airborne dust are including the Mediterranean areas of Europe and Africa, northern Sahara, Arabian Peninsula, Central Asia and China, southwest United States, South Africa, and Southern Australia. Most of these areas create a so-called dust belt.

The frequency of sandstorms, particularly across the lands bordering deserts, has increased dramatically over the past five decades. Thus, the possibilities for aircraft to encounter SDS when operating in these regions have been increasing.

Sandstorms move particles by three mechanisms:

1. Suspension – dust and fine sand particles are lifted to great heights by the wind.
2. Saltation – sand particles vibrate, then bounce and dislodge other particles as the wind increases. Thus, particles become negatively charged, which increase the dislodged dust and sand particles.
3. Creep – blowing large particles along the ground.

The mechanisms of SDS movement, the physical or chemical changes during transport, etc. are governed by the weather and geological features. Particulates in SDS are composed of rock-forming minerals such as quartz and feldspar, and clay minerals such as mica, chlorite, and kaolinite. However, SDS particulate may adsorb anthropogenic atmospheric pollutants during transport.

Global estimates of dust emissions vary between one and three Gigatons per year. The total world dust emission in 2012 was 1,564 million tons. It includes natural sources, such as deserts, depressions in arid land, and anthropogenic sources, such as land-use changes, agricultural practices, water diversion [57]. The distribution of dust is maximum in North Africa 840 million tons (maximum) with the minimum anthropogenic source percentage (8%), while South Africa has the minimum dust 40 million tons with 54% anthropogenic source percentage. Australia has 63 million ton dust and maximum anthropogenic source percentage 76%.

The size of dust and sand particles ranges from submicron to several hundreds of microns. Sandstorms occur within a shallow layer of few meters above the ground surface, while dust storms have finer dust particles lifted much higher into the troposphere, and frequently transported over great distances [58]. Normally, in dust storms, dust concentrations reach 100–1000 $\frac{\mu g}{m^3}$. However, during an Iranian dust storm in January 2017, fine particle concentrations exceeded 10,000 $\frac{\mu g}{m^3}$. The safe air quality for concentration of fine particles set by World Health Organization is at or below 50 $\frac{\mu g}{m^3}$. Dust storms having average diameter of 10 μm can travel thousands of kilometers across continents and oceans, entraining other pollutants on the way. Dust settles when winds drop below the speed necessary to carry the particles, but some dust haze may remain at 5,000–9,000 m downstream for days after a dust storm. Sand storms having average diameter of 60 μm may remain airborne for short distances of few hundred meters [59]. An advancing sandstorm associated with a gust front looks like an advancing wall of swirling sand. The height of this wall may be equal or greater than 1 nm. If there is significant atmospheric instability, dust can reach as high as 20,000 ft [60].

Figure 4.1 illustrates a massive dust storm cloud (haboob) is close to a military camp over Al Asad, Iraq, just before nightfall on April 27, 2005.

SDS impact visibility, health, and flight operations [62]. The visibility is at its worst throughout sunlight hours once the wind is at its strongest. Intense dust storms reduce visibility to near zero in and near source regions, but improving away from the source. From the edge of blowing dust to within 240 km downstream, visibility can range from 800 to 4,800 m.

One of the worst sandstorms was encountered at 1751 local time on July 5, 2011 in Phoenix Sky Harbor International Airport. A 1.6 km high, 161 km wide wall of dust roared from the southeast moving at the speed of 48–64 km/h. It reached the airport

FIGURE 4.1 Dust storm rolling over Al Asad, Iraq, on April 27, 2005 [61].

at 1847 and the visibility dropped to 200 m. Airport was closed for 45 min and the reduced visibility and high wind speed lasted for hours [63]. Visibility is more critical for military aircraft landing on an aircraft carrier, since the runway is short and narrow. The pilots must precisely position their plane and snag a rope to decelerate rapidly [64].

Concerning health problems, inhaling particles smaller than 10 microns cause heart, lung, eye, and skin diseases [65].

Sand and dust contaminants, when combined with rain, can form mud, which affects the values of runway coefficient of friction.

4.4.2 The Main Deserts

A list of the main deserts worldwide are given hereafter [66–69]. Sahara is the world's largest hot desert. Arabian Desert extends from Yemen to the Persian Gulf and from Oman to Jordan and Iraq. Turkestan Desert is in Turkmenistan, South of Aral Sea. Iranian Desert contains the hottest deserts in the world, namely, Kavir and Lut. In East Asia many great deserts exist, including Taklamakan, Gobi Desert, and the Great Indian Desert between India and Pakistan.

Australia Deserts represent 38% of Australia. Kalahari Desert covers parts of many African countries including Botswana, Namibia, Angola, Zambia, and South Africa. Namib Desert covers Namibia and southwest Angola. North American Desert has four primary deserts in North America. Atacama Desert covers 105,000 square km of the Northern Chile, and one of the driest deserts in the world. Patagonian Desert is the largest desert in South America.

4.4.3 Famous Sandstorms

The most famous annual sandstorms are highlighted hereafter [66]:

Haboob: Haboob which means "blowing furiously." Haboob is a SDS that occurs in the deserts of the Sudan region of north-central Africa and in the southwestern region of the United States. It is a tumbling, black wall of dust and sand that may rise to 23,000 ft (7,000 m) above the ground, reducing visibility to near zero. It generally lasts for 30–60 min and travels at the speed of about 30 miles (48 kilometers) per hour across short spans or for great distances.

Harmattan: Harmattan is a SDS that originates in the Sahara Desert due to the harmattan. The harmattan is an easterly or northeasterly wind that produces dust and sand storms up to 20,000 ft high and transports more than 300 million tons of reddish Saharan dust westward across the continent each year. Some 100 million tons of which are deposited into the Atlantic Ocean. Saharan dust also travels 1,600 miles to Great Britain twice a year and falls to the ground as a red precipitation called "blood rain."

Khamsin: Khamsin is an annual hot, dry, southerly wind originating on the Sahara that produces large SDS. It forms over Libya and Egypt. If a storm forms over Turkey, the khamsin blows dust over the northern tip of the Red Sea and into Saudi Arabia, and Jordan. The khamsin blows for about 50 days starting in mid-March.

Saltation: Saltation is the wind-driven movement of particles along the ground and through the air. At wind speed of 10 mph, blown sand reaches a height up to about 49 ft. Heavy sand particles tend only to creep along the surface, moving sand dunes from one location to another. Dust suspended in extreme storms may reach a height of 35,000 ft. Very fine clay particles (0.00004 in diameter) can be lifted by turbulent air and remain in the atmosphere for years.

Shamal: Shamal is a hot, dry, dusty wind that blows for one to five days at various times throughout the year. It produces great dust storms throughout the Persian Gulf, the lower valley of the Tigris and Euphrates in Iraq. The great shamal is normally in June and early July, where the wind blows for 40 days at the speed of 30 mph.

Simoom: Simoom (or the poison wind) is a hot, dry, blustery, dust-laden wind that blows across the Sahara and the deserts of Syria, and the Arabian Peninsula, and depositing its dust on Europe. Its temperatures exceed 130 °F (54 °C) and can cause heatstroke.

4.4.4 Sand/Dust Storms and Fixed-Wing Aircraft

4.4.4.1 Effects

Sandstorm and dust storm SDS activity results in: massive canceling of scheduled flights, reduced visibility which may be close to zero in some circumstances, abrasion of windshield and landing lights, damage or scratch of leading edges of wings and tail, reduction in engine power and complete engine failure in severe cases, blockage of the pitot tubes leading to inaccurate or wrong readings for air speed, clog aircraft instruments, blockage and corrosion in air conditioning packs, reduction of cabin and cockpit air quality, and problems to electrical equipment.

4.4.4.2 Defense

The following procedure is suggested:

- **Awareness**
  Knowing the local climatology and the weather forecast for the route to be flown is essential. Schedules should be designed to avoid the times when sandstorms are most prevalent.
- **Plans**
  Airport authorities will deal with the consequences of a sandstorm, such as blowing sand across aircraft operating surfaces, and so on.
- **Action**
  1. Close an airport if the visibility drops below 500 m for safety implications as the flight crews will be unable to see clearly other aircraft or obstructions [70].
  2. Use a better system for illuminating airports and runways which should be able to penetrate rain, fog, and snow to improve visibility in adverse weather conditions [71].
  3. The following actions are followed if aircraft is on the ground and a sandstorm is approaching: use tension fabric aircraft shelters to protect small aircraft from sand/dust storm, move military aircraft to shelters [72], consider turning the aircraft into wind and tie-down if strong winds are forecasted, cover the intakes, vents, and tubes to prevent ingress of sand and dust, cover any area inside the cockpit where the ingress of dust could interfere with flying controls, for example, throttle quadrant and condition levers, check and clear intakes and vents of any sand and dust before subsequent flight, and if possible, vacuum the flight deck to remove any dust.
  4. If airport is surrounded by deserts, use a windbreak (a row of trees or shrubs placed around the airport) to slow the wind and keep it from blowing away the airport.
  5. Using remote sensing methods with others to quantify atmospheric particulate concentrations [73].
  6. In design phase: use a gravel kit as in Boeing 737-200, for fighters use low-pressure tires (10 bar/150 psi) like the MiG-29, for fighters having intakes close to ground like the MiG-29 use louvers on the upper wing during takeoff and close its air intakes to prevent foreign object damage.

4.4.5 SDS and Helicopters

Helicopters frequently operate to and from desert sites, which can lead to the disturbance of loose sediment from the ground, creating a cloud of dust that will envelope the whole aircraft. When the visual environment has been degraded, the situation is commonly known as brownout [74].

As defined in [75],

> **Brownout** is the term used to describe the result of helicopter rotorwash as it kicks up a cloud of sand/dust while landing. Brownout causes the pilot to loose

vision of the surroundings. Thus it accidents during helicopter landing and take-off operations in desert terrain, dust storms or general vehicle movements.

During brownout, particulate impacts the rotor(s) and is ingested into the engine inlets [76]. It quickly erode the rotors blades made of hard metals like nickel and titanium, as well as other metals used to protect their leading edges [77]. The sand striking titanium also creates bright visible sparks at night, forming a "halo effect," exposing the military helicopters to enemy fires.

4.4.5.1 Effects

Solid particulates lead to: damage of the whole aircraft, low visibility, erosion of main rotor blades which increases the operation costs as the price of a rotor blade varies from US$ 100,000 to US$ 700,000 [77], erosion of laminated plastic of the transparent windshields and canopies, impacts on glass domes of captively carried, optically guided missiles, and difficulty of passengers' breathing, and clogging all instruments.

4.4.5.2 Defense

- **Awareness and plans**
  As described in (Section 4.4.4.2).
- **Action**
  1. If aircraft is on the ground and a sandstorm is approaching: store helicopters inside hangers if available, cover the intakes, vents, and tubes to prevent ingestion of sand and dust, cover the whole helicopter, check and clear intakes and vents of any sand and dust before subsequent flight.
  2. In the design phase the following two methods may be applied [78]: develop a filtration system to remove most of the solid particles from the main air stream and choose materials that are not appreciably affected by solid particle erosion and thus extend the engine life.

4.4.6 Effects on Engines

Air-breathing engines powering fixed- and rotary-wing aircraft are subjected to erosion and fouling (deposition) due to SDS and volcanic ash. Also, they are subject to rain erosion and icing due to rainfalls and supercooled large water particles. All of these phenomena will be discussed in detail in Chapter 10. Helicopter engines are especially susceptible to large amounts of dust and sand ingestion during takeoff, hover, and landing [79]. During the Vietnam field operations some helicopters had as low lifetime as 100 h of operation [80]. During the Gulf War field operations, helicopter engine suffered a loss in power and compressors' surge margins led to engines removal after fewer than 20 h [81].

Operation of aircraft during strong SDS leads to erosion of the cold section (fans/compressors) which results in significant changes in its geometry or profile as described in [82–87]. Ingested particulates into the hot section (combustion chamber and turbines) cause both corrosion and fouling (deposition) and may even clog the cooling holes in turbine stators [4].

Engine air particle separators and screens or barrier filters are used to separate any FOD (SDS, wrench, bird, large piece of ice, a rag or foliage, etc.) from the compressors in turboshaft engines powering helicopters [88]. However, fine particles will continue its route into the cold section of engines [89, 90]; thus, many helicopters sustained severe damage after few operating hours (20–100 h) in dusty environments [91].

As will be discussed in detail in Chapter 10, penalties of operation of compressors in dusty environments include [92]: blunting of leading edge and changes in the incidence angle and the leading edge radius, thinning of the blade trailing edge, increase in surface roughness, and so on.

SDS will also affect the hot section of aeroengines including combustors and turbines [84]. The main deterioration (refer to Chapter 10) includes: deposition (fouling) of combustion chamber and turbine nozzle guide, blockage of turbine blade cooling passages, imbalance in inner shaft causing an unwanted vibration, etc.

Deteriorations in the cold and hot sections of aeroengines lead to several changes during operation [93] including: drop in aerodynamic performance, increase in exhaust temperature of the cycle (E.G.T.), fuel consumption, engine stall, flameout, and so on.

As will be thoroughly described in Chapter 8, designers apply several precautions to reduce the effects of particulates on fixed- and rotary-wing aircraft.

4.5 VOLCANIC ASH

4.5.1 Introduction

Volcanoes are proof that the Earth is alive, active, and ever-changing. The word volcano comes from the little island of Vulcano in the Mediterranean Sea off Sicily. When a volcano erupts, it sends vast amounts of hot gases, solid (rocks), and ash up into the atmosphere. The size of these emissions can vary from bombs measuring over 64 mm to fine ash particles less than 0.063 mm. Large solid material precipitates from the atmosphere near the volcano. Small particles (<15 microns) and gaseous material can be transported to high altitudes. These small particles pose the greatest threat to aircraft [90] as they may remain in the atmosphere for days, or even weeks, and can affect commercial air traffic routes.

Volcanic ash influence animates humans, animals, birds, and plants. For humans, it may irritate eyes and throat, reduce lung function, especially in young children, and increase risks of heart attacks. For animals, if inhaled or ingested may cause fluoride poisoning which would result in internal bleeding, long-term bone damage, and teeth loss. Birds may face difficulty in flying since they are weighed down by the ash on their wings, and it may irritate their eyes. Plants and trees may die from high concentrations of CO_2 gas in the soil beneath them.

The strongest volcano in human history was the eruption of Mount Tambora in April 10, 1815. It had a volcanic explosive index (VEI) of 7. It killed 92,000, and another around 117,000 deaths owing to starvation and diseases. Also a Tsunami with a wave height of 10 m was reported. The ash from the eruption dispersed around the world and lowered global temperatures.

Volcanoes continue to erupt around the world, from Iceland to Ecuador and from Chile to Africa, causing disruption to air travel [94]. An ICAO survey [95] confirmed 83 encounters from 1935 to 1993 and nearly 17 from 1994 to 2000 [96]. ICAO displayed the locations of volcanoes in the last 10,000 years [97]. The famous encounter of an aircraft with a volcanic plume was on June 24, 1982. A British Airways B747-200 aircraft flying from Kuala Lumpur, Malaysia, to Perth, Australia, lost power on all four engines at an altitude of 11,300 m (37,000 ft) due to volcanic ash from Mt Galunggung volcano (Indonesia). For 16 min, the aircraft flew as a glider and descended from 11,300 to 3,650 m, at that point the pilot was able to restart three of the engines and make a successful emergency landing at Jakarta, Indonesia [95].

In the 20th century, nearly 550 strong volcanoes erupted. Annual volcanoes are 60, and some are so weak and even not felt. A total of 253 aircraft encounters with volcanic ash or gas clouds between 1953 and 2016 were identified, of which 122 occurred in the period 2010–2016 [98]. In the United States, 50 strong volcanoes were registered. Moreover, the 2017 volcanic eruption in Bali and the eruption in Hawaii in 2018 resemble high-profile volcanoes.

The countries subjected to strong volcanoes are Philippines, Japan, Indonesia, and the United States. Volcanoes with highest number of encounters (more than five) [96] are: Pinatubo, Philippines (1991), Sakura-jima, Japan (1977–1998), St. Helens, United States (1980), Augustine, United States (1976), Redoubt, United States (1989–1990), Galunggung, Indonesia (1982), Avachinsky-Koriaksky, Russia (2008), Colima, Mexico (2015), Etna, Italy (2013), Galeras, Colombia (2010), Nyiragongo, Congo (2002), Santa Maria/Santiaguito Guatemala (2011–2015), and Santorini, Greece (2019) [99]. Some features of strong volcanoes in the 20th century are listed in Table 4.1.

4.5.2 Earth and Under Water Volcanoes

Volcanoes are either on the Earth or under the water (submarine). Earth volcanoes are about 1,500 worldwide. About 500 of the 1,500 volcanoes are located along what is called the "Ring of Fire" [100]. It is about 40,000 km horseshoe shape in the basin of the Pacific Ocean (Figure 4.2). Other active regions are in Iceland, along the great rift valley in Central and East Africa, as well as in countries around the Mediterranean. Many volcanoes worldwide are not hazardous. Volcanoes become destructive when they emit hot magma tied with ash and gases.

Submarine volcanoes are underwater vents which represent 20% of total volcanoes. Though most of submarine volcanoes are located in the depths of seas and oceans, some exist in shallow water, which discharges material into the atmosphere during its eruption. An example for underwater volcanoes is the Kolumbo submarine volcano in the Aegean Sea. Submarine volcanoes may threaten both air traffic and sea planes (Figure 4.3). Such submarine volcanoes may endanger sea or amphibious aircraft or amphibious aircraft carrier.

4.5.3 Regions of Active Volcanoes

Hot spot volcanoes occur somewhat randomly around the globe. In fact, there are over 100 hot spots that have been active sometime during the last 10 million years or so. A daily updated map for the currently erupting volcanoes is available from Volcano Discovery [103]. For any moment, Ref. [104] provides a map for the presently active erupting volcanoes worldwide and the specific date is printed on the map.

TABLE 4.1
Some Features of Volcanoes in the 20th Century [99]

Year	Volcano	Features
1902	Mont Pelée, Martinique	Completely destroyed the town of St. Pierre, killing nearly 30,000
1912	Novarupta, Alaska	Largest US volcanic eruption of the 20th century, produced 21 cubic kilometers of volcanic material
1914-1917	Lassen Peak, California, United States	Pyroclastic flows, debris flows, and lava flows covered over 16 square kilometers
1980-1986	Mount St. Helens, Washington, United States	Killed 57 people and triggered debris flows that temporarily stopped shipping on the Columbia River and disrupted highways and rail lines
1982	Mt Galunggung, Indonesia	Indirectly killed 18 people in traffic accidents and by starvation
1989–1990	Redoubt, Alaska, United States	Disrupted aviation operations in south-central Alaska
1990s	Pinatubo, Philippines	Injected enormous clouds of volcanic ash and acid gases into the stratosphere to altitudes in excess of 100,000 ft. Based on different sources, number of displaced people varies from 250,000 to 2,100,000 and number of deaths ranges from 560 to 1,300
1992	Cerro Negro, Nicaragua	Number of displaced people is 28,000
1997	Soufriere Hills, Montserrat, UK	Number of displaced people is 7,000
1999	Mayon, Philippines	Number of displaced people is 6,000

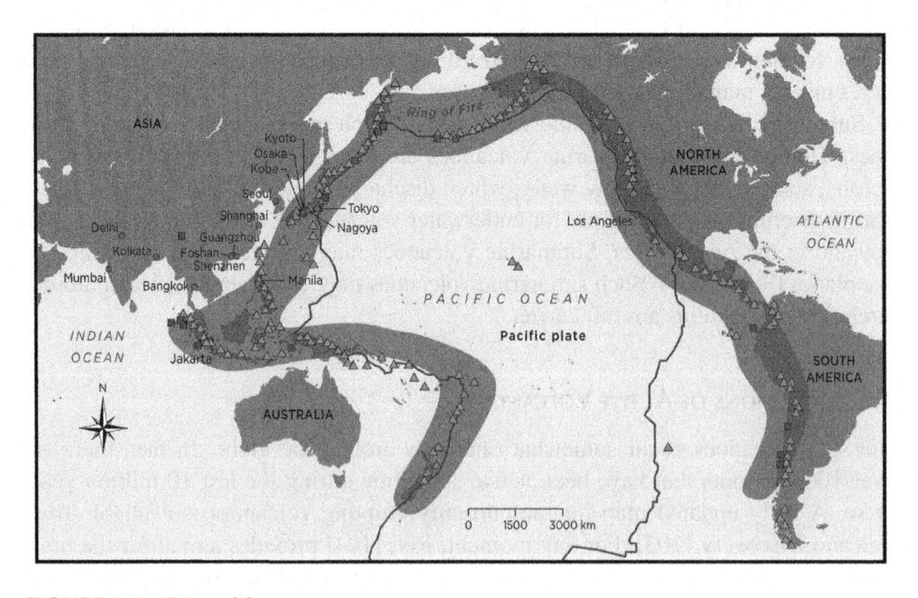

FIGURE 4.2 Ring of fire.

Courtesy: USGS [101].

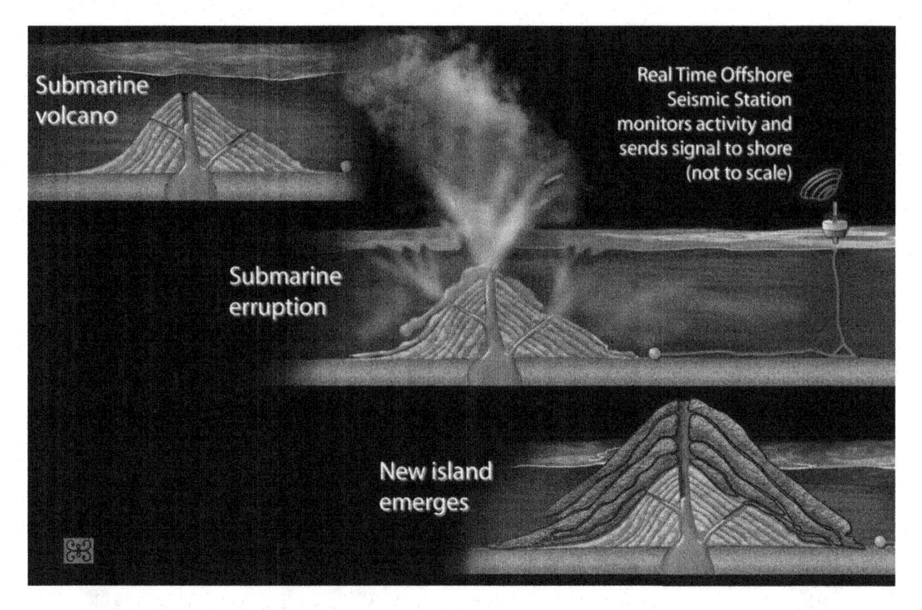

FIGURE 4.3 Submarine or underwater volcano [102].

Courtesy: National science foundation.

4.5.4 Volcanoes Constituents

A volcano is defined as a vent at the surface of the Earth through which lava and other volcanic materials are ejected from the Earth's interior. The following materials are the constituents of a volcano (Figure 4.4): Magma/Lava, Bombs, Ash, and Gases.

1. Magma

Magma is the molten material (or rocks) below the Earth's surface. It can be erupted as lava or pyroclasts. The temperature of magma ranges between 700 °C and 1,300 °C. When the magma reaches the surface, it can extrude as lava. Most lava erupted onto the Earth's surface is basalt.

2. Bombs

Volcanic bombs are larger than 65 mm in diameter (Figure 4.5). They can acquire aerodynamic shapes during their flight. During the 1935 eruption of Mount Asama in Japan, bombs measured 5–6 m in diameter were thrown up to 600 m from the vent [102].

3. Volcanic ash

Volcanic ash is a tiny, light, grey to black sharp-edged, hard glass particles of pulverized rock, primarily basalt which is comprised mostly of silicates. Its diameter ranges from less than 2 mm to less than 0.063 mm. Its melting point is below jet

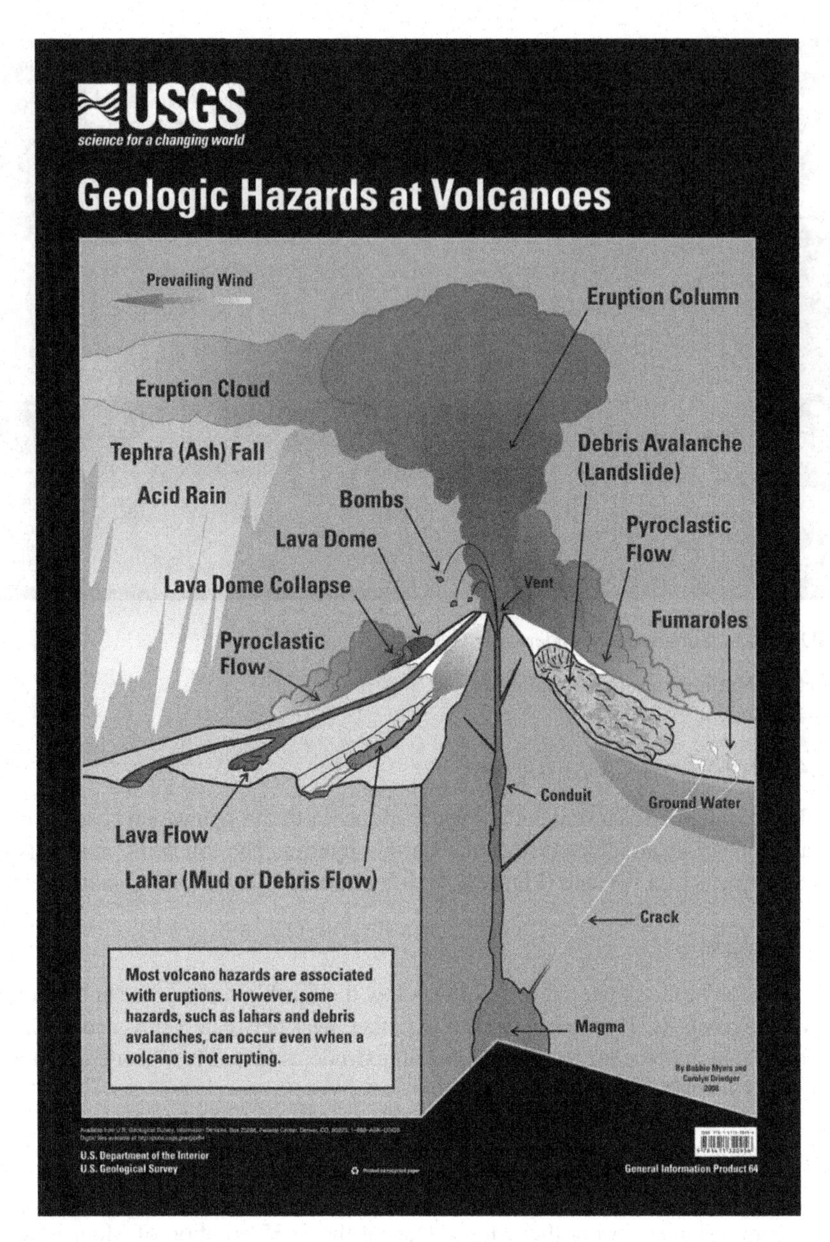

FIGURE 4.4 Constituents of volcano [105].

Courtesy: USGS [106, 107].

engine burner temperature in cruise condition which results in additional hazards to its hot sections [108]. Volcanic ash can be spewed tens of thousands of feet into the air, reaching aircraft cruising altitude (6–11 km). It can be transported hundreds to thousands of kilometers downstream of the source of the eruption. The global annual

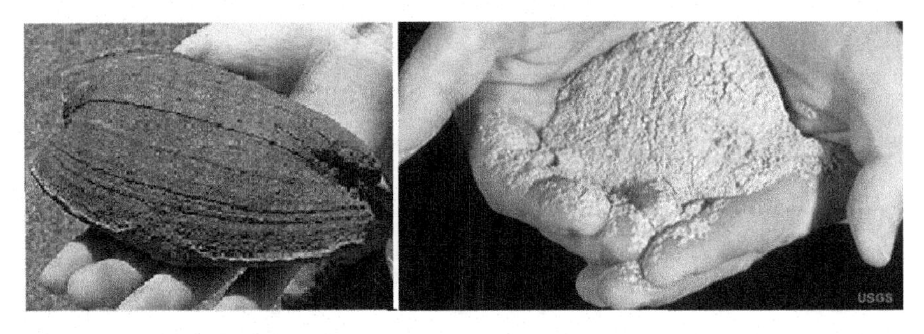

Volcanic Bomb Volcanic Ash

FIGURE 4.5 A volcanic bomb and ash.

[106].

emitted and transported volcanic ash (<63 μm) is estimated to be approximately 33 Tg/year [109, 110].

4. Volcanic gases

The principal components of volcanoes are water vapor (H_2O) (=60%), carbon dioxide (CO_2) (=10%–40%), sulfur either as sulfur dioxide (SO_2) or hydrogen sulfide (H_2S), nitrogen, hydrogen, helium, neon, methane, carbon monoxide (CO), and argon.

Other compounds are hydrogen chloride (HCl), hydrogen fluoride (HF), hydrogen bromide, nitrogen oxide (NO_x), sulfur hexafluoride, carbonyl sulfide (OCS), and organic compounds.

Figure 4.6 illustrates the different gases emitted from volcanic eruption.
The gaseous volcanic toxins that pose the greatest concern to human health include: sulfur dioxide gas (SO2) [112], fluoride acids, and hydrochloric acid.

4.5.5 Volcanic Ash Advisory Centers

One of the results of Eyjafjallajökull volcanic event of 2010 was the establishment of nine Volcanic Ash Advisory Centers (VAACs) [97] worldwide. The VAACs provide an important link among volcano observatories, meteorological agencies, air traffic control centers, and operators. These nine centers are located at: Anchorage VAAC – Anchorage, AK, United States, Buenos Aires VAAC – Buenos Aires, Argentina, Darwin VAAC – Darwin, Australia, London VAAC – London, United Kingdom, Montreal VAAC – Montreal, Canada, Tokyo VAAC – Tokyo, Japan, Toulouse VAAC – Toulouse, France, Washington VAAC – Washington, DC, United States, and Wellington VAAC – Wellington, New Zealand.

Each center monitors and reports any volcanoes activities at a definite region of the globe.

They provide advisory information on the extent and movement of volcanic ash in the atmosphere.

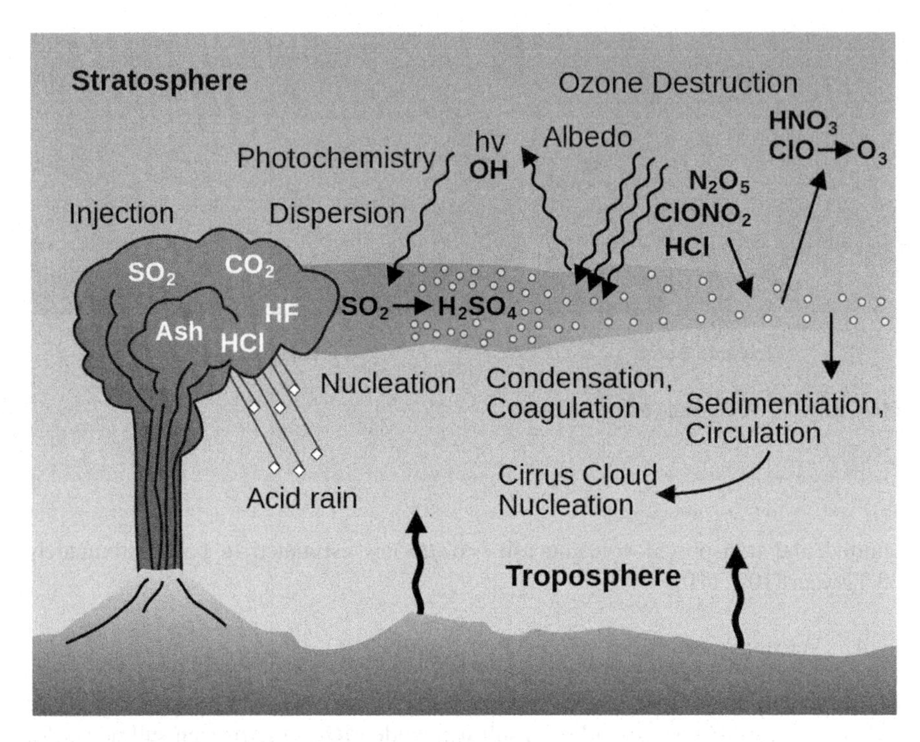

FIGURE 4.6 Gases emitted from volcanic eruption.

Courtesy: USGS [111].

4.5.6 Volcanic Ash Resources

Volcanic ash resources are available worldwide which provide immediate eruption and volcanic ash updates to operators by fax, email, telephone, or teletype. Many information is available on the following sites and references [113–115, 116]:

- The Smithsonian National Museum of Natural History Global Volcanism Program: www.volcano.si.edu/gvp/
- The US Geological Survey: www.usgs.gov/themes/volcano.html
- The Airline Dispatcher Federation
- The Committee on Earth Observation Satellites Disaster Management Support
- Project/Volcanic Hazards Management; the US National Oceanic and Atmospheric Administration
- The Aerospace Industries Association of America (AIA) ad hoc propulsion committee (PC-334-1)
- The Istituto Internazionale di Vulcanologia (a summary of volcanoes in Italy): www.iiv.ct.cnr.it/
- The Nordic Volcanological Institute (information about volcanoes in and around Iceland): www.norvol.hi.is/index.html

- The Volcanological Society of Japan:
  http://hakone.eri.u-tokyo.ac.jp/kazan/VSJ1E.html
- Current Eruptions in Japan:
  http://hakone.eri.u-tokyo.ac.jp/vrc/erup/erup.html

4.5.7 Volcanic Explosivity Index (VEI)

The VEI is described as a semiquantitative measure of the explosiveness of volcanic eruptions, the total ejected volcanic material, eruption column height, and eruption duration [104, 117]. It ranges from 0 to 8.

VEI = 0 is a nonexplosive eruption, while a VEI = 8 is the most disruptive eruption [118]. This index identifies two main eruption styles: effusive and explosive eruptions. Effusive eruptions have constant low viscous magma flows (mainly fluid basaltic lava) and strong gas emissions. Explosive eruptions blast several cubic kilometers of glass particles, pulverized rock (volcanic ash), and corrosive gases high into the atmosphere and over a wide area for several days. Explosive eruptions are divided into three types: Strombolian, Vulcanian, and Plinian eruptions. These eruptions are defined in [95] as follows:

Strombolian eruption – *An eruption consisting of short, discrete explosions which may eject pyroclasts for a few tens to a few hundreds of feet into the air. Each explosion may last for only a few seconds and there may be pauses of tens of minutes between explosions.*

Vulcanian eruption – *A type of volcanic eruption characterized by the short duration, violent explosive ejection of fragments of lava. Vulcanian eruption columns may attain the heights of 45,000 ft (14 km) or more.*

Plinian eruption – *A large explosive eruption that ejects a steady, turbulent stream of fragmented magma and magmatic gas to form an eruption column that may reach altitudes in excess of 100,000 feet (30 km).*

Stromboli volcano in Italy is an example for the Strombolian eruptions and have index (VEI 1-2). Vulcanian eruptions explode more violently and have index (VEI 2-4). Mt. Galunggung eruption in 1982 was VEI = 4.

Plinian eruptions are the most vicious explosive eruptions and have index (VEI 4-8). They can transport large amounts of ash and gases into the upper troposphere or even the stratosphere within a short amount of time. Mt. Vesuvius in 79 AD and Mt. St. Helens eruption in 1980 were classified as VEI = 5. Krakatau in Java in 1883 and Mt. Pinatubo in the Philippines in 1991 were both VEI = 6. Tambora in the Lesser Sunda Islands (Indonesia) in 1815 was VEI = 7.

Explosive eruption presents a direct threat to aircraft in flight and major operational difficulties to aerodromes located downwind of the resulting ash cloud.

4.5.8 Volcanic Ash Encounter Severity Index

Severity index was formulated in 1994 by Tom Casadevall and Karin Budding (in consultation with engine and airframe manufacturers and the Air Line Pilots Association) and endorsed by ICAO [95, 96]. Based on the database, for encounters of aircraft with volcanic ash assembled by the US Geological Survey, six severity classes are defined and described in Table 4.5. The corresponding encounters in the periods 1953–2016 and 2010–2016 [98] are added in Table 4.2.

Table 4.3 lists the Volcanic Ash Encounter Severity Index per aircraft type in the period 1953–2016 [98].

TABLE 4.2
Volcanic Ash Encounter Severity Index and Number of Encounters in the Period 1953–2016 [95, 96, 98]

Class	Effects on Aircraft	Number of Encounters between 1953–2016 (...) between 2010–2016 [...]
0	• Acrid odor or sulfur gas in cabin • Anomalous atmospheric haze in cabin • Electrostatic discharge (St. Elmo's fire) on windshield, nose, or engine cowls • Ash reported/suspected by flight crew	(82) [60]
1	• Light dust in cabin (no oxygen used) • Ash deposits on exterior of aircraft • Fluctuations in exhaust gas temperature (EGT) and return to normal values	(53) [40]
2	• Heavy dust in cabin (dark as night) • Contamination of air handling and air conditioning systems requiring use of oxygen • Pitting, frosting, or breaking of flight deck windscreen or passenger windows • Minor plugging of static pitot tube, without changing instrument reading • Abrasion of exterior surfaces, engine inlet, and fan blades • Deposition of ash in engine	(67) [14]
3	• Engine damage • Vibration or surging of engine(s) • Contamination of engine oil or hydraulic system fluids • Plugging of static pitot tubes leading to erroneous instrument readings • Damage to electrical or computer systems • Interference of navigation or communication systems • Physical impairment of flight crew (due to volcanic ash or gas)	(24) [8]
4	• Temporary engine failure requiring in-flight restart of engine • Engine failure requiring in-flight permanent shutdown of engine(s)	(9) [0]
5	• Engine failure or other damage leading to crash	(0) [0]

TABLE 4.3
The Severity Class Index for Different Aircraft Type in the period 1953–2016 [95]

	Conventional Fixed-Wing Aircraft				Helicopter	
Class	Jet Engine	Turboprop Engine	Piston engine	Unknown	Turboshaft	Unknown
0	44	13	5	8	1	0
1	25	5	3	2	7	6
2	60	2	2	1	1	0
3	23	0	0	0	0	1
4	8	1	0	0	0	0
5	0	0	0	0	0	0

TABLE 4.4
Number of Encounters According to Severity Index and Distance [98]

	Distance (km)				
Class	≤ 100	$100-1000$	$1000-2000$	$2000-3000$	>3000
0	2	11	27	14	17
1	4	4	6	6	1
>2	27	26	5	3	2

The severity of aircraft encounter volcanic ash is related to the distance from the source volcano and listed in Table 4.4.

The degree of destruction of these volcanoes [119] depends upon the speed and height of eruption, physical and chemical properties of volcanic ash particles, availability of gases such as carbon dioxide, sulfur dioxide (SO2), hydrogen sulfide, hydrogen fluoride, hydrogen chloride, and so on, duration of deadly eruption, and nature of human habitation around, etc.

4.5.9 Aviation Color-Code

Aviation color-code notifications provide concise information about potential volcanic ash hazards [94]. Pilots, dispatchers, air traffic controllers, and airline flight planners can use this information to route flights away from potential ash clouds. Features of these colors are given in Table 4.5.

4.5.10 Eyjafjallajökull Volcano in 2010

On April 14, 2010, the eruption Eyjafjallajökull volcano in Iceland entered a dramatic phase, producing an ash plume that rose 9 km into the sky and drifted over the North Atlantic into the path of planes flying to and from Europe, which is the busiest airspace in the world.

TABLE 4.5
Color-Code for Volcanic Ash hazards [98]

Green	Yellow Alert	Orange Alert	Red Alert
Volcano is either in: 1 Normal, non-eruptive state. 2 Volcanic activity has ceased after higher alert activities	Volcano is either: 1 Has elevated unrest level signs 2 Has decreased activities after a higher alert level (should be monitored for possible renewed increase)	Volcano is either: 1 Exhibiting unrest with increased eruption possibilities. 2 Erupting with no or minor ash emission (specify ash-plume height)	Eruption is either: 1 Imminent with significant emission of ash into the atmosphere 2 Underway with significant emission of ash into the atmosphere (specify ash-plume height)

This disruption led to cancellation of 108,000 flights, stranding of 10 million passengers, and a revenue loss of US$ 1.7 billion in the first 6 days of the eruption based on IATA estimates.

4.5.11 Volcanic Ash and Aviation

Volcanic ash is composed of 57%–58% silicon dioxide by weight fine glassy ash and $Al_2O_3, Fe_2O_3, FeO, CaO, Na_2O, MgO$. Volcanic ash particles have less than 2 mm in diameter.

Volcanic clouds and ash can reach thousands of miles away from volcanoes and stay in the atmosphere for long periods. Volcanic ash affects aviation as follows:

- It can reach heights between 6 and 11 km, which is same cruise height of most aircraft.
- It has a melting point of approximately 1000 °C, which is less than maximum temperatures in combustors, so may cause partial or total engine damage.

4.5.11.1 Observation and Detection

Observation/detection and forecasting movement of volcanic ash in the atmosphere is carried out via three methods: namely, ground, airborne, and space-based.

a. Ground methods are visual and radar ones. It is difficult to differentiate between volcanic and regular clouds neither by eye nor by ground radar. Ground-based radar may be located within 100 km from one or more volcanoes. Doppler and polarization of the radar signal could provide important information on particle size, shape, and velocity. However, these radars are a rather expensive units, especially as none of the volcanoes monitored may erupt for decades or even longer [95, 120].
b. Airborne observation from the aircraft cockpit enhances closer observation. Airborne observation is conducted by pilot and airborne radars operating in the X band. Pilots report only what they see in their "pre-eruption volcanic

activity" report. Next, the relevant area control center decides if it is necessary to issue a NOTAM (notice to airmen) or not.

c. Space-based observation uses two basic types of satellite: polar-orbiting and geostationary. Polar-orbiting satellites orbit the Earth at an altitude between 700 and 1,200 km. They pass over the same points on the Earth at approximately the same time every 24 h. Geostationary satellites orbit at the same speed as the Earth's rotation and remain stationary to an observer on Earth. They are located above the equator at an altitude of 36,000 km. Both types can detect volcanic eruptions and volcanic ash cloud.

The polar-orbiting satellites carry sensors which detect volcanic ash, sulfur dioxide and sulfuric acid aerosol, and thus help in identifying and tracking volcanic clouds.

Geostationary meteorological satellites are operated by Europe, China, India, Japan, the Russian Federation, and the United States. They provide lower resolution than similar data from polar orbiting satellites. However, they image the same area of Earth at least every hour, or even every 15 min (Meteosat-8).

4.5.11.2 Effect of Volcanic Ash on Aircraft

4.5.11.2.1 Airborne Personnel

Volcanic ash influences flight crew and passengers. It may cause health problems on breathing and speech and dryness of throat.

4.5.11.2.2 Airframe

The volcanic ash influences both airframe and engine. The damage to airframe modules and systems [95, 121] include: abrasion of the cockpit windows, passenger cabin windows, leading edges of the flight surfaces and the tailfin, "sandblast" the airframe's paint, scratch of landing gear lights, damage to wing leading edges, block pitot-static and fuel and cooling system holes, contamination of fuel, oil, and cooling systems, electrical and avionics units, cargo fire-warning system, etc.

A model for predicting the visibility in the dust storms of Northeast Asia was proposed in [122]. For visibilities less than 3.5 km, the relationship between visibility (V) in (m) and the dust concentration (C_{dust}) in (mg/m^3) is given by:

$$V = \frac{4900}{C_{\text{dust}}^{1.19}} \tag{4.1}$$

Thus for $C_{\text{dust}} = 0.05\,mg/m^3$, the visibility will be greater than 50 km, while for $C_{\text{dust}} = 4\,mg/m^3$ visibility will be only 400 m. For a high concentration say $C_{\text{dust}} = 2000\,mg/m^3$, visibility will be less than 1 m.

4.5.11.2.3 Engine

The engine is subjected to the following damage mechanisms [123, 124, 125]: St. Elmo's glow at the engine face, erosion of the leading edges and blade tips of the cold section (fan/compressors) and a decrease in its efficiency, reduction of compressor surge margin, deposition on the hot section components, reduction of the throat area of high-pressure turbine (HPT) inlet guide-vane causing the outlet static pressure of combustor and compressor to increase rapidly, blockage of fuel nozzles and cooling holes of the HPT, contamination of oil system and bled air for cabin pressurization, loss of thrust and power, and engine flameout.

4.5.12 Procedure to be Followed by Flight Crew When Aircraft Encounter with Volcanic Ash

The following procedures have to be followed [95, 94].

1. Exit the ash cloud as quickly as possible. If possible, turn out the aircraft a 180° of the ash cloud, using a descending turn.
2. Reduce thrust to idle immediately if the flight conditions permit.
3. Turn off auto throttles, engine and wing anti-ice (WAI) devices, all air-conditioning packs, the continuous ignition.
4. If engine fails try restarting again immediately.
5. If all engines flameout try to start the auxiliary power unit (APU), which provides an extra source of electrical power.
6. Monitor the air speed, pitch attitude, and engine exhaust gas temperature (EGT) as they may sharply exceed if engine is contaminated.
7. The crew must use oxygen masks in the presence of an acrid odor, or dust in the flight deck.
8. Do not deploy the masks in the cabin unless there is a loss of pressurization.

The Air Line Pilot Association (IATA) Safety Department adds the following recommendations:

1. Do not attempt to climb out of the ash cloud.
2. Perform engine restart, if necessary.

Finally, each operator's flight operations manual should include more specific instructions.

4.5.13 Inspection After Aircraft Encounter with Volcanic Ash

Aircraft manufacturers list special full inspection procedures to be performed for the aircraft in any of the following cases:

1. After a flight encounter volcanic ash conditions.
2. If there was a sulfur odor during a flight in volcanic ash conditions.
3. After volcanic ash contamination when the aircraft is on the ground.

In any of the above three cases, it is required to examine the aircraft surfaces, structures, systems for contamination, erosion/abrasion, dents, etching, delamination, or other damage. All these inspections are visual unless other procedure is stated by the manufacturer. For example, if nondestructive procedures is necessary, operators use the Nondestructive Testing Manual. Also, if there is damage to the aircraft structure, the Structural Repair Manual (SRM) is used. SRM defines the approved damage limits and approved paint/protection repair procedures [126].

Maintenance staff should follow the following precautions [126]:

- Do not breathe volcanic ash.
- Do not get volcanic ash in eyes.
- Put on protective clothes, eye goggles, and a respiration mask.
- Make sure that the respiration mask can remove volcanic ash from the air.

4.5.13.1 Preparation for Inspection

1. The following recommendations are stated for each case:
   a. If contamination occurred in any flight phase, get information from the crew about the flight conditions and compare them with the aircraft Post Flight Report.
   b. If there was a sulfur odor during a flight in volcanic ash conditions, and no volcanic ash was seen on the external surface of the aircraft, replace the recirculation filters.
   c. If contamination occurred when the aircraft was on the ground with preservation from volcanic ash, no checks are needed for the systems with protective covers/devices if there is no volcanic ash behind them.
2. Install the safety devices on the landing gears.
3. Extend the flight control surfaces (flaps, slats, and spoilers).
4. Open the doors of the main and nose landing gears (MLG) and (NLG) and install the safety devices.
5. Depressurize the hydraulic systems.
6. De-energize the aircraft electrical circuits.
7. Put warning notice(s) in the cockpit to avoid operating systems during maintenance.
8. Put warning signs on ground near the inspection area.

4.5.13.2 External Inspection

4.5.13.2.1 Dry Ash on Aircraft Surfaces

Remove the volcanic ash with a soft brush or vacuum cleaner or a blower. Take care that such removed ash does not cause contamination of other components like probes.

4.5.13.2.2 Wet Ash

If a wet volcanic ash exists, use a low-pressure water supply to remove wet contamination.

4.5.13.2.2.1 Probes and Sensors Concerning the static probes, pitot probes, ice detectors, angle of attack (AOA) sensors, and total air temperature (TAT) sensors follow the following procedures:

1. Remove the tape-adhesive from the bottom side of the film-polyethrlene (if installed before the volcanic ash condition) to let the water drain.

2. If no film-polyethrlene was installed on the probes/sensors before the volcanic ash condition, then remove the contamination from the protection covers/devices with a textile-lint free cotton.
3. If no protection covers/devices were installed on the probes/sensors before the volcanic ash condition, then remove the contamination from the probes/sensors and their adjacent structure with a textile-lint free cotton and install the protection covers/devices.
4. After the installation of the protection covers/devices on the sensors and probes, apply the film-polyethrlene on the protection covers/devices on the probes/sensors. Seal the film-polyethrlene at the top and on the sides with the tape-adhesive. Do not seal the bottom side of the film-polyethrlene to let the water drain.
5. Record the locations where you apply the film-polyethylene and the tape-adhesive.

4.5.13.2.2.2 External Surface A low-pressure water-supply equipment is used for cleaning the top of the aircraft first and then the bottom to ensure that the contamination is fully removed from the aircraft and clean all the cavities on the aircraft (e.g., slats, flaps, brakes, landing gears, and landing gear bays). After cleaning, inspect all the recesses, empty spaces, and outlets on the top of the aircraft (e.g., precooler exhausts). Make sure that there is no contamination on the inner surfaces of these parts.

Remove all the tape-adhesive and the film-polyethylene installed during the preservation procedure.

4.5.13.2.2.3 Inspection of the Fuselage Check any abrasion of the flight cockpit windshield, main cabin windows, and outer surface of the aircraft's fuselage.

4.5.13.2.4 Inspection of the Wings

Examine the top and bottom skin of the wings, all the control surfaces. Examine the flap tracks, flap track fairings, flap inter connecting strut, track sensor, roller mechanisms, steady bearings, and universal joints for contamination. Examine the hinge areas of the spoilers and the ailerons for contamination. Examine all the static dischargers. Examine the piston rods of each servo control of the ailerons and the spoilers for contamination and clean them if contaminated.

Make sure that there is no ash deposit on the roller stops.

4.5.13.2.5 Inspection of the Vertical Stabilizer

Examine for contamination the skin of the vertical stabilizer and all the static dischargers. For the rudder, examine its skin specially the leading and the trailing edges, hinge areas, and piston rods of each servo control. Clean if there is contamination.

4.5.13.2.6 Inspection of the Horizontal Stabilizer

Examine for contamination the skin of the horizontal stabilizers and the elevators, specially the leading and the trailing edges, the hinge areas of the elevators, the piston rods of each servo control of the elevators, and all the static dischargers. Clean the piston rods of the servo controls if there is contamination.

4.5.13.2.7 Inspection of the External Lights and Corrective Actions

Examine for any damage in all the external lights, antennas, and the drain masts. Replace any damaged component(s).

4.5.13.2.8 Inspection of Probes and Sensors

Maintenance personnel must flush all the pressure lines if there is a report from flight crew or maintenance crew stating that there is a very dense cloud of volcanic ashes and the probe covers/devices are not in position, or a general inspection shows signs of heavy contamination.

Examine for damage and contamination of all the probes (pitot and static) and sensors (TAT, AOA), the ice detectors (if installed), and lighted icing indicators. Make sure that the AOA sensors move freely and that the drain holes of the pitot probes are not clogged. Flush the principal static- and total-pressure lines. Drain and flush the standby static- and total-pressure lines.

4.5.13.2.9 Inspection of the Landing Gear Doors, the Landing Gears, and the Landing Gear Wells

If there is contamination remove the contamination from surface, the moving joints and the chromed areas of shock absorbers and actuators before the next flight. Fully lubricate the landing gears before a maximum of ten flight cycles. Examine all the doors parts (skin, hinges, proximity sensors, actuator bare rods, and door recesses) for damage and/or contamination. Examine all the landing gears parts (attachment points of the landing gears, proximity sensors, actuator rods, chromed areas of the shock absorbers, mechanical linkages and the moving joints) for damage and/or contamination. Examine all the landing gear wells and other installed components in them for damage and/or contamination.

4.5.13.3 Internal Inspections

4.5.13.3.1 Inspection of the Fuel System

Examine for contamination the NACA vent intakes for contamination, the refuel/defuel control panel and coupling cap for contamination, the overwing refuel adaptor/cap, fuel drain mast. Clean if necessary.

4.5.13.3.2 Inspection of the Anti-ice System

- Examine for contamination the air inlet and the outlet of the precooler. If there is a contamination, then there are two possibilities:
  1. The anti-ice system was not operating during or after volcanic ash contamination.

     Disconnect and remove the duct immediately upstream of the wing anti-ice (WAI) valve and examine the valve for contamination. If there is a contamination, clean the ducts upstream of the WAI valve, clean the valve, and replace the valve filter

2. The anti-ice system was in operation during or immediately after volcanic ash contamination.

   Remove the valve filter and install a clean filter, disconnect the anti-ice duct from the telescopic duct, and examine it for contamination. If there is a contamination, clean all the anti-ice ducts from the pylon interface to the telescopic duct outlet, and clean the WAI valve.

- Examine the slat anti-ice duct outlets for contamination, and clean them if necessary. Disconnect the slat interconnecting ducts and if there is a contamination remove the slats for maintenance, and make sure that the telescopic duct operates correctly. Examine the telescopic duct, and if contaminated, remove it for maintenance.
- Examine the nacelle anti-ice valves for contamination.
- Inspect the following systems for contamination as described in [126]: the Air Conditioning System, the Hydraulic System, the Ram Air Turbine (RAT), the Potable Water System, and the Inerting Gas Generation System (IGGS).

4.5.13.4 Inspection of the Nacelles and the Engines

- Examine the damage in the air intake cowl, fan cowl doors, and the thrust reversers.
- Examine the fire extinguisher outlets for contamination.
- Check of the distribution piping for blockage if there is a contamination.
- Visual inspection of the first-stage engine fan blades.
- Borescope inspection for the blades of fan, booster (or low pressure compressor, intermediate pressure compressor of a three-spool turbofan engine (like Trent family), high-pressure compressor, combustion chamber, high-pressure turbine (HPT), and low-pressure turbine (LPT) for two-spool engine.
- Eddy Current Inspection for Turbine Exhaust Case.
- Teardown the engine components to locate any traces of volcanic ash within the engine.
- Inspection of HPT hub.
- Examine engine oil, oil filters, and heat exchanger filters.

4.5.13.5 Inspection of APU

Examine compressor blades and other internal parts for erosion. Inspect hot section parts for glassy deposits. Check the clogging of turbine blade cooling channels. Examine corrosion of metallic parts. Inspect oil circuit, hydraulic, and pneumatic systems for contamination.

4.5.13.6 Close-up Inspection and Corrective Actions [126]

1. Clean the aircraft internally and externally if necessary.
2. Retraction of the flight control surfaces (flaps, slats, and spoilers).
3. Make sure that the work area is clean and clear of tools and other items.
4. Close the nose-landing gear (NLG) and main-landing gear (MLG) doors.
5. Close all the access doors, and the access panels opened during the inspection procedure.
6. Remove the access platform(s) and the warning notice(s).

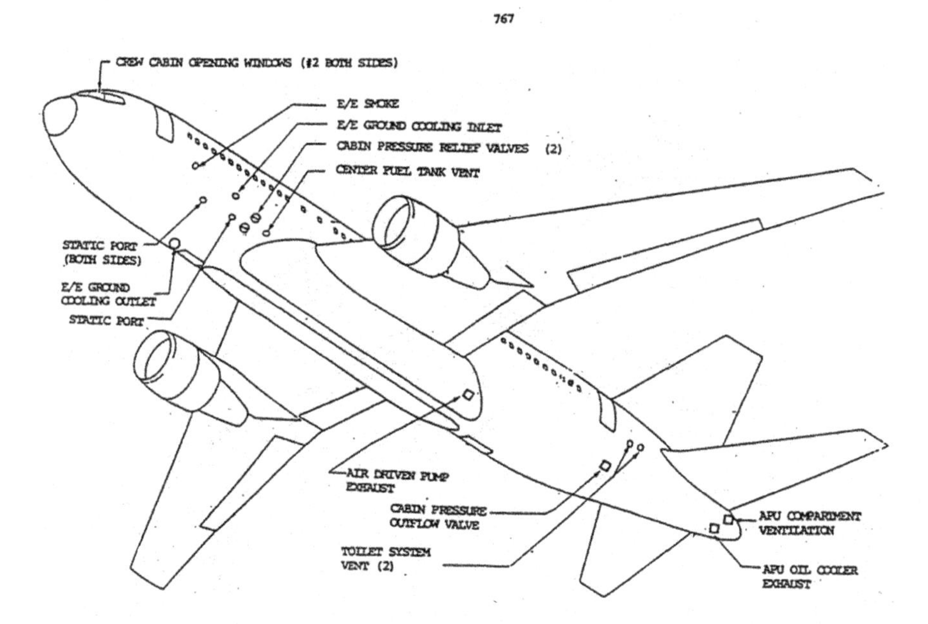

FIGURE 4.7 Openings in Boeing 767.

Courtesy: USGS [128].

7. Put the aircraft back to its initial configuration.
   a. De-energize the aircraft electrical circuits.
   b. Remove the access platform(s) and the warning notice(s).

4.5.14 Engine Protection

For engine protection from volcanic ash and sand storms, use inlet/outlet covers overnight or during sand storms. It is strongly recommended to all operators in Middle East/North Africa regions. Covers should be properly fixed from the front to the rear of the engine to avoid any volcanic ash or sand accumulation.

4.5.15 Stored and Parked Aircraft

For stored and parked aircraft, in case of impending or active ash fallout threat, all the openings of the aircraft have to be closed [127]. Figure 4.7 illustrates the openings of Boeing 767 aircraft. These opening should be covered using appropriate covers. However, if not available, plastic rolls and tapes may be used to close them.

Figure 4.8 illustrates sweeping ash off an airplane at Quito's International Airport, which was covered with ash due to the eruption of Reventador volcano on November 3, 2002 [129].

4.6 RAINFALL

Rainfall is defined as the amount of precipitation falling rain over a given area in a given period of time: it is stated in terms of the depth of water that has fallen into a rain gauge. The amount of precipitation is usually measured by the depth in inches [130].

FIGURE 4.8 Cleaning aircraft [129].

Courtesy: USGS.

4.6.1 Heavy Rainfall

It is characterized by the rate of 100 mm/h or greater. It can impart severe aerodynamic penalties on aircraft which was first identified in 1941 [131]. High velocity water droplet impact can generate pressures of many tons per square inch, which led to several flight accidents and threaten aircraft flight safety [132]. If an aircraft encounters rainfalls during takeoff or landing, it will be in a great risk [133].

4.6.2 Characteristics of Rains

For assessment of the potential hazards of rain on aircraft flight operation, the following parameters should be defined: the raindrop size distribution, the rain intensity, the terminal velocity of raindrop, the range of rainfall rate, the frequency of rainfall occurrence.

- **Drop size**

As defined in [134], the formula of the drop size distribution has an exponential form where $N(D_p)$ $(m^{-3}.\,mm^{-1})$ is equal to the number of rain drops per unit volume per unit size interval having equivolume spherical diameter D_p (mm) is expressed by:

$$N\left(D_p\right) = N_0\, exp\left(-ID_p\right) \quad 0 \leq D_p \leq D_{pmax} \tag{4.2}$$

D_{pmax} is the maximum drop diameter, and $I\left(mm^{-1}\right)$ varied with rainfall rate R $(mm\,h^{-1})$. Thus, $I = 4.1 \times R^{-0.21}\left(mm^{-1}\right)$ is associated with

$$N_0 = 8000 \;\; (m^{-3}.mm^{-1})$$

For drizzle

$$N_0 = 30,000\,(m^{-3}.mm^{-1})\,\text{and}\, I = 5.7 \times R^{-0.21}\left(mm^{-1}\right)$$

For thunderstorm [135, 138]
$N_0 = 1,400\,(m^{-3}.mm^{-1})$ and $I = 3.0 \times R^{-0.21}\left(mm^{-1}\right)$

- **Liquid water content (LWC)**

Liquid water content is the density of liquid water in a cloud expressed in grams of water per cubic meter $(g\,/\,m^3)$.

For drizzle type rain the liquid water content is expressed as:

$$\text{LWC} = 0.0892 \times R^{0.084} \tag{4.3}$$

For thunderstorm type rain [136]

$$\text{LWC} = 0.054 \times R^{0.84} \tag{4.4}$$

- **Terminal velocity** (V_T)

It is assumed that raindrops have a uniform velocity (i.e., without acceleration) before hitting the aircraft surface. So it is important to determine the terminal velocity (V_T) of raindrops. It was developed by Mar-kowitz [21] as V_T:

$$V_T = 9.58\left\{1 - exp\left[-\left(\frac{D_p}{1.77}\right)^{1.147}\right]\right\} \tag{4.5}$$

- **Rates of rainfalls (R)**

Rainfall rate is the linear accumulation depth at ground level per unit time and is typically used to characterize rainfall at ground level [137]. Light rate is 5 mm/h, while heavy rate is 100 mm/h or greater. The greatest rainfall rate record of airborne measurements is approximately 2,920 mm/h, as measured by an instrumented F-100 airplane in 1962 [138, 139].

For an airplane, rainfall rates ranges from light rain of 5–10 mm/h up to very large rainfall rates.

4.6.3 Hazard of Heavy Rainfall

Rainfall influences the airline and airport operation. Even nowadays, people are still unable to resolve the threat of heavy rain to flight; the best means is to stop flying in heavily rainy days.

4.6.3.1 Fixed-Wing Aircraft

Heavy rainfalls have a great influence of aircraft as will be discussed in Chapter 10. A brief description will be given here:

1. Reduce visibility either by scattering of light [140] from a large number of liquid droplets (visibility will be less than 400 m) or water film and splashing of rain drops on the windshield of an aircraft.
2. Decrease of the accuracy of measurement instruments like the α-vane instrument that measures aircraft AOA [141] and pitot tube which measures the airspeed of an aircraft [142].
3. Influence the accuracy of radar scatterometer measurements on the aircraft [143].
4. Erosion of aircraft surface and the carried missiles in defense aircraft [144, 145].
5. Delay or cancelation of flights.
6. Severe aerodynamic penalties to aircraft (lift decrease and drag increase) [141].
7. Affect aircraft material: cracking of brittle material and deformation of ductile materials.
8. Increase the surface roughness of the wings, which in turn induces a change in the pressure distribution, which adversely affects the control input of the aircraft.
9. Increase the aircraft mass.
10. During takeoff or landing configurations, when leading edge slats are extended, a water film is developed in the gap between the leading edge slat and the main airfoil section which may clog the gap causing premature trailing edge separation and aircraft stall.
11. Additional downward force due to rain on the frontal upper surface of wing and front fuselage generating a moment and nose pitching down.
12. Reduction of flight safety particularly during takeoff, approach, and landing.
13. Influence an aircraft's takeoff and landing performance due to the wet runways.
14. If a runway receives over 3 mm of accumulated water, aircraft will encounter the "hydroplaning" or "aquaplaning" phenomenon [130]. This leads to aircraft sliding and ultimately increasing the required runway length for takeoff or landing.

15. Flame out and decrease of thrust of aircraft engine.
16. Water ingestion into the cockpit/cabin/engine of light nonpressurized aircraft [3].

The first study for rain erosion was in 1941. If a Douglas DC-3 aircraft was flying at an altitude of 5,000 ft and encountered a rain cloud with a water mass concentration of approximately 1,270 mm/h, drag increase caused an 18% reduction in airspeed [158]. Another study for the effect of rains on the roughness of Boeing 747 transport aircraft [159] outlined a 2%–5% increase in drag, a 7%–29% reduction in maximum lift, and reductions in stall angle from 1° to 5° due to rainfall rates from 100 to 1,000 mm/h.

4.6.3.2 Airports

Rainfalls affect ground operations such as baggage handling, catering, and so on. They slow all vehicular movement and foot traffic on the airfield. It also leads to runway flooding. Areas of standing deep water will affect braking action and may result in asymmetric braking and possible sliding off runways.

4.6.4 Experimental Investigation

Experimental studies for the impact of rainfalls on aircraft performance are reviewed in this section. Numerical and experimental studies will be described in detail in Chapter 10.

4.6.4.1 Full-Scale Flight Test

Flight test must be made on an airplane during a severe rainstorm. Aircraft should be equipped with measuring devices to record the environmental parameters to explore the safe flight envelope of the aircraft [146]. However, many difficulties arise in these cases including the hazard to the test pilot, the difficulties for extracting accurate performance measurements and environmental parameters, and the extreme difficulty (if not impossible) to obtain the same characteristics of natural rain in several test.

However, some actual flight tests to study rain erosion at subsonic speeds were carried out by the US Air Force and by Cornell Aeronautical Laboratory [147]. An F106 airplane flew at supersonic and high subsonic speeds and altitude of 40,000 ft through thunderstorms. The aluminum nose cone, the cockpit canopy frame, and the plastic laminate antenna covering showed severe erosion due to water drops and/or tiny ice crystals [148]. Also, the Adverse Weather Branch of the Weather Bureau in their Project Rough Rider performed flight tests for an F-100F aircraft [149]. It flew into thunderstorms from April 19 to May 30, 1967 over an area with a radius of 100 nautical miles in Norman, Oklahoma, to gather information for identification and prediction of damage of aircraft modules. The indicated airspeed was 275 knots (317 mph) at altitudes ranged from 10,000 to 35,000 ft depending on the height of the storm system.

Fifteen coating materials were applied to the leading edges of the wing tips, horizontal stabilizers, vertical stabilizer, and the noseboom, including electroplated nickel, advanced polyurethanes, neoprene, adhesively bonded neoprene, urethane, and nitrile rubber boots.

The pilot's comments were recorded during each traverse to specify when rain, hail, and/or slush, were encountered and the severity of the encounter. Moreover, an ice detector system provided information on ice crystal encounter, and other test gear was utilized to determine the magnitude of the vertical wind gusts.

The following forms of deterioration after 163 penetrations at 300 mph for 440 min in storms with varying degrees of rain and hail intensity were observed: the electroplated nickel coating and the brush applied clear polyurethane, and pigmented polyurethane coatings were in good to excellent condition, and the neoprene coating, the urethane tape, and all three types of performed rubber boots either were completely eroded away or had suffered severe adhesion failures.

Another full-scale flight testing was performed from August 1977 through June 1978 by Boeing commercial airplane and Avco companies [150]. The objective was examining the effect of coatings on rain erosion resistance of Boeing 727-200 airplane. Several coatings including CAAPCO B-274 and Hughson Chemglaze M313 were applied to Boeing 727 leading-edge areas on wing and horizontal tail as shown in Figure 4.9.

The condition varies greatly between airlines. Some airlines report severe erosion problems that, in extreme cases, affect low-speed handling characteristics. Several airlines report mild erosion problems; most report no erosion problem. The tested coatings exhibited the highest rain erosion resistance of all other tested materials [151].

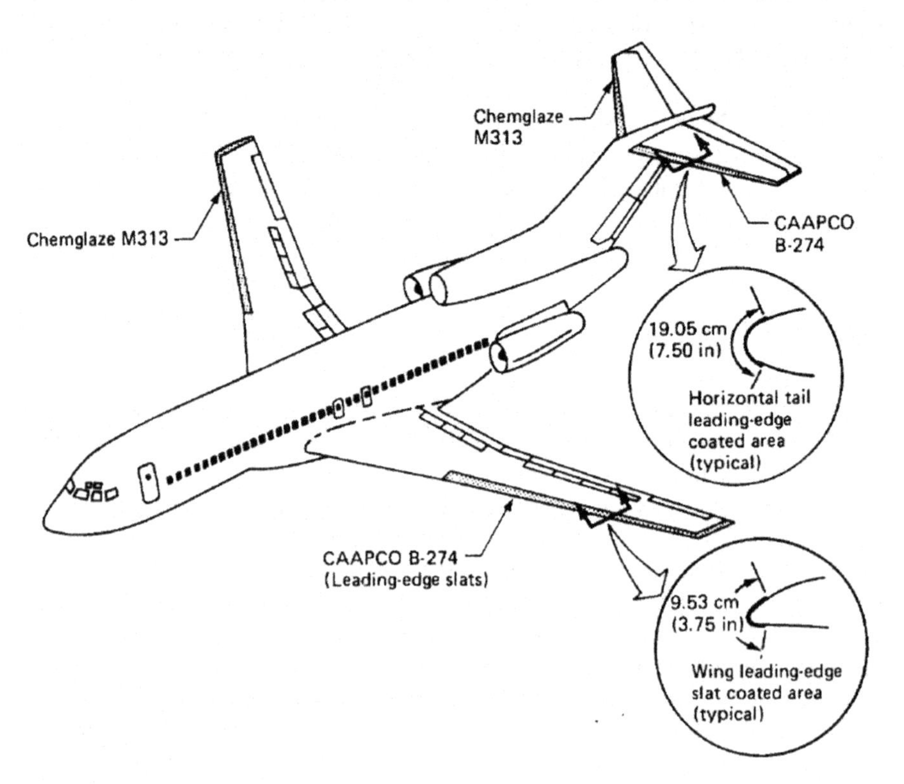

FIGURE 4.9 Coated areas on Boeing 727-200 [150].

Courtesy: NASA.

In general, few flight tests were performed, and unfortunately, most results for flight tests were classified, and a very few is public.

4.6.4.2 Scale Model Test

The performance of a complete scaled model of aircraft or some of its modules like wing and fuselage are tested under simulated scaled rain environment. There are three types of scale model tests: namely, the rotating arm model [152], traditional wind tunnel, and a track in which a model is propelled down a straight track segment.

4.6.4.2.1 Rotating (or Whirling) Arm Experiment

Rotating "whirling" arm (Figure 4.10) facilities have been quite useful for studying single-drop impact dynamics. It consists of a blade or propeller which rotates in a horizontal or vertical plane. Specimens of various airfoil geometries are mounted on the leading edges near the blade tip. The blades with the mounted specimen are rotated through an artificial rainfield usually consisting of uniform water drops, ranging from 0.1 to 1 inch (25 mm) per hour. In general, most of these tests operate at speeds up to Mach number 1.0.

Test studies are digitally recorded and available to the customer for later study. The specimens can be rotated at variable velocities between 100 and 650 mph. Drops have diameters ranging from 1.8 to 2.2 mm and frequency from 6 to 7 drops per second.

The vertical rotating arm apparatus [152] is illustrated in Figure 4.11. A wing section model (say NACA 0024) is placed at the end of a counterbalanced rotating beam. This array is driven by a 15 hp variable speed drive unit. Water drops fall under gravity and impact an airfoil. Experimental tests outlined that the normal or near normal impact of rain drops produced an ejecta cloud directed along the impacting surface [152].

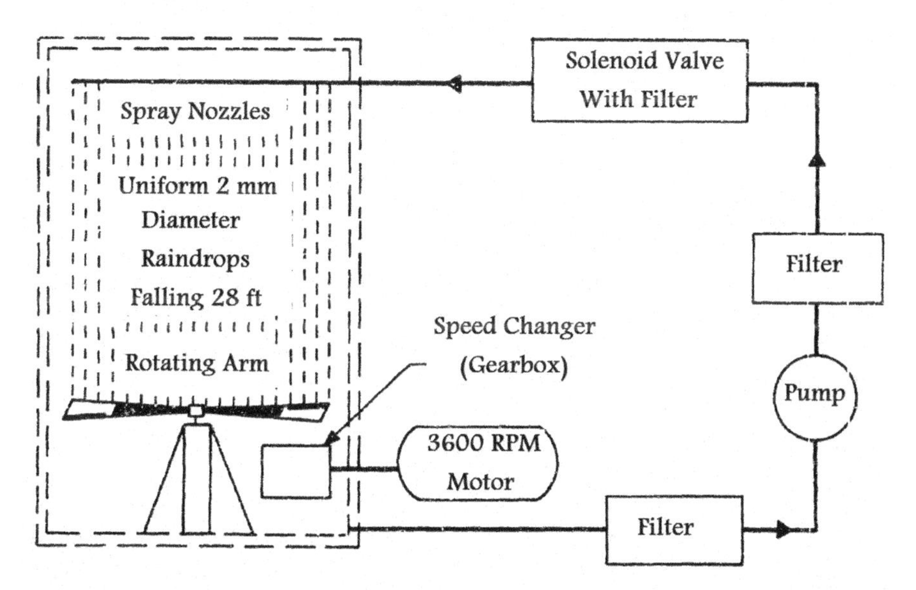

FIGURE 4.10 Horizontal rotating arm testing [152].

Courtesy: NASA.

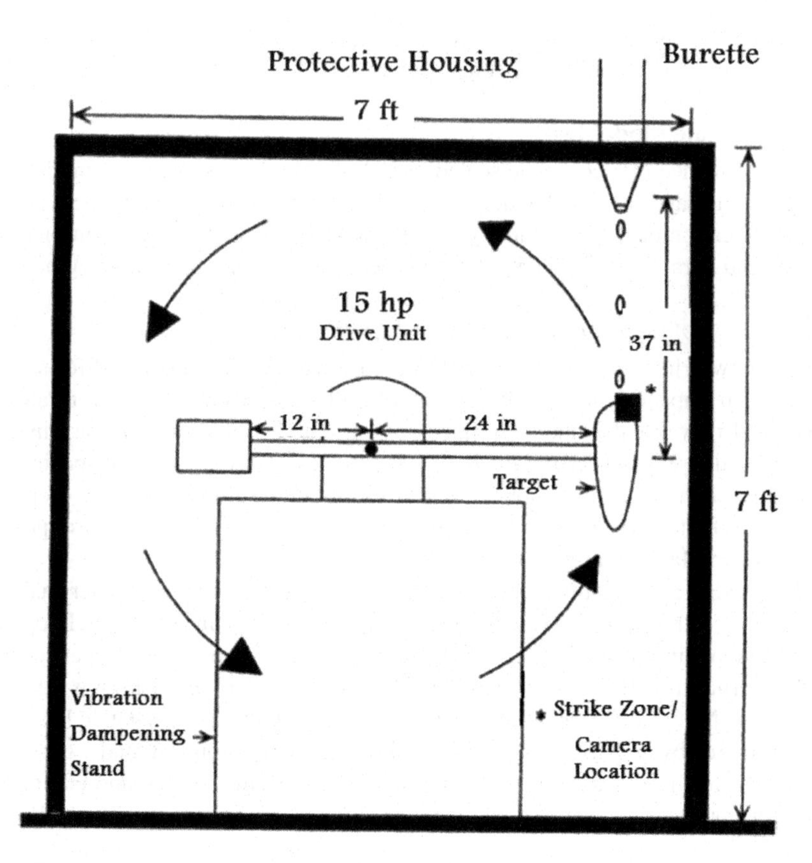

FIGURE 4.11 Vertical rotating arm testing [152].

Courtesy: NASA.

A detailed analysis of the droplets splashback of the ejecta cloud was studied in the INTA (Instituto Nacional De Tecnica Aeroespacial) [153]. Droplets splashback is contained in an annular cone with large cone angle. A splash results from a droplet having a diameter and velocity of (D_i, V_i) respectively, and impacting a rigid surface at angle β. It is assumed that the ejecta droplets are reflected at angle α_r in a cone of angle α_s and the droplets are found in the spray cone annular angle $\Delta\alpha$. The tested materials were Methyl-methacrylate plastic (Plexiglas) and soft aluminum (1100). At an impact velocity of 250 mph, Plexiglas needed 24 h to initiate erosion while soft aluminum (1100) showed erosion in 15 h. The main conclusions were:

- The rebound velocities (V_r) increased several times compared to their incoming velocity ($V_r > V_i$), thus, the rebound particles are ejected at large cone angles and can therefore very effectively interact with the boundary layer.
- The splashed back droplets loose some 30% of their original total mass. The remained mass on the airfoil formed an irregular film on its surface and increased

its surface roughness. This led to degradation of the lift force and increase the drag force.

- The time required to produce a given amount of erosion for most types of materials is inversely proportional to some high power of the impact velocity.

Whirling arm test apparatus are found in the United States [154–157], England, West Germany [158], and Ireland [159].

4.6.4.2.2 Wind Tunnels

The majority of rain studies are obtained in wind tunnel tests [160]. The scaling parameters which control the aerodynamic force generated by an airfoil immersed in a rain environment are the following [161]: Reynolds number based on air, Reynolds number based on water, Weber number, the ratio of mean droplet diameter to model chord, the ratio of the density of air to the density of water, and the AOA.

The tests involve placing a spray distribution system upstream of the aircraft or its modules and directing the spray at the model and measuring the aerodynamic forces (lift, drag, and pitching moment) on it. Figure 4.12 illustrates the 14- by 22-Foot

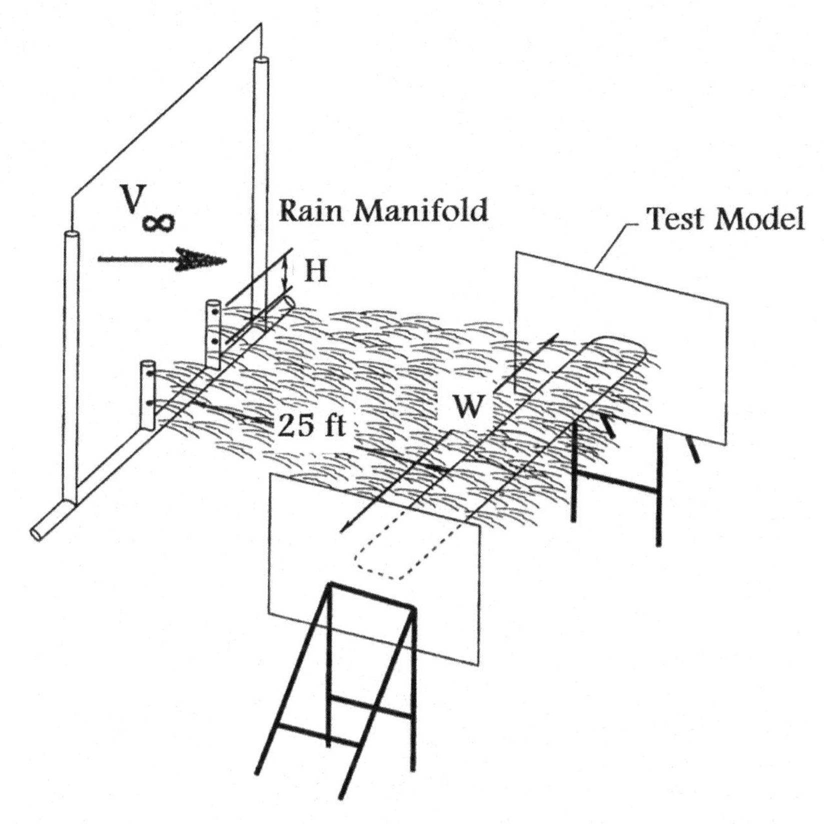

FIGURE 4.12 NASA LaRC wind tunnel facility [162].

Courtesy: NASA.

Subsonic Tunnel in National Aeronautics and Space Administration's NASA Langley Research Center (LaRC) [162].

4.6.4.2.2.1 *NASA Langley Research Tunnel* A NACA 64-210 cambered airfoil simulating a typical commercial transport wing sections was studied to determine the aerodynamic penalty associated with a simulated heavy rain encounter [162, 163]. The model has a chord of 2.5 ft, a span of 8 ft, and was mounted on the tunnel centerline between two large endplates.

4.6.4.2.2.2 *McKinley Climatic Laboratory* The McKinley Climatic Laboratory is an Arnold Engineering Development Complex facility located at Eglin Air Force Base, Florida [164]. It has five testing chambers which include the Main Chamber (MC); the Equipment Test Chamber; the Sun, Wind, Rain and Dust Chamber; the Salt Fog Chamber; and the Altitude Chamber.

The MC is the largest environmental chamber in the world. Its test section is nearly 252 ft wide, 260 ft deep, and 70 ft high. It is used for testing large items and systems for aircraft such as the B-2 Bomber and the C-5 Galaxy. It can simulate all climatic conditions, including heat, snow, rain, wind, and sand. The Sun, Wind, Rain, and Dust Chamber handles wind-blown rain at rates up to 25 inches/h and heavy SDS. It has a cross section 50 × 50 ft and height 30 ft. The temperature ranges between −170 °C and ambient temperature and altitude 80,000 ft.

4.6.4.2.3 Large-Scale Track Experiment

NASA and the FAA (United States) developed a large-scale, outdoors ground-based test capability at the Langley Aircraft Landing Dynamics Facility (ALDF) in 1987 [165] to assess the effect of rain on airfoil performance. A model towing facility in which the model (large-scale wing) is translated down a straight track segment. The ALDF consists of three main components: the test carriage, the rain simulation system, and the data acquisition system.

The ALDF test carriage (which is 70 ft long, 30 ft wide, and 30 ft high) can transport a wing mounted between circular endplates and is propelled along a 3,000 ft track at full-scale aircraft approach speeds (from 100 to 170 knots). The wing had an NACA 64-210 cambered airfoil section to simulate commercial transport wings and equipped with leading-edge and trailing-edge highlift devices deployed to simulate landing conditions. The wing had a rectangular platform of 10 ft chord and 13.1 ft span (geometric aspect ratio of 1.3). A 15-ft diameter, circular endplates were mounted on either side of the wing model.

An overhead rain simulation system produced by a series of nozzles suspended above a 525 ft section of the track. The ALDF rain system simulates rainfall conditions of intensities 2, 10, 30, and 40 in/h, LWC of 2, 9, 26, and 35 $\frac{g}{m^3}$, and rain drop size distribution ranging from 0.5 to 4 mm, which is consistent with realistic rainfall data. Aerodynamic data were measured with and without the rain simulation system. The facility carriage speed ranged from 100 to 170 knots. The wing AOA, which could be changed between test runs, ranged from 7.5° to 19.5° in 2° increments. ALDF data acquisition system consisted of a 28-channel, 12-bit telemetry system with a frame rate of 1,066 frames per second. The data were transmitted from the test carriage to a telemetry receiver that was linked to a personal computer. The transducer outputs necessary for calculating the lift and drag for aircraft model were applied for vertical and longitudinal force, vertical and longitudinal acceleration, roll-, pitch-, and yaw-rate, and free-stream dynamic pressure. This technique is more realistic than wind tunnels.

4.6.5 Helicopter

Helicopter performance is not affected by rainfall as fixed-wing aircraft. Rain does not affect the thrust created by their rotor blades. However, it generally obstructs visibility and makes takeoff and landing more difficult. Rain may also make conditions on the tarmac or landing pad slippery, which can be problematic. However, heavy rain may lead to water accumulation in the engine inlet and cause engine flame out.

4.7 SNOW, ICE, OR SLUSH

4.7.1 Introduction

Icing is defined as any deposit or coating of ice on an aircraft caused by the impingement and freezing of supercooled liquid droplets. Thus, ice collects on the exterior surface [166] of airframe of aircraft (wings, horizontal and vertical stabilizers, fuselage, radomes, etc.) and engines (intakes, fan blades, propellers, helicopter rotor, inlet of APU, exhaust, and drainpipe). Ice accumulation affects both the performance, stability, and weight. It led to severe accidents and numerous incidents. Normally aircraft flies at the troposphere layer ranging from 0 to 11 km. The air temperature decreases within the troposphere. For altitudes ranging from 2.2 to 8.4 km, water droplets are supercooled with temperature below freezing (0 °C to −40 °C) [167]. Supercooled liquid droplets will freeze almost instantly when they touch something solid and cold such as an airplane or its engine(s).

Ice can form even when the outside air temperature is above 0 °C (32 °F). An aircraft equipped with wing fuel tanks may have fuel that is at low temperature, thus it lowers the wing skin temperature to below the freezing point. This phenomenon is known as cold-soaking [168].

As aircraft climbs to near the top of the troposphere, the temperature becomes colder than −40 °C, and all liquid water is frozen into ice crystals so there is no icing hazard at these altitudes. Cumulus clouds with strong vertical motion contain the largest proportion of supercooled droplets. Also, high concentrations of supercooled liquid water may exist at the top of stratocumulus cloud.

Actual weather changes from location to location and from season to season. In summer, flying aircraft may be subjected to icing hazard in the top half of thunderstorms and other deep convective clouds. While in winter, freezing rain can reach the ground, forming ice on parked aircraft. In fall and spring, the supercooled water hazard is often associated with weather fronts such as warm ones [169]. In brief, aircraft flying at altitudes ranging between 2.2 and 8.4 km, supercooled liquid clouds prevail and thus icing of different parts of aircraft occurs. At altitudes less than 2.2 km or greater than 8.4 km, no icing is expected.

TABLE 4.6
Numbers of All Accidents and Their Percentages Caused by Weather from 1997 to 2006 [174]

Year	#all	#weather	#icing	%all	%weather
1997	8,018	332	52	0.65	16
2000	7,998	232	38	0.48	16
2003	8,143	266	32	0.39	12
2006	5,196	188	11	0.21	9

The conditions for icing accretion are [170] supercooled water droplets, non-protected surfaces, and failure of anti-icing system.

For present aircraft, ice can be detected by the following methods: visual, one or more ice detector sensors, an annunciator light that comes on, and all of the above methods. If icing is detected, system automatically turns on the WAI systems.

The annual costs associated with general aviation accidents in the United States range from US$1.64 billion to US$4.64 billion [171]. Weather accounts for roughly 25% of the accidents occurring in general aviation operations [172]. During 32-year period (1982–2013), there were 58,687 general aviation accidents in the United States, of which 11,354 (19.3%) were fatal, producing a total of 20,660 fatalities, with an average of 355 fatal accidents and 645 fatalities per year [173]. Weather was the cause in 25%, or 15,439, general aviation accidents over the period of record. However, the FAA [174] found that weather was associated with 20% of general aviation accidents from 2003 to 2007. There were 3,972 fatal weather-related accidents, or 8,052 fatalities, from 1982 to 2013.

A database for all types of aircraft accidents is recorded by Aviation Safety Reporting System and was analyzed by the NASA Langley Research Center [175]. Table 4.6 shows the number of all accidents from 1997 to 2006 and the percentage of accidents caused by weather and those related to icing [174].

The factors that affect the icing hazard are [176] particle size, liquid water content, environmental temperature, shape of aircraft surfaces, aircraft speed, and aircraft surface temperature (which must be 0 °C or colder). The rate of ice accretion is greatest for an aircraft with thin wings flying at high speed through a cloud with large droplets and high liquid water content. The liquid water content versus the mean effective drop diameter for pressure altitudes ranging from sea level to 22,000 ft is given in [177].

4.7.2 Types of Icing

There are three main types of ice that form when supercooled water freezes: clear, rime, and mixed.

- **Clear ice or Glaze ice**

It is a heavy coating of glassy ice. It forms when aircraft flies in areas/altitudes having a high concentration of large supercooled water droplets, such as cumuliform clouds and freezing rain, and temperature below the freezing temperatures (Figure 4.13a and 4.13b). It spreads unevenly over aircraft surfaces (wing, tail, propeller blades, antennas, etc.). When clear ice forms the temperature of the aircraft, skin rises to 0°C with the heat released during that initial freezing. As more ice accumulates, the ice builds up into a single or double horn shape that projects ahead of the wing, tail surface, antenna, etc., on which it is collecting. When mixed with snow or sleet, clear ice may have a whitish appearance. Clear ice is very hard, heavy, and tenacious. Its removal by de-icing equipment is especially difficult.

- **Mixtures of clear and rime ice**

Mixed ice can form on an aircraft when water drops vary in size or when liquid drops intermingle with snow or ice particles. It is more frequent when the aircraft is flying in clouds where air temperature is –5 to –15 °C. Sometimes it has a mushroom shape on leading edges (Figure 4.13d). Carburetor icing is different as it can occur during warm weather with no visible moisture present.

- **Rime ice**

It forms when small droplets such as those in stratified clouds or light drizzle freeze instantly upon hitting the leading edge of the wing, tail, fuselage, etc., of the aircraft (Figure 4.13c). Droplets in regions of –15 to –20 °C air temperature freeze faster. Ram ice looks white or milky. Rime ice is lighter in weight than clear ice. However, its irregular shape and rough surface decrease the effectiveness and efficiency of airfoils, reducing lift and increasing drag. It is relatively brittle and does not have much strength, so it can break off easier [178].

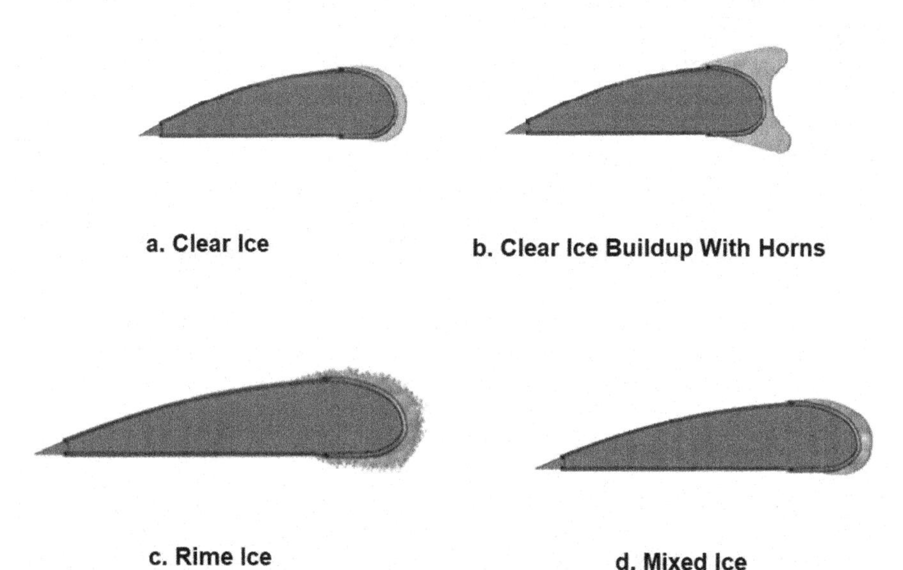

FIGURE 4.13 Type of icing [176].

Courtesy: NASA Glenn Research Center.

TABLE 4.7
Type of Icing in Cumuliform Cloud Based on Temperature [179]

(0°C) – (–10°C)	Clear
(10°C) – (–15°C)	Clear, rime, and mixed
(15°C) – (–20°C) and colder	Rime

- **Frost**
  It forms when water vapor (not liquid water) deposits on the aircraft and forms a "fuzz" of small ice crystals on the aircraft. It has little or no effect on flying. But if deposited on windshield, it will affect pilot visibility. It may also interfere with radio by coating the antenna with ice. Aircraft parked outside on clear cold nights may be coated with frost by the morning which must be removed before takeoff.

Table 4.7 lists the different type of icing in cumuliform cloud based on temperature [179].

4.7.3 Fixed-Wing Aircraft Icing

Ice, snow, and slush have a direct impact on the safety of flight. When a flying aircraft starts to accumulate ice on the aircraft, it has serious negative effects on both the power plant and the aerodynamic performance of aircraft. As little as 0.08 mm of ice on a wing surface can increase drag and reduce airplane lift by 25%.

Two types of icing are critical in the operation of aircraft: induction icing and structural icing [180]. Induction icing is all icing that affects the power plant operation. Its main effect is power loss due to blocking the air inlet of the engine, thereby interfering with the fuel/air mixture. For turbine-based engines, induction icing includes air scoops and lips of intake (or inlet). For reciprocating engines, it includes air scoops, scoops inlet, and carburetor inlet screens. For turboprop engines, induction icing includes the air scoops, propeller hub, and blades. Induction icing for turboshaft engines includes air inlet, air intake screen, and main and tail rotor blades.

Next, structural icing for fixed wing aircraft includes radome, wing, tail surfaces, canopies, windshield, sensors, and antenna. For rotary wing aircraft it includes the fuselage, windshield, main rotor head assembly (swash plates, push–pull rods, bell cranks, hinges, etc.)

4.7.3.1 Hazards During Taxiing from a Contaminated Taxiway

1. Contamination of the underside of the aircraft, especially the landing gear and wing flaps, by spray thrown up from the runway.
2. Snow or slush thrown up onto the landing gear may not prevent retraction but may prevent subsequent extension. It is recommended to recycle the landing gear several times before final retraction to shed as much contamination as possible.

3. Ice may prevent free movement of control surfaces.
4. Wet contamination may freeze after takeoff preventing normal operation of flaps and other control surfaces.
5. Contamination of sensors (e.g., the pitot head or static vents) will cause blockage and erroneous readings of aircraft instruments.

4.7.3.2 Hazard During Takeoff from a Contaminated Runway

In addition to the hazards listed above during taxiing, the following hazards may be encountered:

1. Reduction of aircraft acceleration, delaying the time taken to reach takeoff speed.
2. Difficulty for maintaining directional control using nosewheel steering alone, especially if there is a crosswind.
3. Reduction of the effectiveness of brakes, thus if the takeoff has to be abandoned, then the aircraft may not stop within the runway.

4.7.3.3 Icing Deficiencies

- Reduction of lift (coefficient and force) of wing, horizontal, and vertical stabilizers (Figure 4.14). Thus leading to stall of wing, elevator, and rudder.

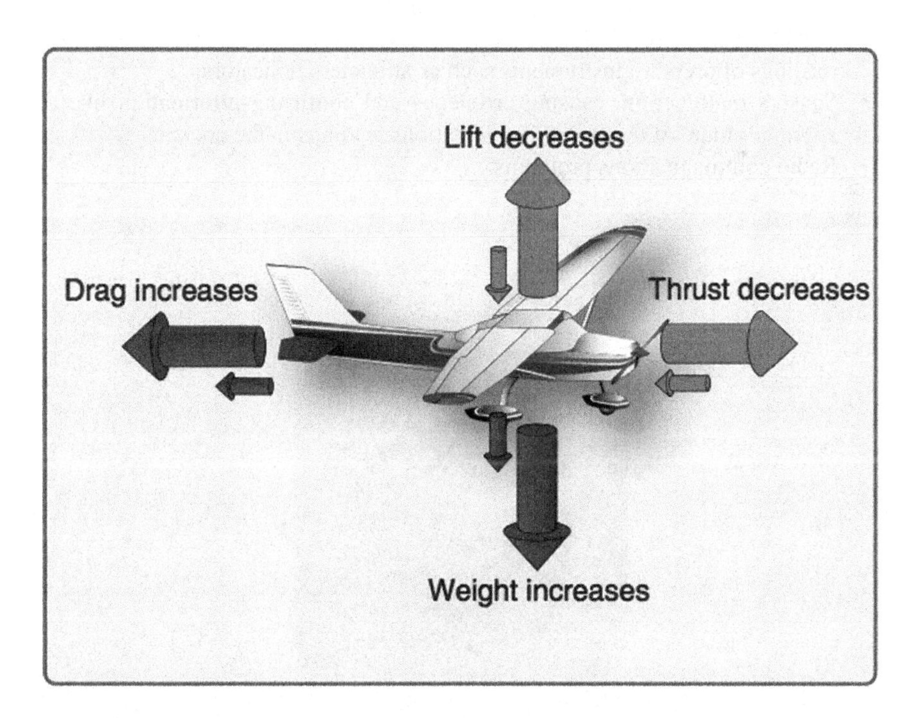

FIGURE 4.14 Effects of structural icing [178].

Courtesy: US Department of Transportation.

- Increase of drag due to ice accumulation on many parts of the aircraft and thus slow down the aircraft, increases fuel consumption, reduces the airplane's range and making it more difficult to maintain speed.
- Shift of the airfoil center of pressure, which may affect the longitudinal stability and pitch trim requirements.
- Increase of weight due to the accumulation of large mass of ice in a short time, which decreases its rate of climb.
- Decrease of the efficiency and thrust of propellers for shaft-based engines.
- Decrease of the power of carbureted engines, due to icing in the carburetor, partially or totally block flow of the fuel–air mixture into the engine.
- Decrease of thrust force of jet engines.
- Reducing the inlet air flow for jet engine when ice forms on air intake, and the stall speed becomes higher than normal (Figure 4.15).
- Causes airflow disturbances that excite harmonic compressor or fan blade frequencies, which become large enough to cause serious engine damage.
- Causes damage due to reduced clearance between rotating and stationary components of turbomachinery of engines (fan, compressor, turbine).
- Reducing the inlet air flow and possibly choking the piston engines due to ice form on air intake filters and manifolds.
- Reduction of visibility for pilots due to icing on the windshield
- Prevent the movement of some control surfaces (ailerons, elevator, rudder, flaps), thus pilot cannot control the aircraft.
- Partial or complete blockage of the pitot static system can produce errors in the readings of pressure instruments such as altimeters indicators.
- Sensors malfunction, causing erroneous and confusing information like airspeed, altitude, AOA, or engine indications readings in the cockpit.
- Radio communications problems.

FIGURE 4.15 Engine with frozen contamination coating the exterior and inlet [181].

Courtesy: NASA.

- Destructive vibration due to the unequal loading on the wings and on the blades of the propeller, which hampers true instrument readings.
- External coverings of ice in locking mechanisms, hinges and seals which may result in inability to open doors after landing which is extremely hazardous in an emergency situation.
- Contamination of retractable landing gear, doors, bays and micro-switches by snow, or wet slush.
- If ice on the wing breaks off during takeoff (due to its flexibility) and goes directly into the engine (particularly fuselage aft-mounted engines), it will lead to engine surge, vibration, and even complete thrust loss.
- When chunks break off and shed, they cause fuselage structural damage.

Even half an inch of ice accumulation may lead to 50% reduction of lift and drag increase by the same amount [118]. On the space shuttle, ice may break off during launch and hit and damage the protective tiles on the wings, which causes the shuttle to burn and explode during reentry.

4.7.4 Icing Intensity

The severity of icing is determined by its operational effect on the aircraft. Based on the flight information handbook, the amount of icing has four categories: namely, trace, light, moderate, and severe. Icing intensity is related to the rate of accumulation of ice on the aircraft, the effectiveness of de-icing/anti-icing equipment, and the amount of ice that would actually be measured (or collected) on small probes, and the actions needed to be taken by pilots.

A summary for the physical properties if different icing category and the immediate reaction of pilots are summarized in Table 4.8.

4.7.5 Representative Incidents to Icing

Many icing mishaps have been encountered when aircraft was not de-iced before takeoff. Since the late 1950s, 16 accidents and 139 fatalities have been attributed to ice-contaminated tailplane stall. Table 4.9 lists several accidents due to icing [182, 187].

4.7.6 Ice Protection System

Aircraft should be ice-free during all flight phases and very critical during takeoff. De-icing and anti-icing systems are designed to keep atmospheric ice from accumulating on aircraft surfaces. In brief, they are defined as de-icing and anti-icing:

- De-icing is a procedure that removes snow, frost, and slush from the aircraft, engine intakes, and fan blades on the ground prior to takeoff to provide clean surfaces.
- Anti-icing provides some protection of those surfaces from contamination by frost/ice/snow for a limited period of time (known as holdover time).

Once an aircraft has been de-iced, delay before takeoff must be kept to a minimum to ensure the contaminant does not refreeze before takeoff. On the takeoff run, the fluid

TABLE 4.8
Physical Properties of Different Categories of Icing [180, 179]

Icing Intensity	Description	LWC g/m³	Ice Collection Rates on Small Probes (inches per 10 miles)	Pilot Action
Trace	Ice becomes perceptible Rate of ice accretion is slightly greater than sublimation	0.0–0.1	0–0.09	- No action is required, if icing occurs for less than 1 h - If encountered for 1 h or more, pilot chooses any of the following: o Apply de-icing/anti-icing equipment o Diversion o Altitude change
Light	The rate of ice accumulation is 6–25 mm/h, which may cause a problem if it is prolonged for more than 1 h	0.1–0.6	0.09–0.18	Choice any of the following: • Use of de-icing/anti-icing is applied occasionally • Diversion • Altitude change
Moderate	The rate of ice accumulation is 25–75 mm/h, thus may cause hazardous problem even if encountered for a short time.	0.6–1.2	0.18–0.36	Choice any of the following • De-icing/anti-icing • Diversion • Altitude change
Severe	The rate of ice accumulation is so dangerous and de-icing/anti-icing is not successful	>1.2	>0.36	Choice of: • Diversion • Change altitude

is shed from the wings and other surfaces so that it does not affect the aerodynamic performance in flight.

4.7.6.1 De-icing

4.7.6.1.1 Large Transport Aircraft

De-icing can be accomplished by several methods [183] as described below:

- Heating: heated or infrared (IR hangar) or portable heater
- Mechanical: scrapers, brooms, ropes, or other tools

TABLE 4.9
Representative Accidents Due to Icing

Date	Location, Airline, Aircraft Type	Observation
February 6, 1958	Munich, Chartered aircraft from British European Airways (BEA), Airspeed Ambassador aircraft	Runway was covered with slush. Pilot aborted takeoff twice and crashed in the third trial into the fence surrounding the airport and next into a house. 23 of the 44 passengers and crew on board died
January 13, 1982	Washington, Air Florida, Boeing 737	Crashed as the flight crew failed to use engine anti-ice before takeoff and to properly de-ice airfoil surfaces
February 6, 1988	California, N2832J Cessena	Inadequate frost removal
December 23, 1988	Montana, N5570H Pipper	Wing Ice
March 10, 1989	Dryden, Air Ontario F28	Heavy snow
February 16, 1991	Cleveland, Ryan DC-9-10	Light snow
March 22, 1992	LaGuardia, NY, USAIR F28	Heavy snow
October 31, 1994	Roselawn, IN, American Eagle ATR72 turboprop engine	De-icing boots did not protect its wings from icing due to freezing rain (68 fatalities)
November 30, 2001	Skien Airport, Geiteryggen, Norway, European Executive Express, BAe Jetsream 31	Airframe ice accretion, severe aircraft damage, and loss of control
January 7, 2004	Pellee Island, Ontario, Canada, Cessna 208 Caravan, Georgian Express	Ice on the airframe. After takeoff, the pilot lost control of the aircraft and it crashed into a frozen lake
October 3, 2014	Flight from Aberdeen to Sumburgh, UK, Saab 340	Encountered an unexpected severe turbulence and icing after climbing and a temporary loss of pitch control and stick shaker activations. Climb was resumed when the severity of the turbulence reduced.
December 21, 2015	North of Darwin NT Australia, Boeing 787-8	Entered an area of ice crystal. Icing penetrated an area which included maximum intensity weather radar returns
February 23, 2018	Island Express Air, Beechcraft King Air B100	As the aircraft climbed out it experienced an aerodynamic stall as a result of wing contamination, five passengers and pilot were seriously injured and the other four received minor injuries

FIGURE 4.16 Infrared hangar at Newark, NJ, United States [184].

Courtesy: NASA.

- Thermal: hot air
- De-icing fluids
- Electro-impulse de-icing (EIDI)
- A combination of these different techniques.

4.7.6.1.1.1 *Heat* Aircraft de-icing can be achieved using heated or IR hangar or portable heater. A *heated* hangar can be used to melt off frozen contaminants. Warm aircraft should not pull into snow or other precipitation at below freezing temperatures. In these cases, the melted precipitation may refreeze before takeoff, especially on a fully fueled, warm wing. IR hangars (Figure 4.16), exist at a few large airports, where radiant heat in the IR wavelength is used to warm aircraft surfaces and melt frozen contamination [178]. It is used for de-icing only. If there is freezing precipitation, an anti-icing fluid application is still required.

4.7.6.1.1.2 *Manual* Mechanical de-icing includes scrapers, brooms, brushes, ropes, squeegees, fire hoses, or other tools. Scrapers can be used to remove ice, but carefulness is required to avoid damage sensors or plexiglas. Technician should make sure that the ice has been removed, not just smoothed [185].

4.7.6.1.1.3 *Thermal (Hot Air)* Hot air – say from an engine preheater – can be used to de-ice spinners, propellers, engine intake areas, turbine engine fan blades, and landing gear. It can also be used to remove ice from the aircraft skin, but take care not to overheat one spot.

4.7.6.1.1.4 *Fluids (De-/Anti-icing)* De-/anti-icing fluids are used to remove ice, snow, and frost from the exterior surface of aircraft. Figure 4.17 illustrates that a US Army C-37B aircraft gets de-iced before departure [186]. De-icing fluids typically consist of glycol-water solution. Glycol is typically a mixture of propylene, ethylene, or dieethylene. De-icing fluids contain few percentages of corrosion inhibitors,

FIGURE 4.17 De-icing of a US Gulfstream G550 before departing Alaska in January 2012 [186].

wetting agent, and dye [187]. Glycol lowers the freezing point and prevents formation of ice contamination at temperatures below freezing. *Corrosion inhibitors* will protect the skin of aircraft, while w*etting agents* allow the fluid to conform to the aircraft surfaces, and *Dye* visually aids type identification.

There are four standard de-icing and anti-icing fluid types used in aviation: namely, SAE Type I, II, III, and IV. They satisfy material compatibility and aerodynamic acceptance for aircraft. They are applied primarily with a forceful, narrow, spray to maximize the hydraulic force to remove the contamination. It should be sufficiently heated, which is indicated by rising steam.

The SAE Type I fluids are the thinnest fluids and always applied heated and diluted [187]. They can be used on any aircraft, as they shear/blow off even at low speeds. They also have the shortest hold over time (HOTs).

Type II and IV fluids add thickening agents that allow fluid to remain longer on the aircraft to melt the frost or freezing precipitation. They may be applied heated or cold, and diluted or full strength. They provide longer HOT, but need a higher speed to shear off the fluid. In North America, Type IV fluids are applied cold and only for anti-icing. In the UK, Type II or IV fluids are applied heated to accomplish de-icing as well as anti-icing.

Type III fluids are relatively new and have properties in between Type I and Type II/IV fluids.

They contain thickening agents and offer longer HOTs than Type I, but are formulated to shear off at lower speeds. Though they are designed for small commuter-type aircraft, but work also for larger aircraft.

TABLE 4.10
Properties of De-/Anti-icing Fluids [187]

SAE	Color	Minimum Rotation Speed (knots)	Sample HOT for Moderate Snow (minutes)
Type I	Red-orange	No minimum	6
Type II	Clear or straw	100	20
Type III	Yellow-green	60	10
Type IV	Emerald-green	100	30

Ground de-icing does not yield permanent results. There are limitations on HOTs prior to takeoff after de-icing is completed. This time depends on the particular de-icing fluid, air temperature, aircraft temperature, and further icing threats due to frost, freezing fog, snow, freezing drizzle, or freezing rain. Table 4.10 lists the properties of de-/anti-icing fluids Types I, II, III, and IV [187].

ICAO lists the following procedure for fluid spraying for aircraft different modules [188]:

1. Fuselage

Aircraft spraying starts by fuselage. Spray along the top center line and then outboard. Avoid spraying directly on flight deck or cabin windows.

2. Wings and horizontal stabilizers

Spray from their leading edge toward the trailing edge.

3. Vertical tail surfaces

Start at the top and work downward, spraying from the leading edge toward the trailing edge.

4. Landing gear, brakes, and wheel bays

Apply minimum de-icing/anti-icing fluid in this area. Do not use high-pressure spraying. Do not spray directly onto brakes and wheels.

5. Engines/auxiliary power units (APUs)

Avoid spraying fluids into the inlets of engines or APUs. Be sure that engines are free to rotate before start up and that the front and back of the fan blades are free of ice. Air-conditioning bleed systems must be switched off during de-icing/anti-icing operations when engines or APUs are running. Do not spray directly onto exhausts or thrust reversers.

6. Instrument sensors

Avoid spraying directly onto pitot heads, static ports or air stream direction detector probes, and AOA sensors.

7. Vents and outlet valves

Avoid spraying directly onto electronic bay vents, fuel tank vents, air outlet valves, or any other similar type of opening.

8. Wiring harnesses and electrical components

Do not spray directly into receptacles, junction boxes, and so on.

Each aircraft has "no spray" zones, which are identified in its airplane flight manual (AFM), aircraft maintenance manual, or the operator's manual. ANNEX A of reference [185] lists the no spray areas for different aircraft types. Figure 4.18 shows critical areas on a typical transport aircraft that should not be sprayed directly during de-icing or anti-icing. It includes wiring harnesses and electrical components, brakes, wheels, exhausts, or thrust reversers, orifices of pitot heads, static ports, airstream direction detectors probes/AOA airflow sensors, engines, APU, other intakes/outlets, control surface cavities, and flight windshields.

Additional precautions:

- When removing ice, snow, slush, or frost from aircraft surfaces (manually or other methods), care shall be taken to prevent it entering and accumulating in auxiliary intakes or control surface hinge areas.

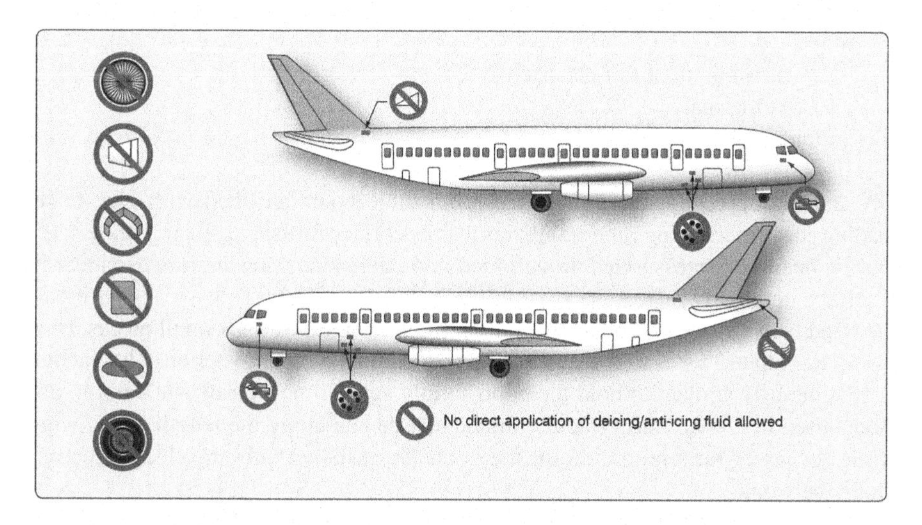

FIGURE 4.18 No direct application of de-icing/anti-icing fluid allowed [177].

Courtesy: US Department of Transportation.

- If Type II, III, or IV fluids are used, all fluid traces on flight deck windshields should be removed prior to departure, particularly if fitted with wipers.

The US Department of Defense has issued the following military specifications concerning both anti-icing and de-icing-defrosting fluids [206]: MIL-A-4823C Type I – standard, MIL-A-4823C Type II – standard with inhibitor, MIL-A-4823D Type I (propylene glycol base), and MIL-A-4823D Type II (ethylene and propylene glycol mix).

4.7.6.1.1.5 *Electromagnetic De-icing (EIDI)* Electromagnetic impulse de-icing (also called Electro-impulse de-icing) is an electric current pulse generated from a capacitor and transmitted through spirally wound, flattened coils made from ribbon wire [189]. The coils are rigidly supported inside the aircraft skin but are separated from it by a gap of approximately 2.5 mm. The EIDI coil current produces a magnetic field, which induces two eddy currents in the thin metal skin, where their fields repel each other. The resultant force causes a movement of the skin adjacent to the coil that shatters the ice and breaks the bond between ice and skin, thus the ice is swept away from the surface by aerodynamic forces. This de-icing method was applied to the bare turbofan inlet of the business jet (Falcoon-20), the inlet of an operating turboprop engine (PW124). Numerous tests were carried out in the NASA-Lewis icing wind tunnel [190] and a one-quarter span of an eight-foot diameter inlet of a high bypass ratio turbofan was tested in France's Centre D'Essais Des Propulseurs R-2 icing wind tunnel. Testing conditions reflected the FAR Part 25 Appendix C icing envelope.

4.7.6.1.2 General Aviation (GA) Aircraft and Turboprop Commuter-Type Aircraft

GA and aircraft powered by turboprop engines use the following methods for de-icing:

- Boots (using hot air or electricity)
- Electrothermal
- Fluids

GA and turboprop powered aircraft use inflatable boots attached with glue to the leading edges of the wings and stabilizers to break off ice formed on them (Figure 4.19). During operation, pressurized air is forced into the bladder causing it to expand with a surface deflection rate of approximately 0.35 in/sec [189]. This expansion breaks the bond between ice and bladders and fractures accreted ice into small pieces. Most boots are inflated for 6 to 8 sec. They are deflated by vacuum suction. The vacuum is continuously applied to hold the boots tightly against the aircraft while not in use. Boots used in GA aircraft typically inflate and deflate along the length of the wing, while for larger turboprop aircraft, the boots are installed only in selected sections along the wing.

Due to the deformation imparted to the wing airfoil during the activated phase of a de-icing cycle, pneumatic boots are limited to low- and mid-velocity applications (up to Mach 0.5).

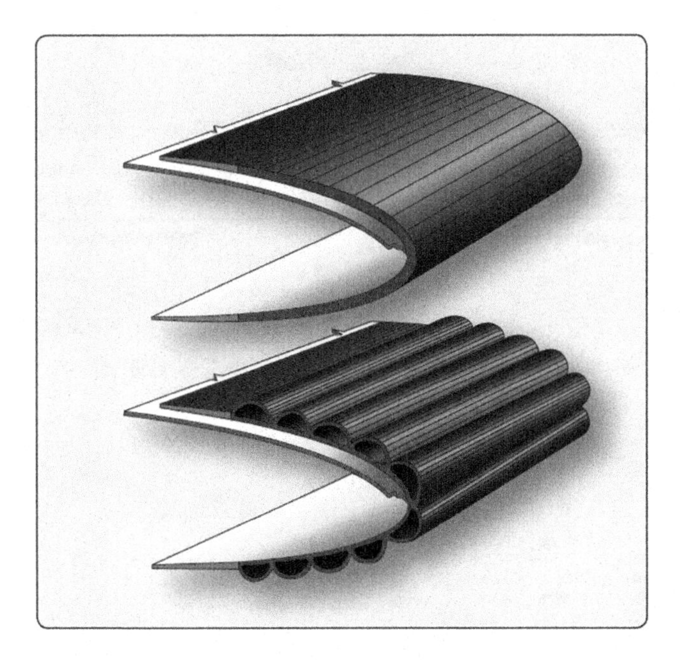

FIGURE 4.19 Cross-section of a pneumatic de-icing boot uninflated (top) and inflated (bottom) [177].

Courtesy: US Department of Transportation.

For reciprocating engine aircraft, a dedicated engine-driven air pump mounted on the accessory drive gear box of the engine is used for inflating the boots to break up the ice formed on the wing and stabilizer leading edges. Also, it is used to hold the de-ice boots tight to the aircraft when they are not inflated. For turboprop engines, bleed air from the engine compressor(s) is used to inflate the boots through switches in the cockpit (Figure 4.20). A relatively low volume of air on an intermittent basis is required to operate the boots. This bled air has little effect on engine power.

Advantages and Disadvantages of Boots

Because de-ice boots use compressor bleed air, aircraft will never run out of de-icing protection. Most aircraft equipped with de-ice boots have manual or automatic modes, which will cycle different sections of the boots for ice removal [191]. However, there are three disadvantages of boots:

- Changing the aerodynamic characteristic of the airfoil when inflated which in turn will increase stalling speed.
- There is a risk of existence of holes in the boots, thus they will not inflate properly, and its ability to remove ice will be decreased.
- The risk of ice forming behind the boot, which cannot be removed by the system.

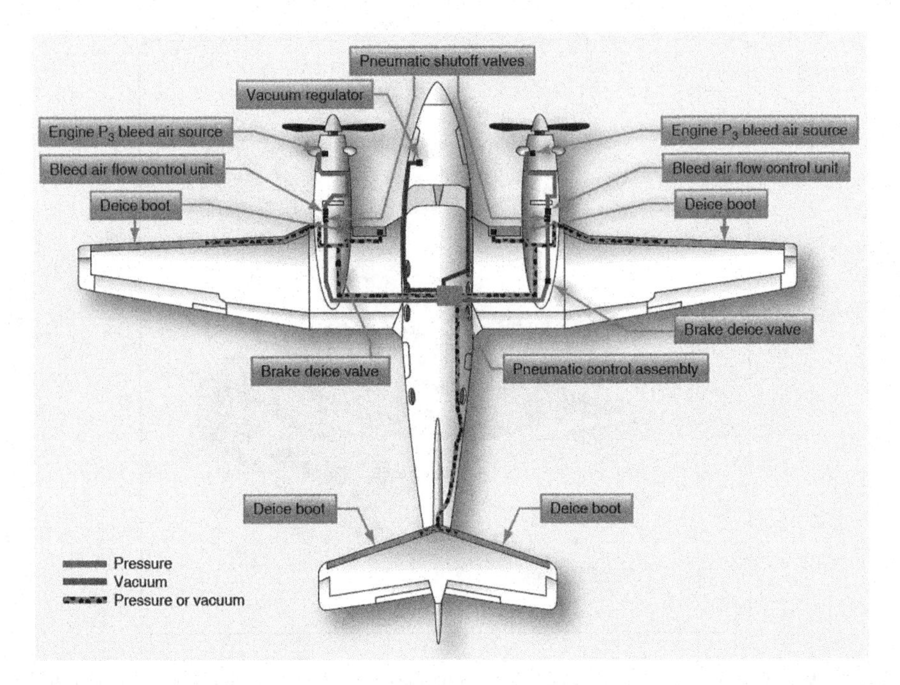

FIGURE 4.20 Wing de-ice system for turboprop aircraft [177].

Courtesy: US Department of Transportation.

A few modern GA aircraft are equipped with:

- Electric de-ice boots on wing and the horizontal stabilizer. These boots contain electric heating elements bonded to the leading edges. When activated, the boots heat up and melt the ice off of leading edge surfaces. The elements are controlled by a timer in a de-ice controller. This system is inoperative when the aircraft is on the ground.
- Electromagnetic expulsion de-icing system consists of an electromagnetic expulsive de-icing boot bonded to the inlet surface. The boot consists of an elastomer polyurethane material (thickness 0.02 inch) having U-shaped ribbon conductors embedded parallel to the boot's surface [189]. A large current from a power unit released through the conductors, which creates opposing magnetic fields in the arms of the U-shaped conductors. The boot deflects 0.09 inch in 200 microseconds, creating a large G-force which is effective in removing ice accretion thicknesses of 0.020 to 1.0 inch. This de-icing system is applied externally and therefore easily retrofitted on existing aircraft.

4.7.6.1.4 Propeller De-ice System

De-icing of the propeller and its components (leading edges, cuffs, and spinner) is performed using electrothermal and chemical methods.

Electrically heated boot firmly cemented on each blade is de-iced by an electric current from a slip ring and brush assembly on the spinner bulkhead. The slip ring transmits current to the de-ice boot. The centrifugal force of the spinning propeller and air flow break the iced pallets from the heated blades.

Some single-engine GA aircraft use a chemical de-icing system for the propellers. The glycol-based fluid may be used for de-icing the propeller system.

4.7.6.2 Checks After De-ice

After de-icing, check that: all ice has been removed, and if there is any ice remaining, de-icing should be continued until all of the aircraft is ice-free, no de-icing fluid has been collected in the control surface gaps, small residual patches of ice may be sufficient to cause roll or pitch upsets, and protective covers have been property removed and stowed [192].

4.7.6.3 Anti-icing

Anti-icing is used to prevent or control ice formation in aircraft. It is turned on before the flight enters icing conditions. It uses the following methods [178]:

1. Heating surfaces with hot air.
2. Heating by electrical elements.
3. Breaking up ice formations, usually by inflatable boots.
4. Chemical application.

Anti-icing of aircraft is accomplished by applying the previously listed SAE Type I, II, III, and IV fluids over the surface. All anti-ice fluids offer only limited protection, dependent upon frozen contaminant type and prevailing weather conditions.

A summary for these methods is listed in Table 4.11.

4.7.6.4 Anti-icing of Aircraft Modules

4.7.6.4.1 Aircraft Powered by Shaft-Based Engines (Piston or Turboprop Engines)

4.7.6.4.1.1 *Wing and Tail Surfaces*

- Weeping wing/tail systems

Pilot activates a switch in the cockpit, then a de-icing fluid is pumped from a reservoir through a mesh screen in the leading edges of the wing and tail surfaces. It can also be applied to the prop and windshield. Weeping wings have the ability to protect the entire airfoil surface since when the fluid is pumped out from the leading edges, it will run back across the top and bottom of the surface, forming a layer of protection against ice. However, since a finite amount of the fluid is carried out, it may run out of it, and the fluid can protect aircraft for only 1.5–2.5 h in normal conditions.

- Heating wings/tails

Heating is used in general aviation today.

TABLE 4.11
Anti-icing Methods

Method	Application	Principle
Thermal		
Hot air bleed	Wings, tail units, engine intakes, windscreens, slats	Hot air from the compressor passes along inside of leading edge structure or blown on outside of windscreen
Combustion heating	As above	Hot air from a separate combustion heater or from a heat exchanger associated with a turboprop engine exhaust gas system
Electrical heating	Wings, tail units, intakes, propellers, rotor blades, wind screens, drain masts, pitot heads, probes, drain masts, control surfaces, carburetors	Electric current (DC or AC) passing through wire, flat strip, or film type elements attached to the outside of the component, or heater elements embedded within the structure (composite material)
Fuel		
De-/anti-icing	Wings, tail units, propellers, windscreens	A chemical which breaks down the bond between ice and water and can be either sprayed over the surface (windshield), or pumped through porous panels along the leading edge of a wing, or allowed to flow by centrifugal force along the leading edge of the propeller

4.7.6.4.1.2 *Propeller* Ice often forms on the propeller before it is visible on the wing. Propeller hubs are treated with chemicals from slinger rings, while leading edges are treated with electrically heated elements.

4.7.6.4.1.3 Windshield

Two systems are used in light aircraft, namely:

- An electrically heated windshield, or a liquid spray bar which includes ethylene glycol, Grade B Isopropyl alcohol, ureas, sodium acetate, potassium acetate, sodium formate, and chloride salts.
- Windshield defroster.

4.7.6.4.1.4 Engines

Most carbureted piston engines heat the carburetor when throttling back from cruise power and may be used during snow or rain and in clouds with near-freezing temperatures. Fuel-injected piston engines use an alternate air door which opens automatically or activated by the pilot if the primary air intake ices.

Some turboprop inlet cowls are also heated with electricity. Table 4.12 presents several aircraft powered by turboprop engines and their different methods for ice protection.

TABLE 4.12
Typical Icing Protection Methods for Some Turboprop Powered Aircraft

		Ice Protection Systems				
Aircraft	**Engine**	**Wing**	**Empennage**	**Windshield**	**Engine Inlet**	**Propeller and Spinner**
C-130	T-56A7	Hot air	Hot air	Electro-thermal	Hot air	Electro-thermal
XC-142	GE T-64-1	Pneumatic boots	Pneumatic boots	Electro-thermal	Hot air	Electro-thermal
Electra	Allison-501-D13	Hot air	Electro-thermal or hot air	Electro-thermal	Hot air	Electro-thermal

4.7.6.4.2 Aircraft Powered by Turbine-Based Engines

Both business jet and large transport aircraft are powered by turbine-based engines and in particular turbofan engines. Thermal pneumatic, thermal electrical, and chemical fluid methods are used for its anti-icing.

4.7.6.4.3 Thermal Pneumatic System

Heated air is ducted spanwise along the inside of the leading edge of wings, leading edge slats, horizontal and vertical stabilizers, engine inlets, and others. The sources of heated air are:

- Hot air bled from the compressors.
- Engine exhaust heat exchangers.
- Ram air heated by a combustion heater.

The hot air bled from the engine compressor is routed through ducting pipe (identified as piccolo), manifolds, and valves for anti-icing of wings (WAI) and horizontal stabilizers (Figure 4.21). Details of numerical analysis of anti-icing systems using piccolo tube will be described in detail in Chapter 10. The piccolo tubes are made of either aluminum alloy, titanium, stainless steel, or molded fiberglass. When the wing leading edge temperature reaches approximately +140 °F, WAI switch is turned on. A pressure regulator is energized, and a shutoff valve opens, and hot air flows into a piping system leading to the piccolo tubes in the wing. If the temperature in the wing leading edge exceeds approximately +212 °F (outboard) or +350 °F (inboard), the red WING OV HT warning light on the annunciator panel illuminates.

For jet engines, one of the anti-icing systems employs a piccolo tube system, where hot air circulates through a tube that extends around the circumference of the inlet. Air exits from holes in the tube to heat the inlet leading edge. In general, hot air bled from the high pressure compressor in a turbofan engine is very efficient in

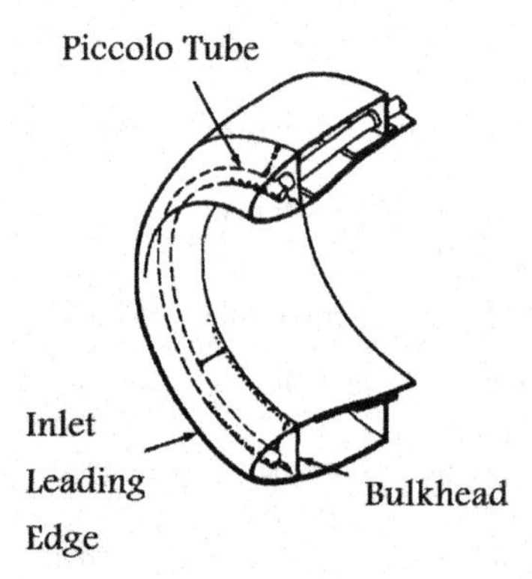

FIGURE 4.21 Anti-icing system for intake of turbine-based engine [177].

Courtesy: US Department of Transportation.

heating the leading edge surfaces and prevent ice formation as long as aircraft engine is running. However, it has the following disadvantages:

- If the heating system for engine cowl is not turned on in time, chunks of ice may break off of the engine cowl and get ingested into the engine.
- Performance penalty due to the bled air mass flow rate.

4.7.6.4.4 Thermal Electric Anti-icing System

This type of anti-ice is limited to small components due to high amperage draw. Thermal electric anti-ice is used on:

- Most air data probes, such as pitot tubes, static air ports, TAT and AOA probes, ice detectors, and engine P2/T2 sensors.
- Water lines, waste water drains.
- Front windshields and side windows.

A current flows through an integral conductive element that produces heat. The temperature of the component is elevated above the freezing point of water, so ice cannot form. Various schemes are used, such as an internal coil wire, externally wrapped blankets or tapes, as well as conductive films and heated gaskets. Electrically heating has the following disadvantages:

- If the heating device is left on during ground operations surfaces, the heated surface can be damaged.
- Inability to heat large areas, such as wings and tail surfaces.

4.7.6.4.4 Chemical Methods

Chemical anti-ice systems prevent and delay the formation of ice on surfaces and windshields of aircraft. The liquid chemical is sprayed through a nozzle onto the outside of the aircraft surfaces. Such chemical systems have a fluid reservoir, pump, control valve, filter, and relief valve.

One-step or two-step procedures are used for aircraft de/anti-icing, depending on:

- The meteorological conditions
- Airport and state practices
- Available equipment
- Available fluids, and
- Hold over time (HOT).

For one-step procedure, Type I only or thickened Type II/III/IV fluid is used for snow/ice removal and protection of the aircraft's surfaces. In some Central Europe airports, which have humid oceanic or humid continental climate, a one-step procedure with Type I fluid is employed for de-icings. This due to the HOT is typically 45 min for clear skies and no precipitation.

For South Europe, a one-step procedures use warm Type II fluids having higher freezing point used for anti-icing and not de-icing. Heated Type II fluid temperature should be significantly lower compared to a Type I application. Operators from the United Kingdom and the Scandinavia and Baltic countries use a one-step method employing Type I fluid. However, British Airways at Heathrow airport is switching toward Type I mixture for the de-icing and Type II or IV for next anti-icing treatment.

In a two-step procedure, the aircraft is cleaned by the Type I fluid first and then Type II or IV fluid is applied within 3 min. To make the cleaning of the aircraft easier, either de-icing fluid is applied hot and under high pressure or hot pressurized water at a temperature of up to −7 °C is applied for cleaning of the aircraft. Protective fluids are less stable and must not therefore be exposed to higher temperatures. They are applied cold.

In the United States and Canada, the two-step procedure is usually used for de-icing on special pads located close to the runway holding points.

4.7.6.5 Hold Over Time (HOT)

It is the time that anti-icing fluid will prevent the formation of ice and frost and the accumulation of snow on the protected (treated) surfaces of an aeroplane. HOT depends on the following factors: de-icing/anti-icing fluid (type, fluid/water ratio, temperature), type and rate of precipitation, ambient temperature, relative humidity, wind direction and velocity, including jet blast, aeroplane surface (skin) temperature.

Tables 4.13 and 4.14 list HOT for SAE Types I and IV for different outside temperature and various weather conditions [193].

With a one-step de-icing/anti-icing procedure, the HOT begins at the commencement of de-icing/anti-icing. With a two-step procedure, the HOT begins at the commencement of the second (anti-icing) step.

4.7.6.6 Checks After Anti-icing

After anti-icing, be sure that there is a uniform layer of anti-icing fluid over the entire aircraft, no fluid has been collected in the control surface gaps, and notify the flight crew.

TABLE 4.13
FAA Holdover Time Guidelines for SAE Type I Fluid on Critical Aluminum Aircraft Surfaces

		Approximate Holdover Times Under Various Weather Conditions (hours: minutes)						
			Snow, Snow Grains, or Snow Pellets					
Outside Air Temperature (C)	**Wing Surface**	**Freezing Fog or Ice Crystal**	**Very light**	**Light**	**Moderate**	**Freezing Drizzle**	**Light Freezing Rain**	**Rain on Cold Soaked Wing**
−3 and above	Aluminum	0:11– 0:17	0:18–0:22	0:11–0:18	0:06–0:11	0:09–0:13	0:02-0:05	0:02–0:05
Below −3 to −6	Aluminum	0:08–0:13	0:14–0:17	0:08–0:14	0:05–0:08	0:05–0:09	0:02–0:05	No holdover time guidelines
Below −6 to −10	Aluminum	0:6–0:10	0:11–0:13	0:06–0:11	0:04–0:06	0:04–0:07	0:02–0:05	
Below −10	Aluminum	0:05–0:09	0:07–0:08	0:04–0:07	0:02–0:04	No holdover time guidelines		

TABLE 4.14
FAA Holdover Time Guidelines for SAE Type IV Fluid

Outside Air Temperature (C)	Type IV fluid Concentration Neat-Fluid/ Water (volume %/volume %)	Approximate Holdover Times Under Various Weather Conditions (hours: minutes)				
		Freezing Fog or Ice Crystal	Snow, Snow Grains, or Snow Pellets	Freezing Drizzle	Light Freezing Rain	Rain on Cold Soaked Wing
−3 and above	100/0	1:50–2:55	0:35–1:10	0:50–1:30	0:35–0:55	0:10–1:15
	75/25	1:05–1:45	0:30–0:55	0:45–1:10	0:30–0:45	0:09–0:50
	50/50	0:20–0:35	0:07–0:15	0:15–0:20	0:08–0:10	No holdover time guidelines
Below −3 to −14	100/0	0:20–1:20	0:25–0:50	0:20–1:00	0:10–0:25	
	75/25	0:25–0:50	0:20–0:40	0:15–1:05	0:10–0:25	
Below −14 to −25	100/0	0:15–0:40	0:15–0:30	No holdover time guidelines exist		

4.7.6.7 De-icing Facilities

At some airports, a safe and efficient aircraft de-/anti-icing [194] is carried out either at:

- Terminal gates, or
- A location away from them as a centralized aircraft de-icing facility.

Such a remote de-icing stands on the aircraft's route to the takeoff point, which permits the collection and ecologically safe disposal of surplus fluid. Some de-icing fluids remain on aircraft after landing and the dried deposits may collect in aerodynamically quiet areas. These deposits should be washed from aircraft with unpowered flying controls as they may freeze at a later point, causing jamming of control surfaces. Once taxiing of an uncontaminated aircraft has commenced, falling snow may build up on the aircraft. This is likely to become dangerous if departure is delayed for any reason. Therefore, pilots should be informed if snow is observed by controllers on taxiing aircraft as it may need the return of aircraft to the de-icing bay for re-treatment.

The use of terminal gates to de-ice/anti-ice aircraft is the most common option in use today. It should handle aircraft de-icing/anti-icing treatments and allow acceptable taxiing times which assures that the time from the gates to the departure runway is less than the HOT.

Centralized aircraft de-icing facilities are facilities along taxi routes leading to the departure runway(s) or on an apron away from the terminal gates. Centralized facilities are preferred than gate facilities for aircraft de-icing in the following cases: gate facilities experience excessive gate delays, the taxiing times to arrive at the departure runway exceeds the HOT, the taxi route encounters a variety of weather conditions, and if retreatment of aircraft de-icing is required.

Centralized aircraft de-icing facilities have many basic components including [194] aircraft de-icing pad(s), mobile de-icing vehicles, environmental runoff mitigation measure, control center building, permanent or portable nighttime lighting system, and, storage tank(s), transfer system(s) for approved aircraft de-icing/

anti-icing fluid(s), and fixed-fluid applicator (turret applicators instead of mobile de-icing vehicle).

Airports during its peak hour departure should balance between using only gates, a combination of gates and a centralized aircraft de-icing facility, or just a centralized aircraft de-icing facility.

Airport authorities should consider the following parameters for using a centralized aircraft de-icing facility: the number of aircraft to be treated for a set time, the number of aircraft requiring re-treatment if exceeding the HOT, the individual times to treat aircraft for various weather conditions (wet snow, pellets, freezing drizzle, cold rain, etc.), the types of aircraft to be treated since the narrow-body aircraft needs the least time, followed by the wide-body aircraft, while aircraft with center fuselage mounted engines (like DC-10) requires the longest time for treatment. If the mobile de-icing vehicles have small tank capacities or require extended periods of time to heat fluids after refilling (around 20 min), additional de-icing pads may be needed.

The number of de-icing pads needed at a facility depends on:

1. Procedures and Methods of Users
2. Variations in Meteorological Conditions
3. Type of Aircraft Receiving Treatment
4. Heating Performance and Volume Capacity of Mobile De-icing Vehicles.

Based on the estimated number of de-icing pads, the dimensions of the facility should be determined. If the estimated number of de-icing pads cannot be managed at a single site, the airport operator should consider additional centralized facilities. It is advantageous for airports serving a wide variety of scheduled flights (main airline, regional, or commuter air carriers) and nonscheduled flights (air taxi, GA aircraft, and charters) to construct a separate de-icing/anti-icing facility for each group.

Figure 4.22 illustrates a centralized aircraft de-icing facility that allows aircraft de-icing while other aircraft to continue unimpeded for departure.

4.7.7 Rotary-Wing (Helicopter) Icing

Ice is one of the most terrifying conditions a helicopter pilot can encounter. Flight of military or civil helicopters at temperatures less than or equal to 0 °C poses a significant hazard and may lead to unacceptable limitations in their available flight envelope. Some engine inlet configurations may ice at ambients slightly above 0 °C. Helicopter may be iced due to supercooled liquid water, snow, freezing rain, freezing drizzle, and freezing fog.

Rotor icing is different from icing of the propellers due to its low rotational speed. Also, icing of the rotor blade is different from that of fixed wing aircraft due to its small chord, variation of airspeed along rotor blade span, cyclic pitch change, and high catch efficiency. Although the slow forward speed of helicopters reduces ice-build up on the fuselage [179], the rotational speed of the main and tailrotor blades produces a rapid growth of ice on certain surface areas.

The most susceptible helicopter surfaces to icing are the leading edge of the main and tail rotors, rotor control rods and gyro, linkages, empennage, engine and engine inlets, stabilizers, antenna, pitot probes, secondary inlets/screens, fuel vent, drain

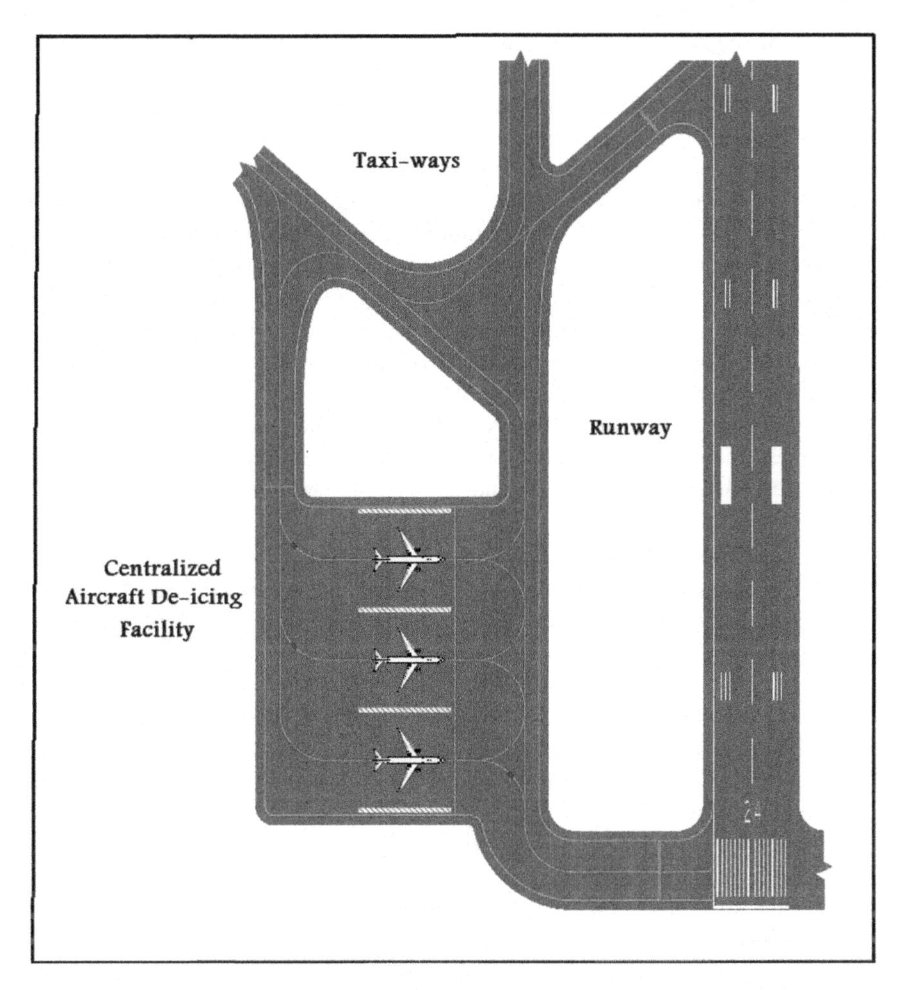

FIGURE 4.22 A centralized aircraft de-icing facility.

Courtesy: US Department of Transportation, FAA.

lines [195]. A 3/16 inch of ice on the rotor may prevent some types of helicopters from maintaining flights. Moreover, tandem rotor blades are subjected to more icing hazards than single rotor ones. Icing of the main rotor has two factors: namely, ice accretion and ice shedding. Ice accretion depends on the rotor aerodynamics and dynamics. It is dependent on freezing rain, snow, mixed liquid/ice, stratiform clouds, and cumuliform clouds.

Ice-shedding is either natural or forced. Natural shedding is either symmetric or asymmetric.

4.7.7.1 Hazard

As stated in [196, 197], ice has the following impact on helicopter modules: jeopardize its safe operation, confine its maneuverability, raise the cost of mission, decrease of rotor lift and increase of its drag, decreases pilot's vision due to windshield icing, increases the required power thus decreasing the available power margin, loss of efficiency and control, blockage of airflow due to icing of the inlet guide vanes and the first rotating stage, damage to guide vanes and rotating blades due to shedded ice from the surfaces ahead of the engine inlet duct such as the windshield, fuselage, or the main rotor blades, rise in engine turbine inlet temperature, decreases blade stall margin, increases rotor blade pitching moment causing high control load, increases of rotor vibration due to asymmetrical ice accretion or shedding, degrades autorotational capability, reduction of helicopter stability or control, reduction of performance (range, endurance, and climb capability), damage of rotor blades, fuselage or engine inlet due to ice shedding, injury of ground personnel due to ice shedding, flame out due to engine blockage or excessive ice or snow ingestion, coating of engine control cables may reduce or prevent throttle movement, blockage of pitot tubes, static ports and others causing inaccurate instruments readings, degradation or vibration of empennage effectiveness, increase of weight due to ice accumulation, locking of wheel brakes and even freeze a wheel or skid to the ground, and more hazard for tandem rotor helicopter

4.7.7.2 Protection Methods

Different methods for icing control (de-icing and anti-icing) are used [196, 197]. A list for the different protection methods for different modules of helicopters are given below:

- Main rotor: heated liquid, electrothermal, bleed air, chemical (alcohol-glycerin mixture: 90%/10%), pneumatic boot system, electrovibratory, electroimpulsive, piezoelectric ultrasonic, ice-phobic, and new centrifugally powered pneumatic de-icing system [198].
- Tail rotor: heated liquid, electrothermal, and chemical methods (glycol, alcohol).
- Engine inlet: electrothermal, compressor bleed air, and hot oil.
- Carburetor: heater.
- Windshield: electrothermal and chemical.
- Empennage: chemical methods, electrothermal, bleed air, pneumatic, and heated liquid.
- Control surface: thermal (bleed air and pneumatic boot system).
- Wing: chemical methods, pneumatic inflatable boots, bleed air, heated liquid, and electrothermal.
- Nose radome: pneumatic and chemical.
- Horizontal and vertical stabilizer: chemical and electropulse.
- Pitot tubes and other flight probes: electrical heat.

Moreover, the following new ice protection technologies are employed: ice phobic coatings, mechanical vibrators, microwave emitters, EIDI systems, and pneumatic impulse systems [189].

4.7.8 Snow and Airports

Contamination of the aircraft movement areas in airports due to snow, slush, and ice is a source of problems for airport administrators and airlines. It may limit or even close its operation. Airport should be closed if the runway is covered by a layer of slush greater than 13 mm or a dry snow of 5 cm thick. Thus, airport authorities must maintain a clear surface of all runways and other movement surfaces of the aerodrome as requested by ICAO Instructions in its Annex 14 [199]. Icing on the aircraft and airport may decrease utilization and disruption of the flight schedule. This leads to loss of revenue for the airlines, airports, and additional costs for clearing the snow and ice. Successful airport operators establish an airport snow and ice control committee that conducts pre- and post-seasonal planning meetings, operates a Snow Control Center, and implements a written plan [199–201].

4.7.8.1 Hazard

A layer of snow on the runway surface causes:

- Additional resistance force during aircraft takeoff run. This resistance force depends on the runway temperature, characteristics of snow (density, temperature, and thickness), and the characteristics of aircraft (undercarriage, speed, and mass).
- Possible damage of aircraft fuselage and other modules due to the snow thrown up from the wheels, particularly the nose one.
- Decrease of the braking force particularly when the contamination is ice. It may lead to a rejected takeoff or increase the possibility of exceeding the available distances for landing.
- Corrosion is exacerbated by the potassium-based de-icers.
- Adverse environmental impact on local environment and water courses.

4.7.8.2 Snow Removal Plan

The snow, slush, or ice must be removed quickly from the movement areas of the airport including the runway(s) in use, taxiways serving them, apron, and holding bays. An example for a winter plan is given in Airport Services Manual, Part 2, Chapter 7 [200]. Preparation of the whole airport for the winter includes preparation of equipment, training of new workers and retraining of fulltime personnel, maintenance of the movement areas, coordination with air traffic control. The basic training of involved workers includes:

- Radiotelephonic familiarization with the operation of the radio station, transmitters, and the radio phraseology.
- Procedures for removing different kinds of snow and ice, as well as the runways in use.
- Knowledge of the characteristics of aircraft and chemicals used for de-/anti-icing of aircraft and runways.
- Knowledge of workers how to operate the equipment in any weather and time (day or night) without affecting safety of operation.
- Knowledge with the airport layout and aircraft movement areas during night or low visibility conditions.

In preparing the snow plan, the following factors are considered for each airport: location, topography, climatic conditions, types of aircraft that use the airport, frequency of operations and physical characteristics of airport movement areas.

The procedure used for the first clearing of the runway depends on several factors including the available equipment, the type of snow (dry, wet, compacted, or slush), and wind speed and direction. The runways that shall be cleared are specified by the airport duty manager in co-operation with air traffic control.

If two runways are available, it is advantageous to open one runway and continue to clear snow from the other, while if only one runway exists, it is necessary to use equipment to clear the snow at high speed.

The FAA define the following two steps for snow and ice removal of airport:

1. The determination of the priority 1 area which consists of the primary runway(s) with taxiway turnoffs and associated taxiways leading to the terminal, portions of the terminal ramp, portions of the cargo ramp, airport rescue and firefighting station ramps and access roads, mutual aid access points (including gates), emergency service roads, access to essential NAVAID, and centralized de-icing facilities [202].
2. Classification of airport as either a commercial service airport (which provide scheduled air carrier service) or a noncommercial service airport (having 10,000 or fewer annual operations). Next, selecting the high-speed rotary plows and snow plows needed for the removal of tonnage of snow in a given time.

Removal of snow and ice from the movement areas may be by mechanical, chemical, or thermal means.

4.7.8.2.1 Mechanical Equipment

Mechanical equipment are preferred for snow and ice clearing than chemical and thermal means, due to their lower operational costs and negligible environmental impact.

The mechanical equipment used on large airports are high speed air blower machines, ploughs, sand/aggregate trucks, rotary blow, snow blow, material spreaders, tankers and loaders, and runway brooms [202]. Using a rotary brush sweeper and air blower is the best way for removing small layers of snow.

According to Airport Services Manual, [200], a large airport with regular air transportation should be equipped with one or more high performance snow blowers, which can throw snow with a specific weight of 400 kg/m^3 a distance of at least 30 m. While for a general aviation airport (which serves aircraft up to 5,700 kg), it should be equipped by one snow blower capable of throwing snow of a same specific weight at least 15 m. For small airports with general aviation operations or with only a few scheduled traffic movements a day, it may be either outsource snow clearing or close the airport for a few hours or even days if snows is occasionally.

4.7.8.2.2 Chemical

Chemicals are used for removing or preventing icing on the runway surfaces. However, after applying the de-icing chemicals, a thin layer of water is formed over the remaining ice. Such a surface is slippery and the braking action is nearly zero. The time needed for a complete dissolving ice depends on the type and concentration of the chemical, the meteorological conditions, thermal condition of the runway, and the thickness of the ice layer.

Chemicals should satisfy the following requirements:

- Must be cheap and effective.
- Should not damage the aircraft structure or the runway.
- Must not be either toxic or harmful to the environment.

Both chlorides and sodium chloride chemicals cannot be used as they are corrosive and may damage the pavement surface. Urea having a chemical composition of $CO(NH_2)_2$ is one of the most common chemicals used at airports. It is cheap, non-toxic and has unlimited shelf life, and can be used up to −5 °C for de-icing or anti-icing. It can be applied in the form of a solution, or by sprinkling dry granules or the granules mixed with water and sprayed. However, it is preferred to clean the runway by mechanical equipment before application of urea as a de-icing fluid. Its effect will take about 30–60 min after application. After the ice has softened, mechanical methods should be used to complete the cleaning of the runway. The disadvantage of using urea is that it may cause damage to asphalt concrete pavements, much more than ordinary road salts.

If its use is unsuitable, more expensive but environmentally friendly acetate or formate-based chemicals are used. Airport operators in United States and several countries in Western Europe use acetates-based de-icing fluids, due to their advantages: more effective at lower temperatures, act for longer time, leave a less slippery surface, easer storage, non-toxic, and do not damage the environment.

Potassium acetate-based de-icing fluids are used in Scandinavian countries. They are effective up to the temperature of −60 °C, but very expensive. Potassium formates are preferred than acetates since formates are more environmental-friendly.

Finally, high pressure water can be used for removing ice from runways. The water is sprayed through nozzles which penetrates the snow or ice layers, disturbs and separates it from the surface of the runway. Next, the runway is mechanically cleaned and treated by a small volume of glycol or any other de-icing chemical to prevent further icing.

4.7.8.2.2 Thermal

Thermal procedures adopting a heated pavement system is used for removing snow and ice as an alternative to mechanical and chemical means. Its disadvantages are high costs and problems of maintenance of some types.

The following methods can be used for removing snow and ice from apron stands and runways: waste heat, geothermal energy, photovoltaic solar energy, and electric heating pavements.

Geothermal resources are preferred due to the following reasons: decreasing the costs of removing snow/ice due to the ever increasing oil price used in mechanical equipment, reducing carbon emissions of mechanical equipment to match the increased public awareness of global warming and environmental protection, and avoiding toxicity of aircraft de-icer and anti-icer chemicals which also create undesirable algae and cause various diseases in fish [203].

Heating runways with geothermal methods can pay for itself in as little as 2–5 years. Its design includes either metallic or plastic pipes cut into the pavement and receive a flow of warm liquids, either from geothermal water (if available) or through heat exchanger systems, hot runoff liquids from local industry, or power plants. Geothermal hot water exists in areas where high-temperature water wells can be drilled for direct use, which mostly found in the western half of the United States.

Nichrome heating wire is embedded at the slab surface level and energized by an outside source to heat the pavement. An anti-icing runway slab was developed by supplying DC energy from a photovoltaic energy system to concrete slabs with surface embedded heat wire [204, 205].

Stainless steel electrodes embedded within the concrete of runways were tested and proved their validity in Des Moines International Airport in Iowa [206].

4.7.9 Pilot's Action in Case of Icing

For a pilot, his/her life and the lives of his/her passengers depend on his/her ability to understand icing and to take the proper preflight and inflight steps to deal with it safely. The following checklists should not replace or supersede AFM or pilot's operating handbook (POH). All types of airplanes are discussed. When a special action is needed for one aircraft type, it will be mentioned. Pilot should perform the following checks [207].

4.7.9.1 Preflight

- Obtain a detailed preflight weather briefing focusing on thunderstorms, types of precipitation, winds, and clouds.
- The weather may be different from that forecasted, thus arrange alternative airports along the route of flight to divert to one of them and choose those with longer runways.
- Define the ceiling and visibility. For mountainous terrain especially if unfamiliar, higher minimums for ceiling and visibility should be considered.
- If there is strong winds aloft, a turbulence is expected and a hazardous downdraft may be encountered.
- Check if there are clouds and what are their types, bases, and tops, then arrange to escape icing conditions (either climb or descend to warmer areas, make a 180° turn).
- Are there icing precipitation, current or forecast?
- Clean all ice, frost, and snow off the aircraft, brakes and wheel fairings in accordance with the POH or AFM. Note that by physically touching the surface, any fine contaminants not easily visible can be detected.
- What is the temperature at cruise altitude? Can descent be safe within his route?
- Are the aircraft de-icing and anti-icing equipment in good conditions? Is the pilot familiar with its operation?
- Cycle any de-icing and anti-icing systems to check for proper operation.
- Check the operation and freedom from contamination of the following: pitot heat and static heat (if installed), warning sensors, AOA, pitot/static openings, fuel drains, and stall warning sensors.
- Check controls, all the airframe, propeller and windscreen systems are operating correctly.
- Check that all boots on wing and tailplane (if equipped) inflate properly.
- If the aircraft is being sprayed with passengers on board close all outside vent.

- Inspect the engines' inlets and remove any accumulated ice from the nacelle inlet as well as around the nacelle drain hole and around the fan blades.
- For turbine-based engines (turbojet and turbofan) the following additional points are recommended in addition to the previous points:
  - Ground de-icing operations should be carefully performed as per the company procedures and FAA circulation [208] and [192].
  - Ensure that de-icing fluids are not sprayed into engines, APUs, pitot inlets, probe openings, or static ports.
  - Do not spray heated fluids onto cold windows.

4.7.9.2 Taxi/Takeoff/In-Flight

- During taxi, carefully use brakes to prevent skidding.
- Check controls for full range of motion.
- Perform regular engine power run-ups to shed any accumulated ice while taxiing, as per each AFM.
- After takeoff cycle landing gear to clear snow or slush from wheel wells if recommended by the manufacturer.
- Anti-ice systems should be activated at the first sign of visible moisture with air temperatures slightly above freezing, while de-icing systems should be activated at the first sign of ice accretion (refer to the AFM or POH).
- If bleed air is used, engine power setting should be set according to the POH or AFM reference section.
- Visually check for any ice formation and check any ice accumulation behind protected areas on the aircraft.
- Carefully identify any ice formations on wings which may cause control problems.
- If there is a need to use wing-de-icing systems, check also the tail as it may be accumulating ice as well.
- Since icing on a tailplane may cause uncontrollable pitch forces if flaps lowered, always be prepared to reverse any flap selection to regain control [209].
- If using an autopilot, if workload permits, periodically disengage and manually fly the aircraft to identify handling changes caused by ice. This is especially important if operated in slow flight or in a holding pattern.
- Use airspeed bug to monitor changes to airspeed.
- Fly out of the cloud (laterally or vertically). Thus, make a U-turn to fly out of the cloud, or climb or descend to clear air above or below the cloud.
- Climb to altitudes where there might be warmer air in the cloud.
- Descend to altitudes where the air is warmer in the cloud or rain (doesn't work in winter when the cold air reaches the ground).
- Since ice buildup on the wing lowers the stall AOA, the stall warning sensor might not provide warning in icing conditions. Pilots therefore should know the manufacturer's published minimum icing airspeeds and treat them as limitations, even if they are not in the limitations section. If flight manual or POH does not have minimum icing airspeeds, add 15–20 KIAS to the normal operating airspeed.
- Restrict maneuvers in icing conditions.
- Propeller anti-/de-ice may help to avoid some performance deterioration.

4.7.9.3 Approach and Landing [207]

- Determine if freezing drizzle or freezing rain is being reported and avoids flying into these areas. A ground observation of any type of precipitation when temperatures are near freezing may indicate freezing precipitation aloft, so be vigilant for severe icing conditions.
- Cycle boots before final approach, if equipped.
- In accordance with the POH or AFM, use a higher approach speed into the landing when carrying an accumulation of ice. Use a longer runway if available.
- Carry some power on flare and flare slightly faster than normal if carrying ice. Use a longer runway if available.
- After touchdown, use brakes sparingly to prevent skidding or in case of ice buildup in brakes.

4.7.9.4 Military Aircraft

Military aircraft do not operate in icing conditions. Aircrews are advised to avoid areas where icing is probable and that extended flight in an icing environment is considered an emergency. They have to check freezing levels and areas of probable icing from weather service. Fighter aircraft has thin, low aspect wings and sharp leading edges. Thus, it is more susceptible to ice accumulation than other kinds of aircraft wings. The primary concern with flying in icing conditions is ice accumulation sufficient to cause engine damage. If inadvertent or unavoidable operation in known or suspected icing conditions has occurred, the following precautionary action should be taken immediately [210]:

- Turn the ANTI-ICE switch to ON.
- Switch the PROBE HEAT to HEAT.
- Make all throttle movements slower than normal to reduce possibility of engine stalls and/or stagnation.
- Frequently monitor the engine instruments: RPM and EGT indications. A reduction of rpm or an increase in EGT accompanied by a loss of thrust is an indication of engine icing.
- Avoid clouds and other areas of visible precipitation.
- Descent below the freezing level if time and fuel permit.
- If unable, fly at altitudes above approximately 25,000 ft or ambient temperatures below −30 °C which are generally free of icing conditions.
- If unable to avoid precipitation, adjust aircraft Mach or altitude as necessary to remain outside of the icing zone.
- Try to eliminate the ice before landing by remaining well below the freezing level for an extended period of time.
- A straight−in field landing is preferred. Minimum power setting after landing is recommended.
- De-ice the aircraft before the aircraft taxi for another take-off to fly [211].

4.8 HAIL

Hail is defined as a form of **precipitation** composed of balls or irregular lumps of ice, always produced by convective clouds which are nearly always **cumulonimbus**

[180]. Hailstones are lumps of ice/water/air mixture with sizes greater than 5 mm that are generally opaque with layered structure [212]. They exist along and near mountains and may fall detached or frozen together into irregular, lumpy masses [213]. **Hail** is formed when raindrops are carried upward by thunderstorm updrafts into extremely cold areas of the atmosphere and freeze.

Even if an aircraft passes through a shower of hail for a short time (order of seconds), it is sufficient to cause damage to aircraft structure such as depression, cracking, and even piercing if crushed with hailstone at a high speed [214, 215]. The exposed parts, such as wheel door, leading edge, radome, windshield, and engine, will be easily damaged during this condition, which severely impact both flight performance and mechanism operating.

When a stationary aircraft on ground is exposed to a hail storm, the velocity of hail impacting is approximately 25 m/s, which is the velocity of free fall for a 0.04 m hail ball.

4.8.1 Properties

4.8.1.1 Size

One way hail that can grow to large sizes is by being recirculated through the storm. For sizes between 5 and 10 mm, they generally appear spherical or conical. Hailstones that have the size of 10–20 mm tend to be ellipsoidal or conical. Larger 10–50 mm hailstones take on ellipsoidal shapes with lobes, while still larger hailstones between 40 and 100 mm appear irregular (including disk shapes) with protuberances [211].

4.8.1.2 Density

For sizes smaller than 20 mm, hail densities range from 50 to 890 $kg\,m^{-3}$. For larger sizes the density ranges from 810 to 915 $kg\ m^{-3}$. The lower density particles are likely to be quite fragile. Since a hailstone may be composed of clear or opaque ice, the limits of its specific gravity are between 0.25 and 0.92 [211].

4.8.1.3 Temperature

Temperatures inside hailstones are determined immediately after the stones strike the ground. It have been found to be at, or less than, 0 °C, and often between −5 °C and −15 °C.

4.8.1.4 Fallspeed

Fallspeed (V) for a sphere is

$$V = \left(\frac{4gD\rho_{\text{hail}}}{3C_d\rho_{\text{air}}} \right)^{\frac{1}{2}} \tag{4.6}$$

where g is the acceleration due to gravity, ρ_{hail} and D are the density and the size of the hail, C_d is the drag coefficient, and ρ_{air} is the air density.

4.8.2 Hail Threat

Due to the increasing flight speed of civil and military aircraft, the probability of surface damage caused by impact with hail, and/or other precipitants, has become a design consideration.

The key requirement from a structural design perspective is that when an aircraft encounters a hailstorm, it is to ensure that it will survive and land safely. Since composites are increasingly used in aircraft manufactures, it is mandatory to design composite aircraft structure to be resistant to typical hail strike energies to minimize the amount of repair required following a hailstorm. The threat depends on the speed of the aircraft and the size of the hailstones. Unlike metallic aircraft structures that often dent and deform during an impact event, carbon fiber aircraft structures leave little to no visual indication of damage on their external surfaces.

The serious problem is that hail is more or less invisible until the aircraft runs into it. The other problem is that hail can be ejected from the bottom of clouds as well as the top of larger storms.

With the growth of aviation, numerous aircraft encounter hail during its routine flights. Hail particles can cause severe damage to all external components of the aircraft.

The threat from hail arises from the multiple impact of hail stones with aircraft surfaces over a prolonged period of time.

The damage of different parts of aircraft is identified hereafter.

- Fuselage: the forward section including radome, cockpit, windshield, and canopy.
- Wings: the leading edges, wing tips and roots, and control surfaces.
- Horizontal and vertical tail: leading edges and control surfaces.
- Engine: cowling and ignition harnesses.
- Propeller: blades and propeller assemblies.
- Wheel doors: lattice hybrid structure.
- Aircraft accessories: radar coverings, antenna loop housings, and landing and navigational lights.

Testing for metallic and composite material of aircraft was performed experimentally.

The CAA conducted experimental tests on the metal portions of typical aircraft (Douglas DC-6 and DC-3 wing sections). Frozen ice spheres having the sizes of 0.75, 1.25, and 1.88 inches in diameter were fired by a compressed-air gun at the wing sections at speeds ranging from 110 to 460 miles/h [213]. The following findings were obtained:

1. The extent of damage depends on the mass of the hailstone, the impact velocity, the impact angle, and the hit aircraft part properties (thickness, strength, and shape).
2. Hailstones less than 0.75 inch in diameter do not cause significant damage at airplane speeds between 200 and 300 miles/h.
3. Hailstone having diameters ranging from 0.75 to 1.88 inch and speeds ranging from 210 to 420 miles/h caused dents in wing leading-edge sufficiently large to require repair.
4. Two-inch hailstones cause extensive damage to airplanes at speeds above 300 miles/h.

Simulated hail impact (SHI) tests were conducted on full-scale carbon fuselage panels at velocities of up to 118 m/sec. It left little to no surface visual damage, except at direct shear tie impacts. Though the SHI located at the middle of the bays did not show any visual damage, there was an extensive interplay delamination and subsurface damage. Moreover, hail impacts directed at the stringer flange and shear tie induced only substructure disbands [216].

Flight tests are also performed by a T-28 and Thunderbolt A-10 [212]. The National Science Foundation refurbished a retired Thunderbolt A-10 warplane, equipping it with scientific instruments and sensors to fly into thunderstorms and study severe-weather systems (winds, hail, and lightning). Thus, researchers could study massive, energetic storms from the inside. Also, it helped meteorologists to predict when and where these thunderstorms will strike. A-10 could not fly directly into hailstones much bigger than golf balls.

An analytical approach for evaluating the denting characteristics of airplane surfaces caused by hailstone is described in detail in [213]. When a hailstone impacts the thin skin of an airplane, a plastic deformation exists near the point of impact and elastic deformation in the surrounding area. If the impact is severe, both bending and membrane action are present in the skin. Consequently, the problem is rather a sophisticated, dynamic, elastic-plastic, large-deformation one. An approximate analysis of the problem for the problem was given in [213], where hailstone was idealized by a crushable spherical projectile impact normal to a flat plate. Separate elastic and rigid-plastic analyses have been applied to the area around the center of impact, which has become plastic. The rigid plastic analysis applied to this central plastic region determines the resulting permanent deformation (dent).

4.8.3 Recent Accidents

Table 4.15 presents some details of recent accidents due to hail.

TABLE 4.15
Representative Accidents Due to Hailstorms

Date	Location, Airline, Aircraft Type	Observation
June 4, 2018	Emergency landing at El Paso International Airport, in Texas, American Airlines Flight # AA1897, Airbus A319 plane	Flight from San Antonio to Phoenix, hail led to cracked windshield, nose damage
May 26, 2019	Beijing, China Southern Airlines flight CZ3101, Airbus A380	Massive chunks of hail destroyed the front windshield and shattered side windows of the plane
October 12, 2020	O'Hare International Airport, United Airlines plane flight # 349, Boeing 767	Flight from O'Hare International airport to Washington, D.C. Hail nearly shattered windshield before it turned back and had an emergency landing at Chicago's O'Hare International Airport.

REFERENCES

[1] Francisco, J.J., Serrano and Kazda, A. Airline disruption management: yesterday, today and tomorrow. Transportation Research Procedia, 28, 3–10, 2017 www.elsevier.com/locate/procedia

[2] Williams, A. Exploring the weather hazards behind 5 deadly, notorious plane crashes, AccuWeather, Updated July 10, 2019, www.accuweather.com/en/weather-news/exploring-the-weather-hazards-behind-5-deadly-notorious-plane-crashes/360542

[3] World Metrological Organization, Aviation Hazards, ETR-20, WMO/TD-No. 1390, Secretariat of the World Metrological Organization – Geneva – Switzerland, June 2007.

[4] Goodman, C.J. and Small Griswold, J.D. Meteorological Impacts on Commercial Aviation Delays and Cancellations in the Continental United States, Journal of Applied Meteorology and Climatology, 58, 479–494, 2019.

[5] NTSB (2010). NASDAC Review of National Transportation Safety Board (NTSB) weather-related accidents (2003–2007). www.asias.faa.gov/

[6] Cook, L., Wood, B., Klein, A., Lee, R., and Memarzadeh, B. Analyzing the share of individual weather factors affecting NAS performance using the weather impacted traffic index, 2009. In AIAA 2009-7017. 9th AIAA aviation technology, integration, and operations conference (ATIO), Hilton Head, SC, September 2009. https://doi.org/10.2514/6.2009-7017.

[7] Rudra, R., Dickinson, W. T., Ahmed, S. I., Patel, P., Zhou, J., and Gharabaghi, B. Changes in rainfall extremes in Ontario. International Journal of Environment Research, 9(4), 1117–1372, 2015.

[8] Rebecca Grant, Storms of War, AIR FORCE Magazine, July 1, 2004, www.airforcemag.com/article/0704storm/

[9] Atkinson, G.D. Impact of Weather on Military Operations: Past, Present, Future, US Army War College, AD911125, 1974.

[10] Aviation Weather Service, FAA Advisory Circular AC-00-45H.

[11] Lara Peck, The Impacts of Weather on Aviation Delays at O.R. Tambo International Airport, South Africa, M.Sc. Thesis, University of South Africa, November 2015.

[12] Mahapatra, P.R. and Zrnic, D.S. Sensors and systems to enhance aviation safety against weather hazards, Proceedings of The Institute of Electrical and Electronics Engineers (IEEE), 79(9), 1234–1267, 1991.

[13] Barrass, J. Cumulonimbus – More Frightening than Bengt's Mother-in-Law, www.skybrary.aero/index.ph p/Cumulonimbus

[14] Types of precipitation, National Weather Service, www.weather.gov/jetstream/preciptypes

[15] World Meteorological Organization (WMO) 2007. Aviation hazards, education and training programme ETR-20, Geneva, Switzerland.

[16] Kulesa, G. Weather and aviation: How does weather affect the safety and operations of airports and aviation, and how does FAA work to manage weather-related effect? The Potential Impacts of Climate Change on Transportation, Washington, DC, Engineering 2003. [17/16] Aviation Weather, FAA Advisory Circular AC 00-6B, 8/23/ 2016.

[17] Obscuration Types, National Weather Services, US Dept of Commerce, National Oceanic and Atmospheric Administration, National Weather Service, www.weather.gov/jetstream/obscurationtypes

[18] Carn, S.A., Krueger, A.J., Krotkov, N.A., Yang, K., and Evans, K. Tracking volcanic sulfur dioxide clouds for aviation hazard mitigation. Natural Hazards, 51(2), 325–343, 2009.

[19] Ismail Gultepe, R. Sharman, Paul D. Williams, Binbin Zhou, G. Ellrod, P. Minnis, S. Trier, S. Griffin, Seong. S. Yum, B. Gharabaghi, W. Feltz, M. Temimi, Zhaoxia Pu, L.N. Storer, P. Kneringer, M.J. Weston, Hui-Ya Chuang, L. Thobois, A.P. Dimri, S.J. Dietz, Gutemberg B. Franc¸A, M. V. Almeida, and F. L. Albquerque Neto, A Review of High Impact Weather for Aviation Meteorology. Pure and Applied Geophysics, 176 (2019), 1869–1921, 2019 Crown https://doi.org/10.1007/s00024-019-02168-6

[20] Wan, T. and Wu, S.W. 2004. Aerodynamic analysis under the influence of heavy rain, 24th International Congress of The Aeronautical Sciences.

[21] ICAO (2013). Annex 3: Meteorological Service for International Air Navigation, 18th edition.

[22] Fultz, A.J. and. Ashley, W.S. Fatal weather-related general aviation accidents in the United States, Physical Geography, 2016, DOI: 10.1080/02723646.2016.1211854

[23] NTSB (1996). Aircraft accident report. Vol. 1. National Transportation Safety Board NTSB/AAR–96/01–PB96–910401, 322 pp.

[24] NTSB (2010). NASDAC Review of National Transportation Safety Board (NTSB) weather-related accidents (2003–2007). www.asias.faa.gov/.

[25] Eick, D. Turbulence related accidents and incidents. Presentation at NCAR Turbulence Impact Mitigation Workshop 2, Sep 3–4, 2014. https://ral.ucar.edu/sites/default/files/public/ events/2014/turbulence-impact-mitigation-workshop-2/docs/eick-turbulencerelatedaccidents.pdf

[26] Sharman, R. and Lane, T. Aviation turbulence: Processes, detection, prediction (p. 523). Berlin: Springer, 2016.

[27] Jenamani, R.K. and Kumar A. Bad weather and aircraft accidents–global vis-a-vis Indian scenario. Current Science, 104(3), 316–325, 2013.

[28] ASC, Taiwan Aviation Occurrence Statistics 2007 to 2016, Aviation Safety Council, Taipei, Taiwan, 2017, www. asc.gov.tw/upload/statistics_files/Taiwan%20 Aviation%20Oc currence%20Statistics%202007-2016.pdf.

[29] Spirkovska, L. and Lodha, S. K. AWE: aviation weather data visualization environment. Computers & Graphics, 26(1), 169–191, 2002.

[30] Bendinelli, W.E., Humberto F.A.J., Bettini, and Alessandro V.M.O, Airline delays, congestion internalization and non-price spillover effects of low cost carrier entry. Transportation Research Part A: Policy and Practice, 85, 39–52, March 2016.

[31] Abdelghany, K.F., Shah, S.S., Raina, S., and Abdelghany, A.F. A model for projecting flight delays during irregular operation conditions. Journal of Air Transport Management, 10, 385–394, 2004.

[32] Allan, S.S., Evans, J.E., and Gaddy, S.G. *Delay Causality and Reduction at the New York City Airports Using Terminal Weather Information Systems*, Project Report ATC-291, Massachusetts Institute of Technology, 2001.

[33] Pejovic, T., Noland, R.B., Williams, V., and Toumi, R. A tentative analysis of the impacts of an airport closure. Journal of Air Transport Management, 15, 241–248, 2009.

[34] Robinson, P.J. The Influence of weather on flight operations at the Atlanta Hartsfield International Airport. Weather and Forecasting, 4, 461–468, 1989.

[35] Goodman, C.J. and Small Griswold, J.D. Meteorological Impacts on Commercial Aviation Delays and Cancellations in the Continental United States. Journal of Applied Meteorology and Climatology, 58, 479–494, March 2019.

[36] Joint Economic Committee (2008). "Your flight has been delayed again": A report by the Joint Economic Committee Majority Staff. 12 pp., http://online.wsj.com/public/resources/documents/ jecreport05222008.pdf

[37] European Commission (2011). Flightpath 2050 Europe's Vision for Aviation. Report of the High-Level Group on Aviation Research. [Online] pp. 11. Available: https://ec.europa.eu/transport/sites/transport/files/modes/air/doc/flightpath2050.pdf

[38] Sasse, M. and Hauf, T. A study of thunderstorm-induced delays at Frankfurt Airport, Germany. Meteorological Applications, 10(1): 21–30, 2003.

[39] Massimiliano Zanin, Yanbo Zhu, Ran Yan, Peiji Dong, Xiaoqian Sun, and Sebastian Wandel. Characterization and Prediction of Air Transport Delays in China. Applied Science, 2020, 10, 6165, www.mdpi.com/journal/applsci

[40] Pramono, A., Middleton, J.H., and Caponecchia, C. Civil Aviation Occurrences in Indonesia. Journal of Advanced Transportation, 2020, Article ID 3240764, 17 pages, https://doi.org/10.1155/2020/3240764

[41] The effect of Australian aviation weather forecasts on aircraft operations: Adelaide and Mildura Airports, Australia Research ATSB Transport Safety Report Aviation Research Report AR-2013-200 Final –July 10, 2017

[42] Evans, J.E. 1995. *Safely reducing delays due to adverse terminal weather.* In: Modelling and simulation in air traffic management, Bianco, L., Dell'Olmo, P., Odoni, A.R. (eds), Springer: Berlin Heidelberg, 85–202.

[43] Belvedere, M.J. and LeBeau, P. Weather flight disruptions cost $1.4 billion: Data, CNBC, Wed, Jan 8, 2014 www.cnbc.com/2014/01/08/weather-flight-disruptions-cost-14-billion-data.html#:~:text=The%20massive%20flight%20delays%20and,masFlight%2C%20an%20airline%20consulting%20firm

[44] Gultepe, I., Heymsfield, A.J., Field, P.R., and Axisa, D. Ice-phase precipitation. Meteorological Monographs, 58, 6.1–6.36, 2017 https://doi.org/10.1175/Amsmonographs-D-16-0013.1.

[45] Möller, D., Wieprecht, W., Hofmeister, J., Kalass, D., Elbing, F., and Ulbricht, M. 2001. *Fog dissipation by dry ice blasting: process mechanism*, Paper presented to the 2nd Fog Conference.

[46] Klein, A., Kavoussi, S., and Lee, R.S. 2009. *Weather forecast accuracy: study of impact on airport capacity and estimation of avoidable costs*, Paper presented to the eighth USA/Europe Air Traffic Management Research and Development Seminar.

[47] Robinson, P.J. The Influence of weather on flight operations at the Atlanta Hartsfield International Airport. Weather and Forecasting, 4, 461–468, 1989.

[48] Liu, Y., Yin, M., and Hansen, M. 2016. Estimating Costs of Flight Delay for Air Cargo Operations. Institute of Transportation Studies, University of California, Berkeley.

[49] Borskya, S. and Unterberger, C. Bad weather and flight delays: The impact of sudden and slow onset weather events. Economics of Transportation, 18, 10–26, 2019.

[50] Gayle, P.G. and Yimga, J.O. How much do consumers really value air travel on-time performance, and to what extent are airlines motivated to improve their on-time performance? Economics of Transportation, 14, 31–41, 2018.

[51] Forbes, S.J. The effect of air traffic delays on airline prices. International Journal of Industrial Organization, 26, 1218–1232, 2008.

[52] Bureau of Transport Statistics (2018). On-Time performance- flight delay at a Glance, United States Department of transportation. URL: www.transtats.bts.gov/HomeDrillChart.asp

[53] Sand and Dust Storms: Subduing a Global Phenomenon (2017) https://wedocs.unep.org/bitstream/handle/20.500.11822/22267/Frontiers_2017_CH4_EN.pdf?sequence=1&isAllowed=y

[54] Ginoux, P., Prospero, J.M., Gill, T.E., Hsu, N.C., and Zhao, M. Global-scale attribution of anthropogenic and natural dust sources and their emission rates based on MODIS Deep Blue aerosol products. Reviews of Geophysics, 50, 2012. http://onlinelibrary.wiley.com/ doi/10.1029/2012RG000388/epdf

[55] Middleton, N. and Kang, U. Sand and Dust Storms: Impact Mitigation, Sustainability, 9, 1053, 2017.

[56] Lekas, Th. I., Kushta, J., Solomos, S., and G. Kallos. Some considerations related to flight in dusty conditions. Journal of Aerospace Operations 3 (2014), 45–56, DOI 10.3233/AOP-140043
[57] www.skybrary.aero/index.php/Sand_Storm
[58] https://en.wikipedia.org/wiki/Operation_Eagle_Claw#/media/File:Sandstorm_in_Al_Asad,_Iraq.jpg
[59] NCMS warns of active winds, low visibility. Emirates 24/7 News, August 4, 2016. www.emirates247.com/news/emirates/ncms-warns-of-active-winds-lowvisibility-2016-08-04-1.637979
[60] Johnson, W.B. Arizona's biggest weather stories of the past 10 years: Heat, floods and dust storms, weather, January 2, 2020 www.azcentral.com/story/news/local/arizona-weather/2019/12/30/arizonas-biggest-weather-stories-decade-2010-2019-dust-storms-hottest-years-biggest-floods/4125203002/
[61] Khalifa, E.M. Conscious Study of Impact of Dust Storm on Aviation and Airport Management. International Journal of Science Research and Technology, 2(2), 51–57, June 25, 2016.
[62] What is Particle Pollution? Particle Pollution and Your Patients' Health, EPA; United States Environmental Protection Agency, www.epa.gov/pmcourse/what-particle-pollution
[63] Sand storm, Encyclopedia.com, September 2020, www.encyclopedia.com/earth-and-environment/atmosphere-and-weather/weather-and-climate-terms-and-concepts/sandstorms
[64] Overview of Dust Storms, atmospheric dust, produced by the comet program, 2012 http://kejian1.cmatc.cn/vod/comet/mesoprim/at_dust/print.htm
[65] Atmospheric Dust: Impact of Air Traffic, http://kejian1.cmatc.cn/vod/comet/mesoprim/at_dust/navmenu.php_tab_1_page_1.3.0_type_text.htm
[66] Shao, Y., Yang, Y., Wang, J., Song, Z., Leslie, L.M., Dong, Ch., Zhang, Z., Lin, Z., Yutaka, K., Yabuki, S., and Chun, Y. Northeast Asian dust storms: real-time numerical prediction and validation. Journal of Geophysical Research: Atmospheres, 108 (D22) (November 2003) 2003JD003667, https://doi.org/10.1029/2003JD003667
[67] Bojdo, N., Filippone, A., Parkes, B., and Clarkson, R. Aircraft engine dust ingestion following sand storms, Aerospace Science and Technology, 2020, http://doi.org/10.1016/j.ast.2020.106072
[68] TH. I. Lekas, G. Kallos, J. Kushta, S. Solomos, E. Mavromatidis – Dust Impact on Aviation‖, 6th International workshop on Sand/Dust Storms and Associated Dust fall, September 7–9, 2011, Greece.
[69] David Cenciotti, Take a Look at These Photos Of Luke Air Force Base F-35s Engulfed By Sand Storm, The Aviationist, July 31, 2018, https://theaviationist.com/2018/07/31/take-a-look-at-these-photos-of-luke-air-force-base-f-35s-engulfed-by-sand-storm/
[70] Bojdo, N. and Filippone, A. A Simple Model to Assess the Role of Dust Composition and Size on Deposition in Rotorcraft Engines, Aerospace 2019, 6, 44 doi:10.3390/aerospace6040044
[71] Introduction to Helicopter Engine Inlet Protection, Pall Aerospace (2009). www.docin.com/p-714184439.html
[72] https://en.wikipedia.org/wiki/Brownout_(aeronautics)#/media/File:U.S._Soldiers_with_Bravo_Troop,_3rd_Squadron,_71st_Cavalry_Regiment,_3rd_Brigade_Combat_Team,_10th_Mountain_Division_prepare_to_board_a_CH-47_Chinook_transport_helicopter_after_completing_their_mission_140313-A-YK672-002.jpg
[73] Sand Erosion, HONTEK, 2020, https://hontek.com/products/sand-erosion/

[74] Wood, C.A., Slater, S.L., Zonneveldt, M., Thornton, J., Armstrong, N. and Antoniou, R.A. Characterisation of Dirt, Dust and Volcanic Ash: A Study on the Potential for Gas Turbine Engine Degradation, Australian Government, Aerospace Division Defence Science and Technology Group DST-Group-TR-3367, May 2017.
[75] Smeltzer, C. E., Gulden, M. E., and Compton, W. A. Mechanisms of Sand and Dust Erosion in Gas Turbine Engines, Quarterly Technical Progress Report, Solar R. P. 2-27S2-7, 1969.
[76] Ahmed F. El-Sayed, FOD in Intakes – A Case Study for Ice Accretion in the Intake of a High Bypass Turbofan Engine, RTO-EN-AVT-195
[77] Mund, M.G. and Guhna, H. Gas Turbine Dust Air Cleaners, American Society of Mechanical Engineers, Paper 70-GT-104, August 1970.
[78] Sirs, R. C. The Operation of Gas Turbine Engines in Hot & Sandy Conditions-Royal Air Force Experiences in the Gulf War, AGARD-CP-558, Paper No. 2, May 1994.
[79] Giovanni Maria De Pratti, Aerodynamical Performance Decay Due to Fouling and Erosion in Axial Compressor for GT Aeroengines, E3S Web of Conferences 197, 11002, 75° National ATI Congress (2020), https://doi.org/10.1051/e3sconf/202019711002
[80] Boynton, S.L. et al. Investigation of Rotor Blade Roughness Effects on Turbine Performance in Transactions of the ASME, Journal of Turbomachinery, 1115, 614–620, 1993.
[81] Calvert, W. Prevent damage to gas turbines from ice ingestion, in Power, Oct.'94, pp. 73–75, 1994.
[82] Lakshminarasimha, A.N. et al. Modeling and Analysis of Gas Turbine Performance Deterioration, in Transactions of the ASME, Journal of Gas Turbine and Power, 116, 46–52, 1994.
[83] Natole, R. Gas Turbine Components-Repair or Replace, in IGTI Global Gas Turbine News, pp. 4–7, May/June 1995.
[84] Tabakoff, W. et al. Simulation of Compressor Performance Deterioration Due to Erosion, in Transactions of the ASME, Journal of Turbomachinery, 112, 78–83, January 1990.
[85] Duffy, R.F. et al., Integral Engine Inlet Particle Separator. Volume II. Design Guide, AD-A015 064, 1975.
[86] Bojdo, N. and Filippone, A. Comparative Study of Helicopter Engine Particle Separators,. Journal of Aircraft, 51(3), 1030–1042, 2014.
[87] Charlie Page, How pilots deal with volcanic ash encounters, September 12, 2020, https://thepointsguy.com/news/pilots-deal-with-volcanic-ash/
[88] Bojdo, N. and Filippone, A. Effect of Desert Particulate Composition on Helicopter Engine Degradation Rate, 40th European Rotorcraft Forum, Southampton, September 2014
[89] Impact of Air Traffic, Atmospheric Dust (2012) http://kejian1.cmatc.cn/vod/comet/mesoprim/at_dust/index.htm
[90] Calero, "ADF (ARMY) A17-036 Kiowa Failure Investigation of Bell 206B-1 Rolls Royce 250-C20 Series Engine," DSTO Client Report, DSTO-CR-2005-0001, 2005.
[91] Bojdo, N., Filippone, A., Parkes, B., and Clarkson, R. Aircraft engine dust ingestion following sand storms. Aerospace Science and Technology, 106 https://www.sciencedirect.com/science/journal/12709638/106/supp/C, November 2020, 106072. www.sciencedirect.com/science/article/abs/pii/S1270963820307549?via%3Dihub
[92] Kihm, D. and Macer, D. Safe, Efficient Flight Operations in Regions of Volcanic Activity, Boeing Aero magazine, 11, qtr_03, 5–11, 2011

[93] Manual on Volcanic Ash, Radioactive Material and Toxic Chemical Clouds, ICAO Doc 9691, AN/954, Second Edition – 2007.

[94] Guffanti, M., Casadevall, T., and Mayberry, G. Encounters of Aircraft with Volcanic-Ash Clouds: An Overview, USGS, https://pages.mtu.edu/~gbluth/Teaching/GE4150/lecture_pdfs/L7a_aircraft_hazards

[95] *Handbook on the International Airways Volcanic Watch, ICAO* Document 9766-AN/968, 10/3/2020.

[96] Christmann, C., Nunes, R.R., Schmitt, A.R., and Guffanti, M. Flying into Volcanic Ash Clouds: An Evaluation of Hazard Potential, NATO STO-MP-AVT-272.

[97] Catherine Annen and Jean-Jacques Wagner, The Impact of Volcanic Eruptions During the 1990s, Natural Hazards Review, 4(4), November 1, 2003. ©ASCE, ISSN 1527-6988/2003/4-169–175

[98] How many active volcanoes are there on Earth? USGS, www.usgs.gov/faqs/how-many-active-volcanoes-are-there-earth?qt-news_science_products=0#qt-news_science_products

[99] Mount Katami, http://jrschmidt3.tripod.com/mountkatmai/id1.html

[100] www.nsf.gov/news/news_images.jsp?cntn_id=108985&org=NSF

[101] Daily updated map of currently erupting and restless volcanoes; www.volcanodiscovery.com/daily-map-of-active-volcanoes.html

[102] https://blogs.agu.org/magmacumlaude/files/2011/04/gip64-copy.jpg

[103] https://en.wikipedia.org/wiki/Volcanic_bomb#/media/File:Volcanic BombMojaveDesert.JPG

[104] Flight Safety and Volcanic Ash Risk: Management of flight operations with known or forecast volcanic ash contamination, ICAO Doc 9974 AN/487, first Edition-2012.

[105] What Are the Special Hazards From Volcanic Ash? https://chis.nrcan.gc.ca/volcano-volcan/haz-vol-en.php

[106] Vogel, A. Volcanic Ash: Properties, Atmospheric Effects and Impacts on Aero-Engines. Ph. D. Thesis, Department of Geosciences, University of Oslo, 2018.

[107] https://en.wikipedia.org/wiki/Volcanic_gas#/media/File:Volcanic_injection.svg

[108] Martin, E. Volcanic Plume Impact on the Atmosphere and Climate: O- and S-Isotope Insight into Sulfate Aerosol Formation. Geosciences, 8, 198, 2018 doi:10.3390/geosciences8060198 www.mdpi.com/journal/geosciences

[109] Casadevall, T.J., ed. 1994. The First International Symposium on Volcanic Ash and Aviation Safety: Proceedings Volume: U.S. Geological Survey Bulletin 2047.

[110] Casadevall, T.J., Thompson, T. B., and Fox, T. 1999. World map of volcanoes and principal air navigation features. U.S. Geological Survey Map I-2700.

[111] Przedpelski, Z.J. AIA Recommendations Aimed at Increased Safety and Reduced Disruption of Aircraft Operations in Regions with Volcanic Activity.

[112] Warneck, P. 1988. Chemistry of the natural atmosphere. Academic Press.

[113] Newhall, G. and Self, S. The Volcanic Explosivity Index (VEI). An Estimate of Explosive Magnitude for Historical Volcanism. Journal of Geophysical Research, 87(1), 1231–1238, 1982.

[114] Andreae, M.O. Climatic effects of changing atmospheric aerosol levels. World Survey of Climatology, 1995, 347–398.

[115] Flight Safety and Volcanic Ash Risk: Management of flight operations with known or forecast volcanic ash contamination, ICAO Doc 9974 AN/487, first Edition-2012.

[116] Rafael Arellano, S. Studies of Volcanic Plumes with Remote Spectroscopic Sensing Techniques, Ph. D. Thesis, Department of Earth and Space Sciences Chalmers University of Technology, 2014.

[117] Blake, D.M., Wilson, T.M., and Stewart, C. Visibility in Airborne Volcanic Ash: Consideration for Surface Transportation Using a Laboratory-Based Method, Natural Hazards, 12, February 2018.

[118] Yang Y. and Wang J., et al. Northeast Asian dust storms: Real-time numerical prediction and validation. Journal of Geophysical Research, 2003, 108. DOI: 10.1029/2003JD003667

[119] Dunn, M.G. Operation of Gas Turbine Engines in an Environment Contaminated with Volcanic Ash. Journal of Turbomachinery, ASME Transactions, September 2012, Vol. 134/051001-(1-18)

[120] Dunn, M.G. and Wade, D.P. Influence of Volcanic Ash Clouds on Gas Turbine Engines, Proceedings of the First International Symposium on Volcanic Ash and Aviation Safety, *Edited by* Thomas J. Casadevall, 1994.

[121] Zygmunt J. Przedpelski and Casadevall, T.J. Impact of Volcanic Ash from 15 December 1989 Redoubt Volcano Eruption on GE CF6-80c2 Turbofan Engines, Proceedings of the First International Symposium on Volcanic Ash and Aviation Safety, Edited by Thomas J. Casadevall, 1994.

[122] Airbus A318/A319/A320/A321, AMM manual, 05-51-25-200-003-A – Inspection after a Flight through Volcanic Ash, after Volcanic Sulfur Odor or after Volcanic Ash Contamination on the Ground, August 1, 2020.

[123] Casadevall, T.J. and MURRAY, T.M. Volcanic Ash Avoidance, BOEING AERO Magazine, QTR_01 2000, 19–27.

[124] Advice for Airport Operators, 2013, https://volcanoes.usgs.gov/vsc/file_mngr/file-110/Advice_for_Airport_Operators.pdf

[125] Casadevall, T.J. Discussions and Recommendations from the Workshop on the Impacts of Volcanic Ash on Airport Facilities, USGS Open-File Report 93-5, Seattle, Washington, April 26–28, 1993 https://pubs.usgs.gov/of/1993/0518/report.pdf

[126] Ash Removal, Volcanic ash fall removal, https://volcanoes.usgs.gov/volcanic_ash/airports_clean_up_mitigation.html

[127] Wan, T. and Wu, S.W. 2004. Aerodynamic analysis under the influence of heavy rain, 24th International Congress of The Aeronautical Sciences.

[128] Rhode R.V. Some effects of rainfall on flight of airplanes and on instrument indications. NASATN-803; 1941.

[129] Luers, J.K. and Haines P.A. The effect of heavy rain on wind shear attributed accidents. AIAA-81-0390, 19th aerospace sciences meeting. St. Louis (MO), 1981.

[130] Luers, J.K. Heavy rain effects on aircraft. AIAA paper 83-0206, 1983.

[131] Marshall, J.S. and Palmer, W.K. The distribution of raindrops with size. Journal of Meteorology, 5(4), 165–166, 1948.

[132] Joss, J. and Waldvogel, A. Raindrop size distribution and sampling size errors, Journal of the Atmospheric Sciences, 26(3), 566–569, 1969.

[133] Bezos, G.M., Dunham, R.E., Gentry, G.L., and Melson, W.E. Wind tunnel aerodynamic characteristics of a transport-type airfoil in a simulated heavy rain environment, NASA TP-3184, 1992.

[134] Dana, L.S., Dunham, J., Earl Dunham, R., Jr., and Bezos, G.M. A Summary of NASA Research On Effects of Heavy Rain on Airfoils, AGARD Conference Proceedings AGARD-CP-496: Effects of Adverse Weather on Aerodynamics, 1991, pp. 15.1–16.16.

[135] Reinmann J.J. Effects of adverse weather on aerodynamics. In: AGARD Conference Proceedings 1991, 496.

[136] Roys, G.P. and Kessler, E. Measurements by Aircraft of Condensed Water in Great Plains Thunderstorms. National Severe Storms Laboratory Publications. TN49-NSSP-19, 1966.

[137] Huffman, P.J. and Haines, P.A. Visibility in heavy precipitation and its use in diagnosing high rainfall rates. AIAA-84-0541, AIAA 22nd Aerospace Sciences Meeting. Reno (NV), 1984.
[138] Cao, Y. Zhenlong, W., and Zhengyu, X. Effects of rainfall on aircraft aerodynamics. Progress in Aerospace Sciences, 71, 85–127, 2014.
[139] Haines P.A. and Luers, J.K. Aerodynamic penalties of heavy rain on landing airplanes. JAircrc, 20(2), 111–119, 1983.
[140] Trammell A. Three unconventional airborne radar clues to severe convective storms. AIAA-89-0705, AIAA 27th aerospace sciences meeting. Reno (NV), 1989.
[141] Clayton, R.M., Cho, Y.I., Shakkottai, P., et al. Rain simulation studies for high-intensity acoustic nose cavities. J Aircraft, 25(3), 281–284, 1988.
[142] Spillard, C.L., Gremont, B., Grace, D., et al. The performance of high-altitude platform networks in rainy conditions, AIAA2004-3220, 22nd AIAA inter- national communications satellite systems conference and exhibit, Monterey (CA), 2004.
[143] Adams, K.J. The airforce flight test center palletized airborne water spray system, AIAA-83-0030, AIAA 21st aerospace sciences meeting, Reno (NV), 1983.
[144] Peterson, G.P. and Maralo, S.A. Rain Erosion Flight Test Program, WADC Technical Report 58-454, May 1959.
[145] Roys, G.P. Operation of an F106A. In Thunderstorms at Supersonic and High Subsonic Speeds, ASD Technical Note 61-97, October 1961.
[146] Schmitt, G.F., Jr. Flight Test-Whirling Arm Correlation of Rain Erosion Resistance of Materials, AFML-TR-67-420, September 1968.
[147] Aircraft Surface Coatings Study: Energy Efficient Transport Program, NASA CR-158954, Contract NAS 1-14742, Task 4.1.3, January 1979.
[148] Hogg, D. Aircraft Composites Damage Through Rain Erosion, Aircraft Composites – Damage Through Rain Erosion, www.azom.com/article.aspx?ArticleID=1740
[149] Bilanin, A.J., Quackenbush, T.R., and Feo, A. Feasibility of predicting performance degradation of airfoils in heavy rain. NASA CR 181842, 1989.
[150] Feo, A. Single Waterdrop Collision Experiments, Instituto Nacional De Tecnica Aeroespacial Nota Technica No. 221/510/85.009, Madrid, Spain, April 1985.
[151] J. Gibson. Testing of Rain Erosion Resistance of Aircraft Coatings. Applied Science Research Laboratory, University of Cincinnati, Ohio. Final Report 15 December 58. FCSTI AD-210 362.
[152] Robertson, R.M., Lobisser, R.J., and Stein, R.E. High Speed Rain Abrasion of Glass Cloth Laminates. Industrial and Engineering Chemistry, 38, 590, June 1946.
[153] Bullis, L.H. Rain Erosion Test of Lockheed ATC-1 Conductive Coating. Materials Lab., WADC. Technical Note WCRT 53-180. September 53. FCSTI AD-30 339.
[154] Air Force Research Laboratory (AFRL), Materials and Manufacturing, Rain Erosion Test Apparatus, Use Policies, Operating Procedures & Specimen Configurations, Updated March 2019.
[155] Wetmore, W.C. German Scientists Investigate Materials, Aviation Week and Space Technology, 110–111, April 27, 1964.
[156] Rhode, R.V. Some Effects of Rainfall on Flight of Airplanes and on Instrument Indications. NACA TN 803, April 1941.
[157] Calarese, W. and Hankey, W.L. Numerical Analysis of Rain Effects on an Airfoil, AIAA-84-0539, AIAA 22nd Aerospace Sciences Meeting, Reno, NV, January 1984.
[158] Tobin, E.F. and Trevor M. Young and Dominik Raps, Evaluation And Correlation of Inter-Laboratory Results From a Rain Erosion Test Campaign, 28th International Congress of the Aeronautical Sciences, ICASE 2012.
[159] Wahl, N.E. Investigation of the Phenomena of Rain at Subsonic and Supersonic Speeds, Technical Report AFML-TR-65-330, October 1965.

[160] Bilanin, A.J. Scaling Laws for Testing of High Lift Airfoils Under Heavy Rain. AIAA-85-0259, January 1985.
[161] Bezos, G.M., Earl Dunham, R., Jr, and Gentry, G.L., Jr. Wind Tunnel Aerodynamic Characteristics of a Transport-Type Airfoil in a Simulated Heavy Rain Environment, NASA Technical Paper 3184, 1992.
[162] Earl Dunham, R., Jr. Influence of Environmental Factors on Aircraft Wing Performance, Von Karman Institute, Lecture Series, 1987-03, February 16–20, 1987.
[163] Arnold Air Force: Test Facility Guide, www.arnold.af.mil/Portals/49/documents/AFD-080625-010.pdf?ver=2016-06-16-100801-260
[164] Bezos, G.M. and Campbell, B.A. Development of a Large-Scale, Outdoor, Ground-Based Test Capability for Evaluating the Effect of Rain on Airfoil Lift, NASA TM 4420, 1993.
[165] FAA Advisory Circular (AC) 20-73A, Aircraft Ice Protection, August 16, 2006.
[166] Aircraft Icing: Learning Goal 3g. Explain how and where supercooled water forms, and explain how ice on aircraft affects flight, ATSC 113: Weather for sailing, flying and Snow Sports, www.eoas.ubc.ca/courses/atsc113/flying/met_concepts/03-met_concepts/03g-Icing/index.htm
[167] Guidelines for Aircraft Ground Icing Operations, Second Edition, TP 14052E (04/2005).
[168] Aviation Weather, FAA Advisory Circular AC No: 00-6B, U.S. Department of Transportation, Federal Aviation Administration, 8/23/16.
[169] Cao, Y., Tan, W., and Wu, Z. Review Aircraft icing: An ongoing threat to aviation safety. Aerospace Science and Technology, 75, 353–385, April 2018 www.sciencedirect.com/science/article/abs/pii/S1270963817317601?via%3Dihub
[170] Capobianco, G. and Lee, M.D. The role of weather in general aviation accidents: An analysis of causes, contributing factors and issues. Proceedings of the Human Factors and Ergonomics Society Annual Meeting, 45, 190–194, 2001. doi:10.1177/154193120104500241
[171] FAA Advisory Circular AC-00-6A, Chapter 10–12, www.faa.gov/documentlibrary/media/advisory_circular/ac%2000-6a%20chap%2010-12.pdf
[172] Fultz, A.J. Fatal Weather-related General Aviation Accidents in the United States: 1982–2013, M.Sc. Thesis, Northern Illinois University De Kalb, Illinois, 2015.
[173] Federal Aviation Administration (2010). Weather-related aviation accident study 2003–2007. Retrieved from www.asias.faa.gov/pls/apex/f?p=100:8:0::NO::P8_STDY_VAR:2
[174] Nikisha Maria Nagappan, Numerical Modeling of Anti-Icing Using an Array of Heated Synthetic Jets, M.Sc. Thesis, Aerospace Engineering Department, Embry-Riddle Aeronautical University – Daytona Beach, 2013.
[175] Icing, www.weather.gov/source/zhu/ZHU_Training_Page/icing_stuff/icing/icing.htm
[176] Jones, A.R. and Lewis, W. Recommended Values of Meteorological Factors to be considered in the Design of Aircraft Ice-Prevention equipment, NACA TN No. 1855, March 1949.
[177] US FAA Aviation Maintenance Technician Handbook – Volume 2 (2018), FAA-H-8083-31A. AMA. Chapter 15: Ice and Rain Protection.
[178] Ground Deicing Using Infrared Energy, FAA Advisory Circular, FAA AC 120-89, December 13, 2005.http://www.airweb.faa.gov/Regulatory_and_Guidance_Library/rgAdvisoryCircular.nsf/ACNumber!OpenView&Start=1
[179] Peterso, A.A. and Dadone, L.U. Helicopter Icing Review, Boeing Vertol Company, FAA-CT-80-210, AD-A094175, June 1980 https://apps.dtic.mil/dtic/tr/fulltext/u2/a094175.pdf
[180] Lester, P.F. Aviation Weather, 4th edition, Jeppesen, 2013.

[181] A Pilot's Guide to Ground Icing, Module I – Risks, How Ground Icing Can Hurt You, Section: Contamination Penalties, https://aircrafticing.grc.nasa.gov/2_1_2_1.html
[182] Kaushik Das, Numerical Simulations of Icing in Turbomachinery. Ph.D. Dissertation, Cincinnati University, April 19, 2006.
[183] Aircraft Icing Handbook, Version 1, Civil Aviation Authority, June 14, 2000.
[184] A Pilot's Guide to Ground Icing, Module IV – De-Icing Operations, How to Take IT Off, Section: Mechanical De-Icing, https://aircrafticing.grc.nasa.gov/2_4_3_1.html
[185] International De/Anti-icing Chapter, www.faa.gov/other_visit/aviation_industry/airline_operators/airline_safety/deicing/media/standardized_international_ground_deice_program.pdf
[186] https://en.wikipedia.org/wiki/De-icing#/media/File:A_U.S._Army_C-37B_aircraft_transporting_Army_Chief_of_Staff_Gen._Raymond_T._Odierno,_gets_de-iced_before_it_departs_Joint_Base_Elmendorf-Richardson,_Alaska.jpg
[187] A Pilot's Guide to Ground Icing, Module III – Fluid Basics, which fluid is right for your aircraft, https://aircrafticing.grc.nasa.gov/2_3_3_1.html
[188] Rosenthal, H.A., Nelepovitz, D.O., and Rockholt, H.M. De-Icing of Aircraft Turbine Engine Inlets, U.S. Department of Transportation, Federal Aviation Administration, DOT/FAA/cr-87/~, June 1988.
[189] Nelepovitz, D.O. and Rosenthal, H.A. Electro-Impulse De-icing of Aircraft Engine Inlets, AIAA Paper No. 85-0546, January 1986.
[190] Colin Cutler, Types Of Deicing Equipment, And Their Advantages And Disadvantages, Boldmethods, January 30, 2020, www.boldmethod.com/blog/lists/2020/01/types-of-deicing-systems-and-advantages-and-disadvantages/
[191] FAA Advisory Circular (AC) 135-17, PILOT GUIDE Small Aircraft Ground Deicing, December 14, 1994.
[192] Kazda, T. and Caves, B. Airport Design and Operation, Third Edition, Emerald Group Publishing Limited, 2015.
[193] FAA Advisory Circular (AC) 150/5300-14D, Design of Aircraft Deicing facilities, March 17, 2020.
[194] Peterson, A.A. and Dadone, L.U. Helicopter Icing Review, Final Report, FAA-CTMO-210, September 1980.
[195] Ice Protection Investigation for Advanced Rotary-WIN6 Aircraft, AD-771 182, Prepared for Army Air Mobility Research and Development Laboratory, August 1973.
[196] Soltis, J.T., Palacios, J. and Wolfe, D.E. Design and Testing of an Erosion Resistant Ultrasonic De-Icing System for Rotorcraft Blades, Applied Research Laboratory, The Pennsylvania State University, Technical Report No. TR 14-007, August 2013.
[197] Szefi, J. and Palacios, J. Centrifugally Powered Pneumatic Deicing for Helicopter Rotor Blades, NASA Aeronautics Research Mission Directorate (ARMD) FY12 LEARN Phase I Technical Seminar November 14, 2013, https://nari.arc.nasa.gov/sites/default/files/PALACIOS_LEARN.pdf
[198] ICAO – Annex 14 – Aerodromes – Volume I – Aerodromes Design and Operations 8th Edition, July 2018.
[199] ICAO-Airport Services Manual-Part 2: Pavement Surface Conditions, 4th Edition, 2002, ICAO Doc 9137-AN/898.
[200] ICAO-Manual of Aircraft Ground De-icing/Anti-icing Operations, Doc 9640 AN/940, Third Edition, 2018.
[201] FAA Advisory Circular (AC) 150/5200-30D, Airport Field Condition Assessments and Winter Operations Safety, November 14, 2016.
[202] Athmann, T., Bjornsson, R., Borrell, P. and Thewlis, P. Geothermal Heating of Airport Runways- (FAA) Design Competition, http://emerald.ts.odu.edu/Apps/FAAUDCA.nsf/AcevesDADEFullProposal.pdf?OpenFileResource

[203] Heymsfield, E., Daniels, J.W., Saunders, R.F., and Kuss, M.L. Developing anti-icing airfield runways using surface embedded heat wires and renewable energy, Sustainable Cities and Society, Volume 52, January 2020, https://www.sciencedirect.com/science/journal/22106707https://www.sciencedirect.com/science/journal/22106707/52/supp/C101712, https://doi.org/10.1016/j.scs.2019.101712
[204] ICAX™ Solar Runway Systems clear ice from aircraft parking stands using under runway heating, www.icax.co.uk/Solar_Runways.html
[205] Rossen, J. and Bomnin, L. Culd heated airport runways melt away your winter travel headaches?, Today, January 26, 2018 www.today.com/money/could-heated-runways-melt-away-your-winter-travel-headaches-t121729
[206] FAA Advisory Circular (AC) 91-74B, Pilot Guide: Flight in Icing Conditions, October 8, 2015.
[207] FAA Advisory Circular (AC) 120-58, Pilot Guide Large Aircraft Ground Deicing, September 30, 1992.
[208] In flight Icing, for General Aviation Pilots, GA 10, European General Aviation Safety Team (EGAST), July 2015.
[209] Natops Flight Manual Navy Model F−14D Aircraft, NAVAIR 01−F14AAD−1, January 15, 2004.
[210] KALLAX, Sweden, Ice and Snow Removal Is Key Before Fighters Take Off, NATO, https://ac.nato.int/archive/2018/ice-and-snow-removal-is-key-before-fighters-take-off
[211] Hail Threat Standardization, Final Report EASA_REP_RESEA_2008_5 Research Project.
[212] Grunbaum, M. Weather-Studying Warthog: A Fixed-Up A-10 Will Fly Into Thunderstorms, popular mechanics, November 18, https://www.popularmechanics.com/author/7568/mara-grunbaum/2011, www.popularmechanics.com/science/environment/a7485/weather-studying-warthog-a-fixed-up-a-10-will-fly-into-thunderstorms/
[213] Thomson, R.G. and Huydzdk, R.J. An Analytical Evaluation of the Denting of Airplane Surfaces by Hail, NASA-TN D-5363, 1969.
[214] Souter, R.K. and JEmerson, J.B. Summary of Available Hail Literature and the Effect of Hail on Aircraft in Flight, National Advisory Committee For Aeronautics, NACA TN 2734.
[215] Shengze Li, Feng Jin, Weihua Zhang and Xuanzhu Meng, Research of hail impact on aircraft wheel door with lattice hybrid structure. Journal of Physics: Conference Series, 744 (2016) 012102, doi:10.1088/1742-6596/744/1/012102
[216] Neidigk, S.O., Roach, D.P., Duvall, R.L., and Rice, T.M. Detection and Characterization of Hail Impact Damage in Carbon Fiber Aircraft Structure, DOT/FAA/TC-16/8, Department of Transportation, FAA, September 2017.

5 Drones, UAVs, and Space Debris

5.1 INTRODUCTION

Unmanned Aircraft Systems (UAS) or drones are no longer science fiction but became a part of our daily life. They have gained great importance both in civil and military applications. For civil UAS, there are favorable applications, but there is also the hazard of its strike with civil aircraft. At the same time, military UAS work as a team with military manned aircraft. The civil unmanned aerial vehicles (UAVs) market is developing at exponential growth with business applications and hobbyist market. The global Unmanned Aircraft Systems (UAS) market size is projected to reach US$ 6,320 million by 2026, from US$ 4,542 million in 2020, at a CAGR of 5.7% during 2021–2026 [1].

There are many beneficial civilian applications of commercial and public small unmanned aircraft systems (sUAS) in uncontrolled low-altitude airspace, including the film industry, aerial photography, mapping, surveying, environmental conservation, an inspection of technical infrastructure like wind farms, real estate, precision agriculture and private and public security, communications, and parcel delivery. They proved to be a great benefit in humanitarian efforts, such as searching and rescuing, and delivering medical supplies in developing nations especially during the COVID-19 pandemic [2, 3].

Also, it proved a reasonable success in deterring birds and ground-based animals and observing patterns of wildlife on or near airports [4].

Though business professionals have no prior knowledge of aviation, they follow the required training and observe all the safety procedures when operating their UAV. Hobbyists also do not have prior knowledge of aviation nor professional liabilities, but they fly UAVs that can fly at several thousand feet in height. Thus, millions of such UAVs have been sold and are in the hands of people unaware of the damages they could cause to civil aviation.

Drones or small UAVs can pose the following risks to aviation (including airports and aircraft):

- Safety risk to aviation due to the growing number of amateur drones, including near misses and collisions with buildings, personnel on the ground, passenger of aircraft and helicopters

DOI: 10.1201/9781003133087-5

- Disruption of air transport due to prompt closure of runways is repeatedly done with busy airports and hubs like Dubai and Gatwick airports [5, 6].

The collision between a UAV and a manned aircraft could lead to the damage of its radome, front exterior of cockpit, wing and empennage leading edges, flaps, stabilizers, and the engines or propulsion system of the latter.

However, drones have several advantages in process improvements and cost efficiencies [7]. Applications include aircraft checks, calibration of navigational equipment (like ILS), ground lighting, runways, taxi ways inspections, pavement surface condition, obstacle surveys, construction work surveys, areas and perimeters surveillance, improved situational awareness during emergencies, and 3D mapping.

5.2 DEFINITIONS

1. Unmanned Aircraft System (UAS) – An unmanned aircraft and the equipment necessary for the safe and efficient operation of that aircraft. An unmanned aircraft is a component of a UAS. It is defined by statute as an aircraft operated without the possibility of direct human intervention from within or on the aircraft (Public Law 112-95, Section 331(8))
2. sUAS are defined as less than 55 lbs. It is categorized into three classes:
    a. Class A (under 4.4 lbs)
    b. Class B (under 20 lbs)
    c. Class C (under 55 lbs)
3. Hazard – Any real or potential condition that can cause injury, illness, or death to people; damage to or loss of a system, equipment, or property; or damage the environment. A hazard is a prerequisite to an accident or incident. For unmanned aircraft weighing less than 300 lbs, damage to the unmanned aircraft itself is not considered
4. Accident – An unplanned event or series of events that result in death, injury, or damage to, or loss of, equipment or property
5. Incident – An occurrence other than an accident that affects or could affect the safety of operations
6. Cause – One or several mechanisms that trigger the hazard that may result in an accident or incident (the origin of a hazard)
7. Airspace closure – for an airport is a period during which no aircraft is permitted to operate to and from that airport
8. Incident's categorization:

Incidents are categorized based on its close encounters or sightings as follows:

A. Close encounters

It poses a hazardous threat to a manned aircraft and includes two cases:

- Near Mid Air Collision (NMAC)

  As defined by the Federal Aviation Administration (FAA) [8], NMAC is an incident associated with the operation of an aircraft in which a possibility of a collision occurs as a result of the proximity of fewer than 500 ft to another aircraft, or a report is received from a pilot or flight crew member stating that a collision hazard existed between two or more aircraft. The pilot declares an NMAC and takes evasive action, or when the pilot uses descriptive language that indicates the drone as being dangerously close (like: “almost hit” or “passed just above”)

- Mid Air Collision (MAC)
  As defined by SKYbrary [9], MAC as an accident where two aircraft come into contact with each other while both are in flight.

B. Sightings

A UAV is in the vicinity of an airport or manned aircraft as seen by a pilot or traffic control but does not pose an immediate collision threat. It also includes two cases:

- Jetliner sighting
  As defined in [2], it is the action of taking pictures and videos with the camera embedded in a UAV for an aircraft, typically a jetliner, during its landing and taking off flight phases. Those pictures and videos are next posted on social media.
  - Airport indoor sighting
  As defined in [2], it is the action of taking pictures and videos with the camera embedded in a UAV flying inside an airport hall.

9. Accident characteristics

Accidents are characterized by:

- Serious injury or fatality to any person or
- Substantial damage to another aircraft

Any mishap that is not an accident is an incident.

10. Controlled and uncontrolled airspace

In controlled airspace, air traffic controllers (ATC) are responsible for separating aircraft flying under instrument flight rules (IFR) from one another. In uncontrolled airspace, ATC does not provide that service. Airliners normally fly on an instrument flight plan, IFR, and in controlled airspace. In rare cases, especially in remote areas, air carrier flights operate in uncontrolled airspace or under visual flight rules. In either cases, pilots should check any other air traffic and avoid it.

In the United States, Classes A, B, C, D, and E airspace are controlled. Class G airspace is uncontrolled [10].

11. Controlled airspace

The FAA categorizes the airspace into five classes of controlled airspace, coded A to E [11].

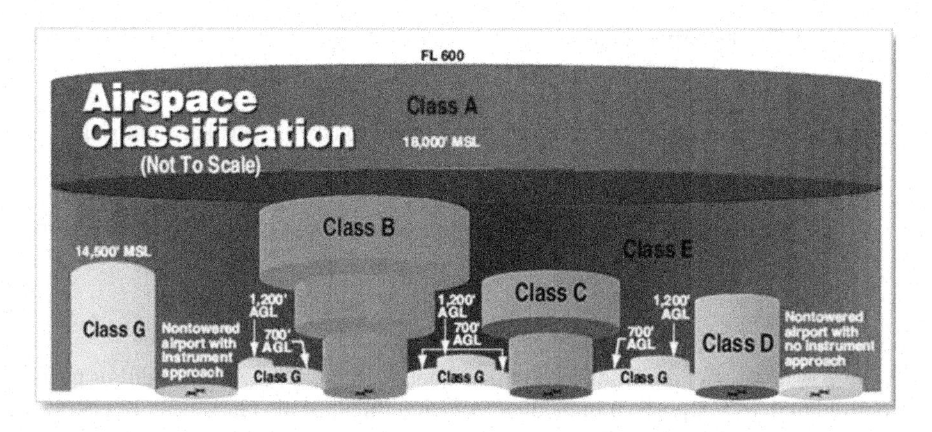

FIGURE 5.1 US Airspace Classification [12].

Courtesy: Department of Defense.

Class A airspace starts Flight Level (FL) 180 (18,000 ft mean sea level [MSL]) to FL 600 (60,000 ft MSL). Altitudes that drones typically do not reach. This airspace is mostly used by commercial airlines for long flights between different cities or countries (Figure 5.1). All operations are subject to ATC clearance, and all flights are separated from each other by ATC.

Classes B, C, D, and E are mainly differentiated by the level of activity of their included airports.

Class B airspace is generally airspace from the surface to 10,000 ft MSL surrounding the country's busiest airports, such as those in the major air travel hubs like New York and Los Angeles. Class B usually has the widest extent – a radius of around ten nautical miles measured from the airport's location. One can think of the "B" in Class B airspace as standing for "big city" airports. All operations are subject to ATC clearance, and all flights are separated from each other by ATC.

Class C airspace is generally airspace from the surface to 4,000 ft above the airport elevation (AGL) surrounding those that have an operational control tower, serviced by a radar approach control, and have a certain number of IFR operations or passenger enplanements. You can think of the "C" in Class C airspace as standing for "cities," sizable city airports.

Class D airspace is assigned to the areas surrounding the smallest airports (generally up to 2,500 ft AGL) with a functioning control tower. In most cases, the extent of Class D airspace is only within a 5-mile diameter with the airport as the center. One can associate the "D" in Class D airspace with "diminutive" or "dime-sized."

Thus, to flight in controlled airspace Classes B, C, and D are prohibited to avoid close encounters between drones and manned aircraft unless one gets an authorization from the relevant ATC facility.

Although Class E airspace can extend to the surface, it generally begins at 700 ft AGL, 1,200 ft AGL, 10,000 ft MSL, or 14,500 ft MSL and extends upward until it meets higher class airspace (A–D). Class E airspace runs into the upside-down wedding cake airspace of Class B or Class C. For a drone pilot to operate in Class E airspace, he must get permission.

Uncontrolled airspace in Class G may extend to 14,499 ft MSL but generally exists below 1,200 t AGL and Class E airspace. ATC separation is not provided. Traffic information may be given as far as is practical in respect of other flights.

5.3 STATISTICS FOR INCIDENTS OF CIVILIAN DRONES/UAVS

A quantitative and qualitative analysis of the population of worldwide UAV incidents is performed in several worldwide studies to identify the characteristics of the UAV incidents at civilian airports. Sources for such analysis were the study in [13] and the FAA and NASA databases [8, 14].

5.3.1 THE PERIOD 2010–2015

A detailed analysis for the mishaps (accidents and incidents) of sUAS in the period from 2010 to 2015 was assembled in a database. A summary of its details is given hereafter:

- 100 mishaps have complete data (96 incidents and four accidents)
  - 15 incidents due to flight controls
  - 14 mishaps (11 incidents, 2 accidents, and 1 fatal accident) due to flight crew which includes remote pilot-in-command, another pilot manipulating controls, and any visual observers designated by the pilot-in-command to see and avoid other air traffic or objects
  - 9, 8, and 6 incidents due to propulsion, lost the link and software, respectively
  - Two incidents for each of the sensors: wind shear and remote control
  - 10 for other reasons
  - 31 incidents and 1 fatal accident due to undetermined reasons
- Different configurations for UAVs resulted in the following mishaps
  - 33 incidents and 2 accidents for multi-rotor
  - 33 for fixed-wing
  - 7 incidents and 2 accidents for helicopters
  - 5 for hybrid
  - 1 for thrust vector
  - 17 not reported
- Categorization by mission
  - 34 Research & Development
  - 23 incidents, 2 accidents, 2 fatal accidents for personal use
  - 9 incidents for aerial photography
  - 6 incidents for each of aerial survey/observation, law enforcement, and training
  - 2, 3, and 7 incidents for illegal activities, unknown
- The outcome of the mission
  - 19, 18, 3, and 4 incidents due to collision with terrain, obstacle, person, and ground vehicle, respectively

- 2 non-fatal accidents as the UAS struck and injured people after a collision with terrain and obstacle
- 13 incident for each of the uncontrolled descent and crash in the landing area
- 10, 6, 5, 3, and 2 incidents occurred when returning to base, flight termination, landing without further incident, air conflict, and unknown

5.3.2 Drones in the USA in the Period December 17, 2013–September 12, 2015

A comprehensive and detailed analysis of unmanned aircraft incidents and manned aircraft in the U.S. National Airspace System is given in [13]. The Federal Aviation Administration (FAA) and Department of the Interior assembled the reports submitted by pilots and air traffic controllers for drones and manned aircraft in the national airspace.

According to these criteria, the following findings were deduced:

- 921 incidents
  - 327 (35.5%) were Close Encounters
  - 594 (64.5%) were Sightings
- 213 incidents in 2014 (70 in the first 7 months)
- 707 incidents in 2015 (585 in the first 7 months)
- 137 reports in October 2015 alone
- 90.2% of incidents occurred above 400 ft, though the FAA's ceiling for unmanned aircraft is 400 ft
  - 61.6% of incidents occurred between 400 ft and 4,000 ft
  - The highest recorded altitude for an incident was 29,000 ft, a Sighting at Alma, Georgia, on May 11, 2015
- 665 incidents have known distance from an airport
  - 391 (or 58.8%) occurred within 5 miles from the nearest airport (FAA prohibits the use of unmanned aircraft within 5 miles of any airport in the United States without permission from air traffic control (ATR))
  - 273 (41.2%) occurred beyond 5 miles from the nearest airport
- Incidents within 5 miles of airports occurred at lower altitudes than incidents beyond 5 miles of an airport.
- Most accidents occur in areas where manned air traffic density is high and where drone use is prohibited.
- 241 reports that indicated a drone-to-aircraft proximity 158 incidents in which a drone came within 200 ft or less of a manned aircraft (two-thirds of all Close Encounters in which a concrete drone-to-aircraft proximity is given)
- 51 incidents in which the proximity was 50 ft or less
- 28 incidents in which a pilot maneuvered (taking evasive action) to avoid a collision with a drone
- 116 cases of the Close Encounters involved multiengine jet aircraft, with 90 of which were commercial aircraft having 50 or more passenger's capacity
- 38 Close Encounter incidents involving helicopters

- The reports do not always clearly identify the type of drone involved in incidents
- Out of the 340 drones identified, 246 were multi-rotors (quadcopters, hexacopters, etc.) and 76 were fixed-wing, and 17 involved helicopter-type drones
- For close encounters aircraft:
  - 116 Close Encounters involved multiengine jet aircraft (small corporate jets to large commercial airliners)
  - 125 involved single-engine propeller aircraft
  - 23 involved multiengine propeller aircraft
  - 38 involved helicopters
- Metropolitan areas were subjected to the highest number of incidents
- 877 incidents for which a time was reported
  - 283 (32.3% percent) occurred between 10 a.m. and 2 p.m. local time
  - 298 (34%) of incidents occurred between 2 p.m. and 6 p.m.
  - 194 (22.1%) occurred between 6 p.m. and 10 p.m.
  - The remaining 11.7% of incidents occurred between 10 p.m. and 10 a.m.

5.3.3 Drones in Europe in the Period from May 2014 to May 2018

The two FAA and NASA databases are adopted here, and only accidents where the manned aircraft pilot declared NMAC were selected. A population of 139 UAV incidents in the vicinity of worldwide civilian airports (less than 35 km) has been considered [2].

The findings of this study can be summarized as follows:

- From the 139 UAVs incidents: 91 cases from the FAA database, 11 from the NASA database, 36 from online news, and 1 from another database
- Incidents distributed over the number of involved UAVs
  - 132 one UAV
  - 5 two UAVs
  - 1 three UAVs
  - 1 four UAVs
- Incident may be classified as:
  - 2 MAC
  - 124 NMAC
  - 2 jetliner sighting incidents
- Types of involved manned aircrafts
  - 46 multi-jet engine aircraft
  - 41 mono–propeller (private) aircraft
  - 9 helicopter private
  - 8 multi-propeller private aircraft
  - 5 multi-propeller airliner
  - 1 military
  - 3 helicopter military
  - 8 unknown

TABLE 5.1
Incidents Distributed Over the Height Interval [2]

Height (ft)	0–500	500–1,000	1,000–1,500	1,500–2,000	2,000–2,500	2,500–3,000	3,000–3,500	3,500–4,000	4,000–5,000	5,000–7,000	7,000–9,000	9,000–12,500
Incident	12	12	16	8	10	14	1	9	9	7	6	4

TABLE 5.2
UAV Incidents Distributed Over the Distance Intervals Between Manned and Unmanned Vehicle

Distant (ft)	Collision	1–25	26–50	51–75	76–100	101–200	201–300	301–400	10,00
Incident	2	20	15	6	21	14	4	4	1

- The most economically impacted airports due to UAVs incidents are Dubai, Gatwick, Sharjah, Chengdu Shuangliu, Chongqing, and Auckland
- Airports with the biggest UAV incidents are:
  - 13 for total London airports
  - 8 for total New York airports
  - 4 Miami International Airport (United States) Nombre
  - 3 for each of the international airports of Dubai, Logan (Boston), Los Angeles, and Philadelphia
- Table 5.1 displays the distribution of incidents based on their height. The worst height is in the range of 1,000–1,500 ft, followed by 2,500–3,000 ft.
- Table 5.2 presents the UAV incidents over the intervals of the distance between the UAV and the manned aircraft (for known cases).
- Regarding the distance between UAV and the nearest airport:
  - Average 9.47 km: not available
  - Minimum range: 0 km (above the runway)
  - Maximum range: 31.5 km
- Concerning action taken by the pilot:
  - 100 no evasive action
  - 26 evasive action.

5.3.4 Drones in Australia in 2017

There have been 1,596 reported drone incidents in 2017, of which 131 were deemed concerns to aviation safety

5.4 EXAMPLES FOR UAV MISHAPS

- In February 2015, Boeing 757, operated by Delta Air Lines Flight 1559 flight from Honolulu to Los Angeles LAX airport, was on final approach at an altitude of 3,000 ft AGL when a drone flew by passing 150 ft to the right of the jet [15].

The first officer informed the tower that a drone had been sighted within the final approach to LAX.

- In 2016, Dubai International Airport (DXB) was closed three times due to illegal drone activities [16]. The airport was closed for 30 min in two incidents (June and September), while the third intrusion in October led to a 115-minute closure. With the financial losses of AED 350,000 ($95,368) a minute according to Emirates Authority for Standardization and Metrology estimates, the total loss, therefore, is estimated at AED 61 million ($16.62 million) [6].
- On Sunday July 9, 2017, a drone flying close to UK's busiest Gatwick Airport led to the closure of the runway, grounding and forcing flights to be diverted [17, 18].
- UK Airprox Board, the authority which monitors near-miss incidents, said 70 drones came close to aircraft over the UK in 2016, more than double the number for 2015. Thirty-three incidents have been reported in the period January–May 2017.
- Boeing 737-887 operated by Aerolineas Argentinas flight # AR1865 at 07:41 local time on November 11, 2017 collided with a drone while on approach to Buenos Aires-Jorge Newbery Airport, Argentina. The aircraft safely landed but suffered minor damage to the left-hand fuselage below the left-hand flight deck window [19].
- A Canadian Slyjet passenger plane on October 12, 2017 hit a drone while descending on Quebec City's Jean Lesage International Airport. The incident occurred around 3 km from the airport at an altitude of 450 m [20].
- On March 27, 2018, Boeing 777-200 operated by Air New Zealand flight # NZ92 flight from Tokyo was landing at Auckland airport in New Zealand and carrying 278 passengers came within 17 ft of hitting a hobby drone which was so close to the plane that the pilots thought it had been sucked into an engine. The pilots spotted the drone at a point in the descent where it was not possible to take evasive action.
- On March 6, 2017, all flight operations were halted for 30 min after an Air New Zealand pilot reported a drone flying within the airport's controlled airspace [21].
- On December 17, 2018, a Boeing 737 departed from Guadalajara, Mexico, to Tijuana. However, just before landing at Tijuana International Airport, a drone struck the aircraft at its nose causing damage to the radio and communications equipment inside the aircraft's nose [22].
- On September 17, 2017, a DJI Phantom 4 UAS collided with a US Army UH-60 Black Hawk at 300 ft AGL and 2.5 miles from the drone operator [23]. There was damage to the Black Hawk's main rotor blade, windshield, and fairings. Several components of the UAS were lodged in the Black Hawk. The helicopter landed safely. The UAS pilot did not hold a remote pilot certificate and was intentionally operating beyond the visual line of sight (BVLOS), which is prohibited under current regulations.
- Only three collisions between drones and manned aircraft have been confirmed in the United States. The above Black Hawk, a DJI Mavic piloted by an amateur drone operator, collided with a hot air balloon in Driggs, Idaho, and another

TABLE 5.3
Details of Some Mishaps [2, 3]

Date	Aircraft	Mission	Location	Phase of Flight	Occurrence
June 19, 2012	RQ-11B	Aerial Survey	Angeles, WA	Enroute	Collision w/ Obstacle Collision w/ Person
July 11, 2013	Gaui-FX7	Personal Use	Luzern SR U A	Unknown	Collision w/ Person
September 4, 2013	T-Rex700N	Personal Use	New York, NY	Maneuvering	Collision w/ Person
July 4, 2014	Unknown	Photography	Key West, FL	Unknown	Collision w/ Person
November 15, 2014	Unknown	Personal Use	Tuscaloosa, AL	Unknown	Collision w/ Person
April 7, 2015	Unknown	Photography	Australia	Maneuvering	Collision w/ Person
May 25, 2015	Unknown	Photography	Marblehead, MA	Unknown	Collision w/ Person
September 2016	Unknown	Unknown	Saint Paul Minneapolis International Airport (United States)	Take on	Near Midair Collision

Mavic collided with a Eurocopter AS350 helicopter as both filmed an off-road race in California.

Few mishap cases are described in Table 5.3.

5.5 UAS STRIKE VERSUS BIRD STRIKE

The impact energy for a collision of sUAS with a manned aircraft depends on the UAS mass and impact velocity. Normally, sUAS collisions cause greater structural damage than bird strikes for equivalent impact energy levels.

The differences between drone and bird strike with manned aircraft can be thoroughly identified as follows [24, 25]:

1. For the same volume, birds are lighter. In other words, drones are more densely packed and will cause more damage on impact to aircraft.
2. Drones have much sturdier metal parts. Those parts can damage wings, windshields, and other elements of the plane. However, not all drones are equal – small ones do not possess a lot of punching power.

3. The birds are more flexible and less dense and hence likely to retain their integrity after the impact and will most likely get deflected. This would lead to lesser damage to aircraft.
4. Drones have moving parts that are traveling at a much higher speed. The damage caused during impact is directly proportional to the square of relative speed. The relative speed is likely to be very high in the case of drones and hence will cause more damage to aircraft.
5. Most bird or drone strikes occur during the takeoff and landing phase of aircraft. Birds with strong survival instincts can spot the approaching aircraft and initiate avoiding action which further minimizes the damage they are likely to cause.
6. Large commercial jet engines include design features that ensure they can shut down after ingesting a bird weighing up to 1.8 kg. Engine shut down is a serious event because it fails multiple aircraft systems requiring emergency action by the flight crew.
7. Commercial aircraft structures are certified to withstand bird collision but not certified against drone strikes and their effect.
8. The drone strike may cause fire or explosion. As drones continue to explode in popularity, more costs and injuries may be enhanced.

A drone accident must be reported if there is a serious injury requiring hospitalization or a fatality; if there's the damage of more than $500 to any property other than the unmanned aircraft; or if the aircraft has a maximum gross takeoff weight of 300 pounds or more and sustains substantial damage.

5.6 UAS REGULATIONS

Since commercial and private enthusiast drone usage is escalating, the aviation industry is ringing alarm bells. Worldwide aviation authorities have issued numerous laws for flying drones that prohibit the usage of drones around airports. The main goal of such regulations is to ensure a reasonable level of safety. However, since consumers purchase millions of drones every year, it is hard to find the pilots of the unmanned aircraft and enforce the law even with the help of radars that detect small drones. Consequently, a lot of incidents are impossible to investigate further to determine the troublemakers.

5.6.1 Worldwide

- Three types of unauthorized UAS flights pose a threat to safe airport operations: clueless, careless, and criminal.
- Strict instructions are released by aviation authorities worldwide.
- In Canada, Transport Canada recommends that drones should be flown no higher than 90 m and at least 5.5 km away from anywhere aircraft may takeoff and land. Anyone wanting to fly a drone in Canada is subject to a set of safety regulations. Anyone caught endangering the safety of an aircraft risks getting hit with a CA$ 25,000 (US$ 12,000) fine [20].

- New Zealand's current rules state that drones must be kept at least 2.4 miles from any airdrome, a location from which aircraft flight operations take place. Anyone breaching these rules can be fined up to NZ$ 5,000 ($3,600) [21].
- In the UK, drone pilots could face on the spot fines of up to £1,000 for offenses such as not having or displaying a flyer ID on drones weighing over 249 g, not being able to provide proof of permissions and exemptions and, of course, for flying dangerously and in restricted locations [26].
- The FAA established special security instructions (on May 17, 2019) that restrict drone operations in airspace up to 2,000 ft MSL near US territorial and navigable waters. These new restrictions specifically prohibit drone flights in this airspace within a stand-off distance of 3,000 ft laterally and 1,000 ft above any US Navy vessel [27].
- In August 2019, the US military gave its bases permission to shoot down any drones they feel may threaten security or aviation safety.
- In Australia, fines of up to $1,110 can be issued per offense, and if matters are sent to court, he may be convicted of a crime, have demerits added to his license or certificate, and fined up to $11,100 [28]. Moreover, operating a drone in a hazardous way to other aircraft, the penalty can be up to two years in prison and a fine up to $26,640 for an individual. Moreover, it is illegal to shoot down or interfere with a drone, even if flying over his home or backyard. The penalty can be up to two years in prison and a fine of up to $26,640 for an individual. From January 28, 2021, a fine of up to $11,100 will be charged for flying an unregistered drone or without an operator accreditation (or remote pilot license) for business use or as part of your job.
- In Russia, the general rules for flying a drone are [29]:
  - All drones weighing more than 250 g (0.55 pounds) must be registered.
  - Fly only during the day with clear weather and visibility.
  - Drone pilots must maintain a direct visual line of sight with their drones while flying.
  - Do not fly over people or congested areas.
  - Do not fly over sensitive areas such as military installations.
  - Flights are not allowed on the Moscow Kremlin and Red Square.
- In China [30], according to China's national aviation authority, the Civil Aviation Administration of China (CAAC), flying a drone is legal in China, provided well awareness and compliance with the following drone regulations:
  - Any drones weighing 250 g (0.55 pounds) or more must be registered with the CAAC.
  - Licensing is required for commercial operations and in other scenarios.
  - Do not fly beyond your visual line of sight.
  - Do not fly above 120 m (394 ft).
  - Do not fly in densely populated areas.
  - Do not fly around airports, military installations, or other sensitive areas such as police checkpoints or sub-stations.
  - All drones are subject to China's "NO-Fly-Zones" or NFZs. Beijing is an NFZ.
  - Do not fly in controlled areas unless you have approval by the CAAC in advance.

5.6.2 FAA

FAA follows the vast increase of UAS and drones' numbers in the last decades closely. A snapshot of the current state of drones in the United States [31]:

- 869,517 Drones Registered
  - 371,091 Commercial Drones Registered
  - 498,426 Recreational Drones Registered
- 220,076 Remote Pilots Certified

1. In 1981, the FAA issued Advisory Circular 91-57, *a set of voluntary guidelines for model aviation hobbyists, such as staying under 400 ft and at a safe distance from airports [32]. Given the relatively small and controlled scale of model aircraft flights, there was little need for the federal government to police the modelling community closely. Under this arrangement, incidents involving remote control aircraft in the national airspace were exceedingly rare.*
2. In September 2015, the FAA updated Advisory Circular 91-57 with stricter *guidelines, including a provision affirming its right to "pursue enforcement action against persons operating model aircraft who endanger the safety of the National Airspace System." The FAA regulations identify 400 ft as the ceiling for remote control aircraft as an airspace rule to avoid the flight path of a manned aircraft. Drone operators without knowledge of airspace norms may not be aware that the sky directly above them is frequented by aircraft flying at low altitudes. Moreover, regulators and air traffic controllers are concerned that drone operators may not see and avoid other aircraft, an essential capability for any pilot.*
3. In June 2016, the FAA published its Small UAS Rule (14 Code of Federal Regulations (CFR) Part 107). While FAA Part 107 broadly authorizes low-risk commercial small UAS operations in the United States, the rule contains several key operating restrictions to maintain the safety of the National Airspace System and ensure that small UAS do not pose a threat to national security. Such key operational restrictions in Part 107 include the following [33]:
   - *Unmanned aircraft must weigh less than 55 pounds (lbs) (25 kg) including payload*
   - *Visual line of sight (VLOS) operations only*
   - *Daylight-only operations (official sunrise to official sunset, local time). Civil twilight operations (30 min before official sunrise to 30 min after official sunset, local time) are approved when the small UAS is equipped with lighted anti-collision lights*
   - *Must yield the right-of-way to other aircraft, all manned or unmanned public and military aircraft*
   - *UAS may not operate over any persons not directly involved in the operation*
   - *Maximum airspeed of 100 mph (87 knots).*

- *Maximum altitude of 400 ft above ground level unless flown within a 400-ft radius of a structure and no higher than 400 ft above the structure's immediate uppermost limit*
- *Minimum weather visibility of 3 statute miles from a control station*
- *No operations are allowed in Class A (18,000 ft and above) airspace*
- *Operations in Class B, C, D, and E airspaces are allowed with the required ATR) permission. ATR permission comes in the form of an airspace authorization*
- *Operations in Class G airspace are allowed without ATR permission*
- *UAS that weigh 55 pounds or more and do not hold an airworthiness certificate will require a Part 11 exemption from various sections of the CFRs and an accompanying Certificate of Waiver or Authorization (COA) as well as a Special Authority for Certain Unmanned Systems exemption (49 U.S.C. §44807). These exemptions were previously obtained under "Section 333" exemptions; however, Section 347 of the FAA Reauthorization Act of 2018 repealed Section 333 of the FAA Modernization and Reform Act of 2012.*

5.7 STUDIES OF UAV STRIKE WITH MANNED AIRCRAFT

5.7.1 Introduction

The collision between a drone and an aircraft becomes a daily threat to the public and aviation authorities at all levels. This collision depends mainly on the drone's weight, the type of aircraft, and the module impacted.

A light drone would do only light damage to an airliner, which can be quickly repaired and checked before the next flight. In contrast, a heavy drone collision with an airliner may be catastrophic due to solid-to-solid impact. A drone weighing several dozens of pounds and moving at a speed of 150 knots could severely damage the aircraft parts. For example, if a drone penetrated the cockpit windows, it may injure the pilots. General aviation (GA) aircraft are exposed to more risk due to their light structure and flight at lower height closer to the drones. Besides, if they are powered by only one piston engine fitted with a propeller, the impact of a drone in the propeller will force the plane to an emergency landing without an engine.

While the effects of bird impacts on airplanes are well documented, little is known about the effects of more rigid and higher mass UAS on aircraft structures and propulsion systems. For this reason and during the last decade, airborne hazard severity thresholds for collisions between unmanned and manned aircraft have been analyzed by some researchers worldwide, including the United States and Singapore [33–36]. A detailed study for the impact of a small UAS with manned aircraft is performed in the last couple of years by the FAA Center of Excellence ASSURE (ASSUREuas.org). The study handling UAS-Five US universities conducted aircraft collision. The study included the impact of two types of sUAS (multi-rotor small quadcopter and a small fixed-wing UAS) with two manned aircraft (commercial transport jet and a typical business jet aircraft) certified under *14 CFR Part 25* or *Part 23* requirements [34]. Ingestion into the engines of jet engines represents the most critical case. Detailed studies for sUAS ingestion into the high bypass turbofan engine are described in detail in [35–38].

5.7.2 Case Study 1

An extended project supported by FAA and performed through collaboration between some US universities is carried out to criticize the hazard of strikes between UAVs and manned aircraft.

The finite element (FE) model of the small UAV (quadcopter and fixed-wing) was developed for the airborne collision studies [39–41]. This research evaluates sUAS (under 55 lbs) collisions on commercial and business jet airframes and propulsion systems. Two sUAS were selected, namely, the DJI Phantom 3 quadcopter and the Precision Hawk Lancaster HawkEye Mark III is a lightweight fixed-wing UAS. The DJI Phantom 3 was constructed with a polycarbonate plastic body/casing that mounts four electric motors, a Lithium-Polymer (LiPo) battery, and a camera with a metallic casing. Its mass, dimensions, and speed were 1,216 g, 290 × 289 × 186 mm, and 16 m/s (31 knots), respectively. The Precision Hawk Lancaster HawkEye Mark III consists of a forward fuselage structure, polystyrene wings, vertical tail, horizontal stabilizer, and carbon/epoxy composite wing spars and tail booms. Its mass, wingspan, length, and speed were 1,800 g, 1,500, 800 mm, and 19.5 m/s (37.9 knots), respectively. Concerning commercial aircraft, the narrow-body single-aisle aircraft such as the Boeing 737 or the Airbus 320 families are chosen to be the most popular commercial transport jets in use throughout the world. Learjet 31A aircraft was selected as a representative for the business jet category.

The high-velocity impact may be either at landing/takeoff or at holding flight phases. For these cases, the maximum flight speed is limited to 208 knots at an altitude of 2,500 ft. The worst-case scenario is encountered when a frontal impact between the aircraft and UAS is encountered. Then, the impact velocity will be the sum of speeds of both bodies, namely 250 knots. FE models are developed for both quadcopter UAS and the wing leading edge FE model. Simulation is validated through simulation and testing by NIAR in a previous project for bird strikes [38].

A parametric study with over 140 impact scenarios was carried out. Figure 5.2 categorizes the damage severity into four levels.

- Level 1 generally corresponds to a minimal amount of localized damage.
- Level 2 represents significant visible damage to the external surface of the aircraft with minor internal damage but no appreciable skin rupture.
- In Level 3, the aircraft's outer surface allows ingress of foreign objects into the airframe, with some damage to the substructure.
- Level 4 includes all the preceding aspects and extensive damage to internal components and possibly compromising part of the primary structure.

5.7.2.1 Airframe Impacts (Case Study 1)

The selected areas for impact on the commercial transport jet were the vertical stabilizer, horizontal stabilizer, wing leading edge, and windshield. Two UAS were

Severity	Description	Example
Level 1	• Airframe undamaged. • Small deformations.	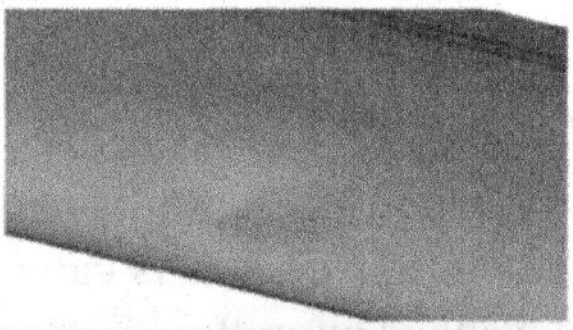
Level 2	• Extensive permanent deformation on external surfaces. • Some deformation in internal structure. • No failure of skin.	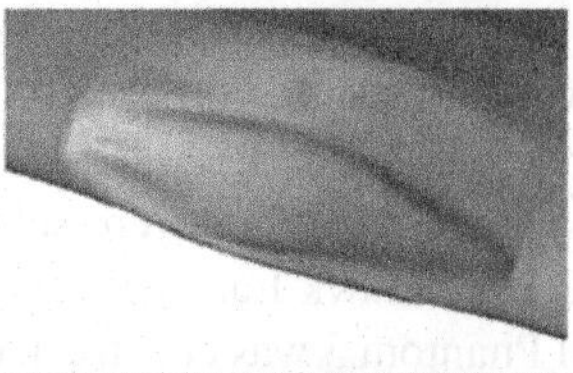
Level 3	• Skin fracture. • Penetration of at least one component into the airframe.	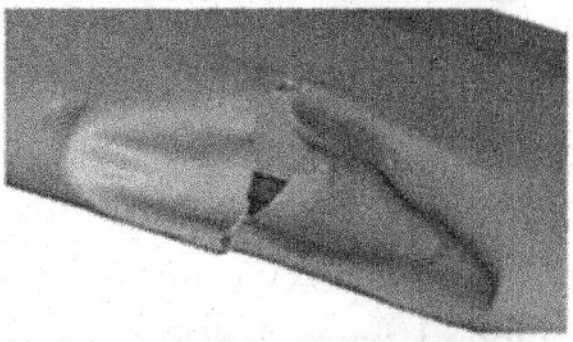
Level 4	• Penetration of UAS into airframe. • Failure of parts of the primary structure.	

FIGURE 5.2 Damage level categories [33].

Courtesy: U.S. Department of Transportation, FAA.

used: namely, 1.2 kg quadcopter and 1.8 kg fixed-wing UAS. Both were having an impact velocity of 250 knots. Sixteen explicit dynamic simulations of impacts were adopted: windshield: 3, wing: 4, horizontal stabilizer 5, and vertical stabilizer:4.

A summary of the severity levels for the quadcopter UAS striking the commercial transport aircraft is as follows:

- Windshield: level 2 severity for all the four impact locations
- Wing: level 2 for one impact close to wing tip, and level 3 for the other three impacts
- Horizontal stabilizer: level 4 for the three impacts close to the stabilizer tip and level 3 for the two impacts close to the fuselage
- Vertical stabilizer: level 3 for all impact locations.

Higher or equal severity levels were encountered for fixed-wing UAS striking the commercial transport aircraft. Same level 2 for the windshield, level 3 for all wing points, level 4 for all horizontal stabilizers, and level 4 for three impacts on the vertical stabilizer and only one point with level 3.

Business jet has the same trend as commercial transport, except that the wind shield experienced the highest severity damage (level 4) when impacted by the fixed-wing UAS.

A FE model of a quadcopter and fixed-wing UAS is developed to analyze the collision simulations with manned aircraft in LS-DYNA software. Moreover, the FE model is employed for global aircraft.

A series of impact scenarios were set up to characterize the dynamic event of a midair collision [39].

A comparison between the damage severity due to the strike of the vertical stabilizer of the business jet with a bird, a quadcopter UAS, and a fixed-wing UAS is illustrated in Figure 5.3. All have the same mass (4 lbs) and impact speed (250 knots). The impact severity for the bird impact was level 2, while both UAS resulted in level 4.

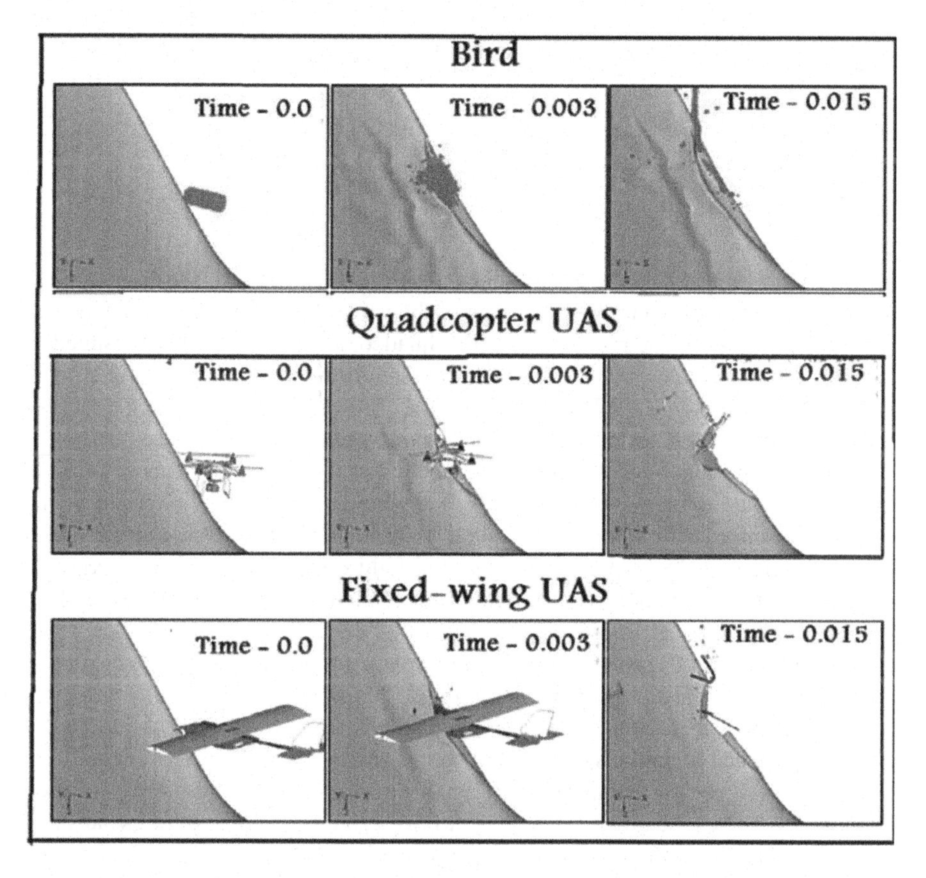

FIGURE 5.3 Comparison of impact kinematics for 4.0 lbs (1.8 kg) projectiles: bird (top), quadcopter (middle), and fixed-wing UAS (bottom) [38].

Courtesy: Department of Transportation, FAA.

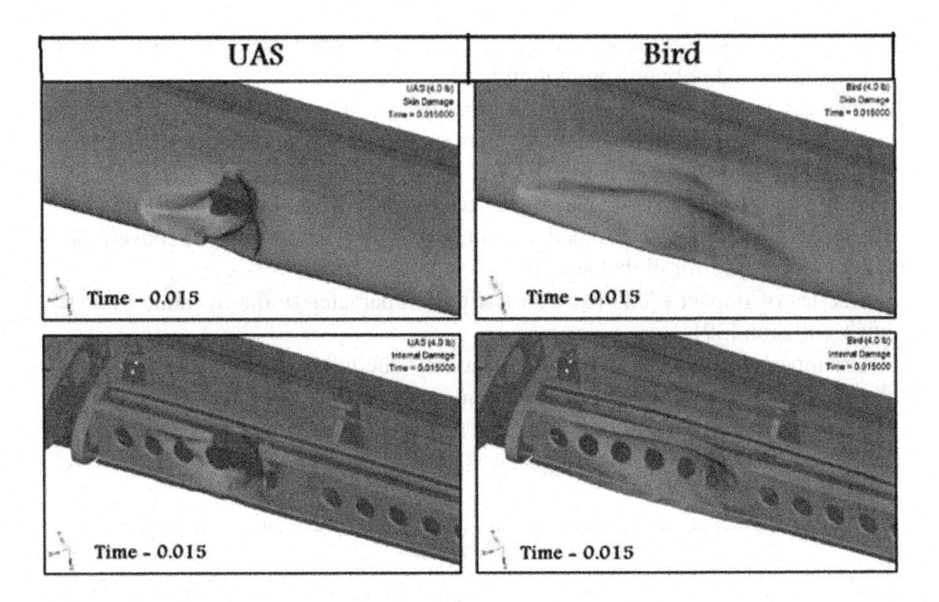

FIGURE 5.4 Comparison of skin and internal damage produced by a 1.8 kg (4.0 lbs) UAS and bird after impact with a commercial transport jet wing leading edge [39].

Courtesy: Department of Transportation, FAA.

As shown in Figure 5.3, the bird impact (top) created a dent in the leading-edge skin, which extended over a longer region of the vertical stabilizer than the two UAS impacts but did not penetrate the skin. The quadcopter impact (middle) resulted in critical damage and intrusion of the UAS into the airframe. Similarly, the fixed-wing UAS collision (bottom) resulted in penetration of the leading-edge skin, significant permanent deformation of the adjoining structure.

Figure 5.4 presents the damage sustained by the commercial transport jet wing leading edge due to both a bird and UAS. The wing leading edge was dented by the bird strike impact with some substructural deformation, which is less critical than the UAS impact. UAS impact resulted in the failure of the leading-edge skin and substructure, which increases the potential for a significant fraction of the UAS to become lodged in the airframe [40].

The windshield of the commercial transport jet was subjected to impact at three different locations (1, 2, 3), as shown in Figure 5.5. Location 3 is on the symmetrical line for the windshield and fuselage (Figure 5.5). A fixed-wing UAS has a mass of 1.8 kg (4 lbs), and an initial speed of 250 knots is used.

Figure 5.6 shows the damage of the UAS due to striking the windshield at location 1. Figure 5.7 shows the breakage of the UAS (top) and the damage of the windshield of the aircraft and surrounding structure (bottom).

5.7.2.2 Ingestion into the Engine (Case 1)

Ingestion of UAVs into jet engines may cause more damage than birds of similar mass. This is due to their denser and harder materials than what birds are composed

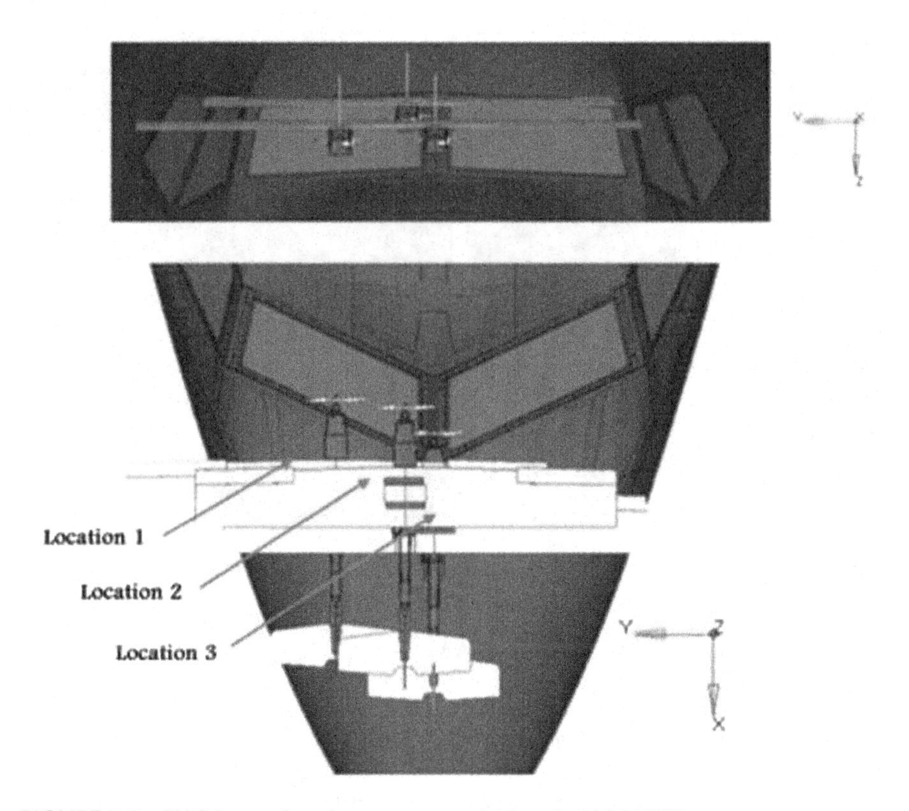

FIGURE 5.5 UAS impact locations – commercial jet windshield [39].

Courtesy: Department of Transportation, FAA.

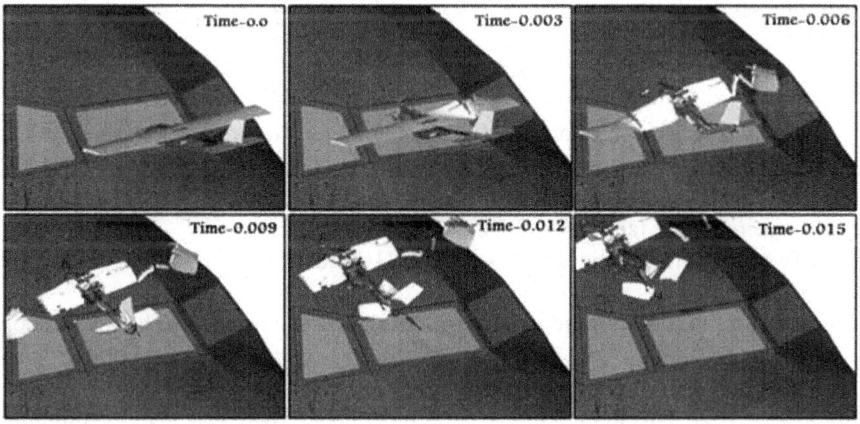

FIGURE 5.6 Time history for impact between a commercial transport jet windshield and a 1.8 kg fixed-wing UAS at 250 knots at location 1 [39].

Courtesy: Department of Transportation, FAA.

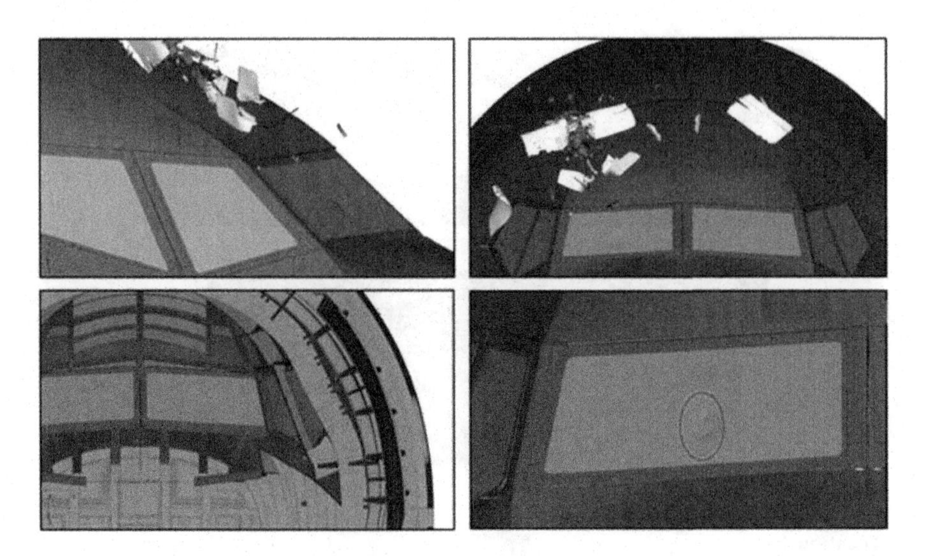

FIGURE 5.7 External/internal damage sustained by a commercial transport jet windshield impacted at location 1 with a 1.8 kg fixed-wing UAS at 250 knots [39].

Courtesy: Department of Transportation, FAA.

of. Damage due to UAS ingestion can be manifested in broken fan blades that immediately cause unbalance of the fan rotor. Also, broken fan blades can travel downstream and damage the next compressor modules. It may penetrate the containment casing. Moreover, the UAVs batteries could cause an engine fire.

A detailed study for the damage caused by the ingestion of UAV into a generic turbofan engine for a mid-sized business jet is carried out [35, 41]. The examined engine modules are the nosecone, shaft, fan disk, fan blades (20 blades), nacelle, and containment ring.

The parameter considered included the following.

- Two UAVs: the quadcopter DJI Phantom 3, popular among hobbyists, and a common fixed-wing UAV, the Precision Hawk. Mississippi State
- Two positions for ingested quadcopter; direct (0^0) and $\left(90^0\right)$ Pitch orientation concerning the fan (Figure 5.8).
- Two positions for ingested fixed-wing UAS, direct and 180° yaw orientation (Figure 5.8)
- Three flight conditions representing takeoff, approach, and flight below 10,000 ft
- Flight speeds are 180 knots for takeoff and approach, 250 for flight below 10,000 ft
- The generic fan adopted had a 40-inch diameter which was in close similarity to the mid-size business jet engines PW306B, TFE731, AE3007, CFE738, and HTF7000 installed to Dornier 328JET, Learjet Gulfstream G100, Embraer ERJ 145 family, Dassault Falcon 2000, Cessna Citation Longitude and having diameters ranging from 34.2 to 44.8 inch

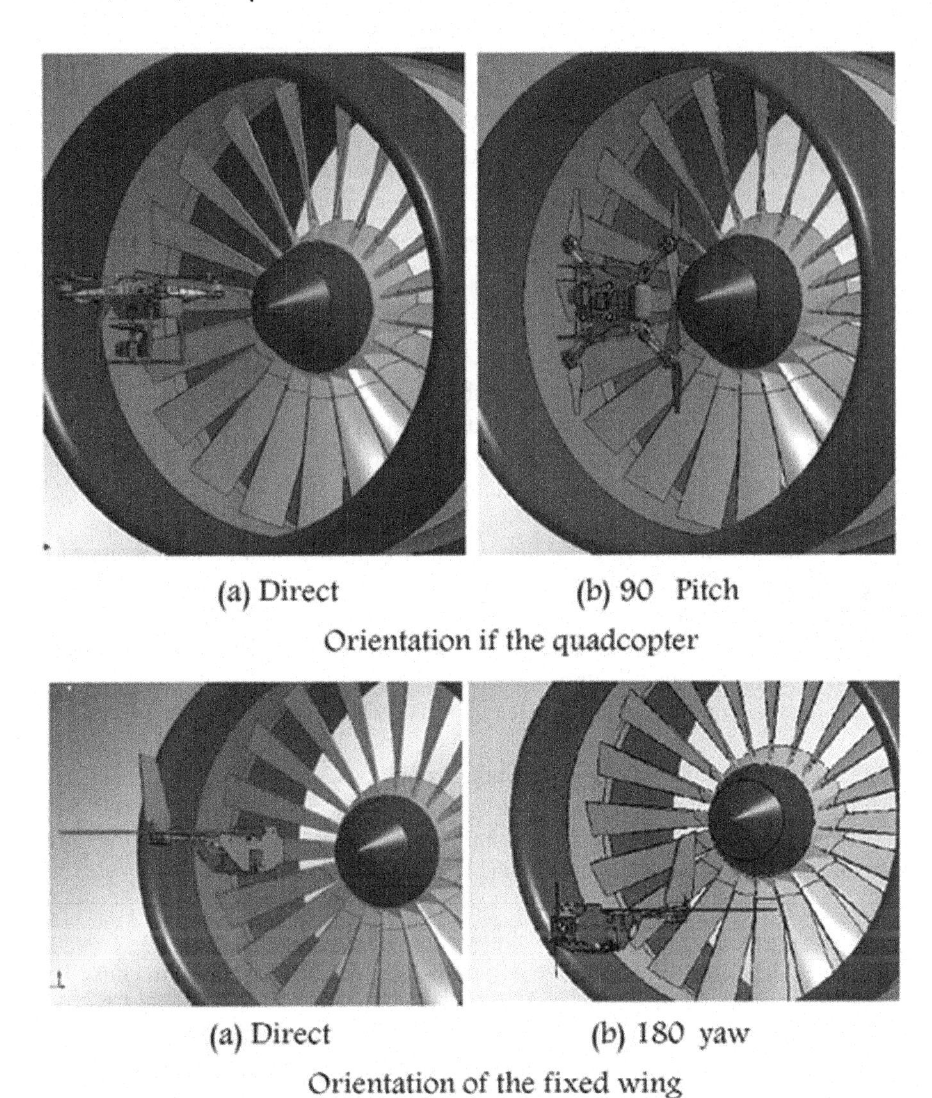

FIGURE 5.8 Orientations of the quadcopter and fixed-wing UAVs [40].

Courtesy: Department of Transportation, FAA.

- Fan rotational speeds were 8,500, 2,000, and 6,000 rpm for takeoff, approach, and flight below 10,000 ft, respectively
- The fan disk geometry was generated using CAD software
- Two sets of fan blades, namely thin and thick ones
- The thin blade has a NACA 4410 airfoil for the root and NACA 4403 airfoil for the tip
- The thick blade has a NACA 4414 profile for the root and an NACA 4405 profile for the tip

TABLE 5.4
Engine Operating Conditions for Three Cases

Flight Phase	Maximum Aircraft Speed (knots)	Fan Rotational Speed (rpm)	Fan Blade Tip speed (ft/s)
Takeoff	180	8,500	1,422
Below 10,000 ft	250	6,000	995
Approach	180	2,000	355

- The FE models were generated for the fan modules
- The nosecone was designed with a bionic shape
- The engine modules have the following material:
    - The nosecone is composed of aluminum 2024
    - The fan blades and disk are composed of the titanium alloy
    - The containment ring is composed of 0.18 inches of aluminum wrapped in 0.968 inches of Kevlar®
- LS-DYNA was used to simulate the impacts
- The FE models were used for the quadcopter and the fixed-wing UAV as well as their components (battery, camera, and motor)
- A comparison for the damage due to birds and UAV was performed

Table 5.4 lists the different engine operating conditions.

For each ingestion simulation, the bird or UAV is placed a few millimeters in front of the fan. The bird or UAV is given an initial translational velocity toward the fan to simulate the relative velocity between the UAV and the airplane.

5.7.2.2.1 Ingestion of Quadcopter into an Engine

Figure 5.9 illustrates the successive motion of quadcopter UAV into the intake and strike with the fan blades.

Figure 5.10 shows the damage caused to the blades. The quadcopter ingestion for the takeoff case results in the loss of multiple blade tips and damage to multiple other blades, but there was no damage to the fan containment case.

5.7.2.2.2 Ingestion of Fixed-Wing UAV into the Engine

The case of a full fixed-wing strike with the fan in the direct orientation is shown in Figure 5.11. The fixed-wing UAV is hitting the outer part of the fan, having a thin blade geometry. Figure 5.11 depicts the kinematics of the ingestion. Figure 5.12 shows the damage caused to the blades.

The fixed-wing ingestion resulted in the loss of multiple blade tips and damage to multiple other blades, but there was no significant damage to the fan containment ring.

Another case was considered for takeoff case where the fixed-wing UAV was also ingested close to the blade tip, but thick fan blades were considered (Figure 5.13). The fixed-wing caused significant damage to the leading edge of the blades that impacted the motor and the camera. There was also some minor damage to many other blades impacting other softer components including the battery.

FIGURE 5.9 Kinematics of the quadcopter ingestion during the takeoff case [40].

Courtesy: Department of Transportation, FAA.

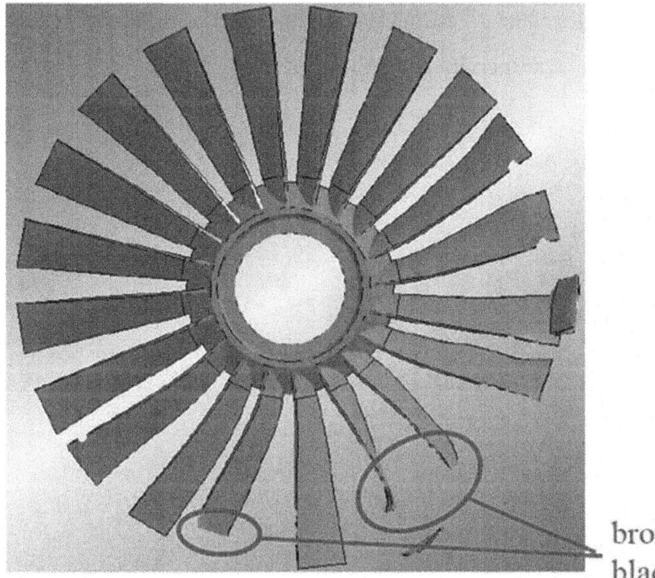

FIGURE 5.10 Damage of the fan due to quadcopter ingestion in the takeoff case [40].

Courtesy: Department of Transportation, FAA.

A third case was studied for the takeoff case, where the fixed-wing UAV was also in the direct orientation but hitting the inner part of the blade near its root instead of the outer part of the blade. The fan blades were of thin geometry. Figure 5.14 shows the damage caused to the blades.

The fixed-wing ingestion at the inner part of the blade resulted in larger plastic deformation in a few blades and some material loss due to the camera and motor impacts. There was no full blade loss.

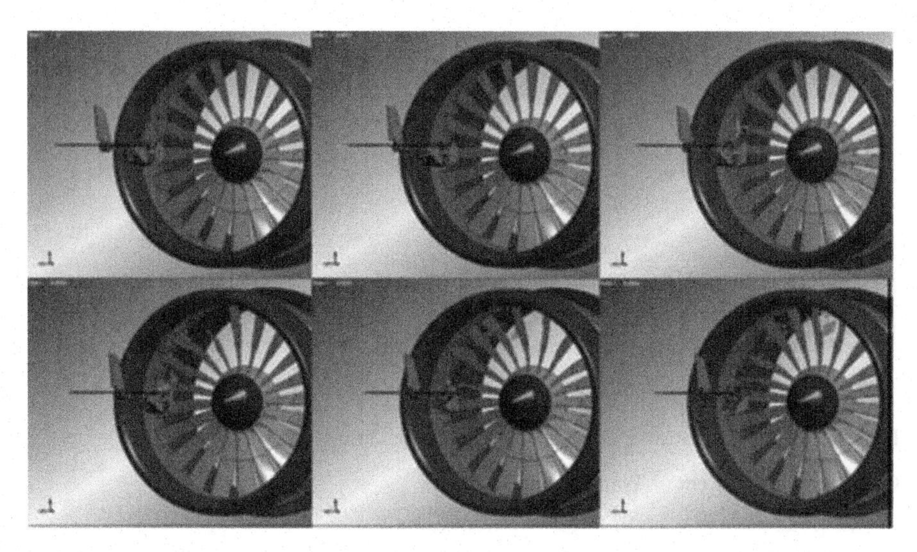

FIGURE 5.11 Kinematics of the fixed-wing UAV ingestion during the takeoff case [40].

Courtesy: Department of Transportation, FAA.

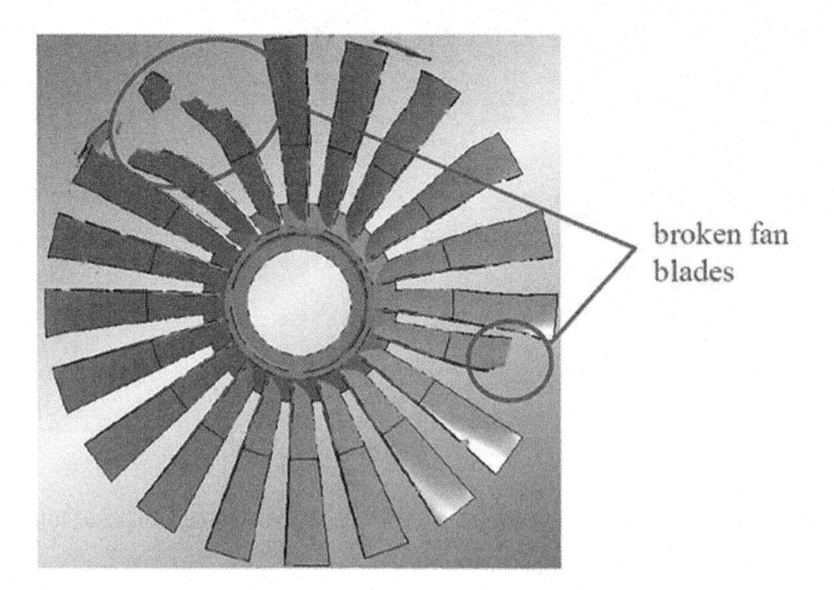

FIGURE 5.12 Damage of the fan (thin blades) due to fixed-wing UAV ingestion close to the blade tip during takeoff [40].

Courtesy: Department of Transportation, FAA.

FIGURE 5.13 Damage of the fan (thick blades) due to fixed-wing UAV ingestion close to the blade tip during takeoff [40].

Courtesy: Department of Transportation, FAA.

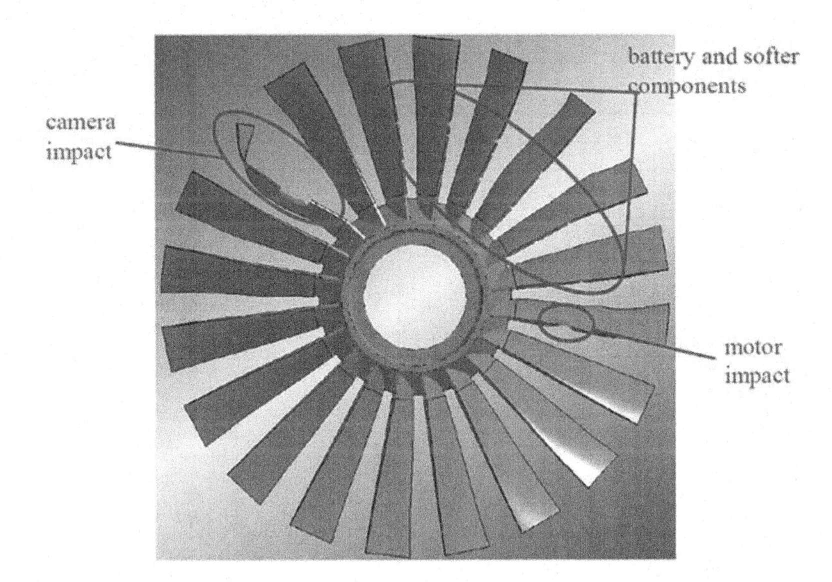

FIGURE 5.14 Damage of the fan (thin blade) due to fixed-wing UAV ingestion close to the blade root during takeoff [40].

Courtesy: Department of Transportation, FAA.

5.7.2.2.3 Damage Levels

A classification of the different damage possibilities is shown in Figure 5.15. A summary of the different cases is briefly described hereafter.

Severity	Description	Example
Level 1	• Deformation of fan blades. • Minor material loss from fan blades. • Dent in nosecone. • No containment failure.	
Level 2	• Significant material loss from one or multiple blades. • Loss of up to one full fan blade. • Crack in nosecone. • No containment failure.	
Level 3	• Loss of multiple fan blades. • UAV penetration of the nosecone. • No containment failure.	
Level 4	• Containment failure due to UAV ingestion.	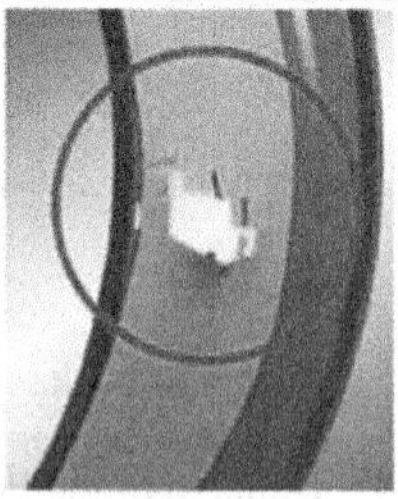

FIGURE 5.15 Damage level categories for nosecone, fan, and containment system [40].

Courtesy: Department of Transportation, FAA.

- A quadcopter in the takeoff case and the direct position striking the thin fan blades close to the tip caused level 3 damage
- A quadcopter in the takeoff case and the 90^0pitch striking the thin fan blades close to tip caused level 2 damage
- A quadcopter in the takeoff case and in the direct position striking the thin fan blades close to the root caused level 1 damage
- A quadcopter in the takeoff case and the direct position striking the thick fan blades close to the tip caused level 1 damage
- The cases of any of the elements of the quadcopter (motor, battery, camera) striking the outer blades or nose cone of thin fan blades configuration the direct position caused level 1 damage
- A quadcopter in both of approach and below 10,000 ft cases and in the direct position striking the thin fan blades close to tip caused level 1 damage
- A fixed-wing UAV in the takeoff case and the direct position striking the thin fan blades close to the tip caused level 3 damage
- A fixed-wing UAV in the takeoff case and the 180^0yaw-angle striking the thin fan blades close to tip caused level 2 damage
- A fixed-wing UAV in the takeoff case and the direct position striking the thin fan blades close to the root caused level 2 damage
- A fixed-wing UAV in the takeoff case and the direct position striking the thick fan blades close to the tip caused level 2 damage
- The cases of the motor and camera of the fixed-wing UAV striking the outer blades of thin fan blades configuration the direct position caused level 2 damage
- In the case of the battery of the fixed-wing UAV striking the outer blades of thin fan blades configuration the direct position caused level 1 damage
- The cases of any of the elements of the fixed-wing UAV elements (motor, battery, camera) striking the nose cone of thin fan blades configuration the direct position caused level 1 damage
- A fixed-wing UAV in the approach phase and the direct position striking the thin fan blades close to the tip caused level 1 damage
- A fixed-wing UAV in flight below 10,000 ft case and the direct position striking the thin fan blades close to tip caused level 2 damage.

5.7.2.3 Ingestion into the Engine (Case 2)

This work examined the ingestion of a UAS of similar size and mass of a typical adult Canadian goose into a modern high bypass engine [36] to quantify the risk arising from the increase in UAS. The following data resembles this case study.

Drone data

- Mass = 5.4 kg
- Maximum diameter = 1.0 m
- Primary components: the frame, motors, batteries, payload, and gimbal
- The UAS frame, including the arms, were modeled with a full composite material model utilizing the Chang matrix failure criterion
- Ingestion speed 92 m/s

Turbofan engine

- High bypass ratio turbofan engine including the inlet fan section, low-pressure compressor (LPC), casings, bearing, and shaft system
- Diameter = 2.9 m
- Fan blades are made from carbon fiber material
- 18 blades
- Rotational speed = 2,160 rpm (226 rad/s)

The ingestion scenario was assumed takeoff flight phase as the propulsion system operates at the most critical conditions of maximum thrust and within heights of flights for UAVs.

Upon ingestion, the UAS impacts 7 of the 18 blades. Blades that were impacted by one or two of the UAS arms experienced little to no permanent damage. Notable damage was seen on the leading edges of blades impacted more UAS arms and their attached motors. The worst case was when blades contacted the UAS hub including the heaviest modules (the battery, payload, and gimbal). These blades were subjected to peak forces up to 82.2 kN in a 0.0125-sec impact duration.

When the damage surpassed the ultimate strength, the fan blade was ejected from the fan assembly resulting in an imbalanced rotation of the entire fan assembly. After the ejection, the blade part was then lodged in between the adjacent fan blade, which escalated the damage on the entire propulsion system. Finally, remain debris from the fan blade can be ingested by the LPC, resulting in a compressor stall.

5.7.2.4 Ingestion into the Engine (Case 3)

The most recent study for the dynamic response due to the ingestion of UAV into the jet engine of a manned aircraft is described in [36]. The engine was simulated based on the combination of finite element method and computational fluid dynamics simulations. This work, for the first time, examined the performance deterioration of the engine in terms of the thrust loss due to UAV ingestion. The damage of the fan blades and the compressor core was evaluated. The damage severity was investigated by considering different collision configurations, different collision positions, and different flight phases. Both the damage of fan blades and the percentage of thrust loss were estimated to reflect the influence of UAV airborne collision on the aircraft operation.

5.8 ENVIRONMENTAL IMPACTS OF DRONES

Drones will create new environmental challenges. The environmental impacts of drones have received comparatively little attention [42, 43]. These include:

1. Noise

It is a key factor that could be a major obstacle to drone integration if it is not carefully managed. Most drones have moderate to quiet noise levels in the range of 20 to 70 decibels. People in cities are accustomed to the ambient noise levels of conventional vehicles. Thus, drone noise are less apparent. However, their proximity to residential areas and the increasing uptake of quieter electric vehicles may make drones more noticeable.

2. CO_2 emission

Most drones use electric batteries as a power source, thus produce zero tailpipe emissions. All drones will consume energy that depends on the design of the drone, its payload, the energy mix used for electricity production, and the method of electricity transmission to the battery. The production and scrapping of drones at the end of their lifespan will also consume energy and produce emissions.

3. Air pollution

Drones help alleviate air pollution in urban cities, as most drones are electric-powered.

4. Wildlife

Drones may also impact wildlife and generate visual disturbance; however, these impacts will need to be carefully managed.

5.9 CLOSURE

Like all new technologies, drones bring both positive and negative impacts identified by their benefits and risks. Recently, scientists and experts examined the acceptability and integration of drones in the transport system [42].

Drones are no longer limited to the small, electric rotor aircraft commonly used by photographers, tourists, and hobbyists. Numerous applications and impacts for human life are as follows:

1. Transport
   a. Networks of small drones that can carry parcels and medical goods over short distances
   b. Larger cargo drones carrying heavier payloads over long distances
   c. Drones transporting passengers within large and sprawling urban areas
   d. Connect cities with isolated regions that are either difficult to access by surface transport or lack the infrastructure for traditional forms of aviation.
2. Economic benefits
   a. Increase productivity and create new manufacturing and technology development
   b. Create jobs throughout the economy
3. Land use and property values
   a. Property values are likely to increase

The above positive impact, together with the negative environmental impact and the hazard of strike with manned aircraft, must be assessed.

The ICAO secretary-general on April 15, 2021, while opening the ICAO's "Drone Enable" Symposium, stressed that COVID-19 had been an accelerator for many UAS and UAS traffic management (UTM) innovations, which helped in saving peoples' lives by the transport of UN COVAX vaccine shipments from ports to hard-to-reach inland communities [44].

She also focused on:

- The increasingly widespread use of unmanned aircraft systems (UAS) as well as related modernization trends associated with digital communications and the emergence of advanced air mobility operators and other new entrants
- The continuous efforts to evolve the remotely piloted aircraft system regulatory framework for international operations. When complete, this will provide the basis for certificated remotely piloted aircraft to operate alongside traditional aircraft, employing similar procedures and separation standards.

5.10 MILITARY UAS

The relation between UAS and manned aircraft is teaming rather than conflict. The employment of UAVs in military operations adds new strike capabilities. In the last century, strike operations were either manned or unmanned aircraft operations.

Technological advances and military adaptation will merge unmanned systems from air, ground, and sea domains into teams of unmanned and manned kill chains (MUM-T) [45].

MUM-T will provide the following key capabilities [46]:

- *Defeating explosive ground surface, sub-surface (tunnel), and sea hazards from greater standoff distances*
- *Assuring mobility to support multiple points of entry*
- *Enabling movement and maneuver for projecting offensive operations*
- *Establishing and sustaining the shorelines of communications required to follow forces and logistics*
- *Protecting austere combat outposts*
- *Providing persistent surveillance to detect and neutralize threats and hazards within single-to-triple-canopy and urban terrain.*

The MUM-T concept was used in the late 1960s when the USAF modified sixteen C-130s to deploy AQM-34 Firebee drones. Each DC-130s had the ability to deploy and control up to four drones simultaneously for reconnaissance and electronic warfare operations [47]. In 2006, Lockheed Martin demonstrated the MUM-T (strike) concept using an AH-64D Longbow Apache helicopter, a UH60 Black Hawk helicopter and an RQ-5B Hunter UAV [48]. In 2014, the US Navy successfully launched and recovered the manned F/A-18 Super Hornet and unmanned X-47B UCLASS aircraft off the aircraft carrier USS Theodore Roosevelt [49]. Figure 5.16 shows a screen grab from a US Navy video of the test [49] posted on the US Navy's official YouTube channel. The tests demonstrated the integration of manned and unmanned aircraft onboard an aircraft carrier and the ability for unmanned aircraft to operate safely and seamlessly with manned aircraft.

FIGURE 5.16 USS Theodore Roosevelt conducts combined manned and unmanned operations [47].

Courtesy: US Navy.

TABLE 5.5
Categories of Military UAS

Group	Size	Weight	Height	Speed	Example
I	Small	0–20 pounds	<1200 ft AGL*	<250 knots	Raven, Bay raktar
II	Medium	21–55 pounds	<3,500 ft AGL	<250 knots	ScanEagle
III	Great	>55 and <1,320 pounds	<18,000 ft MSL**	<250 knots	Shadow and Integrator
IV	Greater	>1,320 pounds	<18,000 ft MSL	Any speed	Fire Scout, Predator, Gray Eagle
V	Greatest	>1,320 pounds	> 18,000 ft MSL	Any speed	Reaper, Global Hawk/Triton, UCLASS

*AGL = Above ground level.
**MSL = Mean sea level.

Several M.Sc. and Ph.D. theses were carried out in Naval Post Graduate School, handling teaming of manned–unmanned aircraft [45, 50] as examples.

Also in Singapore, UAVs are more than just supplements to the fighting force. There is an ever-increasing demand for UAVs to possess higher levels of persistence in providing constant real-time intelligence. Singapore Air Force is planning for a great role in UAV's future fighting force and seeking new methods to steer the application of UAVs [51].

As identified in [52], the Department of Defense classifies UAS into five groups based on weight, height, and speed, as illustrated in Table 5.5.

5.11 SPACE DEBRIS

5.11.1 INTRODUCTION

Space debris is defined as anything in orbit that is man-made and is no longer in use. It includes dead or dying satellites, rocket stages, other discarded hardware (instrument covers, separation bolts), fragments of vehicles that exploded or collided, and countless crumbs of human-made orbital flotsam [53].

Thousands of satellites and space vehicles are now orbiting space. Since hundreds of satellites are uncontrollable or leftover, there will be an increased probability of collision between them or with other space vehicles resulting in hundreds or even thousands of pieces of debris that do not immediately fall to Earth but continue to orbit around it [54].

Space debris can be as big as a large rocket body (over 10 m or 33 ft long) down to microscopic particles that are barely visible. The ones that we are most worried about are those larger than 3 mm. Categories of space debris and its quantities and tracking features are given in Table 5.6.

In 2016, NASA managed to track more than 500,000 pieces of debris (or space junk) orbiting the Earth at speeds of more than 28,000 km/h. In 2012, another study [59] outlined that there are over 20,000 tracked objects (larger than 10 cm diameter) in space with more than 13,000 tracked objects in Low Earth Orbit (LEO).

A reentry is defined as a flight phase when a man-made object in space comes back into the thicker atmosphere. Space hardware reenters at a very shallow angle (<1 degree). There are two kinds of reentry: controlled and uncontrolled. A controlled reentry is when the object uses a propulsion system to put it on a specific trajectory designed to reenter at a known location and time [53]. This location is normally over an ocean so that parts will not burn up land. In this reentry type, the debris is a fully man controlled.

Uncontrolled reentries occur when atmospheric drag slowly causes the object to move to lower altitudes and finally plunge to Earth. This can occur anywhere along the orbit, and thus no one can know where and when its reentry or land will occur. Generally, it is impossible to control objects that are inactive or still active but unable to maneuver – such as the Hubble Space Telescope. Also, many large objects in LEO lack the capability to control reentry location, like if a mission was ended prematurely.

TABLE 5.6
Space Debris [53]

Debris Size	Quantity	Tractable
1 mm	Tens of millions	No
3 mm	Millions	No
1 cm	Hundreds of thousands	No
5 cm	Tens of thousands	Most cannot be tracked
10 cm	Tens of thousands	Most can be tracked
>10 cm	Thousands	Yes, and cataloged

TABLE 5.7
Daily Casualty Expectation [59]

Cause	Daily Casualty Expectation
COVID-19	200 (worldwide 2nd week April 2021), 3,500 (worldwide 1st week January 2021) [55, 56]
Natural Disaster	813 (worldwide 2010), 28 (worldwide 2009) [57]
Work Accidents	12 (Deaths, United States, 2009) [57]
Motor Vehicle Accidents	99 (Deaths, United States, 2009) [57]
Assault (homicide)	45 (Deaths, United States, 2009) [57]
Influenza	8 (Deaths, United States, 2009) [57]
Meteorite Falls	1.1×10^{-4} (Deaths or Injuries, Worldwide, 1800–1995) [58]
Reentry Events*	2.7×10^{-5} (Deaths or Injuries, Worldwide) [59]

*Assumes 100 reentries year, each with Expected Casualty of (1.1×10^{-4}), thus ($100 \times 0.0001/365 = 2.7 \times 10^{-5}$).

Generally, reentries have a positive impact as they reduce the debris population. However, reentry may have a negative impact if it causes damage or injury. Fortunately, reentering objects pose only a marginal risk to people or infrastructure on the ground or aviation.

Objects in the lowest orbits can take few months to reenter. At the same time, objects below about 1,000 km will reenter within a few hundred years in space. Objects in orbits from 1,000 km to 2,500 km reentry may take thousands of years. Above 2,500 km, lifetimes can be much longer.

Typically, 200 to 400 big debris (like dead satellites: UARS, Phobos-Grunt, rocket stages, or fragments) reenter each year. Only a handful of these is controlled reentries. The rest are uncontrolled and can occur anywhere in their respective orbits. Objects of moderate size (1 m or above) reenter about once a week, while on average, two small-tracked debris objects reenter per day. Debris is discovered on the ground only a couple of times each year.

To assure the low hazard of reentry vehicles, a comparison between the daily casualties due to different causes including space debris reentry is given in Table 5.7. For people and property on the ground, the hazards posed by reentering spacecraft or debris are extremely small.

5.11.2 Some Examples of Recovered Space Debris

Large Space debris objects can reach the ground intact as uncontrolled reentry debris.

According to NASA and ESA (European Space Agency), an average of uncontrolled reentry debris (one to two) cataloged pieces of debris has fallen back to Earth each day since 1959.

Statistically, between the first space launch on October 4, 1957 and January 2002, more than 18,000 trackable objects reentered into the Earth's atmosphere [60]. Table 5.8 lists some historic reentry events, which posed a risk either due to their large masses or due to radioactive payloads [61]. The latter ones were two singular events,

TABLE 5.8
Re-entry Objects with Large Masses or Hazardous Payloads [61]

Object	Origin	Mass (kg)	Re-entry Date	Re-entry Type
Mir	CIS	135,000	2001/03/23	Controlled
Columbia (STS-107)	USA	82,000	2003/02/01	Uncontrolled
Skylab	USA	74,000	1979/07/11	Uncontrolled
Salyut 7/Cosmos 1686	USSR	40,000	1991/02/07	Uncontrolled
Salyut 6/Cosmos 1267	USSR	34,000	1982/07/29	Controlled
Cosmos 1443	USSR	19,800	1983/09/19	Controlled
Salyut 2	USSR	18,300	1973/05/28	Uncontrolled
Apollo 5 Nose Cone	USA	17,100	1966/04/30	Uncontrolled
Apollo 8 CSM BP-26	USA	16,700	1989/07/08	Uncontrolled
Compton GRO	USA	14,910	2000/07/04	controlled
Tiangong	China	8,506	2011/09/29	Uncontrolled
UARS	USA	5,900	2011/09/24	Uncontrolled
ROAST	DLR	2,400	2011/10/23	Uncontrolled
Cosmos 954	USSR	4,500	1978/01/24	Radioactive
Cosmos 1402	USSR	990	1983/02/07	Radioactive

both related to the reactor-powered RORSAT satellite class, which was operated until 1989. Details of the Cosmos 954 launched in 1979 will be given in detail below.

Highlights of some space debris events are given below:

1. The Chinese Long March 5B rocket was launched on April 29, 2021 at the Wenchang Space Launch Center in south China's Hainan province. It measured 98 ft long and 16.5 feet wide, and it weighed 21 metric tons. Its mission was to carry into orbit a module containing living quarters for a future Chinese space station. But after completing that task, the body of the rocket circled Earth in an uncontrolled manner. A large segment of the rocket had reentered and disintegrated over the Arabian Peninsula at approximately 10:15 p.m. ET on May 8, and landed at a location with coordinates of longitude 72.47° east and latitude 2.65° north. This impact location is in the Indian Ocean, west of the Maldives archipelago.
2. The US Navy intercepted its defunct spy satellite USA-193 on February 20, 2008, sending a trail of debris that fell over the northwestern United States and Canada. All debris was not larger than a football.
3. A woman in Turley, Oklahoma, got a noggin-knock in January 1997 when she was struck with a Sky junk (lightweight fragment of charred woven material). Luckily, she was not injured. That junk was debris from a Delta 2 booster, which reentered the Earth's atmosphere on January 22, 1997. Other debris from that booster included a steel propellant tank and a titanium pressure sphere (Figure 5.17).
4. The secret Soviet-navy satellite Cosmos 954, which was launched on September 18, 1977, got out of control. Since the satellite antennas each

FIGURE 5.17 A titanium pressure sphere from Delta 2 booster [62].

Courtesy: NASA.

sported a compact nuclear reactor, its reentry was one of the most frightening dates for people on the ground. On January 24, 1978, it reentered over Canada and shed debris across the frozen ground of the Canadian Arctic. Following the crash, the United States and Canada conducted overflights of the area and associated cleanup efforts.

5. On January 21, 2001, a Delta 2 third stage, known as a PAM-D (Payload Assist Module-Delta), reentered the atmosphere over the Middle East. Its titanium motor casing, weighing about 154 pounds (70 kg), slammed down in Saudi Arabia (Figure 5.18). Moreover, its titanium pressurized tank landed near Seguin, Texas, and the main propellant tank plunked down near Georgetown, Texas.
6. On February 1, 2003, during the reentry of Space Shuttle Columbia to Earth, it was disintegrated, killing seven astronauts. The catastrophic, lethal accident shed thousands of debris across 72,520 square kilometers in eastern Texas and western Louisiana. More than 80,000 recovered pieces were stored for follow-up research.
7. Skylab US space station weighing in at 77 tons was launched into orbit on May 14, 1973. Its orbiting operations came to a premature ended on July 11, 1979. When Skylab plummeted through the atmosphere, it sent chunks of debris raining down over a huge area extending from the Southeastern Indian Ocean across a sparsely populated section of Western Australia.

FIGURE 5.18 A titanium motor casing of third stage of Delta 2 rocket in Saudi Arabia [62]. Courtesy: NASA.

REFERENCES

[1] Unmanned Aircraft Systems (UAS) Market 2021 is estimated to clock a modest CAGR of 5.7% during the forecast period 2021-2026 With Top Countries Data, WBOC, Tuesday, April 6, 2021, www.wboc.com/story/43615314/unmanned-aircraft-systems-uas-market-2021-is-estimated-to-clock-a-modest-cagr-of-57nbspduring-the-forecast-period-2021-2026-with-top-countries-data (ACCESSED April 11, 2021)

[2] John Pyrgies. The UAVs Threat to Airport Security: Risk Analysis and Mitigation, Journal of Airline and Airport Management, JAIRM, 2019 – 9(2), 63–96 Online ISSN: 2014-4806 – Print ISSN: 2014-4865 https://doi.org/10.3926/jairm.127

[3] Christine M. Belcastro, Richard L. Newman, Joni K. Evans, David H. Klyde, Lawrence C. Barr, and Ersin Ancel. Hazards Identification and Analysis for Unmanned Aircraft System Operations, AIAA AVIATION Forum, June 5–9, 2017, Denver, Colorado AIAA 2017-3269, 17th AIAA Aviation Technology, Integration, and Operations Conference.

[4] Unmanned Aircraft System: 2019 Guide for Virginia Airports, https://doav.virginia.gov/globalassets/pdfs/policy/suas_-guide-for-virginia-airports-2019.pdf, accessed April 6, 2021.

[5] "Rise of the Drones. Managing the Unique Risks Associated with Unmanned Aircraft Systems", Allianz Global Corporate & Specialty, 2016.

[6] Shutting down Dubai International Airport due to a drone costs $100,000 a minute, Arabian Business, July 9, 2017, www.arabianbusiness.com/content/375851-drone-costs-100000-minute-loss-to-uae-airports

[7] Airports Council International (ACI) Europe (2018). ACI Europe Position on Drone Technology.
[8] FAA (2018). *Federal Aviation Administration UAS Sightings Report.* www.faa.gov/uas/resources/uas_sightings_report/ (Accessed May 1, 2018).
[9] SKYbrary (2018). Initiated by Eurocontrol in partnership with ICAO and other organizations. Retrieved from: www.skybrary.aero/index.php/Main_Page (Accessed April 1, 2021) and www.skybrary.aero/index.php/AIRPROX (Accessed April 5, 2021).
[10] Ben Marcus. Proposed FAA Small UAS Rule – What is Class B, C, D, and E, future of Flight, February 22, 2015, https://medium.com/future-of-flight/proposed-faa-small-uas-rule-what-is-class-b-c-d-and-e-airspace-81e760a36db1
[11] How To Request FAA Airspace Authorization For Class B, C, D, And E Controlled Airspace: A Guide to LAANC and the FAA Drone Zone Web Portal, Drone Pilot Ground School, January 17, 2021, www.dronepilotgroundschool.com/faa-airspace-authorization/
[12] Unmanned Aircraft System Airspace Integration Plan, Department of Defense, Version 2.0 March 2011, https://info.publicintelligence.net/DoD-UAS-AirspaceInte gration.pdf (Accessed April 9, 20211).
[13] Dan Gettinger and Arthur Holland Michel Drone Sightings and Close Encounters: An Analysis, Center for the study of drones, Bard College, December 11, 2015, https://dronecenter.bard.edu/files/2015/12/12-11-Drone-Sightings-and-Close-Encounters.pdf (Accessed April 4, 2021)
[14] NASA (2018). *Aviation Safety Reporting System Database Online.* https://asrs.arc.nasa.gov/search/database.html (Accessed April 3, 2021).
[15] NASA, Aviation Safety Reporting System, Incident Report, (ACN: 1242105), February 2015. Accessed November 29, 2015. http://asrs.arc.nasa.gov/
[16] Dubai airport airspace closed due to "unauthorized drone activity" The National Staff, September 28, 2016, www.thenationalnews.com/uae/dubai-airport-airspace-closed-due-to-unauthorised-drone-activity-1.200601
[17] NATS (2017). *Drone disruption at Gatwick.* https://nats.aero/blog/2017/07/drone-dis ruption-gatwick/ (Accessed April 1, 2021).
[18] NATS2 (2017). *NATS Gatwick drone incident, YouTube.* www.youtube.com/watch?v=V0rfFFZ332k (Accessed April 1, 2021).
[19] Aviation Safety Network, ASN Wikibase Occurrence # 201160https://aviation-safety.net/wikibase/wiki.php?id=201160
[20] Drone becomes FOD, FOD Prevention, https://fodprevention.com/drone-becomes-fod/
[21] David Brennan. Drone Almost Sucked Into Jet Engine As Passenger Plane Comes into Land, Newsweek, March 27, 2018, www.newsweek.com/drone-17-feet-being-sucked-passenger-jet-engines-dramatic-near-miss-861255
[22] Rytis Beresnevicius. A Drone Hit a Boeing 737 in Mexico, Aerotime Extra, www.aerotime.aero/22945-drone-hit-boeing-737-mexico
[23] Jim Moore. Drone Far Beyond Sight During Black Hawk Collision, December 14, 2017 www.aopa.org/news-and-media/all-news/2017/december/14/drone-far-beyond-sight-during-black-hawk-collision
[24] Anil Goyal Private communication, www.linkedin.com/in/testpilotanilgoyal/
[25] Mike Collins, Bird Strike, or Drone Strike? Drone Operators Learn UAS Accident Reporting Rules, AOPA, August 27, 2020, www.aopa.org/news-and-media/all-news/2020/august/27/bird-strike-or-drone-strike

[26] James Abbott. UK drone laws: where can and can't you fly your drone? Techarder, December 31, 2020, www.techradar.com/news/uk-drone-laws-where-can-and-cant-you-fly-your-drone#:~:text=UK%20drone%20laws%3A%20police%20powers,-Police%20forces%20across&text=Drone%20pilots%20could%20face%20on,and%2For%20in%20restricted%20locations.
[27] FAA Establishes Restrictions on Drone Operations Near U.S. Navy Vessels, May 17, 2019, www.faa.gov/news/updates/?newsId=93772
[28] Australian Government, Civil Aviation Safety Authority, Enforcement and penalties, www.casa.gov.au/drones/rules/enforcement#:~:text=home%20or%20backyard.-,The%20penalty%20can%20be%20up%20to%20two%20years%20in%20prison,will%20be%20up%20to%20%2411%2C100.
[29] Russia Drone Regulations, https://uavcoach.com/drone-laws-in-russia/
[30] China Drone Regulations, https://uavcoach.com/drone-laws-in-china/
[31] UAS by the Numbers, FAA, March 30, 2021, www.faa.gov/uas/resources/by_the_numbers/
[32] Federal Aviation Administration, Advisory Circular 91-57, Model Aircraft Operating Standards, Misc. Doc. (1981). Accessed December 6, 2015. www.faa.gov/documentLibrary/media/Advisory_Circular/91-57.pdf
[33] National Academies of Sciences, Engineering, and Medicine 2019. *Airports and Unmanned Aircraft Systems, Volume 2: Incorporating UAS into Airport Infrastructure Planning Guidebook,* ACRP Research Report 212, Washington, DC: The National Academies, Press. https://doi.org/10.17226/25606
[34] Gerardo Olivares, Thomas Lacy, Luis Gomez, Jaime Espinosa de Los Monteros, Russel J. Baldridge, Chandresh Zinzuwadia, Tom Aldag, Kalyan Raj Kota, Trent Ricks, Nimesh Jayakody, Vol. I: UAS collision severity Evaluation, Executive Summary – Structural Evaluation, DOT/FAA/AR-xx/xx, July 2017.
[35] Troy Lyons and Kiran D'Souza. Parametric Study of a Unmanned Aerial Vehicle Ingestion Into a Business Jet Size Fan Assembly Model, Journal of Engineering for Gas Turbines and Power July 2019, Vol. 141/071002-1.
[36] Song, Y. and Schroeder, K., Horton, B., and Bayandor, J. Advanced Propulsion Collision Damage due to Unmanned Aerial System Ingestion, ICAS 2016.
[37] Hu Liu, Mohd Hasrizam Che Man, Kin Huat Low, UAV airborne collision to manned aircraft engine: Damage of fan blades and resultant thrust loss, Aerospace Science and Technology, 2021-04-06, DOI: 10.1016/j.ast.2021.106645
[38] Olivares, G. "Simulation and Modeling of Bird Strike Testing," NIAR Report No. 09-039, National Institute for Aviation Research, 2009.
[39] Gerardo Olivares, Thomas Lacy, Luis Gomez, Jaime Espinosa de Los Monteros, and Russel J. Baldridge. Chandresh Zinzuwadia, Tom Aldag, Kalyan Raj Kota, Trent Ricks, Nimesh Jayakody, Volume II – UAS Airborne Collision Severity Evaluation – Quadcopter, DOT/FAA/AR-xx/xx, July 2017.
[40] Gerardo Olivares, Thomas Lacy, Luis Gomez, Jaime Espinosa de Los Monteros, and Russel J. Baldridge. Chandresh Zinzuwadia, Tom Aldag, Kalyan Raj Kota, Trent Ricks, Nimesh Jayakody, Volume III – UAS Airborne Collision Severity Evaluation – Fixed-Wing, DOT/FAA/AR-xx/xx, July 2017.
[41] Kiran D'Souza, Troy Lyons, Thomas Lacy, and Kalyan Raj Kota. Volume IV – UAS Airborne Collision Severity Evaluation – Engine Ingestion, DOT/FAA/AR-xx/xx, July 2017.
[42] Ready for Take-Off? Integrating Drones into the Transport System, International Transport Forum (ITF) Research Reports, OECD Publishing, Paris, 2021, www.itf-oecd.org/integrating-drones-transport-system

[43] Intaratep, N., W.N. Alexander, W. Devenport, S. Grace, and A. Dropkin, "Experimental Study of Quadcopter Acoustics and Performance at Static Thrust Conditions," 22nd AIAA/CEAS Aeroacoustics Conference, 30 May–1 June 2016, Lyon, France, 2016, https://doi.org/10.2514/6.2016-2873
[44] Exploring the latest developments with unmanned and remotely piloted aircraft systems Uniting Aviation, April 15, 2021, https://unitingaviation.com/news/safety/exploring-the-latest-developments-with-unmanned-and-remotely-piloted-aircraft-systems/ (Accessed April 16, 2021)
[45] Joong Yang Lee, Expanded kill chain analysis of manned-unmanned teaming for future strike operations, M.Sc. thesis, Naval Postgraduate School, Monterey, California, September 2014.
[46] Unmanned Systems Integrated Road Map FY2013-2038, US Department of Defense, Reference Number: 14-S-0553, January 2014, https://apps.dtic.mil/dtic/tr/fulltext/u2/a592015.pdf Accessed April 13, 2021.
[47] C-130 Hercules. (2000, February 20). Federation of American Scientist. [Online]. Available: http://fas.org/man/dod-101/sys/ac/c-130.htm. Accessed April 10, 2021.
[48] Department of Defense, "Unmanned systems integrated roadmap, FY2013 – 2038," Department of Defense, Washington, DC, Tech. Rep. 14-S-0553, 2013.
[49] U.S. Navy. (2014, August. 17). "USS Theodore Roosevelt conducts combined manned, unmanned operations." [YouTube video]. Available: www.youtube.com/watch?list=UUKuSaHewQKWjR2wFuqfkMEA&v=RqiOzO8yV4A. Accessed April 11, 2021.
[50] Fatih Sen, Analysis of the Use of Unmanned Combat Aerial Vehicles in Conjunction with Manned Aircraft to Counter Active Terrorists in Rough Terrain, M.Sc. Thesis, Naval Postgraduate School, Monterey, California, June 2015.
[51] Lance King, DoD Unmanned Aircraft Systems Training Programs: Brief to ICAO, March 24, 2015
[52] LTA Chan Jing Yi, Manned-Unmanned Teaming—An Analysis Of UAVs And Their Interoperability With Manned Aircraft, Pointer, Journal of The Singapore Armed Forces, 66–76, 42(1), January 3, 2016.
[53] Space debris 101, Aerospace, https://aerospace.org/article/space-debris-101.
[54] Debris: Who Is Liable when Two Satellites Collide in Space? http://blogs.esa.int/cleanspace/2016/02/12/who-is-liable-when-two-satellites-collide space/
[55] Daily Updates of Totals by Week and State, CDC Center for Disease Control and Prevention, www.cdc.gov/nchs/nvss/vsrr/covid19/index.htm (accessed April 15, 2021).
[56] Julie Reed Bell and Seth Borenstein, "2010's World Gone Wild: Quakes, Floods, Blizzard," Associated Press, December 19, 2010 (see www.msnbc.msn.com/id/40739667/ns/us_news-2010_year_in_review/t/s-world-gone-wildquakes-floods-blizzards/).
[57] Kenneth D. Kochanek, et al., "Deaths, Preliminary Data for 2009, Vol. 59, No.4," National Vital Statistics Report, U.S. Department of Health and Human Services, March 16, 2011.
[58] John S. Lewis, Table, pp 176–182, "Property Damage, Injuries, and Deaths caused by Meteorite Falls," 6Rain of Iron and Ice, Perseus Publishing, 1996.
[59] William Ailor, Space Debris Reentry Hazards, Presented to Scientific & Technical Subcommittee of the United Nations Committee on the Peaceful Uses of Outer Space, International Association for the Advancement of Space Safety (IAASS), February 2012.

[60] Ailor, W., Hallman, W., Steckel, G., and Weaver, M. (2005). Analysis of Re-Entered Debris and Implications for Survivability Modeling. In *Proceedings of the Fourth European Conference on Space Debris*, ESA SP-587.

[61] Heiner Klinkrad, Space Debris – Models and Risk Analysis, Springer–Praxis, 2006.

[62] Space.com Staff, Worst Space Debris Events of All Time March 8, 2013, www.space.com/9708-worst-space-debris-events-time.html

6 Animate (Biological) Debris

6.1 INTRODUCTION

Animate debris, as discussed in Chapter 1, includes humans, wildlife, insects, and grass. Strike of this animate or biological debris with aircraft will not only damage the flight vehicle but also result in the death and wound of living creatures. Most strikes are caused by wildlife, but in rare cases, humans and grass are sucked into aircraft engines.

As defined by FAA [1],

> **Hazardous wildlife are** species of wildlife (birds, mammals, reptiles), including feral and domesticated animals, not under control that may pose a direct hazard to aviation (i.e., strike risk to aircraft) or an indirect hazard such as an attractant to other wildlife that pose a strike hazard or are causing structural damage to airport facilities (e.g., burrowing, nesting, perching).

A "wildlife strike" is defined as a collision of an animate creature with a flight vehicle. An animate creature may be an avian (bird, bat, or insect) or non-avian (terrestrial mammals, reptiles). Flight vehicle may be an aircraft or missile. Aircraft may be a civilian or military (fixed-wing or rotary-wing) aircraft [2]. A wildlife strike may result in the death of the animal and sometimes major damage to the plane. Also, a bird strike can be a collision of a single bird or a group (a flock) of birds with an aircraft. The Wright Flyer III experienced the first recorded bird strike on September 7, 1905. According to the National Wildlife Strike Database (NWSD), for the 30 years (1990–2019), the number of wildlife strikes was 231,320 for US civil aircraft airplanes: 227,045 in the United States and 4,275 strikes by US-registered aircraft in foreign countries [3]. Globally, wildlife strikes killed more than 292 people and destroyed over 271 aircraft. This assures that though wildlife strikes are not rare, they are hardly ever deadly, and over the years, only a small number of fatal accidents were encountered. Birds and bats strike represent 97% of all strikes; the remaining 3% represents strikes with terrestrial mammals (such as deer, coyotes, skunks) and reptiles (such as turtles, alligators, and iguanas) [3]. White-tailed deer and coyotes are the most struck non-bird species worldwide. Mammals that were most frequently involved in aircraft strikes varied by country. Bats were in Australia and rabbits and dog-like carnivores in Canada, Germany, and the UK. Mammals varied with size

DOI: 10.1201/9781003133087-6

where voles are the smallest and giraffe is the largest. For every million aircraft movements, average mammal strikes per year ranged from 1.2 to 38.7. Mammals' strikes with aircraft are annually increasing by up to 68% [6].

Now, let us remember the repeatable question since Wright Brothers first flight; namely:

Is air travel safe?

The answer is yes. Air travel remains the safest mode of transportation. Reports show that worldwide fatalities in 2018 were for aviation 395, water 697, railroad 815, and highways (passenger cars, buses, trucks, motorcyclists) 36,560 [5, 6]. Moreover, for aviation, the number of damaging strikes is declining. Although the total number of reported strikes has increased, damaging strikes have decreased. In 2000, the number of wildlife strikes reported to FAA increased steadily from 6,002 to 17,228 in 2019, whereas the number of damaging strikes decreased from 764 in 2000 to 720 in 2019 (Table 6.1). The decline in damaging strikes can be attributed in part to increase in the development and use of wildlife hazard management plans at airports.

Two distinguished bird strike accidents that demonstrated to the world the severity of aircraft collisions with birds must be mentioned here:

- The emergency forced US Airways Flight 1549 (Airbus 320), taking off from La Guardia International Airport (LGA), into the Hudson River on January 15, 2009. Migratory Canada geese were ingested into both engines at 3:27:11 p.m. at an altitude of 2,818 ft above ground level (AGL) and 4.5 miles north-northwest of the approach end of runway 22 at (LGA). The aircraft successfully ditched into the Hudson River after 3 min 49 sec. All onboard people (150 passengers and five crew) were safe. It is next called "Miracle on the Hudson."
- The forced landing of Ural Airlines Flight 178 (Airbus 321) with 234 persons aboard in a corn field 3 miles from Zhukovsky International Airport, Moscow, Russia, on August 15, 2019. Gulls were ingested into both engines during takeoff (Aviation Safety Network 2020). Incredibly, none of the 389 people were killed and it was then identified as "Miracle in the Corn Field."
- Both aircraft were damaged beyond repair.

The FAA wildlife strike database for the period 1990–2019 [6–8] reported that:

- Number of species striking aircraft were: 591 species of birds, 51 species of terrestrial mammals, 36 species of bats, and 23 species of reptiles
- 53% of bird strikes occurred between July and October
- 29% of deer strikes occurred in October and November
- 62% of terrestrial mammal strikes occurred at night
- 63% of bird strikes occurred during the day, 8% at the hours of dawn and dusk, and 29% at night
- 62% of bird strikes occur during the landing phases of flight, and 35% during the takeoff run and climb phases, while the remaining 3% occur when the aircraft is en route

- 63% and 86% of terrestrial mammals and bats strikes occur during landing, while 33% and 12% strikes occur during departure.
- Nearly 71% and 72% of bird strikes with commercial and GA aircraft, respectively, occurred at or below 500 ft AGL
- Above 500 ft AGL, the number of strikes declined by 34% and 43% for each 1,000 ft gain in height for commercial and GA aircraft, respectively.
- 1,210 civil aircraft were involved in collisions with deer [4]
- Over these three decades (1990–2019), the estimated cost of damage associated with mammal strikes exceeded $103 million in the United States alone.

Wildlife/aircraft strike has been dramatically increased by the onset of the jet age. The main reasons are as follows:

- Wildlife and airports exist near each other.
- There has been a substantial increase in air traffic worldwide. Commercial air traffic increased from about 23.8 million aircraft movements in 2004 to 38.9 million in 2019 and is expected to reach more than 51 million in 2030 (it was expected to reach 40.3 million in 2020, but due to COVID-19, it dropped to 16.4 million [9]).
- At any instant, there are 35,000 aircraft in the air worldwide, some 6,000–9,000 of which are in the United States.
- Military aircraft have been of a vital role in tactical and logistical operations.
- The increase in aircraft speed and quietness makes it is easier for them to escape the attention of the birds.
- Commercial air carriers are replacing their older three- or four-engine aircraft fleets (Boeing 727, Lockheed L-1011 TriStar, Boeing 747, DC-10, and Airbus 340) with more efficient and quieter, two-engine aircraft (like Boeing 777 and Airbus 330 series). Thus, there is a reduction in engine redundancy which increases the probability of life-threatening situations resulting from aircraft strikes with flocks of birds.
- Modern aircraft are powered by High Bypass Ratio Turbofan HBPR engines (GE90, P&W1000 series and Rolls Royce Trent series), which have large intake diameters, thus even large-sized birds may be sucked into its gas path and damage its modules as they are less resistant to bird strike than piston engines.
- There has been a marked increase in the populations of hazardous wildlife species in many parts of the world in the last few decades. For example, the Canada goose population in the United States and Canada increased from about 1 million in 1990 to over 4 million in 2018. During the same period, the North American snow goose population increased from about 4 million to 15 million birds. Also, white-tailed deer in the United States has increased from 100,000 in 1900 to 30 million in 2019.
- Worldwide environmental protection programs led to considerable increases in populations of many large-bodied species such as cranes, cormorants, gulls, herons, falcons, eagles, vultures, and wild turkeys. In the United States, about

90% of all bird strikes involve species federally protected under the Migratory Bird Treaty Act.
- Billions of birds, bats, and insects use the atmosphere for migration and foraging as civil and military do.

The location of airports can be a significant contributing factor to the number of wildlife strikes, with the following situations being more prone to incidents [10].

- Coastal areas: If an airport is near the sea, gulls can be a problem for aircraft taking off and landing.
- River estuaries: River estuaries are popular grounds for all types of wildfowl, fish, shellfish, and migratory birds.
- Rivers, lakes, and swamps: Popular places for alligators
- Marshland: Popular places for ducks, geese, frogs, toads, turtles, snakes, mammals, and insects
- Landfills: Rubbish (trash) dumps attract birds, like seagulls, various species of mice, voles, shrews, rats, chipmunks, skunks, foxes, feral cats, and dogs
- Farmland: Geese, many mammals including cattle, pigs, poultry, horses, sheep, goats, llamas, donkeys, and rabbits always present
- Considerable areas of grass: Airports with large amounts of grassy areas or open spaces are more likely to attract birds and several mammals.

It is recommended for pilots to:

- Keep departure airspeed at or below 250 knots at altitudes below 10,000 ft
- Maximize the climb rate through 3,500 ft, especially at night in April to May and September through November
- Increase the visibility of their aircraft by keeping exterior lights on when in the airport environment
- Report wildlife strikes
- Get familiar with the FAA's Online Strike Database
- Pilots who habitually fly fast at a low level or bird-rich environment should consider wearing head protection with a visor or goggles.

Strike events may affect passenger safety, airline economics, and local conservation.

6.2 BIRD IMPACT RESISTANCE REGULATION

6.2.1 Fixed-Wing Aircraft

The regulations that cover impact resistance are listed below for different sizes of aircraft [26, 34, 35].

6.2.1.1 Transport Aircraft (Airliners, Civilian, and Military Cargo)

These aircraft are those for which the maximum takeoff weight (MTOW) is greater than 5,670 kg.

6.2.1.1.1 Airframe

The European Aviation Safety Agency (EASA) code (CS-25) necessitates that all airframe modules must be able to withstand without hazard (penetration or critical fragmentation) impact with a bird of 1.8 kg at V_C at sea level or 0.85 V_C at 8,000 ft, whichever is the most critical [26]. The FAA code (Pt 25) has the same regulation for all airframe modules except the empennage. For the empennage, its structure should withstand the impact of a 3.6 kg bird at cruise speed (V_C).

6.2.1.1.2 Engines

EASA European engine ingestion tests (CS-E800) state that for engines:

- *No hazardous engine effect after impact with a single large bird of weight between 1.85 and 3.65 kg.*
- *For a large flocking bird, no more than 50% loss of thrust following ingestion of large mixed birds of between 1.85 and 2.5 kg and for 20 min must be capable of thrust variation*
- *For medium flocking birds, the engine must be capable of ingesting some birds of varying mass, without reducing the thrust of less than 75%*
- *For small birds of weight 0.85 kg, ingestion of some birds must not result in loss of more than 25% thrust.*

6.2.1.2 General Aviation Aircraft

General aviation aircraft include both lightweight executive jets of less than 5,700 kg and planes in the "commuter" category, which are propeller-driven twin-engine airplanes having MTOW of 8,618 kg or less and that have 19 or fewer seats. This group of aircraft has no specific bird strike certification requirements (apart from windshields on commuter aircraft) based on the US Part 23 and European (CS-23).

6.2.1.3 Light Non-Commuter Aircraft

Examples of light non-commuter aircraft are Cesena 206, Socata, and Extra EA-400/500, all of which that a capacity of five passengers. These planes have no bird strike-related certification requirements.

6.2.2 Rotorcrafts

6.2.2.1 Large Rotorcraft

The "large rotor craft" category covers Cat A, with a weight greater than 9,072 kg and ten or more passenger seats, and Cat B, other weights. Based on EASA CS-29 and the FAA Federal Aviation Regulation (FAR) Part 29, the requirements for this type of aircraft are identical [26]. Both require that the aircraft can continue safe flight and landing or safe landing following impact with a single 1 kg bird at the greater of the maximum safe airspeed (V_{NE}) or maximum level-flight airspeed at rated power (V_H) at up to 8,000 ft.

6.2.2.2 Small Rotorcraft

Based on both FAA and EASA requirements, no bird strike-related certification are needed for small rotorcraft (weight 3175 kg or less) [26].

6.3 STATISTICS FOR WILDLIFE STRIKE WITH CIVILIAN FIXED-WING AIRCRAFT

A wildlife strike happens every 15 min in the world. According to the International Civil Aviation Organization (ICAO), 70% of them go unreported. ICAO requests its contracting states to report wildlife strikes [19]. Data are usually collected by the Civil Aviation Authorities (CAAs). Its quality relies on consistent reporting by the pilots, maintenance crews, air traffic control, and wildlife control staff. Recently, wildlife strike reporting has been enforced by many CAAs across the world. European Union (EU) put into force mandatory wildlife strike reporting in 2015 [20, 21]. In the UK, pilots were requested to report all wildlife strikes (damaging and non-damaging) since 2004.

In Australia, it is mandatory to report wildlife strikes for several years. Though in the United States, bird strike reporting (Form FAA 5200-7) is still voluntary, partners are keen to report any wildlife strike. For this reason, the ratio between all reported wildlife strikes and all wildlife strike occurrences increased from 41% to 91% for commercial aircraft in the period from 1990 to 2013 [22]. The reason for this rise is mainly attributed to better reporting, rather than increased bird strike risk.

6.3.1 Wildlife Strikes by Region

ICAO database lists reports for wildlife strikes in 91 states [11]. During the period 2008 to 2015, a total of 97,751 wildlife strikes were reported. The Middle East and South America recorded the least reported wildlife strikes (69 and 78). A moderate number of strikes were reported in Oceania, Central America, and the Caribbean, Asia and the Pacific, as well as Africa (225, 527, 660, and 772, respectively). The maximum numbers of reported strikes were in North America and Europe and North Atlantic (33,221 and 35,775, respectively). The unknown reports were 958.

In a second study by Airbus, for bird strikes by convenient, over 33,000 bird strike accidents to civil aircraft were recorded between 1990 and 2000 [29]. The highest percentage was recorded in Europe (42%), followed by North America (32%), next Asia (19%), then Africa (4%). The lowest figure is the Caribbean and South America (2%) [29].

The most recent study (2020) for the average bird strike rates per 10,000 aircraft movements is reviewed in [12]. The bird strike rates in different countries as listed in [13–18] were 8.19 % in Australia (during the period 2008–2017), 4.62 % in the UK (2012–2016), 4.42 % in Germany (2010–2018), ranged from 4% to 8% in Canada (2008–2018), and 3.95% in France (2004–2013).

6.3.2 Annual Total and Damaging Strikes

Table 6.1 lists an abridged table for the number of reported wildlife strikes in the United States and US-registered civil aircraft in foreign countries, 1990–2019 [6]. A total of 227,045 strikes were registered for the US aircraft. An additional 4,275 strikes were reported for US-registered aircraft in foreign countries resulting in 231,320 strikes in total.

TABLE 6.1
Number of Strikes for US Aircraft within the United States and Foreign Countries in the Period 1990–2019 [6]

	United States		Foreign		Total	
Year	Strikes	Damage Strikes	Strikes	Damage Strikes	Strikes	Damage Strikes
1990	1,816	366	34	6	1850	362
1995	2,717	486	52	11	2,769	497
2000	5,874	743	128	21	6,002	764
2005	7,039	591	180	19	7,219	610
2010	9,660	577	229	18	9,889	595
2015	13,545	610	242	12	13,787	622
2019	17,050	710	178	10	17,228	720

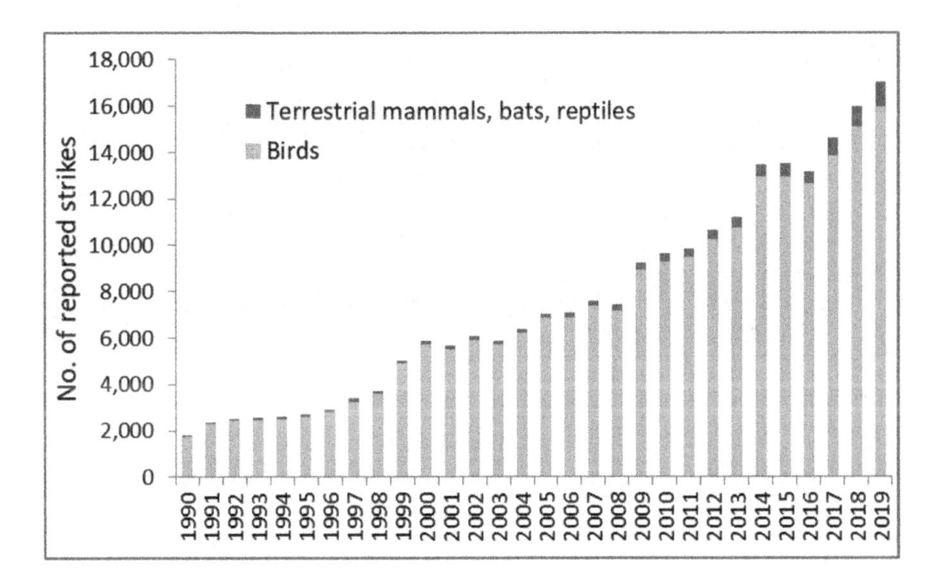

FIGURE 6.1 Number of reported wildlife strikes with US civil aircraft, within United States and foreign countries, 1990–2019 [6].

Courtesy: US Department of Agriculture, Wildlife Services.

Figure 6.1 presents the annual strikes of US aircraft in the United States and foreign countries by birds as well as terrestrial mammals, bats, and reptiles [6].

Table 6.2 lists the number of total and damaging strikes by different species in the period 1990–2019 [6]. Bird strikes were 218,524, while the damaging strikes were 15,768. Bats strikes were 3,269 in total with only 22 damaging ones. Total terrestrial mammals 4,761 and the number of damaging ones 1,165. Reptiles' total

TABLE 6.2
Number of the Total and Damaging Strikes by Different Species in the Period 1990–2019 [6]

Species	Birds	Terrestrial Mammals	Bats	Reptiles
Number of strikes United States (1990–2019)	218,524	4,761	3,269	491
Strikes with damage	15,768	1,165	22	3

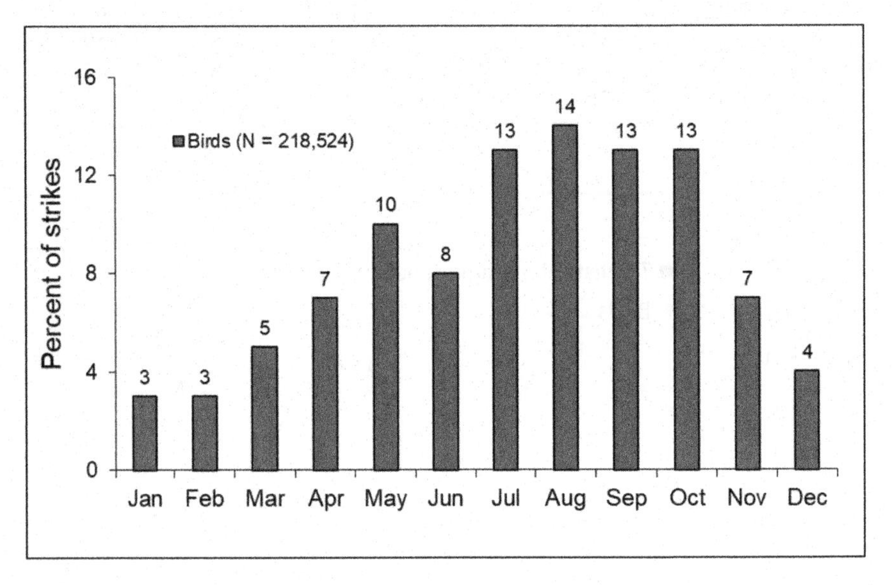

FIGURE 6.2 Monthly percentage of reported bird strikes with civil aircraft by month, United States, 1990–2019 [6].

Courtesy: US Department of Agriculture, Wildlife Services.

and damaging strikes are 491 and 3, respectively. An additional 4,275 strikes were reported for US-registered aircraft in foreign countries for a total of 231,320 strikes.

6.3.3 Monthly Strikes

ICAO database for wildlife strikes outlined that strikes occurred throughout the year [11]. Figure 6.2 presents the monthly percentage of reported bird strikes with civil aircraft by month, United States, 1990–2019 [6]. A total number of strikes within these 30 years reached 218,524.The busiest months are July through October, while the least reported number of wildlife strikes are January and February. August is the month of the highest wildlife strike activity in the periods of 2001–2015. The FAA data for the strikes in the period of 1990–2019 is reported in [6]. Not only a longer period is summarized, but also strikes were categorized for different wildlife species.

The number of strikes with bats was 3,269, with a maximum percentage of 17% in August. In addition, 491 strikes with reptiles were reported of which 57% occurred in May–July. Strikes reported for US-registered aircraft in foreign countries were excluded.

Figure 6.3 illustrates the distribution of bird strikes over the year for different regions in the world. During winters, the risk of collisions between birds and aircraft

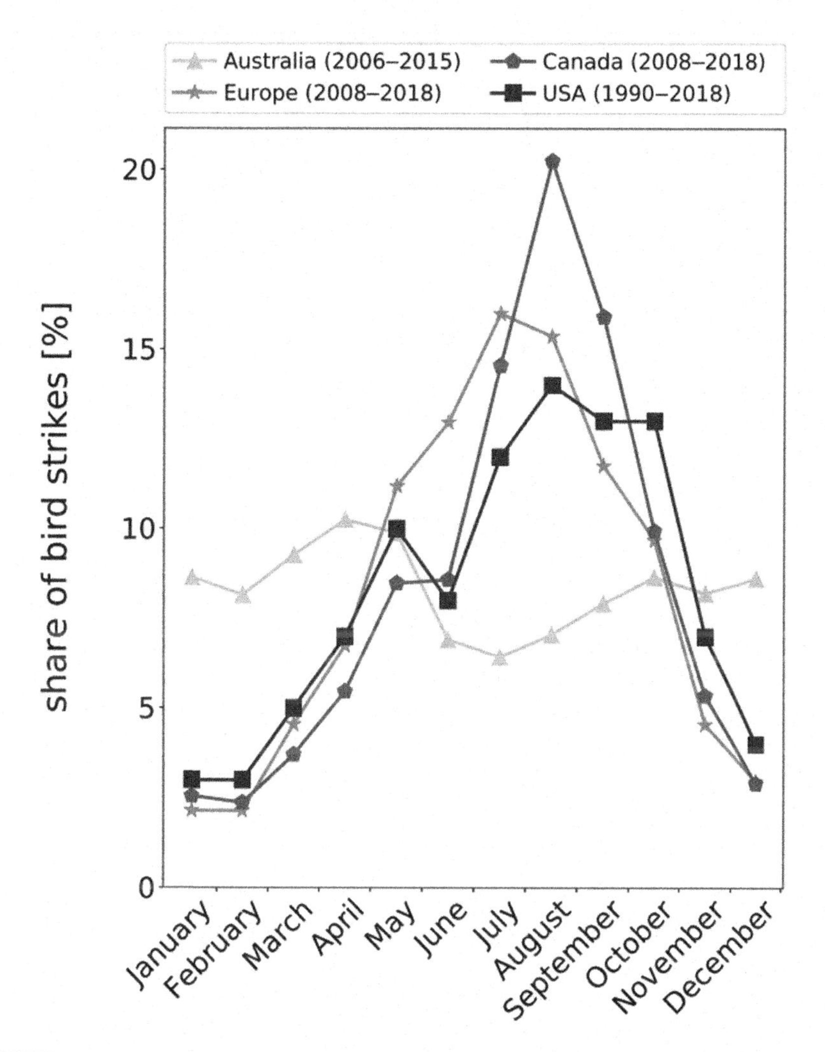

FIGURE 6.3 Seasonal distribution of bird strikes for Australia, Europe, Canada, and the United States [12].

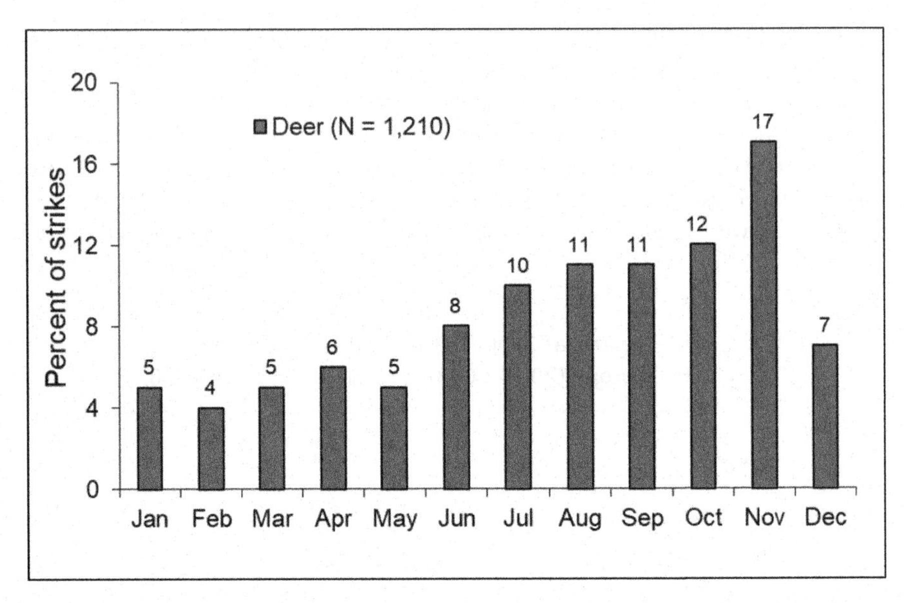

FIGURE 6.4 Monthly percentage of reported deer strikes with civil aircraft by month, United States, 1990–2019 [6].

Courtesy: US Department of Agriculture, Wildlife Services.

is lowest. In contrast, during summertime, the highest number of bird strikes is recorded. During spring and autumn, an increased bird activity due to migration between summer and winter residences leads to more strike.

Figure 6.4 illustrates the monthly percentage of reported deer strikes with civil aircraft by month, United States, 1990–2019. The total reported strikes with deer were 1,210, where strikes with white-tailed, mule, and unidentified species were 1,108, 84, and 18, respectively. The maximum and minimum percentages for deer strike were (17% and 4%) in November and February, respectively.

In addition, 672 strikes with coyotes were reported of which 15% occurred in October.

6.3.4 Time of Occurrence

Most bird strikes occur at day while terrestrial mammals and bats strikes are mostly at night.

The least strikes of all the three groups (birds, mammals, and bats) are at dawn. Table 6.3 lists the reported time of occurrence of known wildlife strikes with civil aircraft, United States, 1990–2019 [6].

The number of unknown times of day accidents was 87,563 cases.

The ICAO analyses for wildlife strike accidents [11] confirmed that during the period of 2008–2015 in 91 countries, 68% of the bird strikes occurred during the day and 25% occurred at nighttime.

TABLE 6.3
Reported Time of Occurrence of Known Wildlife Strikes with Civil Aircraft, United States, 1990–2019 [6]

	Birds		Terrestrial Mammals		Bats	
Time of Day	Number	Percentage (%)	Number	Percentage (%)	Number	Percentage (%)
Dawn	4,886	4	101	4	22	2
Day	84,383	62	633	26	232	22
Dusk	5,945	4	173	7	50	5
Night	39,976	30	1,483	62	770	72

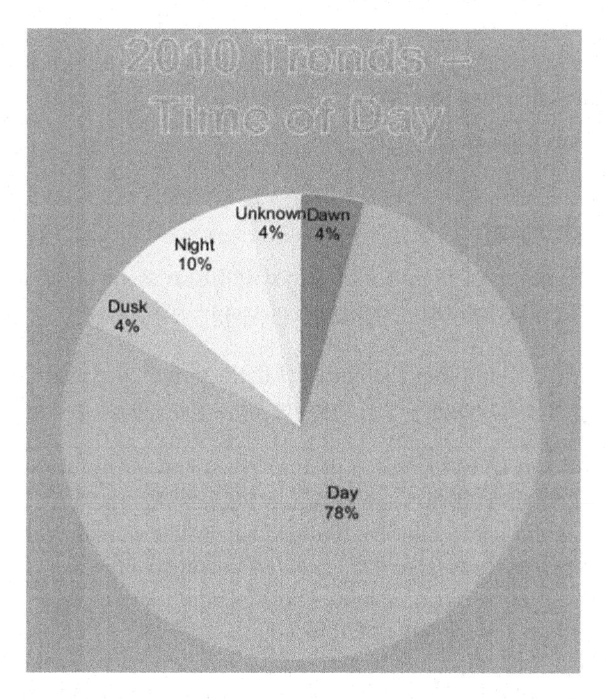

FIGURE 6.5 The number of bird strikes during different parts of the day and night.

Courtesy: Transport Canada [23].

Figure 6.5 shows that some 78% of bird strikes worldwide occur during the day [23].

Figure 6.6 illustrates the hourly bird strike distribution in Canada in the year 1999, which includes both Canadian aircraft overseas and Canadian military aircraft. Most bird strikes occur during daylight hours [23]. Substantial numbers of bird strikes occur in the two periods between 08:00 and 12:00, and 15:00 through 17:00. Unfortunately, these coincide with the periods of maximum scheduled flights.

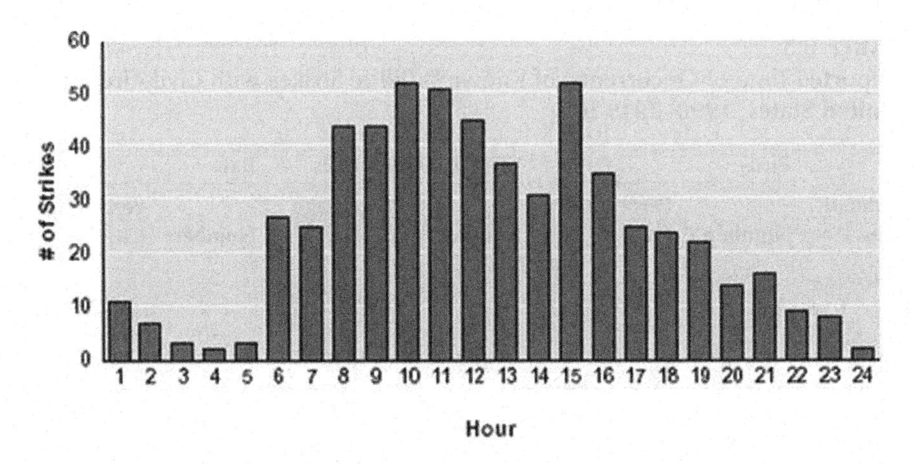

FIGURE 6.6 Hourly bird strike distribution in Canada in 1999 (includes Canadian aircraft overseas and Canadian military aircraft).

Courtesy: Transport Canada [23].

6.3.5 Phase of Flight

Table 6.4 lists the phase of flight for known wildlife strikes with civil aircraft in the period 1990–2019. The following conclusions can be summarized:

1. The asterisk (*) outlines the phase of flight arrival, departure, or local (when pilot conduct "touch-and-go" operations), exact phase of flight could not be identified.
2. A total of "unknown" phase of flight for the three groups were 84,156 strikes. However, 58,950 (70%) were "Carcass Found" reports.
3. The maximum percentage for bird and bat strikes were during approach. The reported percentages were 43% and 71%, respectively.
4. Concerning terrestrial mammals, the maximum strikes were during landing roll recording a percentage of 56%.

Table 6.5 lists the wildlife strikes by several sources including ICAO [11], Canada and United States [23], Australia [24], Panama City [25], EASA [26], and Russia [27].

ICAO database [11] gave the following findings for frequency of strikes per flight phase: parked (0%), taxi (1%), takeoff (31%), en route (4%), approach (33%), landing (26%), and unknown (5%).

As described in [23], the two most critical flight phases in Canada are takeoff and landing where most accidents occur.

Based on the Australian Transport Safety Bureau (ATSB) [24], there were 14,571 birds strikes reported in the period 2004–2013. Bird strikes during takeoff and landing were 38% and 36%, respectively. Strikes during climb and approach were 6% and 18%, respectively. Unknown strikes were 2%.

TABLE 6.4
Reported Known Phase of Flight at Time of Occurrence of Wildlife Strikes with Civil Aircraft, United States, 1990–2019 [6]

	Birds		Bats		Terrestrial Mammals	
Phase of Flight	**Number**	**Percentage (%)**	**Number**	**Percentage (%)**	**Number**	**Percentage (%)**
Parked	96	<1	–	–	2	<1
Taxi	449	<1	–	–	62	2
Takeoff run	24,507	17	39	4	793	31
Climb	23,365	16	58	6	52	2
Departure*	1,736	1	15	2	4	<1
En route	4,169	3	22	2	–	–
Arrival*	550	<1	2	<1	4	<1
Descent	1.881	1	14	2	–	–
Approach	***61,509***	***43***	***655***	***71***	194	7
Landing roll	24,626	17	118	13	***1,449***	***56***
Local*	679	<1	3	<1	28	1

*Phase of flight was determined to be Arrival, Departure, or Local (i.e., pilot conducting "touch-and-go" operations) but exact phase of flight could not be determined.

TABLE 6.5
Reported Known Phase of Flight at Time of Occurrence of Wildlife Strikes with Civil Aircraft, in Canada, UK, Australia, Panama, Russia, and Globally

Phase of Flight	**ICAO 2008–2015**	**Canada 1991–1999**	**UK 1990–2007**	**Australia 2004–2013**	**Panama January 2013–August 2015**	**Russia 1988–1990**
Parked			1.3			
Taxi	1	0.7	0.7		1	
Takeoff	31	30.7	24.5	38	24	12
Landing	26	21.7	33.3	36	27	15
Climb		6.6	10.6	6	6	25
Approach	33	17.1	27.8	18	25	37
En route	4	3.8	1.1			
Descent		1.7	0.7		2	
Unknown	5	17.7		2	15	11

At Panama's Tocumen International Airport, Copa Airlines, the national airline of Panama bird strikes during takeoff and landing are 24% and 27% [25]. Strikes during climb and approach are 6% and 25%, respectively. Strikes during taxi and descent are 1% and 2%, respectively. Unknown strikes are 15%.

The EASA in 2009 [26] and based on data for civil aircraft from the UK and Canada for the period between 1990 and 2007, 24.5% and 33.3% of strikes happened during takeoff and landing, 10.6% and 27.8% during climb and approach, and approximately 1.1% during en-route flight.

In Russia, during a three-year period (1988–1990), the greatest percentage of bird strike was during approach [27]. The climb phase has the second worst percentage. Landing roll is followed, while takeoff run experienced the lowest percentages.

From the above table, aircraft is vulnerable to wildlife strike when it is on or close to ground. Thus, takeoff and climb are the two critical phases when leaving airport, while descent and approach are also the critical phases when aircraft returning to ground. However, an aircraft is much more vulnerable to strikes during takeoff than when landing in most regions. At takeoff, an aircraft's engines are operating at high power settings, and the aircraft is heavier due to a full fuel load.

6.3.6 Bird Strike Versus Altitude

A European study stated that nearly 95% of bird strikes occur at altitudes below 2,500 ft AMSL (above mean sea level), while around 70% occur at altitudes below 200 ft [28].

Moreover, data collected from Directorate General of Civil Aviation of France (French DGAC) and identified by Airbus Industries [29] provide a detailed breakdown for bird strike close to the ground as follows:

- 50%–60% are encountered at altitudes less or equal to 50 ft
- 30% for altitudes ranging from 50 to 500 ft
- 10–20% occur at higher altitudes.

The FAA wildlife hazard management [6] categorized bird strike accidents in United States against altitude for the two groups of commercial and general aviation aircraft, as shown in Figures 6.7 and 6.8.

Figure 6.7 displays the number of reported bird strikes with commercial (top graph) and general aviation (GA) aircraft (bottom graph) in the United States by 1,000-ft height intervals above ground up to 18,000 ft for commercial aircraft and up to 12,000 ft for GA aircraft, 1990–2019.

Above 1,000 ft, the number of reported strikes declined consistently by 34% and 43% for each 1,000-ft gain in height for commercial and GA aircraft, respectively. The exponential equations explained 98% of the variation in number of strikes by 1,000-ft intervals from 501 to 18,500 ft for commercial aircraft and 501 to 12,500 ft for GA aircraft

Figure 6.8 demonstrates that:

- For commercial aircraft, the percentages of total strikes and total damaging strikes below 500 ft are 70.5% and 56.1%, respectively, and above 500 ft 29.2% and 43.9%
- For general aviation aircraft, the percentages of total strikes and total damaging strikes below 500 ft are 71.5% and 48.9%, respectively, and above 500 ft 28.5% and 51.1%.

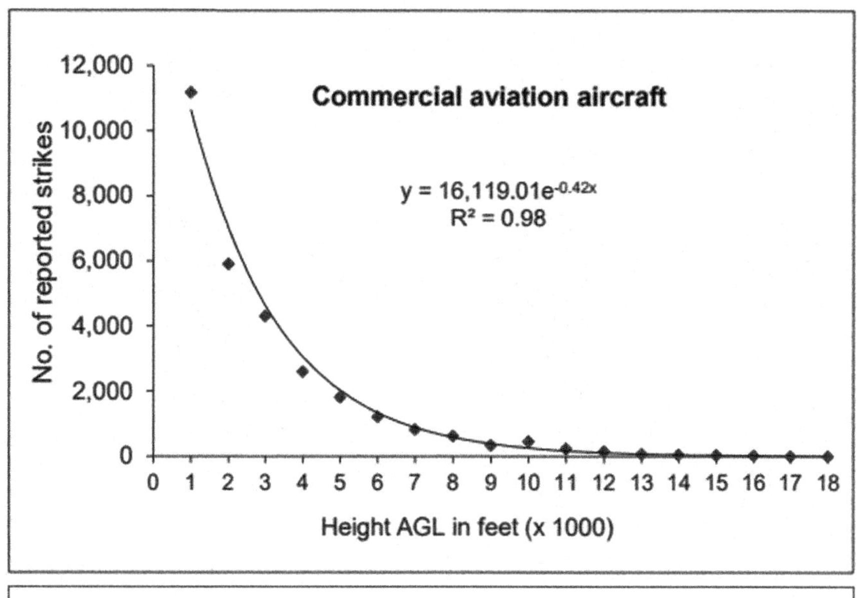

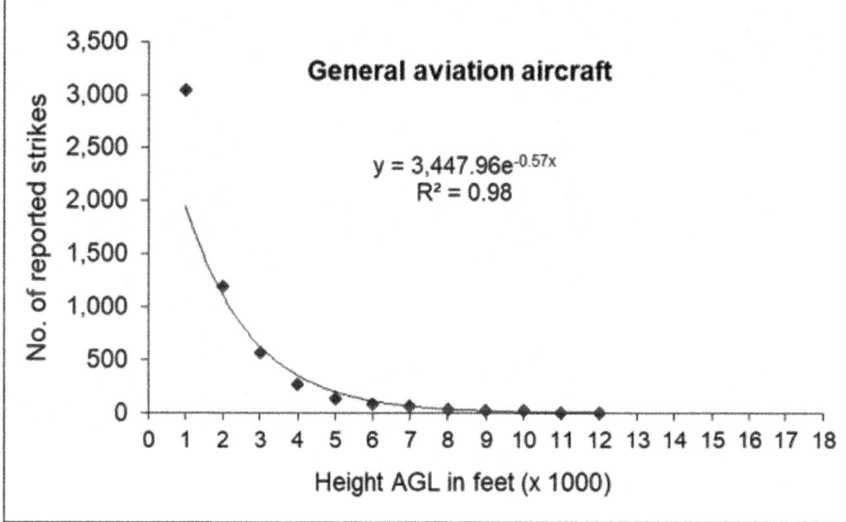

FIGURE 6.7 Number of reported bird strikes with commercial (top graph) and general aviation (GA) aircraft (bottom graph) in the United States by 1,000-ft height intervals above ground level during 1990–2019.

Courtesy: USDA.

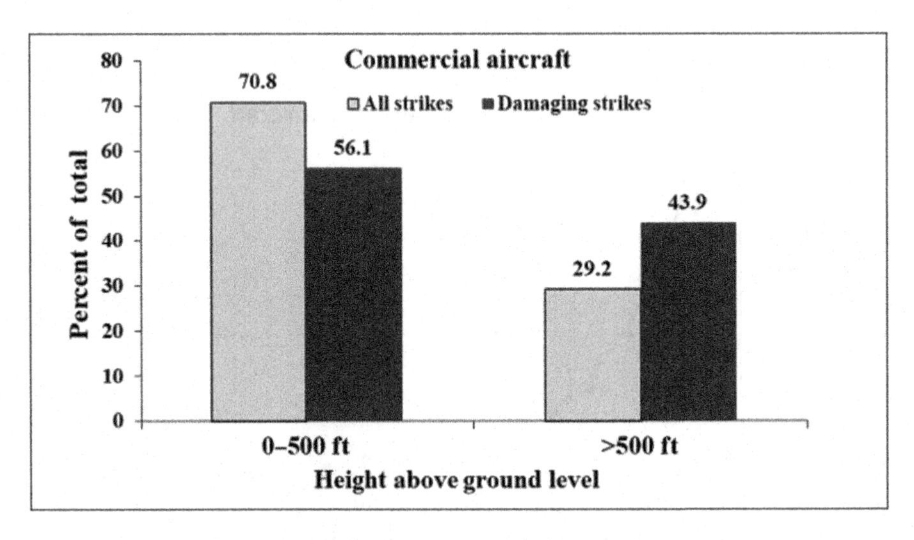

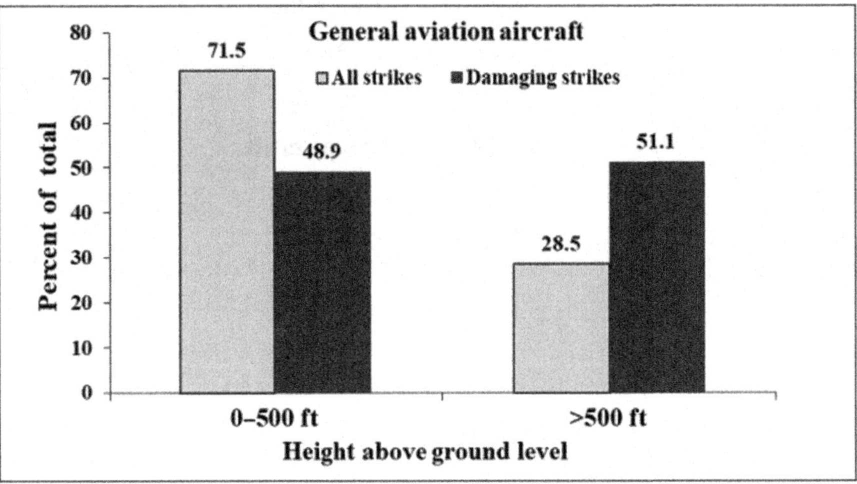

FIGURE 6.8 Percentages of total strikes and total damaging strikes for commercial (top graph) and general aviation (bottom graph) aircraft in the United States, 1990–2019.

Courtesy: USDA.

It is worth mentioning that few accidents occur at high altitudes, some as high as 20,000–30,000 ft. However, two astonishing cases for birds flying at very high altitudes are as follows:

- Bar-headed geese were seen flying at 10,175 m (33,383 ft)
- Rüppell's vulture flying at an altitude of 11,300 m (37,100 ft) collided with an aircraft over the Côte d'Ivoire

6.3.7 Critical Aircraft Parts

Civilian fixed-wing aircraft are categorized as "Commercial Transport and General Aviation" aircraft. Both types are next divided into two groups based on their engine: namely, turbine-powered or shaft-powered aircraft [30]. Turbine-powered aircraft are powered by either turbojet or turbofan engines, while shaft-powered aircraft are powered by either piston engines (normally small aircraft) or turboprop engines.

The parts of civil aircraft (fixed- and rotary-wing) mostly impacted by birds are the nose, radome, windshield, engines, fuselage, wing, helicopter rotor, empennage, landing gear, propeller, and lights. Table 6.6 lists civil aircraft components struck and damaged by wildlife, United States, in the period 1990–2019 [6].

"Others" include pitot tubes, wiper blades, antennae (communication, radar, or global position), total air temperature (TAT) probes, and angle of attack sensors. The number of strikes for these modules are respectively: are 717, 446, 250, 136, and 150.

Many publications focus only on bird strikes including Airbus Industries [29].

Table 6.7 lists the percentages of bird strikes with different parts of the aircraft as reported by ICAO [11], Airbus Industries [29], EASA [31], Boeing Co. [32], and Russian companies [33].

From Table 6.7, engines are subject to the most strikes and damage as registered by Airbus and Boeing companies, as well as European (EASA), ICAO, and Russian airlines. Wing comes next in statistics of EASA, ICAO, Boeing, and Russian airlines. Windshield comes in the third rank, while and propeller are the least parts struck.

TABLE 6.6
Civil Aircraft Components Struck and Damaged by Wildlife, United States, in the Period 1990–2019 [6]

	Birds (30 Years Total)		Terrestrial Mammals (30 Years Total)	
Aircraft Components	Number Struck	Number Damaged	Number Struck	Number Damaged
Windshield	29,563	1,203	9	17
Nose	26,952	1,319	122	114
Wing/rotor	26,560	4,721	336	338
Radome	22,318	1,796	19	17
Engines	21,470	5,042	193	187
Fuselage	21,434	839	162	166
Landing gear	8,332	613	1,502	515
Propeller	3,790	299	357	317
Tail	2,363	793	62	84
Light	1,177	815	55	58
Others	27,612	1,730	459	305
Total	191,571	19,170	3,276	2,118

TABLE 6.7
Summary of Percentages of Bird Strikes with Different Aircraft Parts

Company/Authority Affected area	Airbus	Boeing	EASA	ICAO	Russian Companies	Range
Nose		8%	8%	6%		6%–8%
Radome	41%			7%	4.9%	4.9%–41%
Windshield		13%	13%	5%	6.7%	5%–13%
Engine 1						
Engine 2	41%	44%	44%	21%	48.9%	21%–48.9%
Engine 3						
Propeller				2%		2%
Fuselage	7%	4%	4%	5%	9.0%	4%–9%
Landing gear	3%			5%	6.7%	3%–6.7%
Wing	7%	31%	31%	19%	21.1%	7%–31%
Tail	1%			3%	2.7%	1%–2.7%
Lights				6%		6%
Others				20%		20%

TABLE 6.8
Damaged Modules due to Bird Strike of Aircraft Powered by Turbine-Based and Shaft-Based Engines

Aircraft Engine	Radome and Nose%	Windshield%	Engine%	Propeller%	Wing%	Fuselage%	Landing Gear
Turbine-based	8	13	44	–	31	4	–
Shaft-based	21	14	15–20	7	19	10	9

A comparison between accidents caused by bird strikes with aircraft powered by turbine-based (turbojet and turbofan) and shaft-based engines (piston engines and turboprop engines) are given in Table 6.8.

The noise generated by propellers in either piston or turboprop engines frightens birds and results in lower percentages of bird strike on the engines and wings.

6.3.8 Species of Birds, Terrestrial Mammals, Bats, and Reptiles

In the United States, statistics for wildlife strike in the period of 1990–2019 may be summarized as follows [6]:

- Total bird species were 591 causing a total 124,387 accidents for known birds and 98,366 for unknown birds including:

    - Waterfowl, total strikes: 6,183, damaging strikes: 2,411, and costs: $268,281,376
    - Hawks, eagles, vultures, total strikes: 7,964 and costs: $149,909,175
    - Gulls, total strikes: 12,442, damaging strikes: 1,580 and costs: $64,576,857
    - Pigeons, doves, total strikes: 16,819, damaging strikes: 583 and costs: $25,587,007
    - Owls, total strikes 3,661 and costs: $11,208,435
    - Starlings, mynas, total strikes: 5,249 and costs: $8,132,087
    - Thrushes, total strikes: 2,732 and costs: $6,604,672
- 51 species of terrestrial mammals (mostly artiodactyls [mainly deer] and carnivores [mainly coyotes]). The number of strikes was 1,260 and 1,848, respectively. Damaging strikes were 1,043 and 93, respectively. Both groups cost $62,768,802 and $4,602,067, respectively
- Flying mammals (bats) belong to 36 species. They included megabats and microbats causing 15 and 3,286 strike cases and costs $ 4,678, 299 and $708,520, respectively
- Total reptile species were 23 resulting in 491 cases including 361 turtles, 27 American alligator, and 78 snakes

In conclusion,

- The total known all species were 132,918, where 9,799 causing damage and 7,646 causing NEOF. A total of 17,985 cases with multiple animals, causing 865,763-h downtime and costing $700,108,822.
- Unidentified birds were 98,402 cases, 7,560 only causing damage, 5,297 causing negative effect-on-flight NEOF, cases of multiple animal strikes were 8,983, downtime-hour was 259,406, and total costs were $147,136,268.
- Grand total 231,320, where 17,359 causing damage and 12,943 causing NEOF. A total of 26,968 cases with multiple animals, causing 1,125,169-hour downtime and costing $847,245,090.

According to ICAO [11] analysis for 47,748 cases worldwide, most frequently struck species and its percentages of all cases were:

- Perching birds (22%)
- Shores birds (11%)
- Hawks, eagles, vultures (9%)
- Pigeons, grouse (7%)
- Swift, tree-swift, hummingbird (2%)
- Ducks, geese, swans (2%)
- Owls (1%)
- Mammals (3%)
- Others (39%)

In Australia, based on ATSB [24], 14,571 bird strikes were reported between 2004 and 2013, most of which involved high-capacity air transport aircraft. The four most struck birds are kites, bats/flying foxes, lapwings/plovers, and galahs. Kites had the most significant reported strikes per year, with this species being involved in an average of 129 strikes per year for 2012 and 2013. Galahs were more common birds in strikes involving multiple birds, with more than 38% of galah strikes involving more than one galah. However, larger birds were more likely to result in aircraft damage.

In Panama City, the most critical birds are black vultures, migratory raptors – mostly Swainson's hawks and turkey vultures [25].

6.3.9 Negative Effects for Wildlife Strike On-flight (NEOF)

Both the ICAO and FAA studied the negative effects of wildlife strikes on flight [11, 6].

ICAO [11] analyzed 97,751 reports for wildlife strike, received from 91 countries on strikes occurring in 105 countries for the years 2008 to 2015. A summary of such wildlife strikes reported to the ICAO Bird Strike Information System may be categorized as follows:

- The effect of the wildlife strike on the flight was reported 12,227 times
- Out of those, 2,501 (20%) had a clear indication of an effect on the flight
- There were 1,230 precautionary landings, which account for the highest number of effects (49%). Pilot jettisoned fuel or burned fuel in a circling pattern to lighten aircraft weight or in which an overweight landing was made. An average of 94,358 pounds (13,876 gallons) of fuel was dumped per incident
- Aborted takeoffs 513 cases (21%)
- Engine(s) being shut down 63 cases (3%)
- Others (28%) which includes (but not limited to)
  - Flight delay (211 times)
  - Declaring technical emergency (54 times)
  - Aircraft return (137 times).

During the period 1990–2019, FAA [6] reported 231,320 strikes (227,045 in the United States and 4,275 strikes by US-registered aircraft in foreign countries). The negative effect-on-flight during this period was in 5% and 17% of the bird and terrestrial mammal strike reports, respectively. It is categorized also as [6]:

- Precautionary/emergency landing after striking wildlife in 6,993 incidents (3% of reported strikes). It included 267 incidents in which
  - The pilot jettisoned fuel (each 13,876 gallons) (Figure 6.9)
  - Burned fuel in circling pattern (Figure 6.10)
  - Made an overweight landing (Figure 6.11)
- High-speed aborted takeoff ($\geq 100\,knots$) (Figure 6.12)
- Engine shut down in 470 incidents
- Other reasons in 2,850 incidents like reduced speed because of shattered windshield, flight delays, or crash landing.

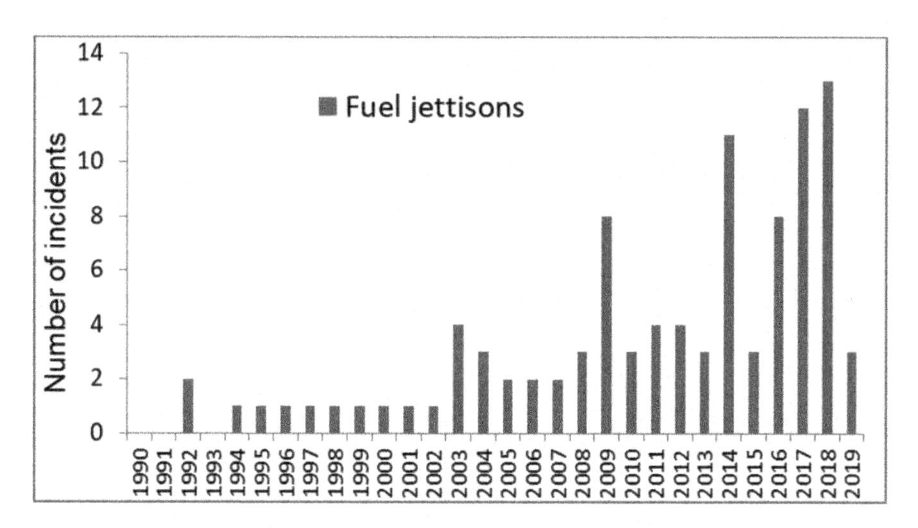

FIGURE 6.9 Number of reported incidents where pilot precautionary jettisoned fuel after bird strike in the period 1990–2019 [6].

Courtesy: USDA.

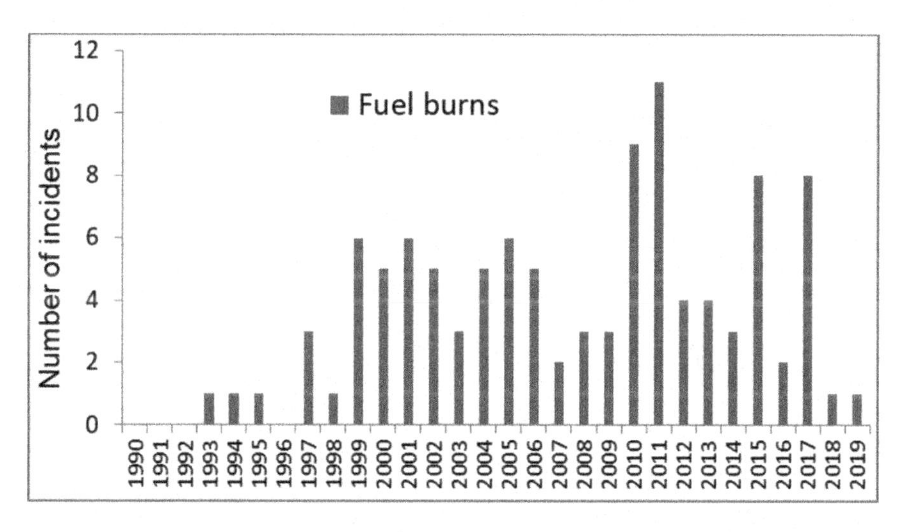

FIGURE 6.10 Number of reported incidents where pilot precautionary burned fuel after bird strike in the period 1990–2019 [6].

Courtesy: USDA.

6.3.10 Hazardous Bird Species

The most hazardous species causing fatal accidents in 100 Years (1912–2012) may be categorized based on the type of aircraft [35]. Two cases are considered: namely,

- Airliners and executive jets
- Aeroplanes 5,700 kg and below

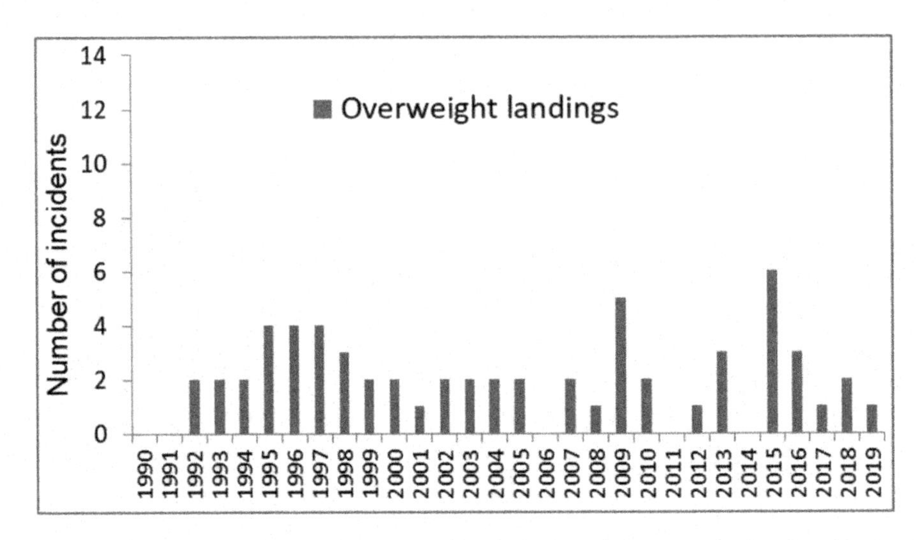

FIGURE 6.11 Number of reported incidents where pilot made overweight landings after bird strike in the period 1990–2019 [6].

Courtesy: USDA.

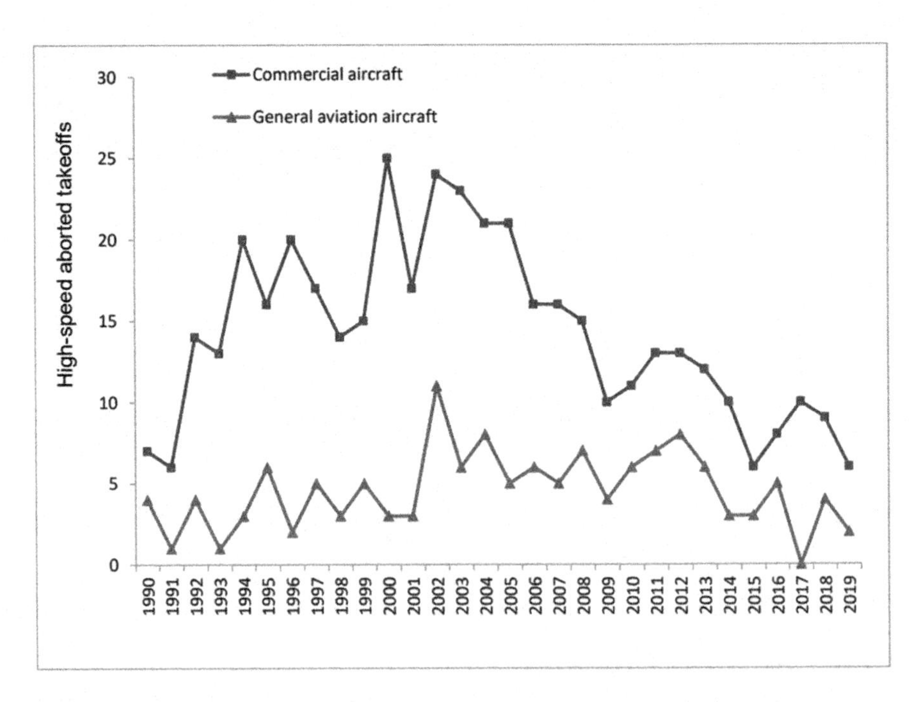

FIGURE 6.12 Number of reported high speed aborted takeoff after bird strike in the period 1990–2019 [6].

Courtesy: USDA.

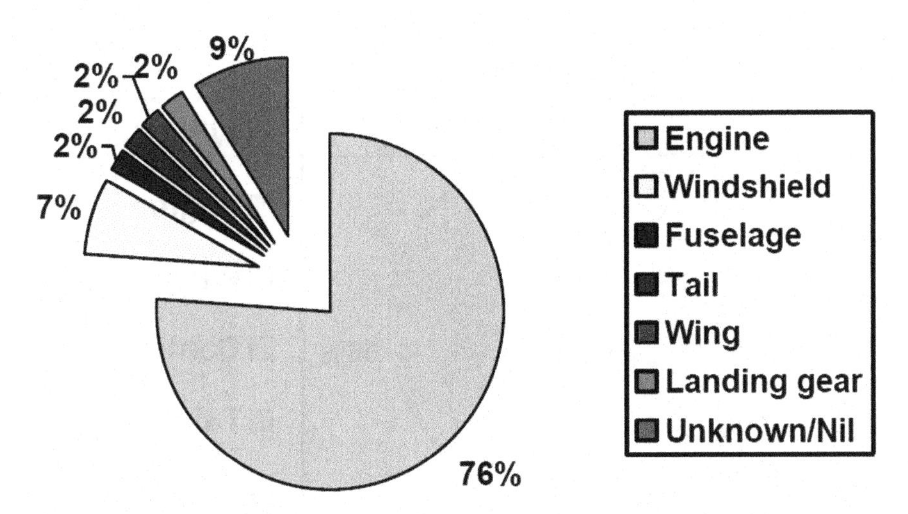

FIGURE 6.13 Part struck by birds for transport and executive jet aeroplanes.

Courtesy: IBSC [35].

TABLE 6.9
Species Causing Fatal Accident for Transport/Executive Jet Aircraft and Small Aeroplanes (5,700 kg or Less)

Bird Species	Gulls	Prey	Pigeon	Waterfowl	Perching Birds	Geese/ Duck	Jackdaw	Pelican	Others
Transport aeroplanes and executive jets	40%	15%	15%	11%	15%	–		–	4%
Aeroplanes of 5,700 kg and below	13%	47%	–	–	–	21%	3%	8%	8%

For Transport Airliners and Executive Jets, there were 45 accidents, among which 16 were fatal ones resulting in 189 deaths (which includes 7 third parties on the ground) and destruction of 44 aircraft.

Figure 6.13 illustrates the parts repeatedly struck by birds. Engine damage was the cause of 76% of the accidents for transport/executive Jet planes, followed by windshields with 7%. Fuselage, wing, tail, and landing gear each contributed 2%.

Small aeroplanes (5,700 kg and below) experienced 32 fatal accidents resulting in 69 deaths and destruction of 56 aircraft [35]. Table 6.9 displays the different species causing fatal accident for transport/executive jet aircraft and small aeroplanes (5,700 kg or less).

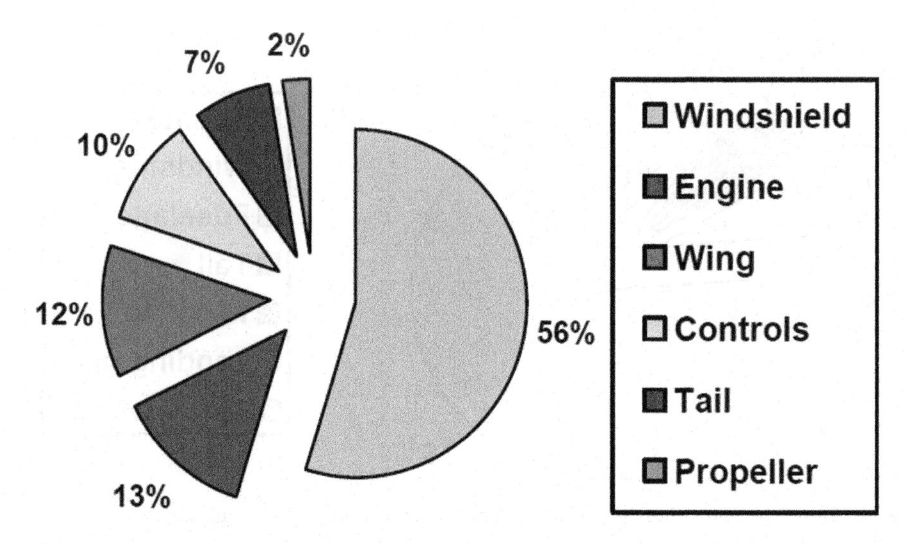

FIGURE 6.14 Part struck by birds for small aeroplanes (5,700 kg or below).

Courtesy: IBSC [35].

As illustrated in Figure 6.14, the windshield is struck in 56% of accidents with the engine and the wing in 13% and 12%, respectively, of cases followed by controls with 10% and the tail at 7%. Propeller for aircraft powered by shaft engines is 2%.

6.3.11 Population Increase of Birds

Wildlife strikes have increased in the past 40 years because of a combination of expanding populations of many wildlife species that are hazardous to aviation and increasing numbers of turbofan-powered aircraft [36].

6.3.11.1 Canada Goose

Resident Canada geese (mean body mass = 9 lbs.) increased fourfold from about 1.1 million to 4.4 million birds in the United States in the period from 1990 to 2019 (Figure 6.15). The resident population now are greater than the migratory population that nests primarily in northern Canada and winters in the United States [6].

6.3.11.2 Snow Goose

North American snow goose (mean body mass = 6 lbs) population increased from about 4 million to 15 million birds during the period from 1990 to 2019.

6.3.11.3 Red-tailed Hawk

The red-tailed hawk population has increased to over 3 million birds since 1990 [37] and expanded into urban environments.

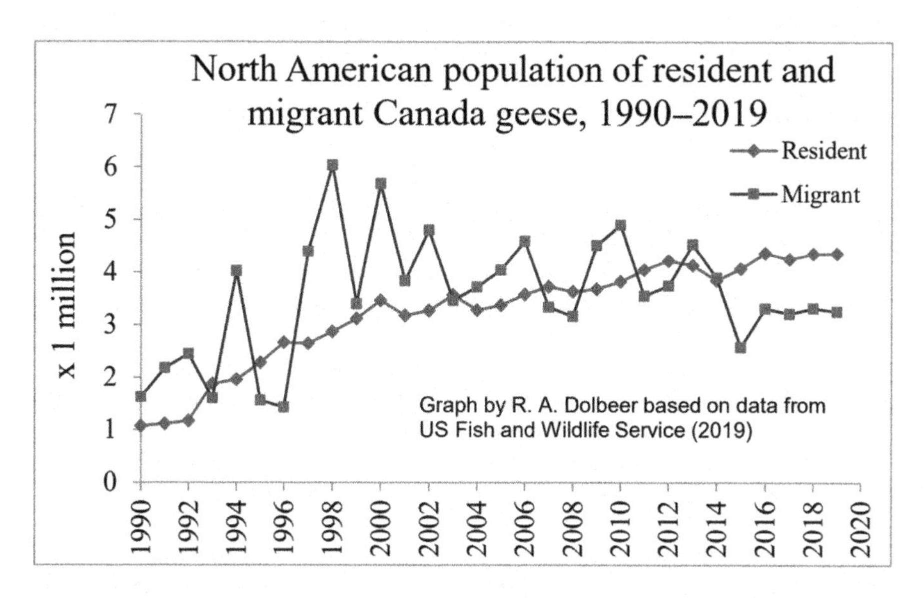

FIGURE 6.15 Population of Canada geese in the period 1990–2019 [6].

Courtesy: USDA.

6.3.12 Terrestrial Mammals

Not only birds are hit by planes, but some of the animals have been hit by planes. Mammals are the most common animals other than birds which are associated with aircraft strikes. Mammal strikes are only a small percent of the total recorded strikes, but strikes with large ungulates including deer are extremely dangerous [38, 39]. In the past decade, airplanes have hit bats, coyotes, raccoons, skunks, opossums, dessert hares, prairie dogs, cats, dogs, foxes, bull snakes, turtles, armadillos, alligators, badgers, moose, at least one woodchuck, an elk, an antelope jackrabbit, and several rather ominous sounding "unknown terrestrial mammals." According to the FAA database [40–45]:

- 25 people were killed due to airplane wildlife strikes (1990–2013)
- 671 planes run afoul due to wildlife strikes with white-tailed deer (1990–2013).
- More than 1,000 deer have been hit by airplanes across the year 1997–2017
- The average repair costs for the plane after hitting a deer is $111,00 (1990–2013)
- In the period from 1982 to 2000,
  - Totally 901 mammal strikes with civil aircraft
  - The number of damaging strikes was 502 (71%)
  - Deer strikes totaled 518 and number of damaging strikes was 425 (86%)
  - Coyote 92 strikes and number of damaging strikes 11
  - Fox 36 strikes and number of damaging strikes 4
  - Rabbit/hare 33 and number of damaging strikes 2
  - 62 deer strikes to civil aircraft carrying over 100 passengers, United States, 12 cases for B737 and 12 cases for DC-9

    - Greatest deer strikes was in November and least in February
    - Most deer strikes occurred at dusk followed by night
    - Greatest deer strikes was during landing roll and least in taxi
    - Aborted takeoff in 75 cases, 10 cases for engine shut down
    - 75% of strikes had a negative effect on the flight. Only 20% of reports indicating damage provided estimates of repair costs. The mean cost for these reports was $74,209 (range $100–$1,400,000).
    - 20 reported human injuries resulting from deer, but the potential exists for a major disaster. For example, aircraft with a capacity of 101–380 passengers were involved in 12% of the reported strikes.
- 1,000 white-tailed deer caused $45 million in damage (1990–2015)
- 152 mammal strikes were reported in Canada as follows: rabbits: 24%, striped skunk: 13%, coyote: 12%, fox: 11%, and white-tailed deer: 7% (1991–1999)
- 27 alligators, 24 green iguanas, and 1 spectacled caiman were struck by aircraft in the United States (1990–2019).

In the most recent study (2021), the following findings are obtained [46]:

- Incidents involving mammals strike with aircraft account for approximately 3%–10% of all recorded strikes. However, relatively little research has been conducted on mammal strikes outside of the United States
- Data on mammal strike obtained from six major national aviation authorities and a global aircraft database identify the following:
    - 40 families were involved in strike events in 47 countries
    - Mammal strike are annually increasing by up to 68%
    - Chiroptera (four families) accounted for the greatest proportion of strikes in Australia
    - Leporids and canids in Canada, Germany, and the UK
    - Chiroptera (five families) and cervids in the United States
    - Most mammal strike occurred during the landing phase
    - Circa-diel strike risk was greatest at dusk
    - Circa-annum strike risk was greatest during late summer
    - The total estimated cost of damage resulting from reported mammal strikes exceeded US$ 103 million in the United States alone, over 30 years.

6.4 STATISTICS FOR WILDLIFE STRIKE WITH FIXED-WING MILITARY AIRCRAFTS

6.4.1 Introduction

For both military and civil aircraft, wildlife–aircraft strikes threaten the safety of aircraft and endanger the lives of both humans and animals. It may result in catastrophic effects, loss of hundreds of millions of dollars per year in aircraft damage as well as lost flight hours [47]. However, military aircraft operations are different from civil

FIGURE 6.16 Bird strike with F-16.

Courtesy: Louis M. DePaemelaere – One Mile High Photography.

ones in several aspects, including mission types, air speeds, and maneuverability [48]. Military aircraft travel at high speeds and close to the ground (Figure 6.16), which likely increases the strike hazard compared to civil aircraft because of the more time spent in bird-rich altitudes [49]. Damage to aircraft is a function of the bird's kinetic energy which is the product of the mass of the bird and the velocity of the aircraft [50–52].

Bird strike reporting is mandatory for marine and naval aviators – even if an aircraft sustains no damage from the incident. Also, all aviation personnel should have a bird/animal aircraft strike hazard (BASH) awareness mindset. Pilots in the cockpit, control tower and ground electronics personnel, aircraft and grounds maintenance personnel, firefighters, and security personnel, even the duty sweeper collecting dead birds and turning them in for identification, all have an integral responsibility to making the entire BASH program effective. This will enable a rather accurate database for military aircraft.

However, until now databases for bird strikes with military aircraft are not as complete as for civilian aircraft [48, 49]. A study covering the period from 1990 to 2013 concluded that the number of strikes with military aircraft is only 10% or even less than those associated with civilian aircraft [53].

6.4.2 Statistics

6.4.2.1 Annual Bird Strike

6.4.2.1.1 Worldwide in the Period 1950–1999

Statistics for accidents caused by bird strikes with military aircraft worldwide in the period 1950–1999 are summarized hereafter [27, 49, 54]:

- A total of 286 serious accidents to military aircraft from 32 countries
- Most accidents were in Europe (east to Russia), plus Canada, the United States, Australia, and New Zealand
- 190 European military aircraft were destroyed because of collisions or attempts to avoid collisions with birds
- At least 63 of them were fatal, with at least 141 fatalities (137 aircrews and 4 on the ground)
- The worst decade was the1990s, with at least 68 fatalities
- Large numbers of bird-related accidents were in Germany (60 aircraft from at least eight countries), the UK (47), and the United States (46+)
- The number of accidents for fighter or attack aircraft with one engine was at least 179 accidents, followed by aircraft with two engines (40+) and jet trainers (34+)
- Seven four-engine heavy transport military aircraft were lost (three of them in the 1990s)
- Some accidents to military aircraft were reported in Asia (especially India)
- Few accidents were reported in Africa and South America.

6.4.2.1.2 US Military Aircraft

6.4.2.1.2.1 *Period 1965 to 1975* The total cost for damaged or lost aircraft exceeded US$ 81 million. From 1965 to 1970, around US$ 20 million was the cost of bird strike. From 1970 to 1975, approximately US$ 70 million has been lost, of which over US$ 61 million was the cost of destroyed aircraft. The actual costs are greater than that US$ 81 million as it does not include the cost for the downtime required to repair the damaged aircraft.

6.4.2.1.2.2 *Period 1990 to 2017* A comprehensive study for bird strikes with the aircraft of the two branches of US military (US Air Force [USAF] and US Navy [USN]) in the period 1990–2017 in foreign and domestic operations was performed in [48, 55]. The main findings are summarized here:

- 104,129 wildlife strike with USAF aircraft
- 6,733 wildlife strike with USN
- Annual average of $20 million in damage and human injuries for the USN
- Annual average of $38 million in damage and human injuries for the USAF
- Costs for USAF from 2008 to 2011 are US$ 11,042,236, $13,084,126, $22,341,664, and $12,560,871
- After removing strikes for unknown avian species occurred in foreign countries, the combined dataset was reduced to 36,979 strikes. Of these, 3,646 strikes were from the USN, and 33,333 strike records came from the USAF

- 923 strikes with substantial damage (Class A/B/C) for the combined dataset, each cost $50,000 to $2,000,000
- 3,024 strikes with minor damage (Class D/E/H) for the combined dataset, with a maximum cost of $87,570
- 33,032 strikes reported with no damage (damage class unknown)
- 186 bird species was involved in 20 or more strikes with USN and USAF aircraft
- There were 40 species or species groups involved in strikes with no substantial damage
- The top three species with substantial damage were snow goose, common loon (*Gavia immer*), and a tie between Canada goose (*Branta canadensis*) and black vulture (*Coragyps atratus*)
- Filtering the strike records to only fighter, cargo, stealth, rotorcraft and species or species groups that were struck over 20 times reduced the total number of strike records from 33,333 to 31,082 having the following breakdown:
    - 9,535 strike records for fighter airframes (65 species struck over 20 times)
    - 20,174 records for cargo airframes (98 species struck over 20 times)
    - 674 strike records for stealth airframes (12 species struck over 20 times)
    - 699 strike records for rotorcraft (16 species struck over 20 times)

Table 6.10 lists the count different mishaps for USAF in 2010 and 2011 [55]. Costs for fiscal 2010 were $10 million more than the surrounding years (2008–2011) due to a single Class A mishap caused by a bird strike.

Figure 6.17 illustrates a comparison of the relative hazard score for the airframe of six avian species and four military aircraft as well as civil aircraft [56].

6.4.2.1.2.3 Period 1995 to 2016 In the period 1995–2016 (21 years), a countdown of the top bird species responsible for USAF wildlife strikes by cost [27]:

- $93,812,397 million due to Canadian geese
- $75,686,764 million due to Black vultures
- $43, 262,092 million, 41,760,459 million, $37,767,636 million, and $29,290,414 million due to strikes by pink-footed geese, American white pelicans, turkey vultures, and mourning dove

TABLE 6.10
Count and Cost of Different Mishaps for USAF in 2010 and 2011 [55]

	Fiscal Year 2010			Fiscal Year 2011		
Class	Count	Percentage of Total	Cost	Count	Percentage of Total	Cost
A	1	0.02	$10,011,204	0	0.00	$0.0
B	2	0.04	$2,082,753	5	0.13	$4,419,861
C	43	0.91	$7,006,128	38	0.86	$5,376,233
E	4,676	99.03	$3,241,579	4,383	99.11	$2,764,777
Total	4,722	100	$22,341,664	4,426	100	$12,560,871

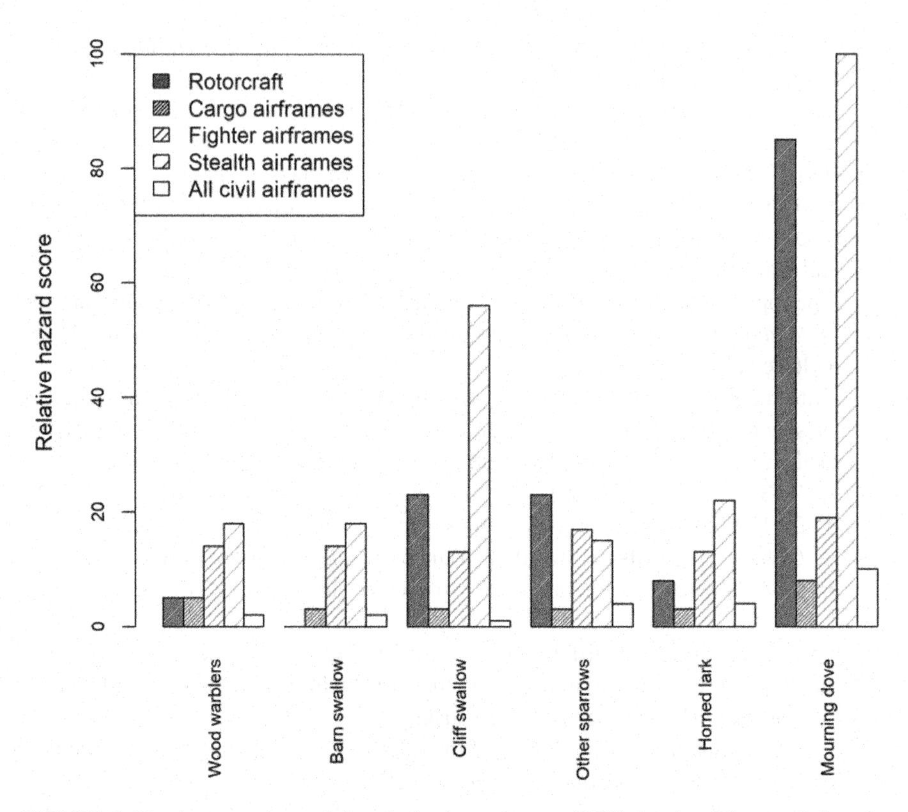

FIGURE 6.17 A comparison of the relative hazard score (RHS) for the airframe of six avian species and four military aircraft as well as civil aircraft [48].

Courtesy: USDA, APHIS, Wildlife Services.

- $24,955,020 million, 17,039,479 million, $10,304,474 million, and $9,400,918 due to strikes by spot-billed duck, red-tailed hawk, mallard, and snow goose.

6.4.2.1.2.4 Period 1995 to 2019 In the period (24 years) [57],

- 27 people killed
- 13 aircraft lost
- Total cost excluding the cost of injuries: $817,546,884

6.4.2.1.2.5 Period 2004 to 2017 In this period (13 years) [59],

- The Marine Corps has recorded roughly 1,540 bird or animal strike incidents. Less than 30 of those incidents resulted in serious damage or a mishap classification of A, B, and C.
- The Air Force recorded more than 69,000 wildlife-aircraft strikes. Those strikes resulted in the deaths of 23 aviators and destroyed 12 aircraft, causing millions in damages.

- On May 7, 2017, F-35 bird strike out of Iwakuni, Japan, caused more than $2 million in damages. It was the second-Class A mishap resulting from a bird strike for the Corps in the past 15 years.
- On April 17, 2017, an Air Force F-16 (49th Wing out of Holloman Air Force Base, New Mexico) struck a hawk during a routine landing.

6.4.2.1.2.6 *Period 2011 to 2017* As reported by *Military Times* [27]

- Air force aircraft costs is nearly US$ 182 million, based on Bird/Wildlife Aircraft Strike Hazard (BASH) team
- Naval aircraft cost is estimated at nearly US$ 64.8 million.

6.4.2.1.3 *European Military Aircraft*

Europe has numerous air bases and military aircraft which confront bird strikes. Gulls and diurnal raptors are the most dangerous bird species [58]. They are followed by swallows, swifts, pigeons, European starlings, Northern lapwings, black-headed gulls, Eurasian kestrels. Herons and vultures are locally dangerous.

Gulls are the cause of many strikes in the British Islands and the coastal part of the continent. Common gulls and herring gulls are the sources of danger in northern Europe. In southern Europe, yellow-legged gulls and vultures pose great danger. In western and central Europe, diurnal raptors, corvids, and storks in addition to gulls are the source of danger. All over Europe, norther lapwings, swifts, black-headed gulls, Eurasian kestrels, and European starlings create real danger.

6.4.2.2 Monthly Strike

Monthly bird strikes in the period 1965–1975 showed that bird strikes are maximum during September, October, and November [49]. It was interpreted as fall (autumn) migration of birds starts in August, and consequently, heavy strikes are encountered in September, October, and November. Another peak for strikes is in April and May. This is due to the spring bird migration. Bird species responsible for major damage of airframe modules (wings, tail assemblies, and the aircraft's nose) are the migrating waterfowl, passerines, and shorebirds.

Bird migration in North America includes the four flyways: Atlantic, Central, Mississippi, and Pacific [48]. The Mississippi flyway causes the lowest probability of aircraft damage while the Atlantic flyway caused substantial damage [48].

6.4.2.3 Bird Strike by the Time of Day

Table 6.11 summarizes the time of occurrence of US military aircraft in 1983. Most bird strikes were encountered during the day (67%) versus 18% during the night [49]. Dawn and dusk strikes made in total 5%. Unknown strike times were only 10% of the total.

The detailed study of the airbase Randolph Air Force Base (AFB; UTM 14N 569946/3266974) [47] concluded that from 2010 to 2014, the peak hours for columbid strikes (pigeons and doves) occurred from 0800–1000, with a smaller afternoon peak from 1500 to 1700. Strikes of white-winged dove strikes were consistent with columbid strikes but relatively fewer during the 0900–1000 interval. However, there were no distinct hourly peaks when considering all wildlife–aircraft strikes.

TABLE 6.11
Reported Time of Occurrence of Bird Strikes with Military Aircraft, United States, 1983 [49]

Time of Day	Dawn/Dusk	Day	Night	Total Known	Unknown	Total
Percentage	5	67	18	90	10	100

TABLE 6.12
The Percentages of Different Parts of Military Aircraft Struck by Birds [49]

Part	Engine/ Engine Cowling	Windshield	Wings	Nose/ Radome	Fuselage	External Tanks	Multiple Hits	Unknown
Percentage	22.3	20.6	19.3	15.1	8.9	6.7	5.2	1.9

If bird strikes with military aircraft (Table 6.11) are compared to civil aircraft (Table 6.3), both have the same trends with different ratios. Daylight is only 62%, while night strikes accumulated 30%, dawn and dusk recorded 4%.

6.4.2.4 Bird Strike by Part

Bird ingestion into the engines of military aircraft is the worst scenario for a bird strike. Next, the windshield is subject to a great share of strikes. Table 6.12 lists the percentages of different military aircraft parts struck by birds in 1950–1999 [49]. Comparing military aircraft with civil ones listed in Tables 6.7 and 6.8, one can conclude that both categories have the same trends.

6.4.2.5 Bird Strike by Altitude

Most bird strikes (80% or more) in the period 1970–1980 occur at altitudes from 0 to 3,000 ft AGL, with some 60% or more occurring below 500 ft [31]. This is due to two factors:

- Birds routinely fly at these altitudes (except when migrating)
- Aircraft must pass through these altitudes when they take off and land

Table 6.13 categorizes bird strikes with military aircraft against the impact altitude from different sources.

6.4.2.6 Bird Strike by Flight Phase

From 2010 to 2014, most bird-strike accidents [47] occurred during landing (49%) and takeoff (26%). Columbid strikes occurred during landing 55% and takeoff 41%, respectively. No records indicated strikes with columbids when aircraft were engaged in low-level flight. Strikes involving white-winged doves only occurred during the

TABLE 6.13
Bird Strikes by Impact Altitude

Altitude Feet (AGL)	0–500	500–1,000	1,000–1,500	1,500–2,000	2,000–2,500	2,500–3,000	>3,000
1970–1980 [49]	61.2	16.1	8.2	8.0	1.6	1.8	3.1
1985–1999 [23]	72	19	unknown	unknown	unknown	5.6	
Randolph Air Force Base 2010–2014 [47]	82	5	13	–	–	–	–

TABLE 6.14
Bird Strikes for Military Aircraft by Flight Phase [60]

Flight Phase	Takeoff	Climb	Cruise	Descent	Range (Munitions Firing)	Low Level	Unknown	Final Approach	Landing
Bird strike %	17.9	1.9	4.7	0.98	7.1	14.6	28.79	9.42	13.0

takeoff and landing phases of flight, with 69% of strikes occurring during takeoff and 31% during landing.

During the period 1950–1999, 53% of the total recorded accidents occurred during low-level flight [27], while 39% took place at or near airfields during takeoff, climb, approach, touch-and-go landings, overshoots, and flight demonstrations – most occurred during takeoff and climb-out.

Figure 6.18 displays the percentages of bird strikes with military aircraft at different flight phases in 1974. Half of the bird strikes occurred during the takeoff and landing phases [54] and 20% in the low-level phase.

A rather detailed statistics for bird strikes with military aircraft during different flight phases are listed in Table 6.14 [60].

6.4.2.7 Bird Strike by Aircraft Type

The percentages of bird strike with different types of aircraft are given in Table 6.15. In the period 1966–1975, fighters and trainers were subjected to the highest percentage (57%) since they spend a great deal of time in the airdrome environment practicing takeoffs and landings [49]. Next comes cargo aircraft (28%), with bombers subject to the smallest number of strikes (13.5%). In the period 1970–1980, fighters and trainers are separated in [60]. Fighters and trainers represent the most collided types (42.2% and 19.1%). Cargo aircraft are in second place, with 28.4%. Bombers follow with 7.9%. In the period 1950–1995, fighters having single- and double-seat

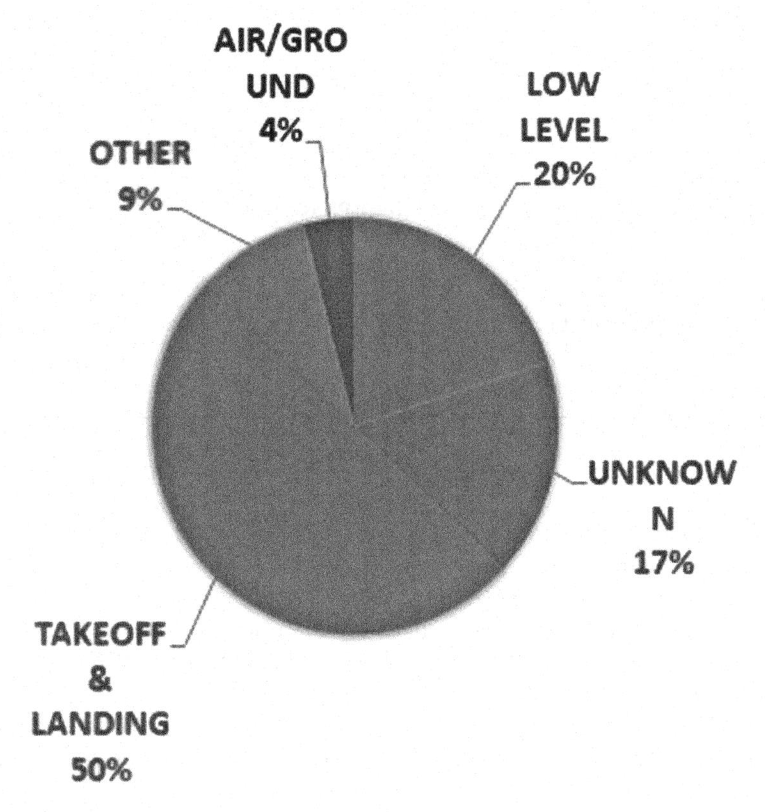

FIGURE 6.18 Bird strike at different flight phases for military aircraft in 1974.

Courtesy: IBSC [49].

together with unknown types accumulate 86.2% of the lost 167 aircraft from 32 countries [23]. Next, trainers and bombers encountered 7.2% and 4.8% of the struck military aircraft. Helicopters encountered 1.8% of total struck aircraft.

6.5 ADDITIONAL DETAILS FOR WILDLIFE STRIKE

Most statistics in Section 6.3 focused on bird strike. Few information was discussed for other wildlife creatures. In this section, other wildlife strike hazards caused by terrestrial mammals will be discussed.

TABLE 6.15
Bird Strikes by Different Military Aircraft Groups

Aircraft Group	Fighter	Trainer	Cargo	Bomber	other
Bird strike % [27]	86.2	7.2	–	4.8	1.8 (helicopter)
Bird strike % [60]	42.2	19.1	28.4	7.9	2.2 (helicopter)
Bird strike% [49]	57		28	13.5	1.5

6.5.1 Reptiles

Several mammal species also pose a significant threat to flight operations and must be considered. Unlike birds that require federal permits to control, many mammalian species require state permits for harassment and control [61]. Every time a pilot flies into a new area, it is important to identify the wildlife hazard before major flying operations begin. This improves aviation safety, adds to the overall integrity of the US Air Force database, and helps in improving aviation safety.

6.5.2 Bats

Bats are the only mammals capable of true and sustained flight.

Bats cause problems to aviation safety as they fly at low altitudes, sometimes migrate with birds in mixed-species flocks. Thus, they may strike and damage aircraft parts or be ingested into engines. Consequently, it will cost the air base the repair or change aircraft parts, interrupt flight operation, and aircraft downtime for repair/maintenance. In the period 1997–2004, there were 126 bat strikes [103]. The red bats and Brazilian free-tailed bats are the most common bat species in USAF strikes, and most North American strikes occur in August (see Figure 6.19).

Examples for damaging bat strikes are the following cases:

- Brazilian free-tailed bats (0.5 ounce) posed a great threat to T-38 and T-37B aircraft, where damage cost nearly $10,000 for each aircraft
- A red bat (0.5 ounce) teamed up with a Mourning Dove and caused $195,707 in damages to a C-130E Hercules
- A grey-headed flying-fox bat struck the right-wing of the one-engine Socata TB-10 Tobago 500 ft AGL and on final approach to Parafield Airport, Australia, on July 21, 2017.

Bats normally do not cause a great threat to aircraft safety in the United States, but with increased overseas, a great threat arises. For example, some fruit bats in Asia have wingspans of up to 6 ft and weigh two pounds.

6.5.3 Terrestrial Mammals

The most hazardous species are mule deer, white-tailed deer, coyotes, foxes, pigs, rabbits, hares, and domestic dogs. While large mammals cause the most damage

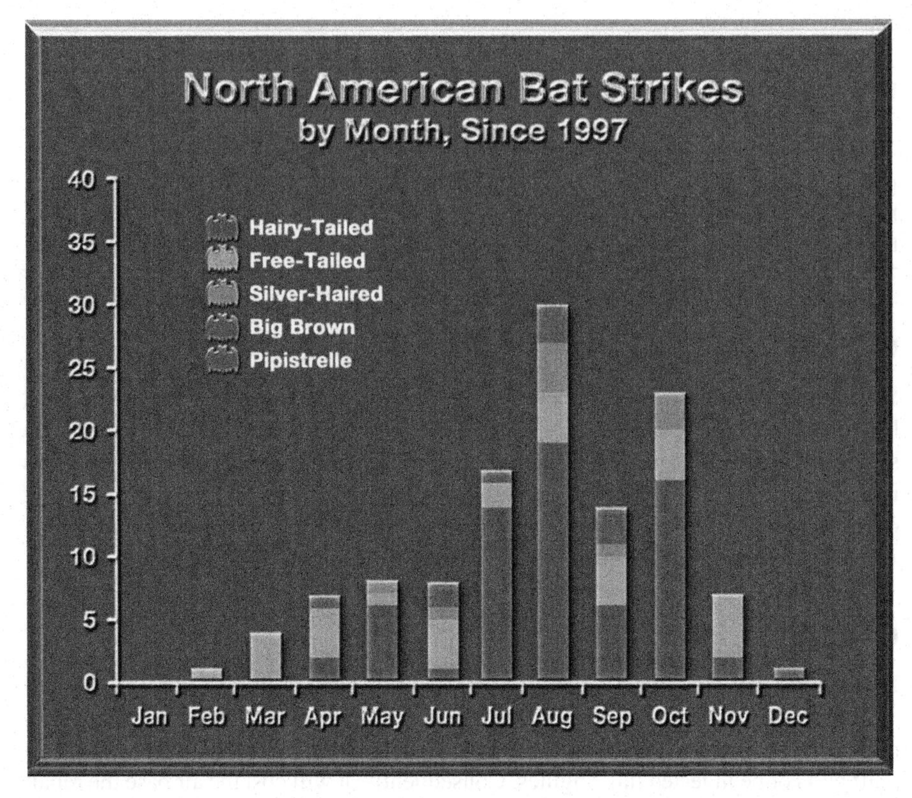

FIGURE 6.19 Different bat species striking USAF aircraft on monthly bases during 1997–2004 [62].

Courtesy: USAF.

when struck, they are involved in less than 3% of wildlife strikes, while more than 97% of wildlife strikes are attributed to birds.

6.5.3.1 Deer

Members of the deer family (including moose, elk, and caribou) may be a rare visitor or a regular visitor to an airfield depending on the airfield location. Depending on local population, deer may pose a severe threat to flight operations.

6.5.3.2 Buffalos and Cows

Aircraft operating from hot areas in Asia and Africa may encounter strikes with buffalos and cows. Examples for these are:

- Boeing B737-800 aircraft operated by Spicejet flight SG622 (Surat–Delhi), on June 11, 2014, during takeoff roll at Surat, the aircraft was at around 350 m

from start of runway, the left engine of the aircraft hit a buffalo. There was substantial damage to the engine (*the engine cowl was badly damaged and there were holes on the sides of engine), but* there was no fire or injury to any person on board
- B737-800 aircraft operated by LION AIR, flight number: LNI892, on August 8, 2013, departing from Sultan Hasanuddin Airport (WAAA) Makassar, at 1203 UTC to Djalaluddin Airport (WAMG) Gorontalo, Indonesia. As the aircraft touched down the runway the aircraft struck two cows which were crossing the runway. The aircraft hydraulic lines of the brake system and weight on wheel sensor were damaged.

6.5.3.3 Feral Pigs

Wild or feral pigs (or hogs) can be attracted to an airfield environment to feed on a variety of foods that may be available in the vegetated areas. These pigs will eat virtually anything from acorns, roots, grain, invertebrates, small mammals, carrion, and other food items.

As an example, for accidents caused by such animal, on July 16, 2010 at South African Express Airways flight SA1107, a de Havilland Dash 8 collided with a wild pig (aardvark/anteater) shortly after touchdown on runway. The accident occurred at night. The impact caused the nose gear to collapse backwards. The pilot managed to maintain runway heading and the aircraft came to rest approximately 1,200 m from the threshold of runway on the centerline.

6.5.3.4 Coyotes and Foxes

These animals are attracted to airfields by rodents, rabbits, and other food sources. Dens may be found in mounds, banks, culverts, and other suitable areas.

6.5.3.5 Rabbits and Hares

In addition to the direct hazard to aircraft, these animals often attract raptors, coyotes, and foxes into the airfield environment.

6.5.4 Rodents (Squirrels, Woodchucks, and Rats)

Rodents are the primary food source for many for the larger species of birds and mammals that present the real threat to aircrews and aircraft. Depredation of the larger rodents (i.e., woodchucks, nutria) is recommended.

6.5.5 Reptiles

6.5.5.1 Snakes, Alligators, and Turtles

They are rarely struck by aircraft. However, a large alligator can present a serious problem to the landing gear of an aircraft. An alligator caused a military airport in Florida to shut down a runway for more than an hour. Other smaller reptiles are a threat when wrapped up in landing gear or tossed into the air by the landing gear.

6.6 HELICOPTERS

6.6.1 Introduction

All rotating wing aircraft (helicopters) are powered by only shaft engines, including piston and turboshaft engines [63]. The continuous increase in the number of helicopters and bird population resulted in a dramatic increase in a bird strike. Surprisingly until now, there are no comprehensive analyses of wildlife strike hazards for helicopters. In 2006, the US Department of Agriculture and the FAA conducted a cursory investigation into reported bird strikes to civil helicopters during 1990–2005 [64]. This analysis indicated that patterns of reported bird strikes to civil helicopters were very different from those involving fixed-wing aircraft [65].

Since helicopters normally fly at low altitudes (and thus may be surrounded by many birds), many strikes with flying avians (birds and bats) may occur. In many cases, birds penetrate the windshield, potentially leading to pilot incapacitation that may lead to a helicopter crash and the death of all persons aboard.

Two-thirds of bird strikes with helicopters occur during the en-route flight phase, where kinetic energy is highest. Only 8%–9% occurred during the approach and 9%–10% during the climb. This contradicts fixed-wing aircraft, where most strikes occur during takeoff/climb and approach/landing flight phases.

In the past, the relatively slow cruising speed, coupled with rotor noise, acts as a sufficient warning for birds to get out of the way. But the new trend toward faster and environmentally quieter helicopters might result in increased problems. Pulsing exterior lights may help to protect helicopters, as they seem to warn birds away. Recent tests have shown that aircraft with pulsing lights are hit less often and suffer less damage when struck. As defined by the FAA, the collision of birds with a helicopter can be categorized as:

- Accidents associated with fatalities
- Incidents that damage the aircraft and create the potential for crashes.

6.6.2 Statistics

In several accidents, birds fractured the windshield and interfered with engine fuel controls. This assured that *improvements are needed in windshield and window design.* Also, the helicopter should be equipped with an audible alarm and warning light to alert the crew to low Main Rotor Speed (Nr). New technologies include integrated visual, aural, tactile, and alert messaging for alerting the flight crew and aiding them in decision-making.

6.6.2.1 Civil Helicopters

In the period 1912–2012 [66]:

- Seven fatal accidents killing 18 people and destruction of 8 helicopters
- 70% of helicopter accidents occur in the United States due to the large number of helicopters operating there

- Windshields were holed in 50% of the helicopter accidents, particularly after collision with heavy birds, including raven (*Corvus corax*), black vulture (*Coragyps atratus*), and red-tailed hawk (*Buteo jamaicensis*)
- Vibration, due to impact on the rotor system, was also a feature in several hull losses.

Wildlife strikes to civil helicopters in the United States, 1990–2011 [70]:

- Helicopters from a variety of public and private organizations, including US federal government agencies, private companies, medical and emergency services, and private citizens, conducted an average of 2,511,227 (±122,922 SE) flight hours annually in the United States
- 1044 wildlife strikes as records of the NWSD
- 51.3% of the reported wildlife strikes occurred during the day (81.3% of the helicopter flight hours occurred during the day)
- 43.7% of reported wildlife strikes occurred at night (only 18.7% of the flight hours for helicopters occurred during the night)
- Approximately 5% of the strikes occurred during dawn and dusk
- Most strikes occurred in September and October, followed by May
- 53.1% of strike occurred during en route, followed by approach (18.2%), climb out 12.2%, terrain flight (11.9%), taxiing (2.6%), takeoff (0.8%), hovering (0.7%), and landing (0.5%)
- 97.2% strikes by birds, whereas 2.8% due to bats and large mammal's strike
- Gulls, waterfowl, raptors, and vultures were the most frequently struck birds, accounting for 26.8%, 19.8%, and 19.6% of all strikes. Doves, pigeons, herons, egrets, and ibises were the next species most frequently struck
- Eleven mammal strikes were reported, including six with bats, two with cattle (cow), and one each with a coyote (*Canis latrans*), a moose (*Alces alces*), and a white-tailed deer (*Odocoileus virginianus*)
- The strike of a cow and the helicopter led to the complete destruction and crash of the helicopter
- 52 human injuries and nine fatalities due to wildlife strike
- The average cost of a bird strike (costs of damaged parts and repair costs) was $41,158 per incident
- The average cost of an on-airfield damaging wildlife strike was $378,539 per incident.

In the period 2015–2017 [69]:

- 665 bird strikes with helicopters in the United States according to Federal Aviation Administration Strike Database
- 79 strikes (12%) resulted in at least some damage
- 41 (6%) produced "substantial damage" – defined as damage to an aircraft's structural integrity, performance, or flight characteristics, normally requiring major repairs or the replacement of the entire affected component

- Twenty-eight incidents resulted in over $10,000 in damage/repair costs, four of which exceeded $100,000
- The total repair and damage costs exceeded $3.7 million not including the revenue lost to an aircraft being out of service while in repair
- Three fatalities in a crash of Bell 407 medical transport due to a bird strike
- Eight injuries (from eight separate incidents) due to medium to large bird strikes where some penetrated the aircraft
- 77 strikes were on government aircraft (including US Coast Guard, US Customs and Border Protection, and other government entities),
- 12 privately owned helicopters
- Most strikes were recorded on helicopters on business flights
- Most damaging strikes were from larger birds, particularly vultures
- Smaller birds in a flock also caused reasonable damage
- Most bird strikes occurred while helicopters are en route
- The windshield is the most struck element with 47% on Part 27 (maximum weights of 7,000 pounds or less and nine or less passenger seats) and 40% on Part 29 (maximum weight greater than 20,000 pounds and ten or more passengers)
- 84%–85% of all bird strikes occurred on components forward of the main rotor mast
- The main rotor experienced 30%–33% of the strikes reported.

Figure 6.20 illustrates the number of bird strikes with helicopters in the period 1990–2013 [66]. A rapid increase is shown:

- 14 cases were in average in the period 1990–1995
- 26 cases were in average in the period 1995–2000

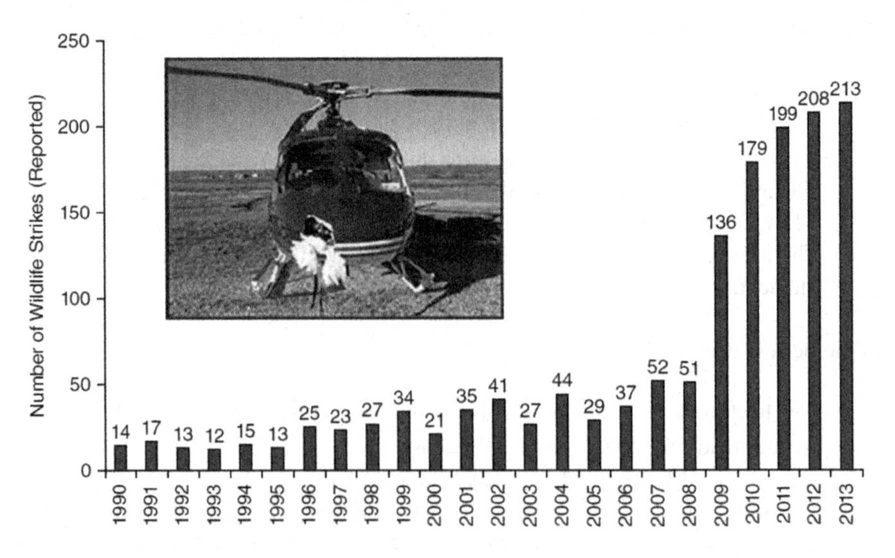

FIGURE 6.20 Number of birds strikes with helicopters in the period 1990–2013.

Courtesy: USDA-APHIS [67].

- 40 cases were in average in the period 2001–2008
- 187 cases were in average in the period 2009–2013
- 213 cases occurred in 2013 only.

The increase in the number of strikes is due to the following:

- The continuous increase in the total number of helicopters. The number of civil and military helicopters in 2020 were 26,466 and 20,519, respectively
- The increasing awareness among pilots regarding the importance of reporting bird strikes since the famous accident involving the ditching of US Airways Flight 1549 in New York's Hudson River in 2009
- The rise in populations of large bird species in North America, including the Canadian goose, American white pelican, and the American snow goose, as well as other worldwide large-bird species with rising including bald eagles, wild turkeys, turkey vultures, double-crested cormorants, sandhill cranes, great blue herons, and ospreys.

Figure 6.21 illustrates the critical parts of the helicopter which are subjected to most bird strikes: namely, the windscreen, main rotor, radome/nose, fuselage, tail, landing gear, and lights. As shown in Figure 6.21, the percentages of bird strike with different

FIGURE 6.21 Percentages of bird strike with different parts of a helicopter [68].

parts are as follows: the windscreen is the most critical part (33.8%), followed by the main rotor (17.9%), and then the radome/nose (13.6%).

6.6.2.2 Military Helicopters

A detailed study for bird strikes with military helicopters in the United States [60] included US Army, Navy, and Air Force, and Coast Guard in the period 1979–2011 resulted in the following findings.

In the period 1979 to 2011 [69]:

A team of scientists reviewed the bird strike records from the US Army, Navy, Air Force, and Coast Guard from 1979 to 2011. Their study included Apache attack helicopters and huge Chinook vehicles transporting troops, supplies and artillery to and from the battlefield.

- The total number of wildlife strikes were 2,511. It is subdivided as follows:
  - 617 strikes in Florida (highest number of incidents)
  - 204 and 192 strikes in New Mexico and Georgia, respectively
  - In 812 of the military's incidents, the type of animal that smashed into the helicopter was described
- Birds were the culprits in 91% of the cases; other animals that struck military craft were bats
- Air Force helicopters were commonly struck by warblers (16.8%) and perching birds (12 %)
- Naval vehicles were mostly hit by gulls (18.2%), followed by seabirds (14.9%), shorebirds (13.4%), and raptors and vultures (12.6%)
- 42% of wildlife strikes occurred in September and November
- 10.4% of wildlife strikes were in December and February
- The cost for these damages ranged from $12,000 to $337,000
- Eight injuries – mostly cuts, lacerations, or bruising – when birds crashed through the windscreen of the aircraft (from 1993 to 2008)
- Two fatalities occurred in 2011 when a red-tailed hawk struck a US Marine Corps in California, which resulted in a crash – costing the life of the pilot and copilot – and the total loss of $24.5 million aircraft.

Figure 6.22 illustrates the monthly bird strikes for civil and military helicopters. It was found that almost 42% of the recorded wildlife strikes occurred between September and November. December and February were less hazardous, with 10.4% of wildlife strikes occurring in those months. Military helicopters encountered a higher number of strikes in October.

During 1990–2010, rotary-wing aircraft (helicopters) within the US Department of Defense (e.g., US Army and US Air Force) have been deployed overseas to conduct various noncombat and combat missions.

The following findings for wildlife strike (birds, bats, insects) are summarized:

- Almost two-thirds (61%) of wildlife strikes to US Army aircraft occurred during deployments in the Middle East (e.g., Iraq), 13% and 12% occurred in Southeast Asia (e.g., South Korea), and Central America (e.g., Panama)

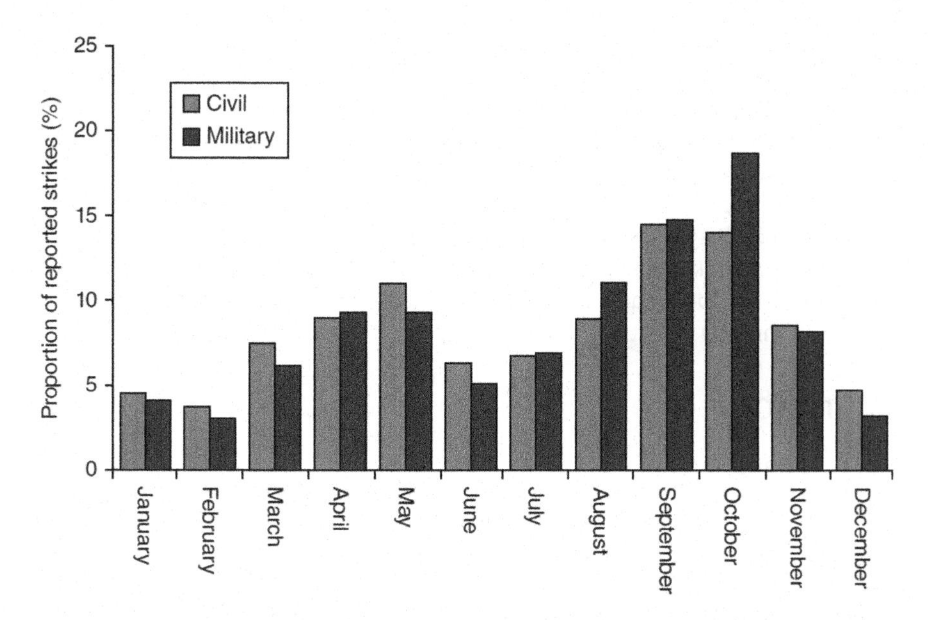

FIGURE 6.22 Monthly record for the proportion of reported strikes, both civil and military helicopters.

Courtesy: IBSC [49].

- Strikes to US Air Force aircraft occurred in Afghanistan (38%), the Middle East (26%), Europe (18%), and Southeast Asia (13%)
- 238 wildlife strikes for the US Army, mostly in Iraq (136), South Korea (28), Germany, and Panama (each 13)
- 463 wildlife strikes for the US Air Force, mostly in Afghanistan (177), Iraq (117), UK (61), and South Korea (35)
- 140 and 15 were the maximum and minimum wildlife strikes with USAF helicopters in 2009 and 1994, respectively
- 48 and 2 were the maximum and minimum wildlife strikes with US Army helicopters in 2007 and 2011, respectively
- 14% and 4% were the maximum and minimum wildlife strikes with USAF helicopters in September and February, respectively
- 13% and 4% were the maximum and minimum wildlife strikes with US Army helicopters in November and July, respectively
- Maximum strikes occurred during terrain flight (forward movement at an altitude less than 305 m AGL), 60.6% for the US Army and 67.5% for USAF
- 24.9%, 7%, and 2.8% were the percentages for wildlife strikes with the US Army during en route, approach, and takeoff
- 16.6%, 4.2%, and 2.7% were the percentages for wildlife strike with USAF during en route, approach, and landing
- Nearly 75% and 50% were strikes during the day for the US Army and USAF, respectively

- Most bird strikes occurred during flight operations off-airfield than on airfield. This contradicts with fixed-wing aircraft, where most wildlife strikes occur with airport field
- Bats accounted for almost all strikes with mammals off airfield (in terrain flight) and resulted in significant damage to aircraft
- Most strikes for US Air Force during off-airfield operations were caused by larks, perching birds, doves and pigeons, warblers, sparrows, and quail
- However, larks, perching birds, thrashers and thrushes, gulls, and waterfowl caused the most damage to US. Air Force aircraft operating within airfield environments.

6.7 INSECTS

6.7.1 INTRODUCTION

Flying insect strikes, like bird strikes, have been encountered by pilots since aircraft were invented.

The first recorded strike was in 1911; young officer Henry H. Arnold nearly lost control of his Wright Model B after a bug flew into his eye while he was not wearing goggles. Generally, insect threats to aviation are less well understood. Insects are generally considered as causing an indirect threat to aircraft, being an attractant for birds that might pose a risk to flight safety rather than a risk in themselves [71]. However, aviation authorities, airport managers, commercial carriers, and private aviators are aware of the threats posed by insects [72]. When flying insects get in the way of an airplane's wing during takeoff or landing, those little blasts of insect's guts increase the surface roughness. Consequently, they disrupt the laminar (or smooth) flow of air over the airplane's surfaces, particularly wings, creating more drag on the airplane as illustrated in Chapter 1 (Figure 1.9).

Insects have the following hazards:

1. Erroneous instrument indications (especially unreliable airspeed and altimeter) due to blockage of pitot tubes
2. Increase in aircraft drag
3. Increase in fuel consumption
4. Loss of engine power
5. Loss of visibility
6. Hurting airport employees and passengers
7. Damage of equipment aboard parked aircraft
8. Delay of aircraft

The most critical insects are mud wasp, honeybees, bugs, locusts, and hornet.

6.7.2 WASPS

Mud wasp (Pachodynerus nasidens Latreille and known as the keyhole wasp) is native to South and Central America and the Pacific Islands, including Hawaii,

Polynesia, Micronesia, and Japan. It is not known how the insects arrived in Australia (possibly in 2006) from Central America, resulting in a novel hazard to aviation. Brisbane Airport, Australia, has been settled by such species which have evaded airport biosecurity to set up nests in the most inconvenient of places, threatening plane security [73]. Surprisingly, wasps can build hives in the probes in less than 20 min.

All cavities in aircraft like TAT probe (front fuselage), vent line to fuel pressure gauge (underwing), probe (on tail), engine tailpipe, engine temperature/pressure probe (rear of engine), and pitot probes are suitable places for mud swap nesting [72]. Wasps nesting in pitot probes may block them and pose a great risk to aviation safety and operational costs. Pitot probes are vital for pilots as they know their airspeed, especially for takeoff and landing. Anomalies between airspeed indications between probes can lead to costly and inappropriate crew action, leading to high speed rejected takeoffs [74] or air turn backs [75].

In rare cases, it led to catastrophic accidents like the Boeing 757-200 aircraft operated by Birgenair flight # 301 on February 6, 1996. The trip was from Puerto Plata in the Dominican Republic to Frankfurt, Germany. The aircraft was sitting unused for 20 days before the trip without pitot tube covers, so a wasp nest was built inside them. Thus, the pitot tubes were blocked, giving different readings for the flight speed indicators of the pilot and first officer, which led to wrong decisions and the crash of the aircraft. All 189 people on board died.

The maintenance manual for the Boeing 737–800 states: "A pitot probe or static port system blocked by foreign objects such as insects may cause large errors in airspeed-sensing and altitude-sensing signals, which may lead to loss of safe flight."

In January–March 2006, five incidents were reported at Brisbane Airport, Australia, in which pitot probes gave inconsistent airspeed readings in the cockpit of Airbus A330s, and flights were either rejected during take-off roll or continued to their destination with an erroneous airspeed indication on one of the three pitot systems [72]. In 2016, a detailed study in Australia [72] to identify the wasp species responsible for:

- Pitot probe blockages
- The seasonal pattern of nesting
- The type of aircraft most likely to be impacted
- The location most at risk.

Replica pitot tubes for Boeing 737-400, 737-800, 747-400, Airbus A330, De Haviland Dash-8, and Embraer 190 were fixed to three panels erected at four locations in Brisbane airport including both domestic and international terminals (Figure 6.23). Panels were fixed so that probes were facing the same direction as aircraft probes when an aircraft was parked at the gate and a roughly equivalent height.

The 3D-printed pitot tubes represented different aircraft have a range of aperture dimensions (2.5–8 mm). Panels were monitored weekly (during the main breeding season) to monthly. Blocked probes were removed and replaced by empty probes of the same type.

The following results for the incidence rate of pitot probe nesting can be concluded from [72, 80]:

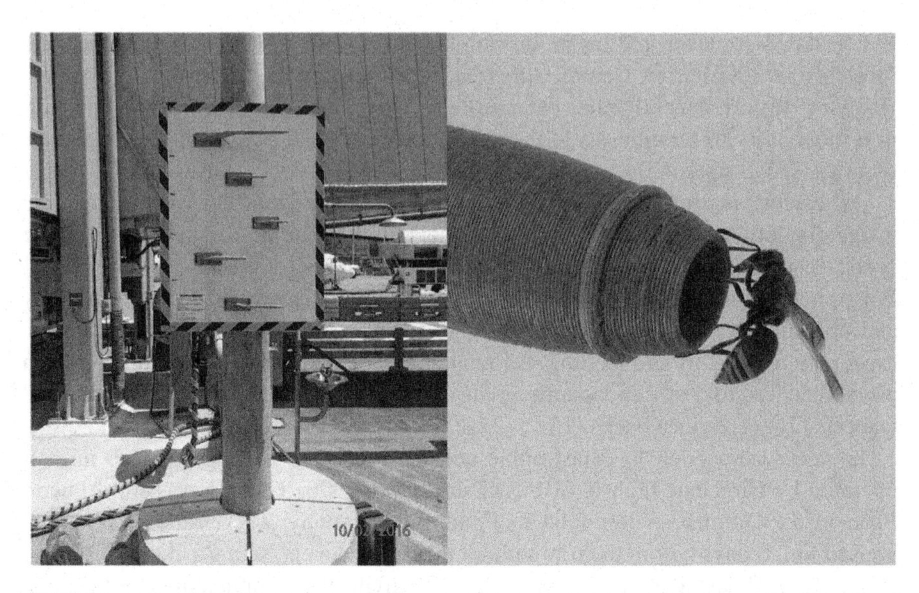

FIGURE 6.23 Left: Pitot probe panel established at Brisbane Airport, with 3D-printed probes of (from top) 747-400, 737-400, A330, 737-800, DHC-8. Right: Female keyhole wasp on 3D-printed DHC8 probe, Brisbane Airport, May 2016 [72].

- Sixty probes were available for nesting over 39 months
- In this time, 93 probe blockages were recorded, with 38 adult wasps emerging (23 males and 15 females), resulting in an average of 2.38 blockages per month, and 0.0397 blockages per probe per month.

Figure 6.24 illustrates the blockage (nesting) of 60 pitot tubes over 39 months [72].

Aircraft manufacturers Boeing [74], Airbus [76], aviation authorities [77], and several carriers at Brisbane recommended (not mandatory) placing plastic covers over gauges upon aircraft landing to prevent wasps from building hives in the probes.

However, such plastic covers incur risks on their own due to failure to remove them. In July 2018, an A330 belonging to Malaysia Airlines bounding for Kuala Lumpur had to return to Brisbane and made a heavy landing because the pitot tube was covered. The covers were not removed before takeoff [78, 79].

6.7.3 Bees

Swarms of flying bees descend on airports and halt operations. The insects arrive en masse at the airport. In some countries like the United States, such bees are protected species, and it is a crime to kill them. In such cases, airport authorities call a master beekeeper to remove them. One can recall the case of a delta flight from Pittsburgh to New York in August 2012, when tens of thousands of honeybees descended on the plane's wing, just when the crew was preparing to fuel it [81].

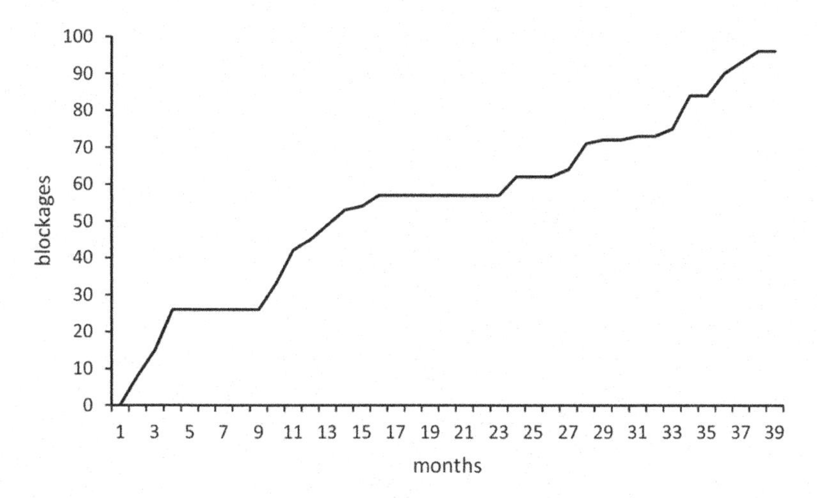

FIGURE 6.24 Cumulative experimental pitot probe blockages due to wasp nesting over 39 months (total of 60 probes available, 93 blockages) [72].

When the bees are not protected by law, they are removed by spraying them with water from the airport fire extinguisher unit. Two examples are listed hereafter:

1. In September 2017, a Citilink Indonesia commuter flight was about to depart from Indonesia's Kualanamu International Airport, North Sumatra province, when thousands of insects descended onto the aircraft's right-hand wing [82]. More than half of the wing was completely covered by the swarm, while other bees hover nearby. For more than 90 min, the airport fire extinguisher unit sprayed the wing with water and dislodged the insects. Maintenance workers carefully checked over the aircraft before being allowed to make its journey.
2. On Sunday, September 16, 2019, an Air India flight was due to leave the eastern city of Kolkata with 136 passengers when a swarm of bees landed on the cockpit window, attacking ground staff when they tried to remove them. Attempts to use the windscreen wipers also failed. The fire extinguisher unit managed to dislodge the honeybees after nearly an hour-long operation [83].

6.7.4 Locusts

Aviation experts say that locusts fly up to 3,000 ft high and can cause huge inconvenience in taking off and landing of an aircraft. However, at cruise level, there is no problem. If a pilot notices locust while he is trying to land at a particular airport, he can abort the landing and fly to an alternate airport. He cannot do anything else. Locusts fly during day only, thus it is easy for pilot to watch them. Almost all air intake ports (engine inlet, air-conditioning pack inlet, etc.) of the aircraft will be prone to ingestion in large numbers if the aircraft flies through a swarm.

Swarms of locust can vary from less than 1 square kilometer (0.38 square miles) to several hundred. Each square kilometer can contain at least 40 million insects, according to the UN's Food and Agriculture Organization.

Such swarms may lead to zero/low visibility, loss of engine power, blocking of an aircraft's pitot tubes, causing inaccurate airspeed readings, as well as drag increase. In 2010, the Australian Civil Aviation Safety Authority issued a warning to pilots about the potential dangers of flying through a locust swarm.

Several incidents have been encountered. One of the old incidents was in the summer of 1986, where a Boeing B-52G Stratofortress flying at 0.8 Mach, on a low-level training mission over portions of Wyoming, Montana, and North Dakota entered a swarm of locusts [84]. The insects' impacted on the aircraft's windscreens. The window wipers were useless, and the crew were unable to see. Thus, they were forced to abort the mission and fly using the aircraft's instruments alone. The aircraft eventually landed safely.

On January 13, 2020, pilots of Ethiopian Airlines B737-700 flight from Djibouti to Dire Dawa were preparing to land their plane, when a swarm of locusts slammed into the aircraft engines, windshield, and nose. They tried in vain to clean the windscreen with the plane's wipers. Thirty minutes later the plane landed safely but in the capital Addis Ababa instead [85]. In the year 2020 East Africa has been hit by the worst locust invasion in 25 years.

6.7.5 Bugs

Tiny insects like bugs, bees, and mosquitoes can cause a lot of troubles for the nowadays airplanes. Bug residue causing drag has been a long-standing challenge for the aviation community.

The impact of insect debris on the leading edges of aircraft wings has an adverse effect on their natural laminar flow during cruise flight.

Accumulating bugs causes the airflow to trip from smooth or laminar to turbulent, resulting in additional drag. An aircraft designed to have laminar wings when flies a long distance can save 5%–6% in fuel usage. Surprisingly, little bugs that trip the flow and thus a part of this benefit is lost.

The seriousness of insect debris contamination is dependent on airplane characteristics, airfoil geometry, mission, place local terrain, humidity, temperature, time of day, and time of year [86, 87]. Extensive research effort started long ago in the 1960s [86], continued in the 1980s [87, 88], and the 21st century [89–91] as examples.

As identified in [92], the threat of insect contamination is typically limited to flight phases close to the ground, where 50%–60% of the insects are collected during the ground run, climb out, final approach, and landing. At altitudes greater than 1,000 ft AGL, contamination is normally negligible.

A group of researchers at NASA's Langley Research Center – the "bug team" – ran several flight tests of coatings that may reduce the amount of bug contamination on the wings of commercial aircraft [88, 91].

Past insect contamination protection techniques included: mechanical devices (paper covering, scrapers, wipers, deflectors) and surface films (soluble films, resilient surface, liquid spray systems, and porous leading edges). However, such insect contamination protection techniques were impractical.The fluid injection was the most promising technique. Fluid may be through slots or through metal skins made porous by election beam or laser beam drilled holes. The holes are about 0.0025 inch in diameter with a spacing of about 0.0205 inch. TKS porous, woven sintered stainless-steel metal leading-edge, as well as mono-ethylene glycol (MEG) and water fluid solutions proved a successful technique. The MEG fluid acts as a solvent for insect body protein content. The drying of this protein content serves as the glue which causes the adhesion of insect debris to the impacted surfaces. Therefore, the effectiveness of such MEG enhances the use of small fluid quantities.

NASA conducted flight experiments to determine insect accumulation rates and protection efficiency under a variety of flight and atmospheric conditions using a Cessna 206 equipped with a TKS porous leading edge [87]. During tests, the TKS system was deactivated on all surfaces except the "right wing leading edge." The insect protection system effectiveness was evaluated by comparing the right, protected wing with the left, unprotected wing. The highest insect population were encountered in the following conditions: flight duration ranged from 10 to 50 min, the airspeed ranged from 80 to 130 mph, temperature ranged from 70 °F to 80 °F, and the altitude was 50 ft.

Good insect contamination protection was achieved using a solution of 80% MEG and 20% water at flow rates between 0.013 and 0.027 $gal / min / ft^2$ of projected leading-edge frontal area.

All insect studies conclude that during takeoff roll/climb and approach flight phases, aircraft speed is high enough to cause a rupture of the insect body. Such insect debris create three-dimensional roughness elements in the boundary layer, resulting in a transition from laminar to turbulent flow behind the surface disruption [90]. Turbulent flow is developed if insect residue exceeds a critical height, which is a function of insect size, impact angle, impact speed, and Reynolds number [93]. Depending on these factors the share of critical insects on overall contamination is about 9%–25% [94].

Engineers at NASA Langley in a more recent study developed and tested more than 200 coating formulations in a small wind tunnel, then took a number of those to flight on the wing of a NASA jet [95] to choose the best that slough bug guts.

They selected five non-stick coatings to the wings of Boeing 757 ecoDemonstrator, in flights from Shreveport Regional Airport. This airport was selected because of its significant bug population, runway length, temperature, humidity, weather, by experts from NASA, Boeing, the US Department of Transportation, and University of California-Davis.

Boeing 757 made 15 flights from the Shreveport Regional Airport, each included several takeoffs and landings. One of the five coating/surface combinations showed especially promising results as it had about a 40% reduction in bug counts and residue compared to a control surface mounted next to it.

Another field study was carried out in Schiphol airport the Netherlands, to analyze the above effects in detail [96]. Weekly visual inspections on eight aircraft were performed for one year to quantify the number of insects on the aircraft leading edges. All aircraft were operated on short-to-medium haul missions (like current state-of-the-art 150 passenger seated aircraft) within a European network. The average contamination rate of all eight aircraft proved to be strongly dependent on season. Contamination was maximum in summer with two local maxima in May and August. While nearly zero contamination was seen in winter.

The mathematical model for insect contamination over the leading edge of a typical aircraft, and transition from laminar to turbulent boundary layer, was correlated to economic implications.

6.8 HUMAN

It is the saddest part in this book to write about human as a debris of the different FODs in aviation. Human can be pulled by engine intake or pushed by engine exhaust. Moreover, human can be struck by aircraft frame especially moving elements.

6.8.1 Civil Aircraft

6.8.1.1 Introduction

As previously mentioned, aeroengines are categorized either as turbine-based or shaft-based engines [30, 63]. Two common types for each group: namely, turbojet and turbofan for the first group, and turboprop and piston engines for the second group. Movement close to running turbojet/turbofan endangers his/her life if sucked into them. Moreover, propellers are usually installed to piston/turboprop engines. Such propellers pose a great threat to personnel whenever the engine is running. It slices human's body into pieces if he/she accidentally gets too close.

For these hazards, all aircraft and engine manufacturers define the dangerous areas around aircraft during its ground operation. Two critical areas are identified:

- Inlet suction area upstream of the intake of aircraft engine(s)
- Exhaust wake area downstream of the nozzle of aircraft engine(s).

Both areas depend on the aircraft type, installed engines and its rotational speed (idle or full power) as well as wind speeds. An entry (or save) corridor is defined also for personnel movement around aircraft when engines are running.

Maintenance and airport staff must be alert when passing beside aircraft with running engines.

6.8.1.2 Humans Sucked into Aircraft Engine

1. Airplane mechanic was sucked into jet engine Boeing 737-500 Continental Airlines Flight 1515 at El Paso Airport, Texas, on January 16, 2006 [97]

2. Boeing 737 plane, owned by Kazakhstan airline Air Astana, jet engine sucked an engineer while it was about to leave Moscow for London on November 4, 2004 [98]
3. Mechanic sucked into engine in Gulf area in 2017.

6.8.1.3 Aeroflot Boeing 737 Fatally Strikes a Man on Moscow Runway

On November 18, 2018, an Aeroflot Boeing 737-800, was performing a flight from Moscow Sheremetyevo (Russia) to Athens (Greece), hit a 25-year-old Armenian citizen on the runway [99].

6.8.2 Military Aircraft

The trainee 21-year-old John Bridget was pulled into the jet engine of the US fighter A63 Intruder aboard the air carrier USS Eisenhower on February 20, 1991 during the Gulf War [100]. While he was moving toward the engine, he was suddenly pulled inside. Luckily, another crewman raced toward the engine to power it off. Incredibly, Bridget was sucked into the engine for 3 min and managed to survive, due to the design of the A-6 engine (the J-52 dual spool turbojet). It has a long protruding cone that extends in front of the first stage of compressor. When he was sucked in, his arm extended above his head which caused his body to wedge between the cone and inside wall of the intake. His helmet and float coat were sucked in first which prompted the pilot to cut the throttle.

6.9 GRASS

As described in [101], turf grass and ditches in airports collect and hold large amounts of light debris such as paper, cardboard, plastic, and various containers that trash often originate in terminal aprons, cargo ramps, and hangar ramps. To avoid their ingestion into engines, a maintenance employee on a small all-terrain vehicle with a litter stick and garbage bags picks up trash in grassy areas and fence-lines.

Also in air bases, soldiers and guards collect any rocks, pebbles, forgotten tools, sticks, and grass in their FOD walks to avoid any damage to the high-precision engines and other systems in aircraft [102].

REFERENCES

[1] Hazardous Wildlife Attractants on or near airports, FAA Advisory Circular AC 150/ 5200 33C, 02/21/2020

[2] El-Sayed, A.F. Bird Strike in Aviation, Wiley 2019

[3] Sexton, C. Mammal strikes with aircraft are becoming much more frequent, Eaerth.com news, February 3, 2021 www.earth.com/news/mammal-strikes-with-aircraft-are-becoming-much-more-frequent/

[4] FAA: Wildlife Management, www.faa.gov/airports/airport_safety/wildlife/management/

[5] Transportation Fatalities by Mode, Bureau of Transportation Statistics, March 2021, www.bts.gov/content/transportation-fatalities-mode

[6] Dolbeer, R.A. Begier, M.J. Miller, P.R., Weller, J.R., and Anderson, A.L. Wildlife Strikes to Civil Aircraft in the United States, 1990–2019, FAA, National Wildlife Strike Database, Report No.26, February 2021

[7] Proactive Airports, Pilots Minimize Wildlife Strike Hazards, National Business Association Aviation (NBAA) https://nbaa.org/aircraft-operations/safety/in-flight-safety/wildlife-strike-response/proactive-airports-pilots-minimize-wildlife-strike-hazards/

[8] FAA, Frequently Asked Questions and Answers, www.faa.gov/airports/airport_safety/wildlife/faq/

[9] Mazareanu, E. Global air traffic – number of flights 2004–2021, Dec 2, 2020 www.statista.com/statistics/564769/airline-industry-number-of-flights/#:~:text=Global%20air%20traffic%20%2D%20number%20of%20flights%202004%2D2021&text=The%20number%20of%20flights%20performed,reached%2038.9%20million%20in%202019.

[10] Finlay, M. What Is A Bird Strike And What Damage Can It Cause?, Simple Flying, August 9, 2020, https://simpleflying.com/bird-strike-damage/

[11] International Civil Aviation Organization (2017). Summary of Wildlife Strikes Reported to the ICAO Bird Strike Information System (IBIS) for the Years 2008–2015. Electronic Bulletin, EB 2017/25, Attachment B. www.skybrary.aero/bookshelf/books/4069.pdf

[12] Metz, I.C., Ellerbroek, J., Mühlhausen, T., Kügler, D., and Hoekstra, J.M. Review: The Bird Strike Challenge, Aerospace, 2020, 7, 26; doi:10.3390/aerospace7030026

[13] Australian Transport Safety Bureau. Australian Aviation Wildlife Strike Statistics 2008 to 2017; Australian Transport Safety Bureau: Canberra, Australia, 2019.

[14] Bird Strike in Canada, https://canadianbirdstrike.ca/history-of-bird-strike-committee-canada/#:~:text=bird%20strikes%20are%20not%20very,only%20a%20few%20cause%20damage (accessed March 6, 2021)

[15] Direction Générale de l'Aviation Civile. Analyse du Risque Animalier en France 2010–2013. 2017. Available online: https://de.calameo.com/read/000687261c8f500e036b0 (accessed on 5 March 2021)

[16] Deutscher Ausschuss zur Verhütung von Vogelschlägen im Luftverkehr e.V. Jahresbericht 2018; DAVVL: Bremen, Germany, 2019.

[17] UK Civil Aviation Authority. Birdstrikes. Available online: www.caa.co.uk/Commercial-Industry/Airports/Safety/Birdstrikes/ (accessed on 5 March 2021)

[18] Dolbeer, R.A., Cleary, E.C., and Wright, S.E. Wildlife Strikes to Civil Aircraft in the United States 1990–2018; Federal Aviation Administration National Wildlife Strike Database, Serial Report Number 25; Federal Aviation Administration, U.S. Department of Agriculture: Washington, DC, USA, 2019.

[19] ICAO. Doc-9332-AN/909—Manual on the ICAO Bird Strike Information System (IBIS), 3rd ed.; ICAO: Montreal, QC, Canada, 1989

[20] European Parliament and the Council. Regulation (EU) No. 376/2014 of The European Parliament and of The Council of 3 April 2014. Off. J. Eur. Union 2014. Available online: https://eur-lex.europa.eu/legalcontent/EN/TXT/PDF/?uri=CELEX:32014R0376&from=EN (accessed on 5 March 2021).

[21] European Parliament and the Council. Commission implementing regulation EU 2015/1018. Off. J. Eur. Union 2015. Available online: https://eur-lex.europa.eu/legal-content/EN/TXT/PDF/?uri=CELEX: 32015R1018&from=EN (accessed on 6 March 2021).

[22] Dolbeer, R.A. Trends in Reporting of Wildlife Strikes with Civil Aircraft and in Identification of Species Struck Under a Primarily Voluntary Reporting System,

1990–2013; Special Report Submitted to the Federal Aviation Administration; DigitalCommons@University of Nebraska–Lincoln: Lincoln, NE, USA, 2015. Available online: https://digitalcommons.unl.edu/cgi/viewcontent.cgi?article= 1190&context= zoonoticspub (accessed on 6 March 2021).

[23] MacKinnon, B. (2004). Sharing the Skies, Chapter 7, Bird- and Mammal-strike Statistics. www.tc.gc.ca/eng/civilaviation/publications/tp13549-chapter7-2144.htm (accessed 2 March 2021).

[24] Australian Transport Safety Bureau (2014). ATSB Transport Safety Report 2014–15. www.atsb.gov.au/media/5366635/ATSB%20Annual%20Report%202014%C2%AD15.pdf

[25] Cheng, E., Hlavenka, T., and Nichols, K. (2015). It's A Bird, It's A Plane, It's A Problem: An Analysis of Bird Strike Prevention Methods at Panama City's Tocumen International Airport. https://web.wpi.edu/Pubs/E-project/Available/E-project-102915-205833/unrestricted/CopaBirdStrikePreventionMethodsInPanama.pdf

[26] Dennis, N. and Lyle, D. Atkins Ltd.; Food and Environment Research Agency (FERA). Bird Strike Damage & Windshield Bird Strike. 2009. Available online: www.easa.europa.eu/sites/default/files/dfu/Final%20report%20Bird%20Strike%20Study.pdf (accessed on 8 March 2021).

[27] Insinna, V. (2018). Aviation in Crisis: Wildlife strikes add to Air Force and Navy's mishap count. *Military Times* www.militarytimes.com/news/your-military/aviation-in-crisis/2018/04/14/wildlife-strikes-add-to-air-force-and-navys-mishapcount (accessed 3 March 2021).

[28] EASA. Bird Population Trends and Their Impact on Aviation Safety 1999–2008; Safety Report; EASA: Cologne, Germany, 2009.

[29] AIRBUS, Flight Operation Briefing Notes, Operating Environment: Birdstrike Threat Awareness, Section IV (Operational Effects of Birdstrikes), part 1 (General), 2014. www.skybrary.aero/bookshelf/books/181.pdf (accessed 8 March 2021).

[30] El-Sayed, A.F. Fundamentals of Aircraft and Rocket Propulsion, Springer, 2016

[31] Reddy, G. Bird Strike. www.slideshare.net/gyanireddy/birdstrike-11316202, 2012, (accessed 8 March 2021).

[32] Nicholson, R. and Reed, W.S. Strategies for Prevention of Bird-Strike Events, *AERO* QTR_03.11, Boeing. www.boeing.com/commercial/aeromagazine/articles/2011_q3/4/, 2011 (accessed 2 March 2021).

[33] Aircraft Damages Caused by Birds, Aviation Ornithology Group, Moscow. www.otpugivanie.narod.ru/damage/eng.html (accessed 12 February 2021).

[34] Maragakis, I. (2009). European Aviation Safety Agency, Safety Analysis and Research Department Executive Directorate, Bird Population Trends and Their Impact on Aviation Safety 1999–2008.

[35] Thorpe, J., 100 Years of Fatalities and Destroyed Civil Aircraft Due to Bird Strike. IBSC30/WP Stavanger, Norway 25th–29th June 2012. www.int-birdstrike.org/Warsaw_Papers/IBSC26%20WPSA1.pdf (accessed 2 March 2021)

[36] Dolbeer, R.A. and Beiger, M.J. 2019. Wildlife strikes to civil aircraft in the United States, 1990–2017. U.S. Department of Transportation, Federal Aviation Administration, Office of Airport Safety and Standards, Serial Report No. 24, Washington, D.C., USA.

[37] Partners in Flight. 2019. Population Estimates Database, version 3.0. (http://pif.birdconservancy.org/PopEstimates).

[38] Wildlife Management, Center for wildlife and aviation, Embry Riddle Aeronautical University, http://wildlifecenter.pr.erau.edu/WildlifeManagement.html#Other%20Wildlife

[39] Jacobson, H.A. and Kroll, J.C. 1994. The white-tailed deer—the most managed and mismanaged species. Presented at Third International Congress on the Biology of Deer: Edinburgh, Scotland, Aug. 28 to Sept. 2, 1994.
[40] Lafrance, A. How Often Do Airplanes Hit Deer? Or alligators? Or bald eagles? Or armadillos? TECHNOLOGY, FEB 16, 2017, www.theatlantic.com/technology/archive/2017/02/what-happens-when-an-airplane-hits-a-deer/516951/
[41] How Often Do Airplanes Hit Deer? Knisley Welding Inc. KWI, March 7, 2017 https://knisleyexhaust.com/blog/how-often-do-airplanes-hit-deer/
[42] Wright, S.E., Dolbeer, R.A., and Montoney, A.J. Deer on Airports: An Accident *Waiting* to *Happen*, Proceedings of the Eighteenth Vertebrate Pest Conference (1998), Vertebrate Pest Conference Proceedings collection, January 1998, University of Nebraska – Lincoln, DigitalCommons@University of Nebraska – Lincoln
[43] Wright, S.E. An Analysis of Deer Strikes with Civil Aircraft, USA, 1982–2001, Bird Strike Committee-USA/Canada, Third Joint Annual Meeting, Calgary, AB Bird Strike Committee Proceedings, August 2001, University of Nebraska – Lincoln, DigitalCommons@University of Nebraska – Lincoln
[44] Namowitz, D. Report Offers Insights on Aircraft-Wildlife Collisions, AOPA, March 18, 2021, www.aopa.org/news-and-media/all-news/2021/march/18/report-offers-insights-on-aircraft-wildlife-collisions, (Accessed March 21, 2021)
[45] Marusak, J. Deer killed in CLT runway latest in series of odd animal collisions at airports, The Charlotte Observer, February 15, 2017, www.charlotteobserver.com/news/local/article133003529.html
[46] Ball, S., Caravaggi, A., and Butler, F. Runway roadkill: a global review of mammal strikes with aircraft, Mammal Review, Wiley, 04 February 2021, https://doi.org/10.1111/mam.12241
[47] Colón M.R. and Long, A.M. Strike hazard posed by columbids to military aircraft, Human–Wildlife Interactions 12(2):198–211, Fall 2018, digitalcommons.usu.edu/hwi
[48] Pfeiffer M.B., Blackwell B.F., and DeVault T.L. Quantification of avian hazards to military aircraft and implications for wildlife management, 2018, PLoS ONE 13(11): e0206599. https://doi.org/ 10.1371/journal.pone.0206599
[49] Richardson, W.J. and West, T. (2000). Serious Birdstrike Accidents to Military Aircraft: Updated List and Summary, International Bird Strike Committee. IBSC 25/WP SA1. Published in IBSC Proceedings: Papers & Abstr. 25 (Amsterdam, vol. 1): 67-97(WPSA1). http://worldbirdstrike.com/IBSC/Amsterdam/IBSC25%20WP SA1.pdf (accessed 12 March 2021).
[50] Clearly, E.C. and Dolbeer, R.A. (2005). Wildlife Hazard Management at Airports: A Manual for Airport Personnel, 2e. https://digitalcommons.unl.edu/cgi/viewcontent.cgi?article=1127&context=icwdm_usdanwrc (accessed 12 March 2021).
[51] Einstein A. The Meaning of Relativity. Princeton: The Princeton University Press; 1922
[52] Lovell C.D. and Dolbeer R.A. Validation of the United States Air Force bird avoidance model. Wildl Soc Bull. 1999; 27: 167–171.
[53] Dan, G. (2013). No. 100: Strike Incidents – Visualizing Data with ggplot2. http://genedan.com/no-100-strike-incidents-visualizing-data-with-ggplot2 (accessed 13 March 2021).
[54] McCracken, P.R. (1976). Bird Strikes and The Air Force. Bird Control Seminars Proceedings, Paper 53. http://digitalcommons.unl.edu/cgi/viewcontent.cgi?article=1052&context=icwdmbirdcontrol (accessed 2 March 2021).

[55] 2ND LT. TIFFANY ROBERTSON, BASH 2011, Wingman, Aviation Safety Special Edition 2012, 2012, www.safety.af.mil/Portals/71/documents/Magazines/Wingman/2012%20Aviation%20Special.pdf?ver=2016-08-19-164502-537 (accessed March 20, 2021)
[56] DeVault T.L., Belant J.L., Blackwell B.F., and Seamans T.W. Interspecific variation in wildlife hazards to aircraft: implications for airport wildlife management. Wildl Soc Bull. 2011; 35: 394–402
[57] Zakrajsek E.J. and Bissonette J.A. Ranking the risk of wildlife species hazardous to military aircraft. Wildl Soc Bull. 2005; 33: 258–264.
[58] Kitowski, I. Civil and Military Bird Strike in Europe: An Ornithological Approach, J Applied Science, 11 (1): 183–191, 2011
[59] Snow, S. This is how 1,500-plus aircraft bird strikes have affected the Marine Corps in the past 15 years, Marine times, May 17, 2019, www.marinecorpstimes.com/news/your-marine-corps/2019/05/17/this-is-how-1500-plus-aircraft-bird-strikes-have-affected-the-marine-corps-in-the-past-15-years/
[60] Payson, R.P. and Vance, J.O. (1984). A Bird Strike Handbook for Base-Level Managers, M.Sc. Thesis, Air Force Institute of Technology, Air University. https://apps.dtic.mil/dtic/tr/fulltext/u2/a147928.pdf (accessed 23 January 2021).
[61] Bird/Animal Aircraft Strike Hazard (BASH) Manual, Commander Navy Installations Command, CNIC M-BASH 1 April 2018
[62] Peurach, S. Bat Strike, Flying Safety Magazine FSM, United States Air Force, pp.18–20, September 2004
[63] El-Sayed, A.F. Aircraft Propulsion and Gas Turbine Engines, Taylor & Francis, Second edition, 2017
[64] Cleary, E.C., Dolbeer, R.A., and Wright, S.E., 2006. Wildlife Strikes to Civil Aircraft in the United States 1990–2005. US Department of Transportation, Federal Aviation Administration National Wildlife Strike Database, Serial Report Number 12. Washington, DC
[65] Brotak, E. FAA: 665 helicopter bird strikes over the last 3 years, Vertical: OCTOBER 5, 2018, https://verticalmag.com/news/faa-665-helicopter-bird-strikes-over-last-3-years/
[66] Thorpe, J. (2012). 100 Years of Fatalities and Destroyed Civil Aircraft Due to Bird Strike. IBSC30/WP Stavanger, Norway 25th–29th June 2012. www.int-birdstrike.org/ Warsaw_Papers/IBSC26%20WPSA1.pdf (accessed 22 March 2021).
[67] Dolbeer, R.A., Wright, S.E., Weller, J., and Begier, M.J., Wildlife strikes to civil aircraft in the United States, 1990–2013. Report of the Associate Administrator of Airports Office of Airport Safety and Standards and Certification. Federal Aviation Administration National Wildlife Strike Database Serial Report 20. Washington, DC: Federal Aviation Administration., 2014
[68] U.S. Coast Guard. https://en.wikipedia.org/wiki/Helicopter#/media/File:R-4_AC_HNS1_3_300.jpg (accessed 23 March 2021).
[69] Chow, D. Bird Strikes Problematic for Military Helicopters, Study Finds, Livescience, March 04, 2014, www.livescience.com/43833-military-helicopters-bird-threat.html
[70] Washburn, B.E., Cisar, P., and DeVault, T.L. "Wildlife strikes to civil helicopters in the US, 1990–2011" (2013). USDA National Wildlife Research Center – Staff Publications. 1247. https://digitalcommons.unl.edu/icwdm_usdanwrc/1247
[71] Hauptfleisch, M.L. and Dalton, C. 2015. Arthropod phototaxis and its possible effect on bird strike risk at two Namibian airports. Appl. Ecol. Environ. Res. 13, 957–965.

[72] House, A.P.N., Ring, J.G., Hill, M.J., and Shaw, P.P. Insects, and aviation safety: The case of the keyhole wasp Pachodynerus nasidens (Hymenoptera: Vespidae) in Australia, Transportation Research Interdisciplinary Perspectives, Volume 4, March 2020, 100096, http://dx.doi.org/10.1016/j.trip.2020.100096

[73] Herald, N.Z. Australian airport's wasp infestation growing safety problem, nzherald.co.nz, 6 Dec 2020, www.nzherald.co.nz/travel/australian-airports-wasp-infestation-growing-safety-problem/OC75EXAB63ASWBPF2AQIVJJAVY/ (accessed March 30m 2021)

[74] Carbaugh, D. Boeing, 2003. Erroneous: flight instrument information – safety and guidance. Aero 29.

[75] Australian Transport Safety Bureau, Rejected Takeoff, Brisbane Airport, Qld 19 March 2006, VH-QPB Airbus A330–303 (Canberra), 2006.

[76] Duquesne, B., Jacquot, A., and Cote, S. Pitot probe performance covered on the ground, Safety First, AIRBUS, 22, 6–13, 2016.

[77] Civil Aviation Safety Authority, 2018. Airworthiness bulletin 02-052 issue 4 – 3 May 2018, wasp nest infestation – Alert. Canberra.

[78] Wikipedia, "Boeing 757–225 Birgenair flight 301 Pitot tube blocked by wasp nest caused it to crash", 1996.

[79] McKenna, J.T. *Blocked Static Ports Eyed in Aeroperu 757 Crash*, Aviation Week and Space Technology, pp. 76, 1996

[80] House, A.P.N., Ring, J.G., and Shaw, P.P. Inventive nesting behaviour in the keyhole wasp *Pachodynerus nasidens* Latreille (Hymenoptera: Vespidae) in Australia, and the risk to aviation safety, bioRxiv preprint doi: https://doi.org/10.1101/2019.12.15.877274

[81] Yoneda, Y. Swarm of Thousands of Honey Bees Delays Delta Flight to NYC, inhabitat, 08/06/2012, https://inhabitat.com/swarm-of-thousands-of-honey-bees-delays-delta-flight-to-nyc/

[82] Draper, J. That'll create a buzz! Packed plane is grounded for 90 minutes after huge swarm of BEES settles on its wing, Daily Mail, 26 September, 2017, www.dailymail.co.uk/travel/travel_news/article-4921404/Packed-plane-grounded-BEES-settle-wing.html (Accessed March 31, 2021)

[83] Buzz off! Angry bees delay Air India flight, 16 September 2019, ARAB NEWS, www.arabnews.com/node/1555136/offbeat (Accessed March 31, 2021)

[84] Low-level locusts: Think through the potential on sequences of any plan, the free Library, www.thefreelibrary.com/Low-level+locusts%3a+Think+through+the+potential+on+sequences+of+any...-a085592518

[85] 'Locust swarm' forces Ethiopian Airlines plane to divert, BBC.news,13 January 2020, www.bbc.com/news/world-africa-51098209

[86] Lachmann, G.K. Aspects of Insect Contamination in Relation to Laminar Flow Aircraft, A.R.C. Technical Report C.P. No. 484, 1960

[87] Croom C.C. and Holmes, B.J. Flight Evaluation of an Insect Contamination Protection System for Laminar Flow Wings, May 1985, www.researchgate.net/publication/23908176

[88] Maddalon, D.V. and Wagner, R.D. Operational Considerations for Laminar Flow Aircraft. Laminar Flow Aircraft Certification. NASA Conference Publication 2413, Langley Research Center Hampton, Virginia, 1986, pp. 247–266.

[89] Young, T.M. and Humphreys B., Liquid anti-contamination systems for hybrid laminar flow control aircraft - a review of the critical issues and important experimental results. Journal of Aerospace Engineering. Vol. 218, 2004, pp. 267–277

[90] Wicke, K., Kruse, M., Linke, F., and Gollnick, V. Impact of Insect Contamination on Operational and Economic Effectiveness of Aircraft with Natural Laminar Flow Technology, 29th ICAS, Saint Petersburgh, Russia, September 7–12, 2014

[91] "NASA researchers to flying insects: 'Bug off!'." ScienceDaily. ScienceDaily, 5 November 2013. www.sciencedaily.com/releases/2013/11/131105122725.htm

[92] Humphreys, B. Contamination avoidance for laminar flow surfaces. Proceedings of the 1st European Forum on Laminar Flow Technology. Hamburg, Germany, DGLR-Bericht 92-06, Deutsche Gesellschaft für Luft- und Raumfahrt, Bonn, pp. 262–273, 1992.

[93] Young, T.M. and Humphreys, B. Liquid anti-contamination systems for hybrid laminar flow control aircraft - a review of the critical issues and important experimental results. Journal of Aerospace Engineering. Vol. 218, 2004, pp. 267–277

[94] Holmes, B.J. and Obara C.J. Observation and implications of natural laminar flow on practical airplane surfaces. Journal of Aircraft, Vol. 20, No. 12, pp. 993–1006, 1983.

[95] Barnstorff, K. and Harrington, J.D. NASA Tests Aircraft Wing Coatings that Slough Bug Guts, Jun 2, 2015, www.nasa.gov/langley/nasa-tests-aircraft-wing-coatings-that-slough-bug-guts, (accessed April 1, 2021)

[96] Elsenaar, A. and Haasnoot, H.N. A survey on Schiphol airport of contamination of wing leading edges of three different aircraft types under operating conditions. Proceedings of the 1st European Forum on Laminar Flow Technology. Hamburg, Germany, DGLR-Bericht 92-06, Deutsche Gesellschaft für Luft- und Raumfahrt, Bonn, pp. 256–261, 1992.

[97] Fox News, Airplane Mechanic Sucked Into Jet Engine, Killed at El Paso Airport, January 16, 2006 www.foxnews.com/story/airplane-mechanic-sucked-into-jet-engine-killed-at-el-paso-airport

[98] Engineer sucked through 737 engine, bluebaron, 4th June 2004, www.pprune.org/engineers-technicians/150865-engineer-sucked-through-737-engine.html

[99] Kaminski-Morrow, D. Departing 737 fatally strikes person on Moscow runway, FlightGlobal, , 20 November 2018. www.flightglobal.com/safety/departing-737-fatally-strikes-person-on-moscow-runway/130364.article

[100] How did he survive? Horrifying footage re-emerges of the moment a Navy trainee was sucked into the engine of a fighter jet during the Gulf War, DAILYMAIL.COM REPORTER, 15 May 2017, www.dailymail.co.uk/news/article-4509288/Video-shows-Navy-trainee-pulled-jet-engine.html (accessed 2 April 2021)

[101] Airport Foreign Object Debris (FOD) Management, FAA Advisory Circular AC 150/5210-24, 9/30/2010

[102] Heaton, D. Doing the FOD Walk, FOD Prevention, https://fodprevention.com/doing-the-fod-walk/

[103] Parsons, J.G., Blair, D., Luly J., and Robson, S.K.A. Bat Strikes in the Australian Aviation Industry, The Journal of Wildlife Management, Vol. 73, No. 4 (May 2009), pp. 526–529 (4 pages), Wiley

7 FOD Prevention and Elimination (FOE) in Manufacturing Processes

7.1 INTRODUCTION

Manufacturing of different modules/parts of fixed- or rotary-wing aircraft differs from other industries by the strict requirements of product integrity and the processes used to assure safety and reliability. However, the potential of introducing foreign objects into the product during fabrication is very high. For example, the number of fasteners used to assemble the structure and install components can reach hundreds of thousands.

Other possible FODs include tools, employee personal items, drill filings/chips, and several others. However, it is impossible to have a FOD free product. FOD controls and prevention practices must be established and enforced throughout the entire build process if product integrity is to be assured.

FOD control, prevention, and elimination are implemented throughout all phases of the *manufacturing process* including design, machining, assembly, testing, inspection, packaging, shipping and receiving, and repair, including facilities and services operations [1].

Component design must ensure their protection from mechanical, electrical, and corrosion damage causing impairment of operation due to rain, snow, ice, sand, grit, and de-icing fluids [2]. However, such threats necessitate the selection of costly material and high technological production methods which in turn reduce the product profitability.

Manufacturing of aerospace vehicles starts with raw materials such as metal alloys, composites, laminates, and carbon fibers. Through numerous processes including machining, stamping, de-burring, grinding, sanding, and cutting, these materials will have the final shapes of wings, fuselage, landing gear, and numerous aircraft parts and systems.

Various size and shapes of debris are generated during the aircraft manufacturing, assembly, and rework processes. Such debris includes machining dust, metal chips, nuts, bolts, composites, and foam insulating materials [3]. Unfortunately, debris can drop into and remain suspended in the radial crevasses of some aircraft modules and remain secure and undetected through preflight inspection. It may jam flight controls, cause an electrical short, overheat avionics and navigation systems, puncture in

DOI: 10.1201/9781003133087-7

the skin of aircraft, or block airflow to critical components, leading to catastrophic accidents.

Military aircraft are especially prone to these circumstances due to their wider flight envelope.

7.2 DESIGN PHASE

It includes several design criteria for both airframe and engine. Aspects of airframe design should keep an eye on FOD protection in addition to the traditional strength and endurance. Thus it must consider all protective techniques to avoid or minimize erosion, corrosion, and icing damage if operating in adverse weather conditions like sandstorm, rain falls, and snowstorms.

Design of engines for protection of its rotating and stationary modules from FOD includes selecting the appropriate installation methods, selecting preferable alloys or composite material for their components and numerous design techniques for avoiding performance deterioration due to erosion, deposition, and icing conditions.

7.3 AIRFRAME OF FIXED-WING AIRCRAFT

Different aircraft modules including the fuselage, wings, tails, air probes, engines, propellers, auxiliary power unit, and radome suffer from various forms of foreign object debris including water droplets, hail, solid particles, as well as birds and insects. These debris will result in erosion, icing, and other serious structural problems.

Solid particles arising from sandstorms or jet blast may lead to erosion of radome, several aircraft parts, and engine modules.

Under certain atmospheric conditions, ice can build rapidly on aircraft wings and tail, air inlets, and engines [4].

7.3.1 RADOME

7.3.1.1 Erosion Protection

Erosion of radome caused by either water droplet or solid particles will cause its wear and detracts from their appearance [5]. It is important to protect it since it houses flight critical instrumentation. A thicker layer of more elastomeric material is applied to prevent erosion. This is achieved by applying several layers of additional paint to add some 250μ or greater of additional elastomeric polyurethane coating. However, it is time consuming to apply, as these layers need to be built up over time to allow solvents to evaporate.

Alternatively, an erosion boot, which is a thin, clear polyurethane film, may be applied to the radome, as a much quicker solution. It acts as a protective barrier between aircraft and the damaging debris. It has a three-dimensional shape that fits exactly over any complex contours of aircraft. This is a simple and easy solution. It ensures a perfect and streamlined fit and due to its ultra-high strength adhesive will guarantee a secure installation for years and years of service.

7.3.1.2 Icing

Since the earliest days of aeronautics, icing, rainfall, and snowfalls were found to be a crucial problem for aircraft flight or in other words the transportation's ancient enemies. The impingement of water and formation of ice on the radomes produces serious effects on the radar performance.

If during flight a thin layer of ice is accumulated on the leading edge of a wing that covers only 2% of the its chord, and has few centimeters thickness, it will drastically influence the wing's aerodynamics. It may cause flow separation, thus destroys lift, increases drag and reduces the control surface effectiveness. In some cases it may decrease engine performance and stability. Therefore, endless researches and deigns are performed on aircraft anti-icing methods for many years [6, 7].

A fluid protection system using ethylene glycol may be used for both anti-icing and de-icing which provide adequate icing protection [8].

7.3.1.3 Bird Strike

The forward-facing edges of an aircraft, especially the radome, windshield, leading edges of wings, and empennage are most crucial. Also, birds usually collide with aircraft engines as well [9].

Based on Federal Airworthiness Regulations (FAR) section §25.571 (e) (1), the radome structure must be designed to assure capability of continued safe flight and landing of the airplane after impact with an four-pound bird when the velocity of the airplane (relative to the bird along the airplane's flight path) is equal to 0.85 VC @ 8,000 ft. Aircraft should safely continue its flight and land safely [10, 11].

7.3.2 Flight Deck Windshield and Passenger Cabin Windows

7.3.2.1 Construction

Flight deck windshields are designed to allow vision of pilots [12] and remain intact despite the massive differential in exterior and interior pressure plus the risks of airborne objects or projectiles hitting the window [13].

Aircraft windshields and cockpit side windows are laminated multiple plies of glass or stretched acrylic material [14]. There are two structural plies (to withstand the pressure differential in the cabin), facing plies, adhesive interlayers, protective coatings, embedded electro-conductive heater films or wires, and mounting structure. On the outside, there is a thin outer "face ply" which also has a de-icing element. The facing and structural plies are laminated together with adhesive interlayer material of poly-vinyl butyral, polyurethane, or silicone.

7.3.2.2 Resistance to Adverse Weather

To keep windshield areas free of ice, frost, and fog, window de-icing, anti-icing, and defogging systems are employed. These can be electric, pneumatic, or chemical depending on the type and complexity of the aircraft:

- Transport category and high-performance aircraft use thermal electric anti-icing in windshields

- Small aircraft use chemical anti-ice systems
- Older aircraft have a space between the plies that allows a flow of hot air bled from engines to keep the windshield warm and fog free.

The laminated construction facilitates the inclusion of electric heating elements into the glass layers, which are used to keep the windshield clear of ice, frost, and fog. The elements can be in the form of resistance wires or a transparent conductive material may be used as one of the window plies. Each windshield incorporates an electrical heating element and three temperature sensors. One sensor is used for normal temperature control and another is used for overheating detection. The third sensor is a spare and is used if one of the other sensors fail. The amount of heat supplied to each windshield is controlled by a temperature controller which automatically regulates power to the heating element.

To remove the rain from the pilot and copilot's windshields, most aircraft use one or a combination of the following systems: windshield wipers, chemical rain repellent, pneumatic rain removal (jet blast), or windshields treated with a hydrophobic surface seal coating [15].

An electrical windshield wiper system consists of independent pilot and copilot systems. Each system consists of a windshield wiper and an electric motor (recent aircraft) or hydraulic motor (old aircraft).

Chemical repellant is applied by two switches or push buttons (for pilot and copilot) in the cockpit. The proper amount of repellant is applied using a solenoid valve to meter the repellent to a nozzle which sprays it on the outside of the windshield. These repellent fluids may be ethylene glycol, propylene glycol, Grade B Isopropyl glycol, urea, or sodium acetate.

Some aircrafts use a surface seal coating called hydrophobic coating on the external surface of the pilot's and copilot's windshield. This coating represents a passive method for icing which eliminates the need for wipers and gives the flight crew better visibility during heavy rain.

The last method employs hot air which either bled from high-pressure compressor (HPC) of the jet engine or heated by an electric blower. This hot air will be used for both removal of rain and the control of windshield icing.

7.3.2.3 Resistance to Bird Strike

According to Federal Aviation Administration (FAA) [14], the strength of glass varies with the rate of loading; the faster the rate of loading, the higher the strength, as is the case for bird impact loading. In addition, glass fracture stress for a load of short duration will substantially exceed that for a sustained load.

For windshields certification, it must undergo bird strike tests. FAA regulations necessitate the windowpanes to "withstand, without penetration, the impact of a four-pound bird at impact speed between 250 to 350 knots. An extensive costly experimentation is needed for designing an optimum impact resistant windshield. However, advanced numerical techniques that simulate the bird strike event are adopted as an effective substitute to the numerous experiments [9, 16–19].

Windshields are also tested for pressure, temperature control, and abrasion including sand or rain erosion.

Next, concerning passenger cabin windows, they are not subject to structural loads or visibility provision requirements as cockpit windows. However, they must preserve the pressure conditions in the cabin. Cabin window are multipaned construction consisting of two structural panes, inner facing panes, protective coatings, and mounting structure. The two structural panes are made from polymethyl-methacrylate and separated by an air gap. The designs with the structural panes separated by an air gap usually are such that the fail-safe pane is not loaded unless the main pane has failed.

Anti-icing and defogging of side windows is achieved by electrically heating them. Each side window incorporates an electrical heating element and three temperature sensors also as described above for windshields.

7.3.3 Wings, Horizontal and Vertical Tails

7.3.3.1 Icing

Ice accretion is expected to form whenever there is visible moisture in the air and the temperature is near or below freezing.

The size and shape of the ice accretion on unprotected aerodynamic surfaces depend primarily on airspeed, temperature, water droplet size, liquid water content, and the period the aircraft has operated in the icing condition. Under a normal icing encounter, most of the ice accretion would occur over the aircraft wings, tail, engine nose, and lips.

FAA requires airplane manufacturers to demonstrate that their aircraft can fly safely in icing conditions as defined by the so-called icing envelopes in the FAA's FAR Part 25, Appendix C [20].

Icing is a cumulative hazard which reduces the efficiency of aircraft by influencing the following parameters (Figure 7.1):

- Reduces lift
- Decreases thrust
- Increases drag
- Increases weight.

Moreover, ice accretion will affect the aircraft safety and performance as it results in:

- Unbalance of the aircraft due to the additional weight and unequal formation of the ice on wings and tail unit
- Reduction of the takeoff performance, rate of climb, and aircraft maneuverability (especially military ones) due to the degraded lift
- Increase of fuel consumption, reduction of the airplane's range and difficulty to maintain speed due to the increased aerodynamic drag

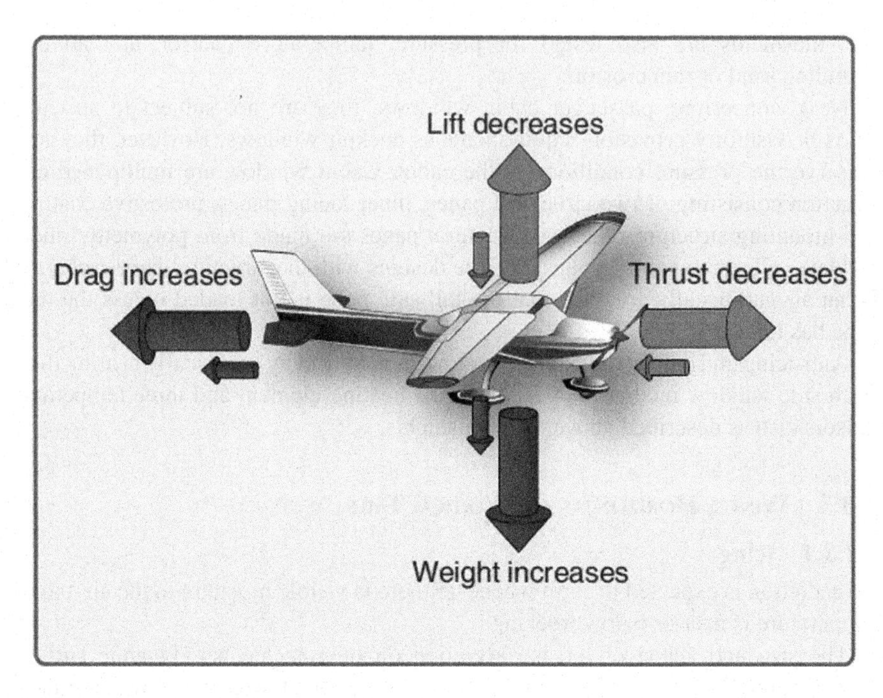

FIGURE 7.1 Effects of Structural Icing [21].

Courtesy: US DOT, FAA.

- Destructive vibration and hampering true instrument readings
- Jamming of the control surfaces
- Damage of aircraft structure as the flexing of the wing may break off and shed of its accumulated ice during takeoff
- Engine's surge, vibration, and complete thrust loss if the broken off ice chunks are ingested into the engines (especially aft-mounted ones)
- Reduction of propeller's efficiency
- Increase of landing distances which may be as twice as the normal distance.

The types of structural icing are clear/glaze, rime, and a mixture of the two [22, 23]. Each type has its identifying features.

Aircraft is designed to have ice and rain protection systems on the following components:

- Wing leading edges
- Horizontal and vertical stabilizer leading edges
- Engine cowl leading edges
- Propellers
- Propeller spinner
- Air data probes
- Flight deck windows

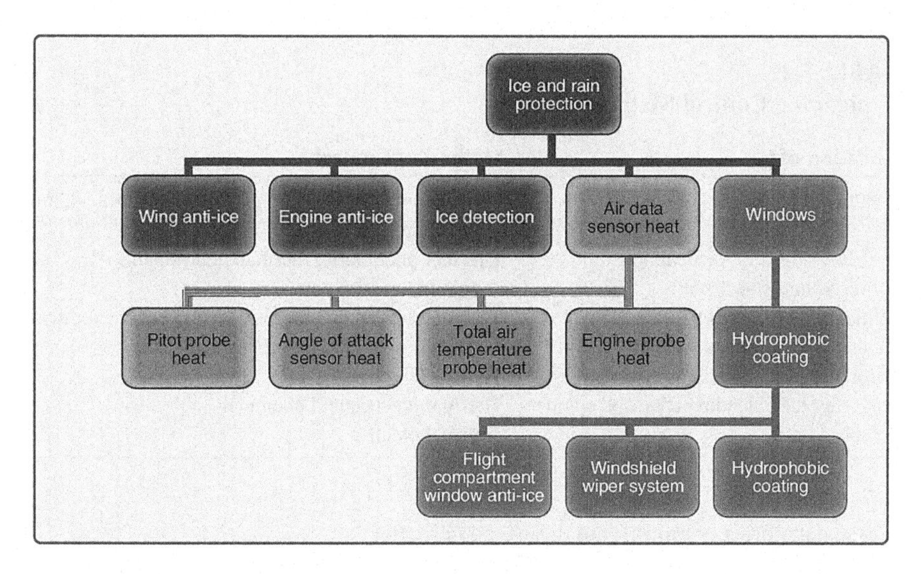

FIGURE 7.2 Ice and rain protection systems [21].

Courtesy: US DOT, FAA.

- Water and waste system lines and drains
- Antenna.

Two types of ice removal are generally employed:

- De-icing means the removal of ice after it had formed. It is a reactive action after significant ice has been build up
- Anti-icing is the prevention of ice formation. It is a preemptive action, which is turned on before the aircraft enters icing conditions.

An overview of ice and rain protection systems installed in a large transport aircraft category is illustrated in Figure 7.2. Ice can be detected visually, but most modern aircraft have one or more ice detector sensors and automatic control by onboard computers that warn the flight crew of icing conditions.

Different methods are employed to prevent icing (anti-icing) or to eliminate formed ice (de-icing) depending on the type of aircraft. These are:

1. Heating surfaces with hot air
2. Heating by electrical elements
3. Breaking up ice formations, usually by inflatable boots
4. Chemical application.

Table 7.1 lists the different modules of aircraft vulnerable to icing and the method for control.

TABLE 7.1
Typical Ice Control Methods [24]

Location of Ice	Method of Control
Leading edge of the wing	Thermal pneumatic, thermal electric, chemical, and pneumatic (de-ice)
Leading edges of vertical and horizontal stabilizer	Thermal pneumatic, thermal electric, and pneumatic (de-ice)
Windshield, windows	Thermal pneumatic, thermal electric, and chemical
Heater and engine air inlets	Thermal pneumatic and thermal electric
Pitot and static air data sensors	Thermal electric
Propeller blade leading edge and spinner	Thermal electric and chemical
Carburetor(s)	Thermal electric

Three categories of aircraft are described hereafter:

- Large transport and regional aircraft are equipped with advanced thermal pneumatic or thermal electric anti-icing systems that are controlled automatically to prevent the formation of ice.
- General aviation (GA) aircraft are mostly equipped with pneumatic de-icing boots, a chemical anti-ice system.
- High-performance aircraft may have "weeping wings."

7.3.3.1.1 Large Transport and Regional Aircraft

Large transport and regional aircraft employ the following systems:

1. Thermal pneumatic for anti-icing and de-icing of wings, leading edge slats, winglets, horizontal and vertical stabilizers, and engine inlets
2. Thermal electric anti-icing for small components like air data probes, ice detectors, as well as windshields and side windows
3. Chemical anti-icing
4. Electro-impulse De-Icing (EIDI)
5. Electro-thermal heaters (Boring 787)

7.3.3.1.1.1 Thermal Pneumatic System

Thermal pneumatic system uses heated air ducted spanwise along the inside of the leading edge of the airfoil and distributed around its inner surface. Heated air may be bled from the turbine compressor, engine exhaust heat exchangers, or ram air heated by a combustion heater.

Thermal wing anti-ice (WAI or TAI) systems for transport aircraft (like Boeing 737 series or Airbus A320) and regional jet aircraft (like Bombardier CRJ700, CRJ900, and CRJ1000) typically use hot air bled from the engine compressor. The hot air is routed through ducting, manifolds, and valves to components that need to be anti-iced. Figure 7.3 shows a typical WAI system for wing and wing leading slat anti-ice

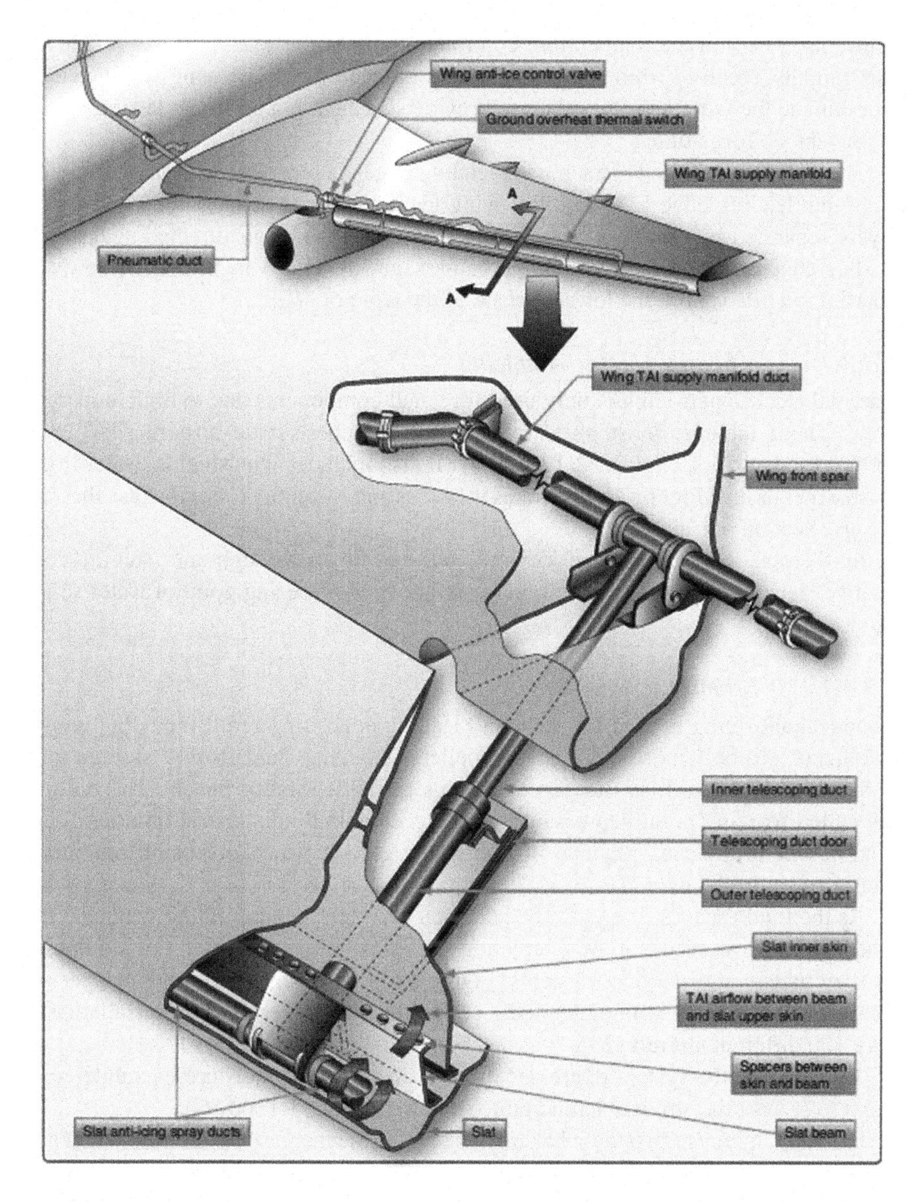

FIGURE 7.3 Thermal wing and wing leading slat anti-ice systems for wing-mounted engine aircraft [21].

Courtesy: US DOT, FAA.

systems. A telescoping duct supplies bleed air to the slats in the extended position. When normal anti-icing is selected, the wing leading edges are maintained at a constant temperature to shed ice and to prevent further ice accumulation. The system is manually activated and is automatically controlled by an anti-ice temperature controller. The anti-ice temperature controller controls the wing anti-ice system by using inputs, received from temperature sensors located on each wing leading edge, to modulate the wing anti-ice valves to provide sufficient hot air to the leading edges to prevent ice formation.

The bleed air is routed to each wing leading edge via piccolo tubes for distribution along the leading edge [25, 26]. The wing leading edge is constructed of two skin layers separated by a narrow passageway (Figure 7.4). The air directed against the leading edge can only escape through the passageway, after which it is vented overboard through a vent in the lower surface of the wing or slat.

7.3.3.1.1.2 Thermal Electric Anti-icing

Thermal electric anti-icing is only used for small components due to high amperage draw. These include: most air data probe (pitot tubes, static airports, TAT, and AOA), ice detectors, engine P2/T2 sensors. It also includes windshields, water lines, wastewater drains. Figure 7.5 illustrates the types and location probes that use thermal electric heat on one airliner.

In thermal electric anti-ice devices, current flows through an integral conductive element to elevate their temperature above the freezing point of water so ice cannot form.

7.3.3.1.1.3 Chemical Anti-icing

Chemical anti-icing is used to protect all leading edges of aircraft including wings, stabilizers, struts, windshields, and propellers. De-icing fluid from a storage tank is fed through micro-filters to several porous metal distributor panels. These fluids are called freezing point depressant fluids. Acceptable fluids for use on aircraft are mixtures of propylene, ethylene or diethylene glycol, water, corrosion inhibitors, wetting agents, and die [27].

As the fluid escapes, it breaks the adhesion between the ice and the wing and tail where the airflow carries it away, creating the so-called weeping wing or are known by their trade name of TKS® systems (Figure 7.6). It is designed for both inadvertent (no-hazard) and flight into known icing conditions. It was certified for installation in over 100 different aircraft [28].

TKS® fluid, the system depresses the freezing point of moisture encountered in flight to at least the ambient temperature or down to −76 °F (−60 °C).

However, there are two main hazards: the first is the increase of aircraft weight and the second is the toxicity of the ethylene glycol, which will result in environmental pollution.

7.3.3.1.1.4 Electro-impulse De-Icing (EIDI)

EIDI method used high-voltage capacitors, which are rapidly discharged through the coils (made of copper ribbon wire) installed just inside the skin of the aircraft leading edge [30], resulting in an electromagnetic repulsive force (on the

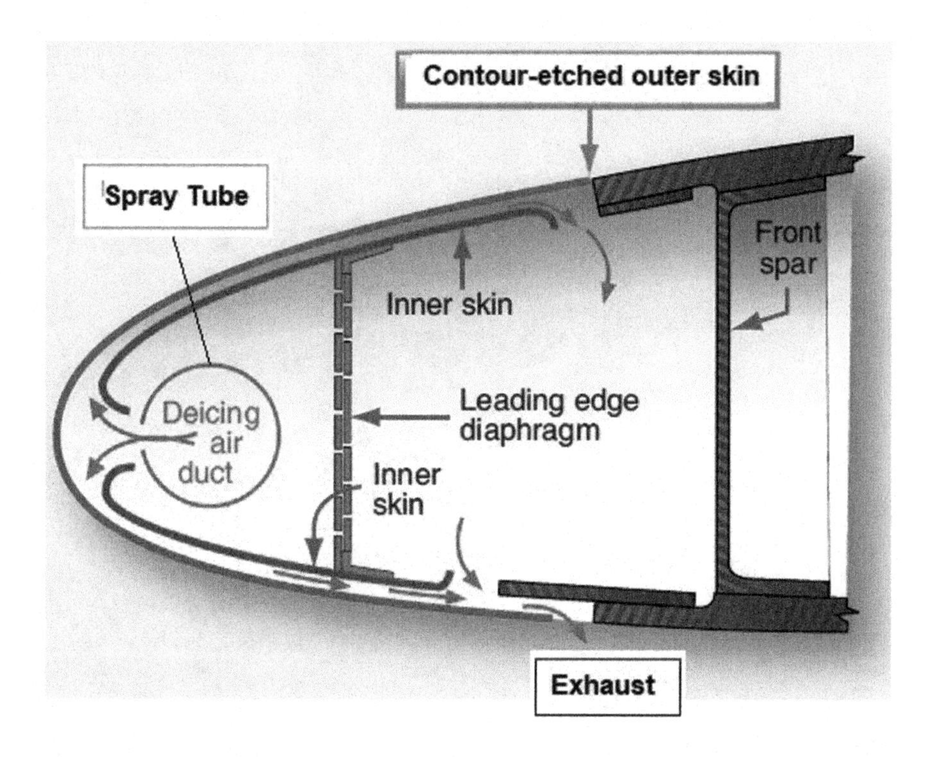

FIGURE 7.4 Heating the wing leading edge using hot air flowing in piccolo tubes [21].

Courtesy: US DOT, FAA.

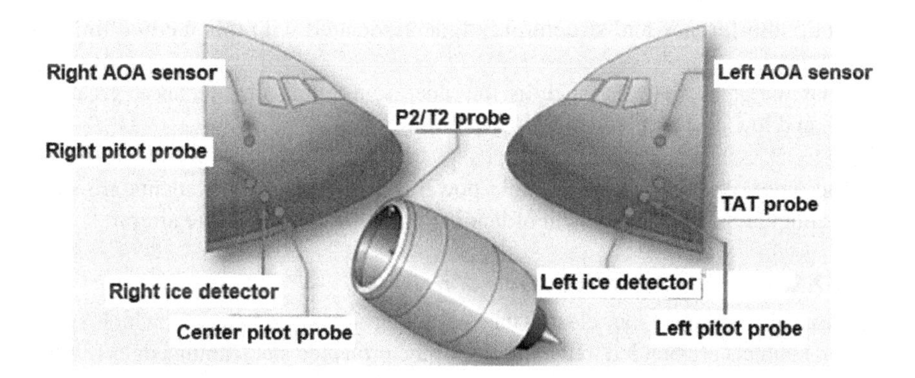

FIGURE 7.5 Thermal electric anti-icing tubes [21].

Courtesy: US DOT, FAA.

FIGURE 7.6 Weeping wing.

Courtesy: NASA [29].

aircraft skin), throwing ice in all directions (Figure 7.7). However, the electromagnetic interference and structural fatigue associated with this method limits its application.

It has major advantages, such as low energy, minimal maintenance, great reliability, and low cost and weight [31–33].

Figure 7.8 shows a wing with coils placed spanwise, separated by about 0.4 m. These are all supplied by a single power unit. Energy requirements are small, being comparable to those typical of landing lights for the same size aircraft.

7.3.3.2.5 Electro-thermal Ice Protection

The Boeing 787 utilizes an electro-thermal ice protection scheme, in which several heating blankets are bonded to the interior of the protected slat leading edges [34, 35]. For de-icing protection, the heating blankets may be energized sequentially, while for anti-icing protection they energized simultaneously.

This method is significantly more efficient than the traditional system – about 35% better efficiency – because no excess energy is exhausted. As a result, the required ice protection power usage is approximately half that of pneumatic systems. Moreover, because there are no-bleed air exhaust holes, airplane drag, and community noise are improved compared to the traditional pneumatic ice protection system.

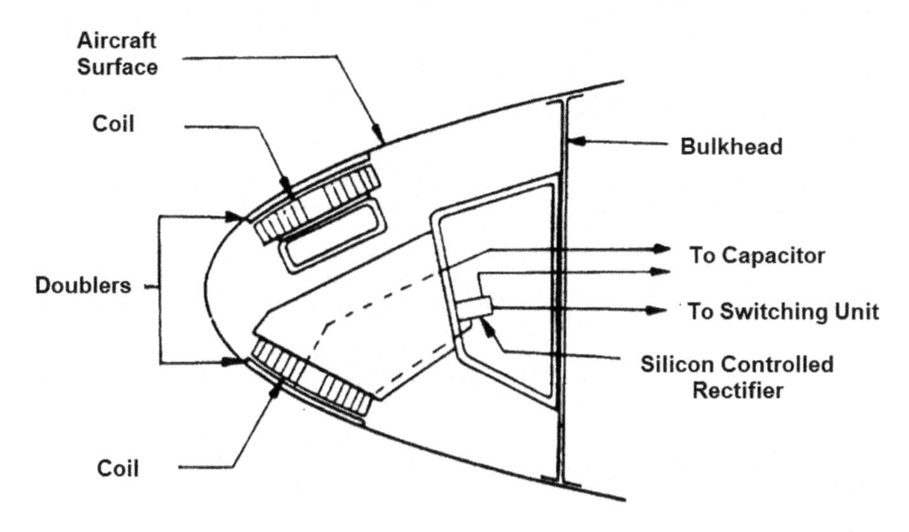

FIGURE 7.7 Impulse coils in a leading edge [30].

Courtesy: NASA.

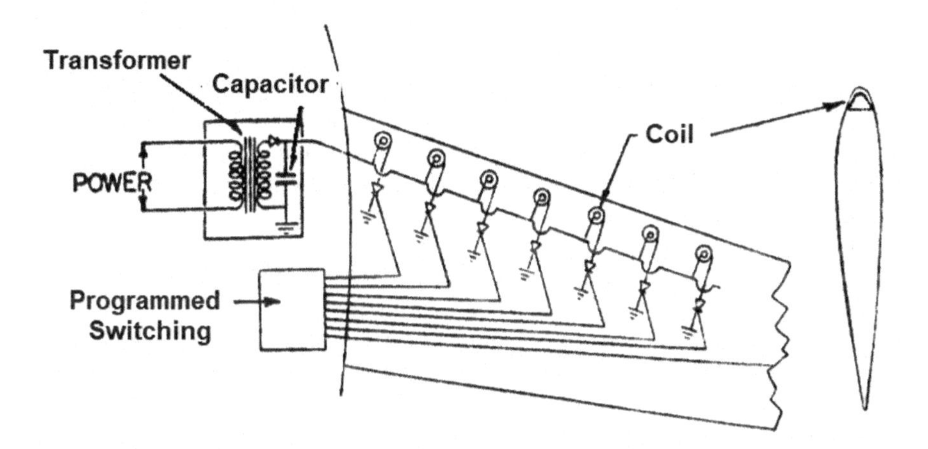

FIGURE 7.8 Electro-impulse coils installed in a wing [30] electro-thermal heaters in case of Boeing 787.

7.3.3.1.2 GA aircraft and turboprop commuter-type aircraft

GA aircraft and turboprop commuter-type aircraft often use:

1. Pneumatic de-icing system
2. Thermal pneumatic system
3. Thermal electric anti-icing
4. Chemical anti-ice system.

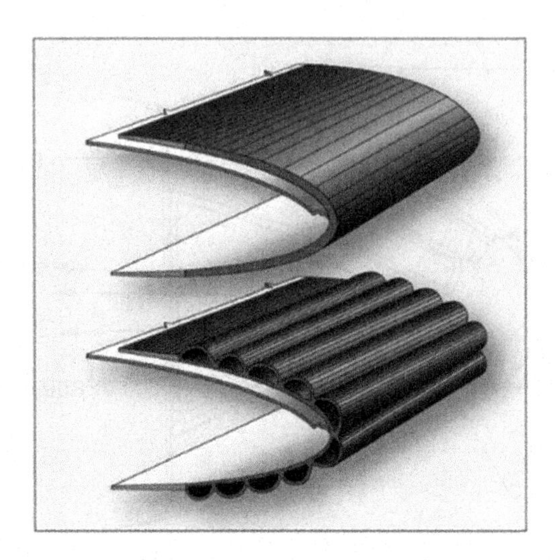

FIGURE 7.9 Pneumatic de-icing boots: top: uninflated and bottom: inflated [36].

Courtesy: US DOT, FAA.

7.3.3.1.2.1 Pneumatic deicing systems Pneumatic boots are commonly used on small airplanes of the type used in many FAR 135 operations (Figure 7.9). The leading edges of the wings and stabilizers have inflatable boots (rubber strips) attached to them [36]. The boots expand when inflated by pneumatic pressure for some 6–8 sec; next, they are deflated by vacuum suction. When these chambers are rapidly inflated and deflated, the adhesive force between the ice and rubber is broken. The vacuum is continuously applied to hold the boots tightly against the aircraft while not in use.

Pneumatic de-icing boots were first invented in 1923 by BF Goodrich.

The pneumatic boot is mostly employed in turboprop aircraft, some larger piston prop aircraft, and smaller jets. Examples for these aircraft are Beech BE200 Super King Air, Beech BE90 King Air, Beechjet 400, Cessna Grand Caravan, and Gulfstream G100/150.

However, such pneumatic boots are subject to corrosion and damage by external objects; therefore, they need to be replaced every two or three years [28].

7.3.3.1.2.2 Thermal Pneumatic System Though it is mostly used by larger jets, it is also used by jets chartered by private aircraft including Citation CJ3, Citation X, Phenom 300.

7.3.3.1.2.3 Thermal Electric Anti-icing Simple probe heat circuits exist with a switch and a circuit breaker to activate and protect the device.

7.3.3.1.2.4 Chemical Anti-ice System Some aircraft types, especially single-engine GA aircraft, use a chemical de-icing system for the propellers (Figure 7.10).

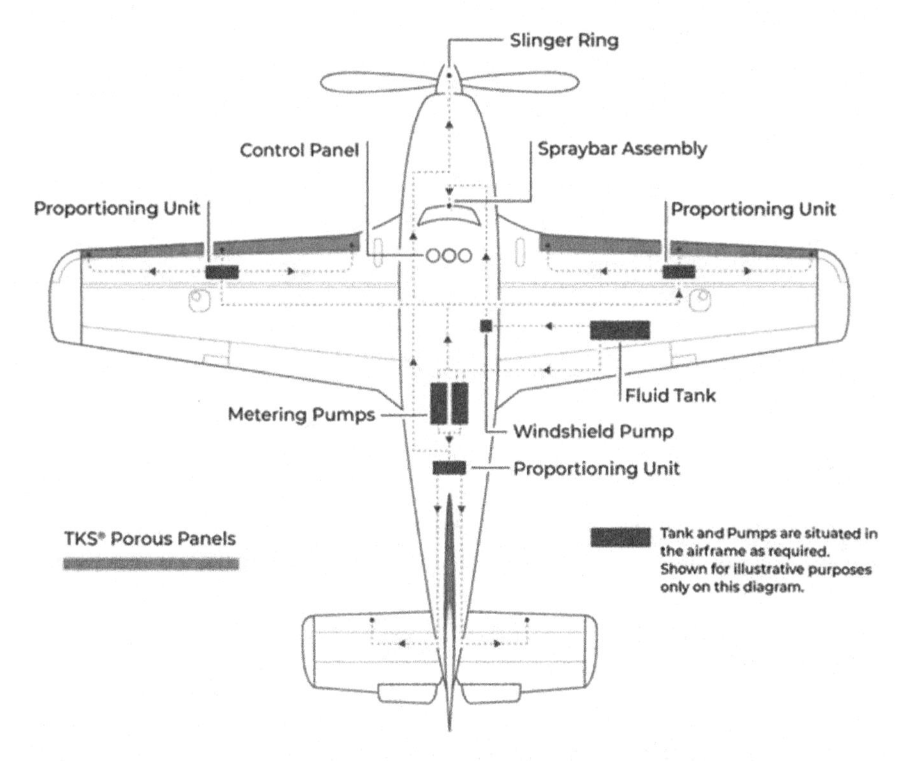

FIGURE 7.10 Chemical de-icing system for propeller, wing, and stabilizer [21]

Courtesy: US DOT, FAA.

Ice usually appears on the propeller before it forms on the wing. The glycol-based fluid is supported by a small electrically driven pump to the slinger rings on the prop hub. The propeller system can be a stand-alone system, or it can be part of a chemical wing and stabilizer de-icing system such as the TKS™ weeping system.

7.3.3.1.2.5 Anti-icing for Inlet Cowls and propellers Some Turboprop Inlet Cowls Are Also Heated with Electricity to Prevent Ice from Forming. Ice can be treated with chemicals from slinger rings on the propeller hub. Graphite electric resistance heater on leading edges of blades can be used. Propeller de-ice boots, which also are used for anti-ice, are also thermal electric.

7.3.3.2 Erosion Protection

A polyurethane straight tape having a thickness of 0.305 mm (0.012″) is used for erosion protection of aircraft leading edges including wing leading edges and horizontal tail. It is available in five different widths and 100 ft per roll. Such coating has several advantages including easiness of installing, cheapness, and availability in different thicknesses.

7.3.3.3 Bird Strike Resistance

The FAA additional regulations (under 14 CFR Part 25-571) state that an aircraft must be able to continue its flight and land safely after its wing structure has been impacted by a 1.8 kg (4 lb) bird at cruise speed (*Vc*) at mean sea level [37].

7.3.4 Thrust Reverser

7.3.4.1 Direction of Flow

The C-17 Globemaster III is powered by four Pratt & Whitney F117-PW-100 turbofan. When thrust reverse is applied, the engines direct their exhausts upward and forward to reduce the chances of foreign object damage (FOD) by ingestion of runway debris into the engines.

7.3.5 Wheels and Tires

7.3.5.1 Front Nose Wheel Downstream of Engine Intake

In fighter aircraft, the front wheel is often the cause of foreign objects being thrown up into the engine. The ingestion of foreign objects into the engine depends on the inlet size (diameter, diagonal, cross-section), the height of the inlet above the ground, and the position of the intake w.r.t the nose wheel.

The engine of F-16 has a low-positioned belly intake and thus more prone to ingesting foreign object debris. To reduce this risk, the intake entrance is designed in front of the nose landing gear, so that debris is less likely to be projected into the intake by the nose landing wheel [38].

7.3.5.2 Nose Wheel with Mudguard

The nose wheel of the Sukhoi Su-27 (NATO codename Flanker) is fitted with a mud-guard (or protective cover) to protect against FOD (Figure 7.11). This mudguard reduces the risk of debris projected by nose wheel into the bottom surface of fuselage and wings especially when aircraft takes off from unpaved runways [39, 40].

7.3.5.3 Nose Gear with Gravel Deflector

Nose gear of BOEING 737-200 is fitted with gravel deflector as explained earlier.

7.3.5.4 Large Number of Tires

Boeing C17 Globemaster has a large number of wide tires. Thus it is designed to allow landing of C17 on rough runways without cuts and tears which generate tire debris. Consequently, it avoids separation of tire treads, breakage into pieces that could impact several areas of the aircraft.

7.3.6 Vertical and Horizontal Stabilizer

7.3.6.1 Anti-icing and De-icing

For large transport-category aircraft, advanced thermal pneumatic or thermal electric anti-icing systems are employed for controlling the formation of ice on horizontal

FIGURE 7.11 Su-27 mudguard fitted to the nose gear [40].

With author permission.

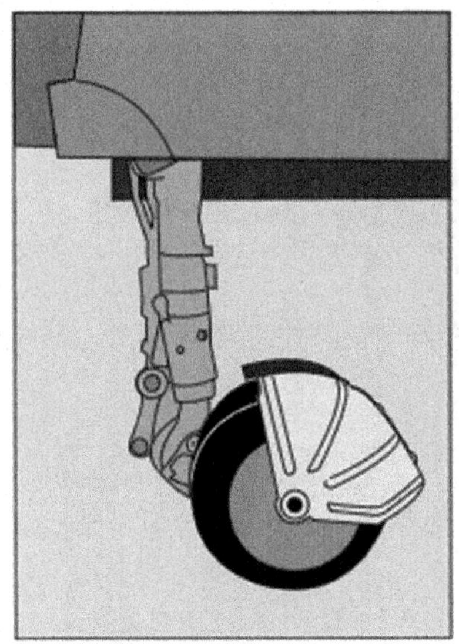

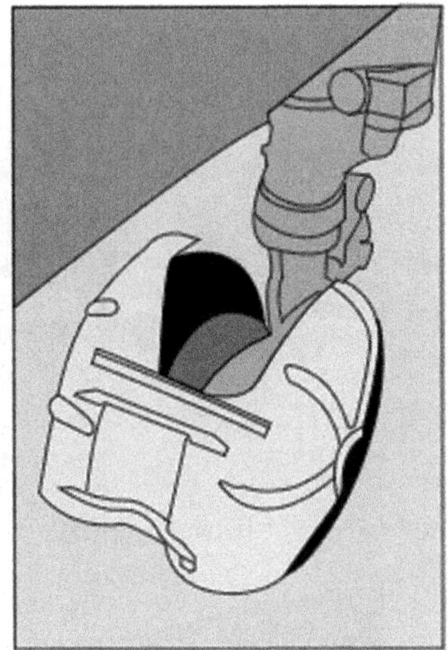

FIGURE 7.11 Continued

and vertical stabilizer. GA aircraft use pneumatic de-icing boots and chemical anti-ice system.

7.3.6.2 Bird Strike Resistance

Based on FAR section §25.631, the empennage structure must be designed to assure capability of continued safe flight and landing of the airplane after impact with an eight-pound bird when the velocity of the airplane (relative to the bird along the airplane's flight path) is equal to V_C at sea level [41].

7.3.7 Composite Materials in Fixed-Wing Aircraft

Composite materials were adopted in the aviation industry since the 1970s due to its high specific strength, high specific stiffness, smooth surfaces features (which reduce drag force), low thermal expansion coefficient, reduction of foreign object damage, and good vibration-damping characteristics [42]. Typical composites are carbon fiber reinforced plastic (CFRP), boron fiber reinforced plastic (BFRP), Kevlar fiber reinforced plastic (KFRP), and glass fiber reinforced plastic (GFRP) [43].

The most widely used composites in airframes are CFRP and the second is KFRP [44]. Due to the high price and difficulty in fabricating BFRP, there is little use of BFRP. Since GFRP has lower strength and stiffness properties compared to CFRP, it is not used for primary load-bearing structures. But it is used in many secondary structures in civil aircraft since it is inexpensive.

Historically, composites were first used on nonload-bearing structures, such as radomes, fairings, and for inner decoration [42]. Since the 1980s, there were used in secondary structures like control surface panels. Since composites indicated excellent in-service performance in commercial aircraft, this encouraged an increased usage of composites in aircraft structures, including small business-type aircraft and large commercial transport aircraft.

In the 21st century, composite materials have been adopted in primary structures in several recent aircraft. For example, in the Airbus 380, 25% of the total airframe weight is made with composites [45]. Many modules of the A380 aircraft are made of the CFRP composites including the Outer Wing, Outer Flaps, Spoilers & Ailerons, Empennage and the unpressurized Fuselage, Rear Pressure Bulk Head, Upper Deck Floor Beams, Engine Cowlings, and Center Wing Box [46].

The Boeing 787 adopts composite materials for the entire fuselage, as well as many components on the wing and nacelle. These composites account for 50% of its weight [47]. The most recent Airbus A350 XWB employs carbon composite resembling 53% of its weight [48, 49].

Concerning military aircraft designers also relied on composites since the 1960s.

Fighter aircraft used carbon fiber composite structural elements in horizontal and vertical stabilizers, flaps, wing skins, and various control surfaces for many years [50]. The Airbus A400M Atlas military transport aircraft has 35% of its structure made of composite material. More recently, the F-35 Lightning II Joint Strike Fighter incorporated carbon fiber composite fuselage and wing structures, while it is employed in the entire structure of the B2 stealth bomber [42].

Coatings are applied to high-erosion areas, such as wing and tail leading edges, to have a maintenance-free service life of about 2 years in normal airline utilization [51].

Two commercially available elastomeric polyurethanes, CAAPCO B-274 and Chemglaze M313, were tested extensively and compared to two coating systems (Astrocoat Type I and Corogard) already in commercial use.

Flight tests were conducted on the NASA B737 terminal configured vehicle to measure the effects of coatings on airplane drag. It was determined that, at a typical cruise condition, CAAPCO reduced airplane drag 0.2% when applied to the bare wing upper surface and that rough Corogard (160 pin roughness) increased drag by about the same amount. The Corogard tested was somewhat rougher than fleet average (130–150 yin). As a result of these tests, efforts are being made to further reduce roughness of surface coatings on new airplane applications. Replacing rough Corogard with CAAPCO on the wing upper surface and replacing enamel with CAAPCO on the empennage surfaces would produce a total airplane drag of about 0.55%.

During one flight in the B737 TCV test series, a severely eroded leading edge was simulated with a 7.6-cm (3-in) strip of No. 50 metallic grit glued onto the leading edge ahead of the wing test section. This strip caused a drag increment of approximately 0.3%. Combining this with the total airplane drag of 0.55% gives the possible total drag reduction, 0.85%. A cost/benefit analysis based on the 0.85% drag reduction indicated a net benefit per airline B737 of up to $10,000 per year with fuel costs at 26.4c/L ($l.OO/gal).

7.4 ENGINES OF FIXED-WING AIRCRAFT

7.4.1 Engine Installation

Engines may be installed in different locations on the wing, fuselage and tail structures. Airframe designers choose the appropriate locations that minimizes the strikes with various FO Debris, or in other words reduce the possibilities of FO Damage.

7.4.1.1 High Wing Aircraft

7.4.1.1.1 Antonov An-225

The jet engines installed to high wing aircrafts have less chance to pick up any foreign objects since engines are high above the runway, for example, Antonov An-225 Mriya [52] (Figure 7.12).

7.4.1.1.2 Antonov An-74

Engines are installed above the wing as shown in Figure 7.13) [53].

7.4.1.2 Rear-mounted Engines

7.4.1.2.1 Fairchild Republic A-10

The two-engine Fairchild Republic A-10 Thunderbolt II aircraft are top-mounted at the rear fuselage as shown in Figure 7.14 [54].

7.4.1.2.2 A-40 Albatross

The Beriev Be-42/A-40 Albatros is a twin-engine amphibious multipurpose jet aircraft developed by the Soviet manufacturer *Beriev Design Bureau*, today *Beriev Aircraft Company* (Russia). The Be-42 is the world's largest amphibious aircraft (Figure 7.15) [55]. Such engine installation above the wings minimizes the injection of water droplets into the engines.

FIGURE 7.12 Antonov An-225 [52].

FIGURE 7.13 Antonov An-74 [53].

FIGURE 7.14 A-10 Thunderbolt II [54].

FIGURE 7.15 Beriev Be-42/A-40 Albatros [55].

Permission GFDL 1.2.

7.4.1.2.3 Lake LA-250 Renegade

The Lake LA-250 Renegade is a six-seat amphibious utility aircraft powered by a highly mounted Lycoming O-540 single engine air-cooled six-cylinder, horizontally opposed engine coupled to a pusher propeller in a pod on a pylon above the fuselage (Figure 7.16) [56].

Since it operates in saltwater which will corrode its aluminum structure in a short amount of time if left unprotected. Aluminum is covered using materials such as polychromate primer or zinc chromate.

7.4.1.2.4 Embraer EMB-145LR (ERJ-145LR)

Embraer EMB-145LR (ERJ-145LR) aircraft is powered by two engines installed to the rear fuselage as illustrated in Figure 7.17 [57]. Such installation minimizes the ingestion of FO Debris generated by tire's motion during ground operation.

7.4.1.3 Engines Buried in the Wing

The B-2 aircraft is powered by four General Electric F118-GE-100 turbofan engines which are buried within the wing for several reasons including reducing the risk of damage to engine fan blades by bird ingestion, reduction of RCS, and minimizing the thermal visibility or infrared signature of the exhaust (Figure 7.18) [58]. Such engine installation reduces the risk of ingesting foreign objects during aircraft ground operation or in flight at low speed and low altitudes.

FIGURE 7.16 LA-250 Renegade Sea Plane [56].

FIGURE 7.17 Embraer EMB-145LR (ERJ-145LR) [57].

GFDL 1.2.

FIGURE 7.18 B-2 [58].

7.4.1.4 Over Wing Podded Engines

Figure 7.19 [59] illustrates a rear view of the Honda HA-420 aircraft having an over-the-wing engine mounts. Engines being installed above the wing and attached to its most rear part are less vulnerable to FO Debris.

7.4.1.5 Top Surface of the Rear Fuselage Section

Northrop Grumman Global Hawk unmanned aircraft for high-altitude, long-duration missions is an example for such engine installation. It is a 44-ft-long Global Hawk powered by a single Rolls-Royce AE3007H turbofan engine [60]. The engine

FIGURE 7.19 Honda HA-420 aircraft [59].

Courtesy: CC BY-SA 4.0.

is mounted on the top surface of the rear fuselage section (tail-unit) with the engine exhaust between the V-shaped tail wings (Figure 7.20).

7.4.1.6 Rear Fuselage Pusher Propeller Turboprop Engine

The General Atomics MQ-9 Reaper (sometimes called Predator B) is an unmanned aerial vehicle (UAV) capable of remotely controlled or autonomous flight operations developed by General Atomics Aeronautical Systems (GA-ASI) primarily for the United States Air Force (USAF). The MQ-9 and other UAVs are referred as remotely piloted vehicles/aircraft (RPV/RPA) by the USAF to indicate their human ground controllers (Figure 7.21) [61]. It is powered by one Honeywell TPE331-10 turboprop, 900 hp (671 kW) with a pusher propeller. Engine is installed to the tail-unit which minimizes FO Damage.

7.4.2 Engine Intake

7.4.2.1 Intake Ovality

Intake ovality is a successful solution to reduce the risk of foreign object ingestion through engine intake which increases the intake-to-ground distance by the opalization of the intake as is the case on the Boeing 737 family of aircraft.

7.4.2.2 Intake Covered with Gratings

The F-117A aircraft is powered by two low-bypass F404-GE-F1D2 turbofan engines from General Electric (Figure 7.22) [62]. It has the following design features:

FIGURE 7.20 Northrop Grumman Global Hawk unmanned aircraft having top surface rear fuselage engine installation [60].

Courtesy: NASA.

FIGURE 7.21 UCAV General Atomics MQ-9 Reaper [61].

FIGURE 7.22 F-117A [62].

1. Each engine has an inlet (or intake), which is covered by a fine, rectangular gratings or grilles
2. An electrical heating system which removes ice during flight
3. Lights on either side of the cockpit to allow the pilot to inspect the icing of intake grilles during flight
4. A large blow-in door fitted atop each engine nacelle to increase airflow to the engine during taxiing, takeoffs, or low-speed flight since the grilles restrict airflow to the engines.

7.4.2.3 Intakes Fitted with Removable Screens

Russian fighters like Mig-29 or the Su-27 have new FOD system for preventing FOD during taxiing, run, and take-off through the use of extendable screens in the air intakes' ducts as shown in Figure 7.23 [63]. Supplementary air inlet ramps are to be added through the lower surface of the air intakes. However, the major danger associated with the use of permanent intake screens is the danger of ice formation on the mesh.

7.4.2.4 Alternative Inlets

MIG 29 has alternative inlets (louvers) on top of the plane which temporarily open during takeoff while the front inlets shut down when it operates from unprepared or bombed runways. It supplies enough airflow to bring the aircraft into takeoff speeds (Figure 7.23).

FIGURE 7.23 Mig-29 [63].

7.4.2.5 Engine Vortex Dissipators

Boeing 737-100/200 is fitted with engine vortex dissipator to minimize the FOD ingestion on ground.

7.4.3 Architecture and Design of Jet Engines

7.4.3.1 Introduction

In brief, all aircraft engines belong to the internal combustion large group, which has two main sub-groups, namely reaction and shaft engines (Figure 7.24). Reaction engines in turn have several types including the turbine-based engines where its turbojet and turbofan engines are the most dominant engines now. Shaft engines include turboprop, turboshaft, and piston engines. Turbojet engines are declining due to its high noise and fuel consumption values. The remaining turbofan and turboprop categories power fixed-wing aircraft while the turboshaft engines power helicopters [64, 65]. Piston engines are employed in small fixed-wing and helicopters.

Turbofan engines are the most available jet engines since the 1970s, for example, Pratt & Whitney PW4000 series, General Electric GE90, Rolls Royce Trent series. Any turbofan engine is composed of the following modules: intake, fan, compressor(s), combustion chamber, turbine(s), hot nozzle, and cold nozzle. A discussion will be given hereafter for each module and how FOD has its impact on its design.

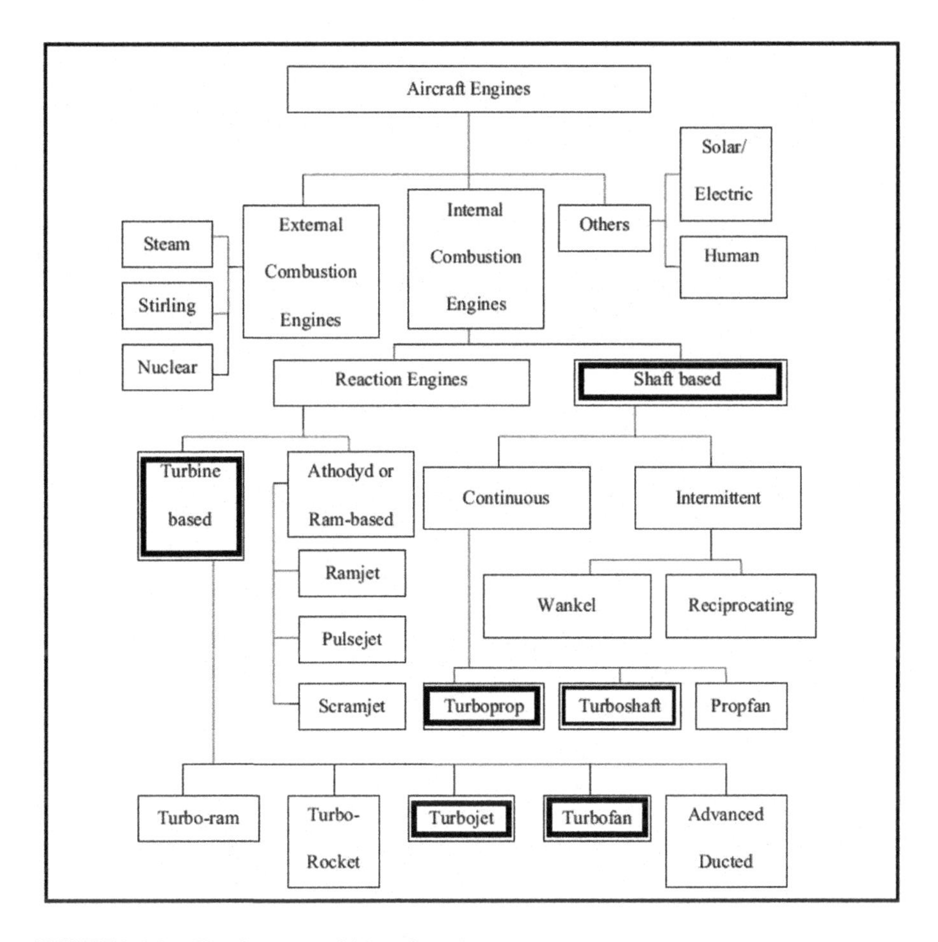

FIGURE 7.24 Classification of Aircraft engines

7.4.3.2 Regulations for Engine Endurance of Bird Strike

New engine design must satisfy the below performance and operation characteristics when a single or a flock of birds are ingested into its core.

Both of the Federal Aviation Administration FAA and the European Union Aviation Safety Agency EASA have their regulations concerning bird strike that must be fulfilled for engines' certification. The EASA issued its CS-E800 for a new engine certification, it must fulfill the following conditions:

- *No hazardous engine effect after impact with a single large bird of weight between 1.85 and 3.65 kg*
- *For a large flocking bird, no more than 50% loss of thrust following ingestion of large mixed birds of between 1.85 and 2.5 kg and for 20 min must be capable of thrust variation*
- *For medium flocking birds, the engine must be capable of ingesting some birds of varying mass, without a reduction of the thrust of less than 75%*

- *For small birds of weight 0.85 kg, ingestion of some birds must not result in loss of more than 25% thrust.*

7.4.3.3 Intake

7.4.3.3.1 Reducing Ice Accretion

Turbine-based engines face serious problems when ice builds up on its inlet. Such icing may restrict the airflow through the engine, affect engine performance, and possibly cause engine malfunction [66, 67]. Numerical and experimental studies for icing of the intake of jet engines were performed since several decades. A numerical analysis for ice accretion of the intake (inlet) of the GE CF6-50 high bypass ratio turbofan HBPR engines, when passing through clouds, is investigated by the author [67]. A parametric study is performed assuming the engine is installed to an aircraft flying an altitude of

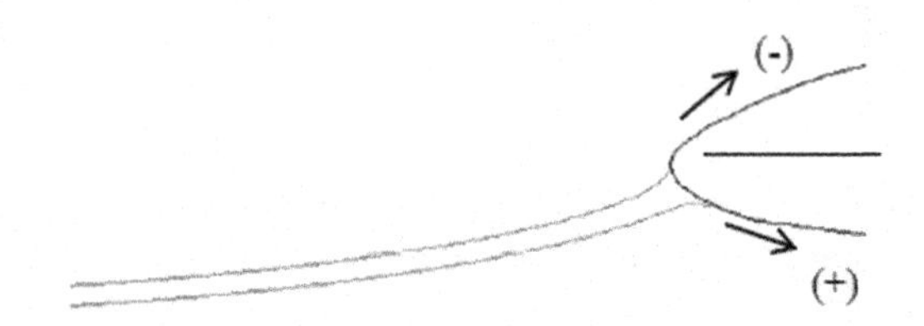

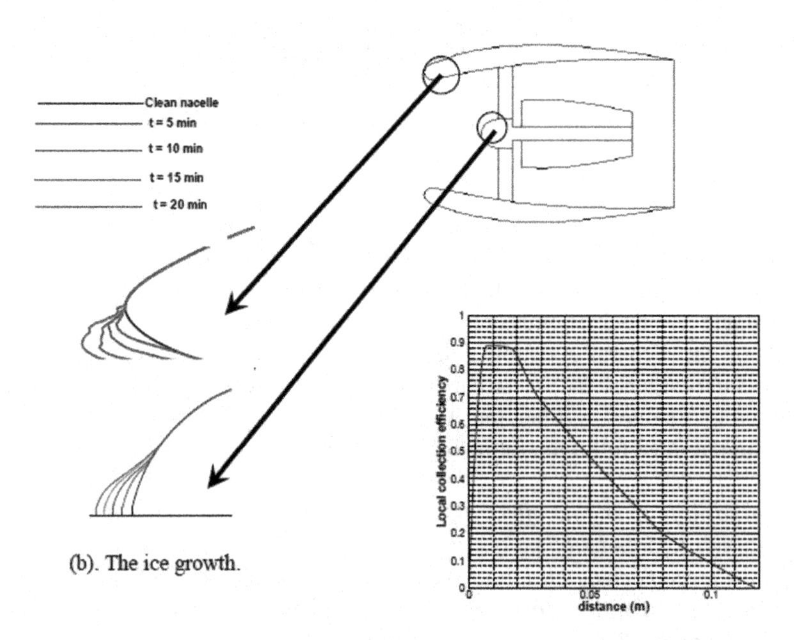

FIGURE 7.25 Ice accretion on the intake and fan's spinner of the General Electric GE CF6-50 turbofan engine (Mach number= 0.8 and water droplet diameter =15-micron).

8,461.5 m, through clouds having a liquid water content of 10 gm/m^3 and water droplet diameters of (15, 25, 35, 50 microns) and different Mach number (0.8, 0.6, 0.4, 0.3).

Figure 7.25 illustrates the ice patterns on the intake and fan's spinner for the case of Mach number equals 0.8 and water droplet having 15-micron diameter. Figure 7.25(a) shows the envelop for droplet trajectories. It outlines that most of the impingements are located in the inside portion of the intake while some of it is above the nose and this is also displaced in the plot of the local collection efficiency. Figure 7.25(b) illustrates the ice growth both on the intake and fan's spinner after 5, 10, 15, and 20 min. Figure 7.25(c) illustrates that the local collection efficiency plot. Full details for ice accretion will be given in Chapter 10.

Furthermore, if the accreted ice on the intake is dislodged, it will be ingested into the engine. It may hit the acoustic material lining the intake duct or impact fan/compressor blades and seriously damage them [68].

Intake design incorporates anti-icing systems if flying into known icing conditions. These systems use either large amounts of engine compressor bleed air (for turbojet and turbofan engines) or electrical power (for turboprop engine) or a combination of both. Electricity may result in an attendant fuel consumption penalty. This penalty may be minimized if de-iced, rather than anti-iced inlets are used.

7.4.3.3.2 Reducing Erosion

The intake shape of high bypass ratio turbofan is designed to reduce the risk of FOD damage. The shape of inlet duct and fan spinner direct any foreign object debris to the bypass or (secondary) duct. This will protect the fan, booster, and HPC from ingestion and damage by foreign object debris [38].

7.4.3.4 Fan

Design for improving FOD resistance is noticed in the following two parts: spinner and fan blades.

7.4.3.4.1 Spinner

Spinner must withstand bird impact, erosion, and ice accretion. To achieve such goals, the spinner is made from glass fiber, laid, and curved to achieve maximum strength. The thickness of the spinner is chosen to meet the heavier bird certification requirements where a double-ply lay-up technique is widely used.

Moreover, the angle of the cone is optimized for both bird impact and ice shedding behavior. The spinner nose also has a rubber front to dislodge ice accretion. A polyurethane coating is employed for protection against erosion.

Spinners may have any of the following three configurations:

- Conical: Provides best ice accretion characteristics
- Elliptical: Provides best hail ingestion capability
- Coniptical: A compromise between ice accretion characteristics and hail ingestion capability.

On an elliptical spinner, the ice accumulation starts at a lower humidity level compared to the other spinner designs. For these spinners, the heating with warm air is often used, because spinners of this shape have a higher risk for ice accretion. One example of this spinner heating is the Pratt & Whitney PW4000 engine. On this engine, warm air from the low-pressure compressor enters the N1 shaft and flows forward into the spinner cavity. The airflow exits the spinner through holes located just in front of the rear edge of the spinner.

7.4.3.4.2 Fan Blades

Fan is the first rotating module of turbofan engine located downstream of the intake. It is subjected to various FOD including water, hail, ice slab, birds (medium and large), and mixed sand and gravel. It is designed to pass different ingestion tests. Impacts of such debris with the fan blades will result in erosion by liquid and solid impact, cracks and even breakage of one or more blades. Consequently, manufacturers of aircraft engine keep their design development from both aerodynamic and dynamic aspects to demonstrate the capability of the engine to operate satisfactorily while ingesting simulated foreign object. As an example, CFM56 engines are designed to withstand the ingestion of the following debris and satisfy the following operating conditions:

- No substantial thrust loss
  - water: 4% (in weight) of total airflow
  - hailstones: 25 × 2 inch + 25 × 1inch stones within 5 sec
  - ice from inlet: 2 × (1 inch × 4 inch × 6 inch) slabs
- With less than 25% thrust loss
  - medium birds: 3 × 1.5 lb +1 × 2.5 lb ingestion into the engine core within 1 sec and operate for a 20-min period
  - mixed sand and gravel: 1 ounce for each 100 inch of inlet area
- With no hazard to the aircraft
  - large bird: 1 × 6 lb at most critical fan blade location.

Concerning erosion due to solid particles, numerical studies have been done by many authors including my group for estimating the erosion rates and lifetime of fan blades of high bypass ratio turbofan engines based on wear limits of its manufacturers [69, 70, 74].

Recent fans have the following design features:

- Low aspect-ratio fan blade like Rolls Royce Trent 1000 (Large Chord and Airfoil Thickness) design leads to a high tolerance to FOD and bird strike [71]
- Even number of carbon fiber composite blades, so that damaged blades can be replaced by a balanced pair and the engine does not need to be rebalanced in the field [72, 73].

General Electric company claimed several advantages for the composite fan blades of its engine GE90 with hollow titanium design, including low fragmentation energy, no low cycle fatigue limit, significant mass savings, high durability, and manufacturing repeatability. However, the fan blades showed FOD resistance disadvantages including erosion and impact properties. The solution to the erosion problem is a metallic leading-edge cover. The impact resistance improvements have been obtained using "second-generation" toughened resins, although 3-D through-thickness reinforcement is also being studied [73].

7.4.3.5 Compressors

Compressors are subjected to erosion when sand, dust, or volcanic ash particles are ingested into their core. Both numerical and experimental studies were carried out by the author and few researchers to evaluate the erosion of axial compressors [75–79] and centrifugal compressors [80–82].

LPC or Low-pressure compressor (or booster) is designed to centrifuge foreign objects and dirt out of the core stream, resulting in reduced compressor airfoil erosion and cleaner airfoils [71].

High Pressure Compressor (HPC_ also uses low aspect ratio airfoils same as the fan blades and thus produces a high tolerance to FOD and improve performance retention [71].

Protective coatings are employed to improve the erosion resistance of compressors, decrease the blade deterioration, and increase their lifetime. For this sake, new protective coatings are developed and tested in [83].

An experimental study used chemical vapor deposition technique to apply the ceramic TiC coatings on INCO 718 and stainless steel 410. It developed one order of magnitude less erosion rate compared to some commercial coatings on the same substrates. Moreover, the coated blade life became one order of magnitude longer.

More recently, physical vapor deposition (PVD) coatings and carbide coatings are used to improve the erosion resistance of HPC.

Nanocoating is recently considered in the design of engine parts and component surfaces to improve their wear and corrosion resistance [84]. An advanced, erosion-resistant (ER) nanocoating protects the compressor blades from erosion for both civilian and military aircraft [85].

It also saves fuel, cuts carbon emissions, reduces operational costs, extends compressor life [86]. It will add more benefits like self-cleaning, improved hardness, and thermal performance and flame retardancy.

Using nanotechnology, multiple very thin layers of material are applied to a compressor blade using PVD. These layers will not impact weight or dimension and provide protection for particle erosion, fluid erosion and corrosion, as well as reduced ice accretion [86].

Nanocoating was applied to CFM56-7 engines used in Boeing 737 and Airbus A320 aircraft of Delta Airline as well as coated CF6-80 engines installed to Boeing 767 and Boeing 747-400 aircraft of Delta Airline.

7.4.3.6 Turbines

High-pressure turbine utilizes cooling flow supply from the compressor. Compressor discharge air cools the stator of the first turbine stage. Air is drawn from the center of the split combustor diffuser for the rotor cooling circuit. The flow reversal into the center of the diffuser separates foreign particles from the air for the rotor cooling circuit [71]. This flow is accelerated tangentially by a radial inflow inducer nozzle prior to entering the shroud.

7.4.4 Geared Turbofan

The Pratt & Whitney PurePower engine PW1217G is an example for a geared turbofan (Figure 7.26) [87], which provides a better FOD resistance since:

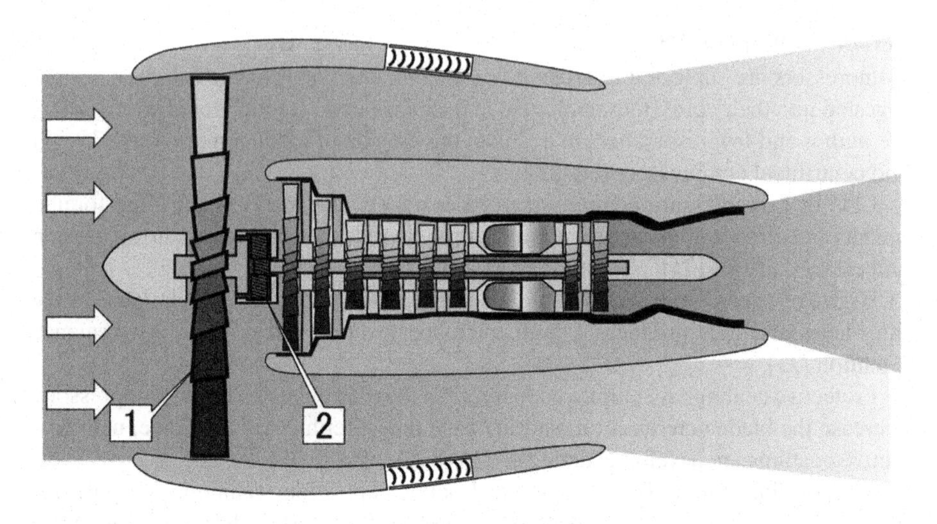

FIGURE 7.26 A schematic of a geared turbofan engine [87]: (1) fan; (2) gearbox.

1. It has a much higher "bypass ratio," and thus a higher chance that most debris sucked by the engine will bypass the engine core. Consequently, wear of the compressors, combustion chamber, and turbines will be reduced.
2. The fan does not run as fast as previous generation engines. Such a Lower Fan Tip Speed will reduce the damage due to foreign objects as its blades will not suffer from severe impact of debris [71].
3. Lower Fan Tip Speed – Reduced damage due to foreign objects.
4. The new engine can be mounted higher (on the wing) than traditional engines, increasing the distance from the ground and from all those objects that could damage it.

7.4.5 Propellers

In the 1960s, glass and fiber polymer composite propeller materials were used for manufacturing propellers. Today, nearly all the large propellers for aircrafts have carbon fiber blades (Lockheed Hercules C-130J). These are light weight and have high specific strength, rigidity, and excellent fatigue resistance. Also, carbon fiber is three times more resistant to erosion than aluminum alloys [88, 89]. Composite material propellers are extensively used nowadays for both erosion and icing protection. Chapter 10 will contain a detailed design for propeller against erosion damage.

7.4.5.1 Propeller De-icers

Propeller de-icing systems ensure safe propeller operation during icing conditions. Therefore, it is important to periodically inspect de-icer or anti-ice boots to ensure they are in an airworthy condition. Check the condition of the bond of the boot to the blade, looking for cracks and disbands. Look also for bumps, loose spots, or wrinkles in the boot. Ensure the propeller can be moved through its entire operating pitch range without placing any tension on wire leads or permitting them to interfere with or rub on nearby parts. Check for propeller resistance values between the de-icer leads and ground.

When conducting a detailed inspection, remove the spinner. Check the wiring leads and harness for looseness and wear. Ensure that wiring clamps are secure. Check slip rings and brushes for wear. Electrically isolate the de-icer circuits from other aircraft wiring and check for intermittent open circuits by moving the de-icer straps slightly. Repairs to propeller de-icers should be made in accordance with the manufacturer's instructions. If the ice protection system uses liquid-based anti-ice boots, check the condition of the slinger-ring and the feed-tubes.

7.5 HELICOPTERS AIRFRAME

7.5.1 Erosion Resistance

Helicopters are subjected to erosion caused by dust and other particulates suspended in the air during its normal flight as well as takeoff, landing, and low altitude hover. Radome erosion boots that applied to fixed-wing aircraft are employed to protect helicopter radomes from erosion [5]. It is simple and easy to install and save thousands on costly maintenance.

Helicopter rotors are even more vulnerable to the detrimental effects of sand, dust, and volcanic ash. Most blades are now constructed from composite structures; the outer layer is wrapped with alloy, and the inner layer is made from a composite. Ti-4Al-1.5Mn, Mg-Li9-A3-Zn3, and Al7075-T6 are commonly used aviation materials and are suitable for simulation experiments with different physical properties [90–92].

7.5.2 Ice Protection

The buildup of ice on rotor blades will affect its performance as it may [93]:

- Change the shape of the airfoil and consequentially
- Reduce lift while increasing drag
- Slow the main rotor (which needs additional power to maintain the speed which might not be available)
- Increase the helicopter's gross weight requiring more power as well
- Rotor imbalance which produces vibrations in the rotor system.

However, these vibrations can cause shedding of the ice. If shedding is symmetrical all the ice may come off, vibration levels, lift, and drag will return to normal. However, asymmetrical shedding can make the vibrations worse.

Few comments will be mentioned here concerning different modules

- Rotors

Helicopter rotors are even more vulnerable to the detrimental effects of ice. Consequently, the NASA Lewis Research Center performed aircraft icing studies almost from its start. Several methods of de-icing or anti-icing are examined and their undesirable aspects regarding energy requirements or effectiveness were analyzed.

The EIDI system has been applied by NASA Lewis Research Center. It is used for ice removing from the main rotor with very low energy, minimal maintenance (no moving parts), great reliability, and weight and cost competitive with existing methods. The tail rotor ice protection system can be set to de-icing mode, which applies power in a scheduled manner or anti-icing mode in which heat is continuously applied to tail rotor heating mats.

- Windscreen

Regarding windscreen anti-icing, electrically heated windscreens provide satisfactory protection even in the most severe conditions.

- Pitot/static systems

Most pitot heads or the combined pitot/static probes are electrically heated and operated satisfactorily in icing conditions.

7.5.3 Composite Materials

The composites content in civil rotorcraft is like that in commercial fixed-wing aircraft, averaging 50%. This is due to the superior material properties of CFRP compared to standard aviation aluminum materials. These properties include corrosion resistance, high fatigue strength, and weight reduction potential [94]. Sikorsky CH-53K and the Airbus Helicopters H160 are examples for the increased use of CRFP [95]. While the CH-53K is the first CH-53 model equipped with airframe side shells made from composite materials, the H160 even introduces the first civil helicopter with a full composite airframe.

The helicopter rotor blade resembles a high level of sophistication in aircraft composite construction [42]. The construction of such a component obviously requires a sophisticated fabrication procedure involving many materials, and some of these fabrication processes will be discussed in the next section.

7.6 HELICOPTER ENGINES

7.6.1 Engine Inlet Particle Separator

Helicopters frequently encounter sand and dusty environments during low-altitude operation due to downwash of rotor. During landing and takeoff, the rotorwash kicks up a cloud of sand or dust. Moreover, at certain areas of operation fine sand and dust particles are found at high altitudes, where the air is less dense and sometime hotter. These particulates can lead to a rapid performance deterioration of the turboshaft engines powering helicopters.

Consequently, particle separators are fitted to helicopter engine intakes to remove harmful dust from entering these turboshaft engines. The use of particle separator is necessary in desert environments to minimize the rapid engine wear and subsequent power deterioration [96]. However, these separators result in the loss in inlet pressure.

There are three main technologies commonly referred to as engine air particle separators. Two are retro-fit devices that may be integrated into the engine intake architecture, while the third type is designed to be an integral component for the engine [97]. These three devices are:

1. Integrated inlet particle separators (IPS) [38, 98, 99] rely on rapid change in curvature of the inlet geometry (Figure 7.27). They are found on many helicopters' turbo-shaft engines as Chinook and Apache.
2. Vortex tube separators (VTS) rely on centrifugal forces created by cyclone-like systems.
3. Inlet barrier filters (IBF) rely on a permeable fabric panel in front of the inlets to arrest the particles [100] (Figure 7.27). Sikorsky UH-60 Blackhawk with Inlet Barrier Filter fitted.

Aerometals designed, manufactured, flight tested and delivered a new generation of engine inlet barrier filter [101, 102]. It is fitted to the Sikorsky S-70i/UH-60 Blackhawk helicopter.

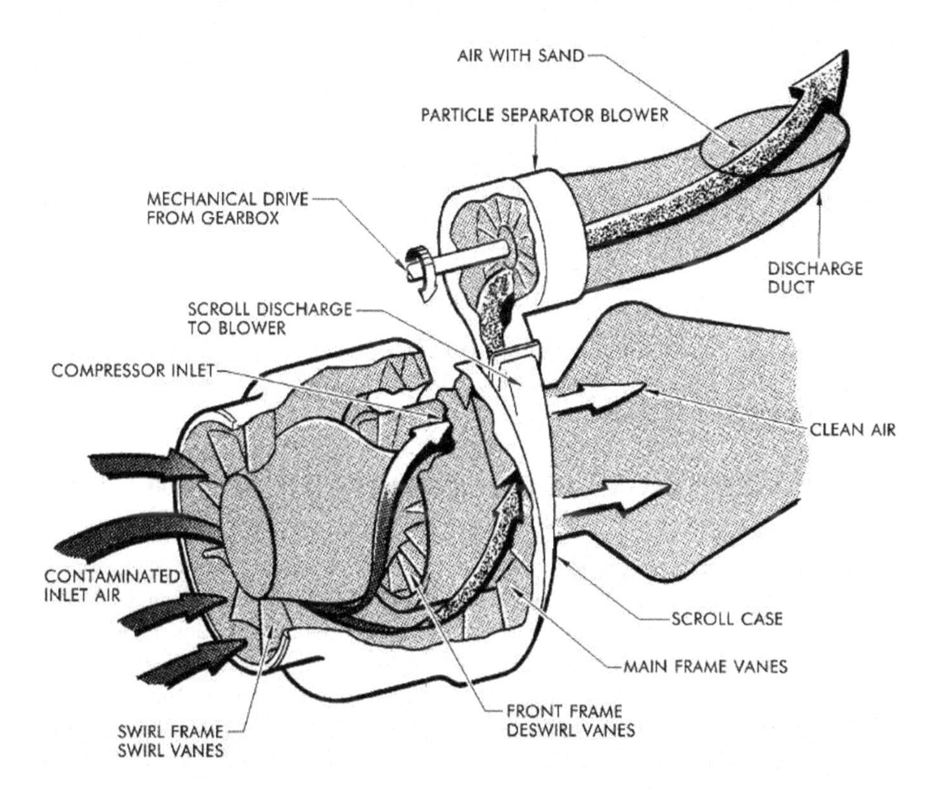

FIGURE 7.27 Integrated inlet particle separators (IPS) [103].

Courtesy: NASA.

A computational study [96] for the performance of these different filtration techniques outlined that:

- The vortex tube separators achieve the lowest pressure drop.
- The barrier filters exhibit the highest particle removal rate.
- The integrated inlet particle separator creates the lowest drag.
- The vortex tube and barrier filter separators are much superior to the integrated particle separator in improving the engine lifetime.

PUREair filtration systems designed by Pall [104] provides the following benefits:

- Protect engine by removing harmful solid and liquid contamination
- Excellent FOD protection
- Reduction of erosion of compressor
- Excellent snow and ice protection
- Self-cleaning and thus yields maintenance free
- Easy installation
- Reduced unscheduled engine removal
- Increased engine mean time between unscheduled removal
- Increased aircraft availability.

PUREair systems are used on a variety of customers which include US Army, UK RAF Royal, Netherlands Air Force, Egyptian Air Force, Royal Australian Navy, NAMSA, Heli Union, Maverick Helicopters, AgustaWestland AW-139, Bell-427, Boeing CH-47 Chinook, Denel, Enstrom, Eurocopter AS332 Super Puma, Eurocopter EC-135, Kaman, Mil Mi-8, Sikorsky SH-3 Sea King, and Aérospatiale SA 315B Lama.

7.6.2 Engine Intake Protection Screen

Permanent or in-flight movable intake protection screens can also be used to protect the engine intake. It can be either a true screen, as on the A109, or a screening obstacle, as on the Kamov Ka-52.

7.7 CONTROLLING OF FOD IN MANUFACTURING PROCESSES

An aviation organization can develop an FOD Prevention Program to eliminate debris as much as possible, reduce its consequences, and prevent debris from contaminating work areas in the first place. Typically, a program is based on guidelines and standards that are issued by authorities such as the FAA and EASA. Major considerations of a prevention program are using area designations (awareness, control, or critical area) to achieve risk mitigation through different levels of control, and training employees to understand which preventative actions to take depending on the area and work task. An effective program greatly reduces the cost of FOD and helps to keep workers safe by maintaining a mindset of keeping areas clean [105].

7.7.1 Control Areas

In aviation manufacturing areas, FOD debris can present a safety and quality risk. Production processes and assembly lines involve a variety of tools and materials that must move quickly through the facility. Any area where products are received, shipped, stored, repaired, inspected, or manufactured can be designated as an area that may contain foreign object debris. Thus workbenches, storage areas, work cells, and other locations must be organized and looking professional.

Three distinct categories for operation are evident: Fabrication, Assembly, and Checkout/Test [106–108].

7.7.1.1 Fabrication

These are the work areas including Sheetmetal, machine shop, tubing, and electrical harness. It is often called "back shops." These shops do not have enclosed panels, so FOD control is less critical than other manufacturing areas.

7.7.1.2 Assembly

In these areas, the fabricated parts are assembled, and cavities are enclosed with fasteners and sealants. Consequently, these areas require greater control, cleaning, and inspection than the fabrication areas. All closed out areas must be inspected for FOD and documented.

Assembly processes have the following sequencing:

- Clean or flush the component after machining operations to assure that it is free of debris
- Cap or seal exposed openings to prevent foreign object entry
- Protect hardware and equipment from splatter during brazing, soldering, or welding
- Inspect components and equipment for any damage before installation and repair if necessary
- Ensure protective devices (dust covers, temporary seals, cushioning, etc.) are present and properly installed
- Carefully clean and cap the ends of fluid and pneumatic lines/tubing after their cutting
- Remove extraneous material as part of the assembly step
- Ensure production tooling is clean, undamaged and FOD free prior to installation and build-up of components or assemblies
- Ditto for work stands, ladders, and special test equipment
- Protect products by using FOD barriers, foam pads, covers, etc.

7.7.1.3 Checkout/Test

This area has the highest level of control as the product is in its final stage prior to delivery or flight. Personnel entering this work area must leave all their personal items like watches, keys, jewelry, and coins. Moreover, all carried tools and hardware should be counted. Workers should be aware with the FOD control requirements for each area.

7.7.2 FOD Management Program

Facilities should establish a comprehensive program that incorporates training, operational guidelines, inspection, maintenance, and communication. The manager should be supportive of and responsible for the FOD Prevention Plan and its implementation. He must demand effective corrective action for FOD incidents and negative trends.

Managers supervise three departments: namely, FOD control, safety, and quality assurance, whose managers should have similar rank/authority.

7.7.2.1 The FOD Department

The number of people required for a full-time FOD effort varies depending on the type of operation, number of aircraft in production, the level of FOD control required, and the available budget. As a rule of thumb, an inspector is assigned for each five aircraft in production.

The FOD department performs the following duties:

- Performs FOD inspections for panel closeouts, manufacturing process
- Investigates FOD incidents/accidents
- Maintains FOD removal equipment
- Collects, records incident data into a database, and sends reports to key personnel
- Plans corrective actions for problems/issues
- Develops FOD training programs
- With all deliveries of aircraft or aeroengine modules, include missing item reports for any items not found or retrieved during the entire manufacturing process.

7.7.2.2 Training Elements

Training should focus on the following elements:

- A description of the methods and housekeeping practices used to control foreign object debris
- Designation of the personnel who are responsible for implementing and managing the FOD program
- A method of accountability for tools and other items at the beginning and end of a work shift
- A visitor protocol: visitors in FOD critical areas must always be escorted
- Reporting procedures for any missing or lost items

7.7.2.3 Awareness and Communication

Awareness and communication are at the forefront of FOD programs. They include:

- Implementing wall signs and floor tape that define which FOD area workers are currently in and remind them to clean as they go
- Employ labels, floor tape, wall signs, and barricades are essential tools to make people aware that FOD has the potential to cause issues.

- Posters and other visual reminders sign that outline FOD controls for each work area
- Arranging visibility boards that display FOD metrics, bulletins, and awareness media in each work center.

7.7.2.4 Implementing 5S in an Organization

Many manufacturing companies have adopted the Five S's of housekeeping as an aid for workplace organization [109]. This methodology enhances organization and efficiency through five phases, which help eliminate unnecessary items and importantly, remove any debris. It makes "clean as you go" second nature during daily operations.

5S is a workplace organization strategy that works on improving the efficiency and effectiveness of the company. It gets its name from the fact that each of the five main areas of focus start with the letter S.

7.7.2.5 The Five S's of Housekeeping

- Sort
  - Eliminate any obstacles that get in the way of production
  - Remove any unneeded items, put things away, and make sure the right people are performing the right jobs
- Set in order
  - Arrange things so that they are located where they need to be used
  - Remove any personal items from clothing before going on/into assemblies
  - Keep food and beverages out of work areas
  - Reduce or eliminate the need for employees to walk to another area to get a tool which helps in preventing wasted time and effort
  - Remove unnecessary items, relocate benches, equipment, storage units and toolboxes to facilitate smooth workflow and logical sequencing
  - Designate spaces for cleaning equipment, trash containers, ladders, air hoses and toolboxes should be outlined and labeled on floors and walls
  - Everything should be returned to its place when the task is completed so that it can be found easily when needed
  - Report missing items.
- Shine

Enforce "Clean-As-You-Go." Clean workplaces employing cleanliness checklists for tools and machines, and responsibilities for each work area. Shop floors should be painted a light color to allow easy identification of dirt/debris and to facilitate cleaning.

- Standardize
  - Set standards for how work should be done to reduce errors and improve efficiency
  - Keep rules concise and displayed prominently
  - Designate areas for cleaning equipment, ladders, toolboxes, hoses, parts, and support equipment

    - Improve standards to have everyone operating in the same way on all shifts to increase production
    - Establish accountability rules for expendables: sandpaper, rags, razor blades, etc.
- Sustain
    - Make sure any improvements implemented will be effective long into the future
    - Perform audits and inspections to sustain the improvements gained through the other four steps.

7.7.2.6 Five (5) S Team

The 5S implementation team should consist of the following people

- **Senior management** – chosen from the upper-level management teams. Their roles are approving changes and bringing all the groups together.
- **Middle management** – for medium and large organizations. They will take a more direct role in helping to coordinate what types of improvements need to be made in the areas they have direct responsibility for.
- **Direct supervisors** – The front-line supervisors are responsible for making the bulk changes and assure the application of any new strategies. Direct supervisors will support the 5S changes and will work hard for the success of the project.
- **Front-line employees** – The front-line employees will be the ones to directly "feel" the changes, so it is important to have them on the team. They will point out improvement opportunities and define the problems that management may not be aware of.

Since the 5S team should know how to work together, keeping everyone in the loop on all decisions and making sure to value his/her input is essential to the success of 5S implementation.

The total number of people on an implementation team can range from 4 to about 12 in most cases, with some of the members acting as advisors or resources to the rest of the group.

7.7.2.7 Auditing

Auditing is necessary to assure FOD free product. If FOD for any reason cannot be removed, it should be secured with sealant to avoid its migration or becoming loose.

Some manufacturers rotate and shake the entire aircraft assembly at different stages to dislodge foreign objects. Mirrors and high intensity flashlights are normally used. Special equipment including borescopes and other are used to verify cleanliness and retrieve FOD especially in engine modules. Since borescope is an expensive and fragile, it should be only used by FOD department personnel. Other tools used for FOD inspection/removal include pneumatic powered vacuum cleaners with suction tubes, magnets, grabbers, and others.

7.7.2.8 Tool Management

Tool management is an important system to improve safety and efficiency and avoiding foreign object debris and FOD. Every tool or group of tools are displayed in color-coded drawers of secure storage trolleys, which enables engineers and technicians to find the needed tool quickly and easily.

7.7.2.9 Product Quality and Customers

Product "Quality" implies not only its reliability, integrity, and safety, but also customer satisfaction. Customers' feedback may identify any foreign objects found in the delivered product and their exact location or zone. Manufacturers will next inform customers with their corrective action specifying what has been implemented to eliminate the error from future deliveries.

Manufacturers also invite the customer to visit the factory on a regular basis, who will give suggestions and reminders.

Successive inspections for FOD must be arranged throughout the entire production cycle. Each work center or department should consider themselves the "customer" of the preceding department. Thus, as the component or assembly moves from one work center to another, two FOD checks are performed, the first before leaving, and second when reached the new work center.

Any debris found after unit's inspection should be documented and reported to the manufacturing manager and the FOD program manager for corrective action. FOD manager will monitor the timeliness and effectiveness of the corrective measures.

Work centers that do not find FOD in their acceptance inspections are then responsible for any FOD found by their customer down the line.

A recognition of work centers having the best customer deliveries is monthly performed. An award should be given to the best work center of the year.

Without an effective FOD control program during the whole manufacturing phases, the product may not meet quality standards and leads to fall of sales, loss of contracts, or even accidents/incidents.

7.7.3 Transportation

All precautions are made during transportation. Fitted tarps and protective padding are used to deliver maximum protection against weather and road debris.

REFERENCES

[1] Foreign Object Debris (FOD) Prevention Procedure https://sapseod.mindtouch.us/Monarch_Global_Industries/Monarch_Aerospace/Foreign_Object_Debris_(FOD)_Prevention_P, Updated: Mon, 27 Apr 2020 20:29:16 GMT
[2] Foreign Object Debris (FOD) Management, FAA AC No: 150/5210-24, September 30, 2010
[3] Central Vacuum System for Advanced Aerospace Manufacturing Foreign Object Debris (F.O.D.) Central Vacuum System, Air Dynamics Industrial Systems Corp., Sep 07, 2011, https://news.thomasnet.com/companystory/central-vacuum-system-for-advanced-aerospace-manufacturing-foreign-object-debris-f-o-d-central-vacuum-system-601463

[4] El-sayed, A.F. FOD in Intakes – A Case Study for Ice Accretion in the Intake of a High Bypass Turbofan Engine, RTO-EN-AVT-195, August 9, 2011

[5] Rowbotham, J. Erosion Protection of Aircraft Radomes and Leading Edges, April 23, 2016 www.linkedin.com/pulse/erosion-protection-aircraft-radomes-leading-edges-your-rowbotham/

[6] Lynch, F.T. and Khodadoust, A. Effects of Ice Accretions on Aircraft Aerodynamics. Prog. Aerosp. Sci. 2001, 37, 669–767.

[7] Anon. "Reduce Dangers to Aircraft Flying in Icing Conditions", Most Wanted Transportation Safety Improvements-Federal Issues: Aviation, National Transportation Safety Board Recommendations and Accomplishments. Available online: www.ntsb.gov/Recs/most-wanted/air_ice.htm (accessed on 17 November 2005)

[8] James P.L. and Blade, R.J. Experimental Investigation of Radome Icing and Icing Protection, NACA RM E52531, 1954

[9] El-Sayed, A.F. Bird Strike in Aviation: Statistics, Analysis and Management, Wiley, July 2019.

[10] Reference F1] Design Airspeeds," Part 25 Airworthiness Standards: Transport Category Airplanes, Federal Aviation Administration, Dept. of Transportation, Sec. 25.571, Washington, D.C., 2003

[11] Mav, R.K. Numerical Analysis of Bird Strike Damage on Composite Sandwich Structure Using Abaqus/Explicit, M.Sc. Thesis, Department of Aerospace Engineering, San Jose State University, 2013

[12] Grether, W.F. Optical factors in aircraft windshield design as related to pilot visual performance, AMRL-TR-73-57, July 1973

[13] Balzo, J.D. Some technical clarity about aircraft windows, JDA Journal, May 23, 2018

[14] Windows and Windshields, FAA AC No: 25.775-1, 1/17/2003 Initiated By: ANM-110

[15] Aircraft Rain Control Systems, Aeronautics Guide, www.aircraftsystemstech.com/2017/05/aircraft-rain-control-systems.html

[16] Dar, U.A., Zhang, W., and Xu, Y. FE Analysis of Dynamic Response of Aircraft Windshield against Bird Impact, International Journal of Aerospace Engineering, Hindawi Publishing Corporation, Volume 2013, Article ID 171768, 12 pages http://dx.doi.org/10.1155/2013/171768

[17] Zang, S.G., Wu, C.H., Wang, R.Y., and Ma, J.R. "Bird impact dynamic response analysis for windshield," Journal of Aeronautical Materials, vol. 20, no. 4, pp. 41–45, 2000.

[18] Samuelson, A. and Sornas, L. "Failure analysis of aircraft windshields subjected to bird impact," in Proceedings of the 15th ICAS Congress, London, UK, 1986.

[19] Boroughs, R.R. "High speed bird impact analysis of the Learjet 45 windshield using DYNA3D," in Proceedings of the 39th AIAA/ASME/ASCE/AHS/ASC Structures, Structural Dynamics, and Materials Conference and Exhibit and AIAA/ASME/AHS Adaptive Structures Forum, pp. 49–59, Long Beach, Calif, USA, April 1998.

[20] Federal Aviation Regulations Parts 25, 21 and 210. "Airworthiness Standards: Transportation Category Airplanes, Normal Category Rotorcraft, and Transport Category Rotorcraft," Appendix C, Washington, D.C. 20591

[21] US FAA Aviation Maintenance Technician Handbook – Volume 2 (2018), FAA-H-8083-31A. AMA. Chapter 15: Ice and Rain Protection

[22] Hassani, A. Flowfield Simulation and Aerodynamic Performance Analysis of Iced Aerofoils, M.Sc. Thesis, Cairo University, 2013

[23] Badry, A.H., El-Sayed, A.F. Salem, G.B., and El-Banna, H.M. Computational investigation (simulation) of flow field and aerodynamic analysis over NACA 23012 airfoil

with leading edge glaze ice shape, Proceedings of ICFD11: Eleventh International Conference of Fluid Dynamics, December 19–21, 2013, Alexandria, Egypt
[24] Aircraft Ice Control Systems and Ice Detector System, Aeronautics Guide, www.aircraftsystemstech.com/2017/05/aircraft-ice-control-systems-and-ice.html
[25] Hassaani, A., Elsayed, A.F., and Khalil, E.E. Numerical investigation of thermal anti-icing system of aircraft wing, International Robotics & Automation Journal, IRATJ, 2020; 6(2):60–66.
[26] Bernardo, B., Alessandra, D., Oronzio, M., et al. Numerical Investigation on a modified "Piccolo Tube" System In Aircraft Anti-Icing. *The American society of Mechanical engineers*. 2017:10.
[27] How Fluids Work, Fluid Basics, NASA Glenn Research Center, https://aircrafticing.grc.nasa.gov/2_3_2_1.html#:~:text=These%20fluids%20are%20also%20known,contamination%20at%20temperatures%20below%20freezing.&text=Wetting%20agents%2C%20to%20allow%20the,conform%20to%20the%20aircraft%20surfaces.
[28] TKS® Ice Protection, www.cav-systems.com/tks/
[29] https://aircrafticing.grc.nasa.gov/media/img/ground-i/I-0113.jpg
[30] Zumwalt, G.W., Schrag, R.L., Bernhart, W.D., and Friedberg, R.A. Electro-Impulse De-Icing Testing Analysis and Design, NASA Contractor Report 4175, 1988
[31] Jiang, X. and Wang, Y. Studies on the Electro-Impulse De-Icing System of Aircraft, Aerospace 2019, 6, 67
[32] Li, Q. Research on the Experiments, Theories, and Design of the Electro-Impulse De-Icing System. Ph.D. Thesis, Nanjing University of Aeronautics and Astronautics, Nanjing,
[33] China, 2012. 8. Li, Q.Y.; Bai, T.; Zhu, C.L. De-icing excitation simulation and structural dynamic analysis of the electro-impulse de-icing system. Appl. Mech. Mater. 2011, 66–68, 390–395
[34] Sinnett, M. 787 No Bleed Systems: saving fuels and enhancing operational efficiency, AERO QTR 4, 2007
[35] Boric, M. Electrification and E-Flight Part 4 Boeing Is on the Way to a (More) Electric Future: The 787 and Zunum Aero are Boeing's pathway to E-Flight Jul 12th, 2018 www.aviationpros.com/article/12414609/electrification-and-e-flight-part-4-boeing-is-on-the-way-to-a-more-electric-future
[36] FAA AC 135-17 PILOT GUIDE Small Aircraft Ground Deicing, DATE: 12/14/1994
[37] Skybrary (2018). Aircraft Certification for Bird Strike Risk. www.skybrary.aero/index.php/Aircraft_Certification_for_Bird_Strike_Risk (accessed 29 December 2018).
[38] W. Beres, Chapter 6 – FOD PREVENTION, RTO-TR-AVT-094
[39] DCS: Su-27 Flanker Flight Manual, https://steamcdn-a.akamaihd.net/steam/apps/250310/manuals/DCS_Su-27_Flight_Manual_EN.pdf?t=1509670012
[40] https://aeroenginesafety.tugraz.at/doku.php?id=5:52:521:5212:5212
[41] Design Airspeeds, Part 25 Airworthiness Standards: Transport Category Airplanes, Federal Aviation Administration, Dept. of Transportation, Sec. 25.571, Washington, D.C., 2003
[42] Gibson, R.F. PRINCIPLES OF COMPOSITE MATERIAL MECHANICS, 4th edition, CRC Press. Taylor & Francis Group, 2016
[43] Wang, B. Composite structural design. Aircraft Design Handbook (Structure Design). Beijing: Aviation Industry Press; 2000. p. 631–685.
[44] Ren, H., Chen, X., and Chen, Y. Reliability Based Aircraft Maintenance Optimization and Applications, Academic Press is an imprint of Elsevier, 2017

[45] Pora, J. and Hinrichsen, J. Material and technology development for the A380. In: Proceedings of the 22nd International SAMPE Europe Conference of the Society for the Advancement of Materials and Process Engineering, Paris; 2001.
[46] Pora, J. Composite Materials in the Airbus A380 - From History to Future, https://pdfs.semanticscholar.org/31ff/a71022ce6bb0549d1711facf792b6b171414.pdf?_ga=2.1193114910.324134500.1589402513-307663090.1589402513
[47] Hale, J. Boeing 787 from the ground up. AERO. Shannon Frew: Boeing Company; 2006. p. 17–20.
[48] A350 XWB Family: Shaping the future of air travel, www.airbus.com/aircraft/passenger-aircraft/a350xwb-family.html
[49] Thévenin, R. Airbus Composite Structures Perspectives on safe maintenance practice, www.niar.wichita.edu/niarworkshops/Portals/0/Wednesday_945_Thevenin.pdf
[50] Harris, C.E., Starnes, J.H., Jr., and Shuart, M.J. 2001. An assessment of the state-of-the-art in the design and manufacturing of large composite structures for aerospace vehicles. NASA TM-2001–210844.
[51] Staff of Boeing Commercial Airplane Company, Aircraft Surface Coatings- Summary Report, NASA Contractor Report 3661, 1983, https://ntrs.nasa.gov/archive/nasa/casi.ntrs.nasa.gov/19850002627.pdf
[52] https://en.wikipedia.org/wiki/Antonov_An-225_Mriya#/media/File:Antonov_An-225.jpg
[53] https://en.wikipedia.org/wiki/Antonov_An-74#/media/File:Antonov-An-74.jpg
[54] https://en.wikipedia.org/wiki/Fairchild_Republic_A-10_Thunderbolt_II#/media/File:A-10_-_32156159151.jpg
[55] https://en.wikipedia.org/wiki/Beriev_A-40#/media/File:Beriev_A-40_Gelendzhik_2Sept2004.jpg
[56] https://en.wikipedia.org/wiki/Lake_Renegade#/media/File:Lake_Seawolf_N64RF.jpg
[57] https://en.wikipedia.org/wiki/Embraer_ERJ_family#/media/File:Embraer_EMB-145LR_(ERJ-145LR),_American_Eagle_AN0969536.jpg
[58] https://en.wikipedia.org/wiki/Northrop_Grumman_B-2_Spirit#/media/File:B-2_Spirit_(cropped).jpg
[59] https://en.wikipedia.org/wiki/Honda_HA-420_HondaJet#/media/File:Paris_Air_Show_2017_HondaJet_rear.jpg
[60] NASA Armstrong Fact Sheet: Global Hawk High-Altitude Long-Endurance Science Aircraft, www.nasa.gov/centers/armstrong/news/FactSheets/FS-098-DFRC.html
[61] https://en.wikipedia.org/wiki/General_Atomics_MQ-9_Reaper#/media/File:MQ-9_Reaper_UAV_(cropped).jpg
[62] https://en.wikipedia.org/wiki/Lockheed_F-117_Nighthawk#/media/File:F-117_Nighthawk_Front.jpg
[63] https://en.wikipedia.org/wiki/Mikoyan_MiG-29#/media/File:MiG-29_Fulcrum_B_Luftwaffe.jpg
[64] El-Sayed, A.F. Aircraft Propulsion and Gas Turbine Engines, Taylor & Francis, Second edition, 2017
[65] El-Sayed, A.F. Fundamentals of Aircraft and Rocket Propulsion, Springer, 2016
[66] Rosenthal, H.A., Nelepovitz, D.O., Rockholt, H.M. De-Icing of Aircraft Turbine Engine Inlets, June 1988DOT/FAA/cr-87 www.tc.faa.gov/its/worldpac/techrpt/ct87-37.pdf
[67] El-Sayed, A.F. FOD in Intakes – A Case Study for Ice Accretion in the Intake of a High Bypass Turbofan Engine, RTO-EN-AVT-195, August, 9, 2011

[68] The Jet Engine, Rolls Royce, © *Rolls-Royce plc 1986,* Fifth edition, Reprinted 1996 with revisions.
[69] El-Sayed, A.F., Gobran M.H., and Hassan, H.Z. Erosion of an Axial Transonic Fan due to Dust Ingestion, American Journal of Aerospace Engineering, Volume 2, Issue 1-1, January 2015, Pages: 47–63.
[70] Hassan, H.Z., Gobran, M.H., and El-Sayed, A.F., Simulation of a Transonic Axial Flow Fan of a High Bypass Ratio Turbofan Engine During Flight Conditions, February 2014, International Review of Aerospace Engineering 7(1):17–24.
[71] Davis, D.Y. and Stearns, E.M. Energy Efficient Engine: Flight Propulsion System Final Design and Analysis, NASA CR-168219, 1984.
[72] Aero engines lose weight thanks to composites (Part 1), Materials today, 6 November 2012, www.materialstoday.com/composite-applications/features/aero-engines-lose-weight-thanks-to-composites/
[73] A3] Saravanamuttoo, H.I.H., Rogers, G.F.C., Cohen, H., Straznicky, P.V., and Nix, A.C. Gas Turbine Theory, 7th Edition, Pearson, 2017.
[74] El-Sayed, A.F. and Gobran, M.H. Performance Degradation and Thrust Assurance Modeling of a High Bypass Turbofan Engine, The Second International Symposium of Fluid Machinery and Fluid Engineering (2nd ISFMFE), October 22–25, 2000, Beijing, China.
[75] Tabakoff, W. Causes for Turbomachinery Performance Deterioration, ASME Paper 88-GT-294.
[76] Hamed, A.A., Tabakoff, W., Rivir, R.B., Das, K., and Arora, P. Turbine Blade Surface Deterioration by Erosion, Journal of Turbomachinery, ASME Transactions, JULY 2005, Vol. 127, 445–452.
[77] El-Sayed, A.F., Lasser, R., and Rouleau, W.T. Effect of Secondary Flow on Particle Motion and Erosion in a Stationary Cascade, Int. J. Heat & Fluid Flow, Vol.7, No.2, June 1986.
[78] El-Sayed, A.F. and Brown, A. Particulate Flow in the Rotor and Stator Elements of Turbomachines, ASME Paper No. 85-GT-215, Presented at the Gas Turbine Conference and Exhibition, Houston, Texas, USA, March 18–21, 1985.
[79] El-Sayed, A.F. and Rouleau, W.T. The Effect of Different Parameters on the Trajectories of Particulate in a Stationary Turbine Cascade, 3rd Multi-Phase Flow and Heat Transfer Symposium Workshop, Miami Beach, Florida, USA, April 1983.
[80] El-Sayed, A.F. and Rashid, M.I.I. Efficiency of Gas Cleanup Systems for Reducing Erosion in Centrifugal Compressors, Proc. of IASTED Energy Symposia, Cambridge, Mass., USA, July 7–9, 1982.
[81] El-Sayed, A.F. and Rashid, M.I.I. Particle Trajectories in Centrifugal Compressors, Proc. of Gas Born Particles Conf., Inst. of Mechanical Engineers, June 30-July 2, 1981, Paper C61/81, I. Mech. E. 1981, Oxford, UK.
[82] El-Sayed, A.F. and Rashid, M.I.I. Particulate Flow in Centrifugal Compressors Used in Helicopters, Proc. 5th Int. Symposium on Air Breathing Engines (5th ISABE), Bangalore, INDIA, Feb. 16–22, 1981.
[83] Tabakoff, W., Hamed, A., and Shanov, V. Blade Deterioration in a Gas Turbine Engine, International Journal of Rotating Machinery, 1998, Vol. 4, No. 4, pp. 233–241.
[84] The Global Market for Nanocoatings in Aerospace, Future Market Inc., February 2018
[85] Varga, G. Erosion-Resistant Nanocoatings for Improved Energy Efficiency in Gas Turbine Engines, U.S. Department of Energy, Industrial Technologies Program,

2013, www.energy.gov/sites/prod/files/2013/11/f4/erosion-resistant_nanocoatings.pdf

[86] Millikin, M. Erosion-resistant nanocoating for turbine blades saves fuel, lowers costs, 15 August 2012, www.greencarcongress.com/2012/08/mct-20120815.html

[87] https://en.wikipedia.org/wiki/Geared_turbofan#/media/File:Geared_Turbofan_NT.PNG

[88] Reference A1] M. Burden, R. McCarthy and B. Wiggins, Chapter 6: Advanced polymer composite propeller blades, In H. Assender and P. Grant (Eds.), *Aerospace Materials*, Institute of Physics, 2012.

[89] FAA PROPELLER] FAA AC 20-37E, AIRCRAFT PROPELLER MAINTENANCE, 9/9/2005

[90] Jackson, R.J. and Frost, P.D. Properties and Current Applications of Magnesium-Lithium Alloys; Technology Utilization Report NASASP-5068; NASA: Columbia, DC, USA, 1965.

[91] Shechtman, D., Blackburn, M.J., and Lipsitt, H.A. The plastic deformation of TiAl. Metall. Mater. Trans. 1974, 5, 1373–1381

[92] Dursun, T. and Soutis, C. Recent developments in advanced aircraft aluminium alloys. Mater. Des. 2014, 56, 862–871.

[93] Aircraft Icing Handbook, Civil Aviation Authority

[94] Brauner, L., *Analysis of process-induced distortions and residual stresses of composite structures* (Doctoral dissertation). Berlin, Germany: Logos Verlag, (2013)

[95] Weber, T.A. and Ruff-Stahl, H.K. (2017). Advances in Composite Manufacturing of Helicopter Parts. International Journal of Aviation, Aeronautics, and Aerospace, 4(1). https://doi.org/10.15394/ijaaa.2017.1153 10.7 Engines

[96] Bojdo, N. and Filippone, A. Comparative Study of Helicopter Engine Particle Separators, Journal of Aircraft, Volume 51, Number 3May 2014, Published Online:19 Jun 2014, https://doi.org/10.2514/1.C032322

[97] Improving Aircraft Performance, Aie particle separator, Flight, www.flight.mace.manchester.ac.uk/air-particle-separator-design/about/about.html

[98] Potts, J.T., Jr. Why an Engine Air Particle Separator (EAPS)?, ASME Paper 90-GT-297

[99] Duffy, R.J. et al, Integral Engine Inlet Particle Separator. Volume II. Design Guide, USAAMRDL-TR- 75-31B, August 1975

[100] Inlet Barrier Filter System for the Bell Helicopter Textron Canada Limited Model 407 Helicopters, AFS-BH407-IBF-KIT-ICA, 18 July 2016, http://docplayer.net/52817940-Inlet-barrier-filter-system.html

[101] AVIATION PROS, Aerometals Introduces Sikorsky S-70i/UH-60 Blackhawk Engine Inlet Barrier Filter, Jul 8th, 2019 www.aviationpros.com/engines-components/aircraft-engines/propellers-rotors-blades/press-release/21087516/aerometals-aerometals-introduces-sikorsky-s70iuh60-blackhawk-engine-inlet-barrier-filter

[102] Aerometals, H-60 / S-70 Engine Inlet Barrier Filter, www.aerometals.aero/filter-detailsh60.html

[103] Delgado I.R. and Proctor, M.P. A Review of Engine Seal Performance and Requirements for Current and Future Engine Platforms, NASA/TM-2008-215161

[104] PUREair, Pall Corporation, 2014, https://aerospace.pall.com/content/dam/pall/aerospace-defense/literature-library/non-gated/Pall%20PUREair%20Introduction%20Presentation.pdf

[105] FOD: How to Control and Prevent Foreign Object Debris, Creative Safety Supply, www.creativesafetysupply.com/articles/fod-how-to-control-and-prevent-foreign-object-debris/

[106] Messenger, P.J. Controlling FOD In a Manufacturing Environment – Part 1 www.fodcontrol.com/controlling-fod-in-a-manufacturing-environment-part-1/
[107] Messenger, P.J. Controlling FOD In a Manufacturing Environment – Part 2 www.fodcontrol.com/controlling-fod-in-a-manufacturing-environment-part-2/
[108] Messenger, P.J. Controlling FOD In a Manufacturing Environment – Part 3
[109] www.creativesafetysupply.com/articles/howimplement-5s/

8 Management of Inanimate (Non-biological) FOD

8.1 INTRODUCTION

Air travel remains a safe mode of transportation. Reports show that although the total number of reported strikes has increased, damaging strikes in the airport environment have decreased [1, 2].

- In 2000, 14% of reported strikes indicated damage to the aircraft, while in 2017, the rate was only 4%.
- The number of wildlife strikes reported to FAA increased steadily from 6,000 in 2000 to 14,500 in 2017, whereas the number of *damaging* strikes decreased from 762 in 2000 to 633 in 2017.

The decline in damaging strikes can be attributed in part to increase in the development. and the use of wildlife hazard management plans at airports.

FOD program will help reduce employee injuries from flying debris as well as save money.

8.2 FOD MANAGEMENT PROGRAM

Several proposals for organizing a FOD management program by different organizations are available in the open literature. The Federal Aviation Administration (FAA) issued numerous circulars in this regard. Also, various literature and circulars were published by other organizations and companies, including the FOD Control Corporation, the US National Research Council, and Boeing Company. Highlights of these proposals will be given hereafter.

8.2.1 The Federal Aviation Administration (FAA)

The FAA, in its Advisory Circular AC No. 150/5210-24, recommends the guidelines and specifications for designing an airport FOD management program [3]. The use of this AC is not mandatory. However, it is mandatory for the acquisition of FOD

DOI: 10.1201/9781003133087-8

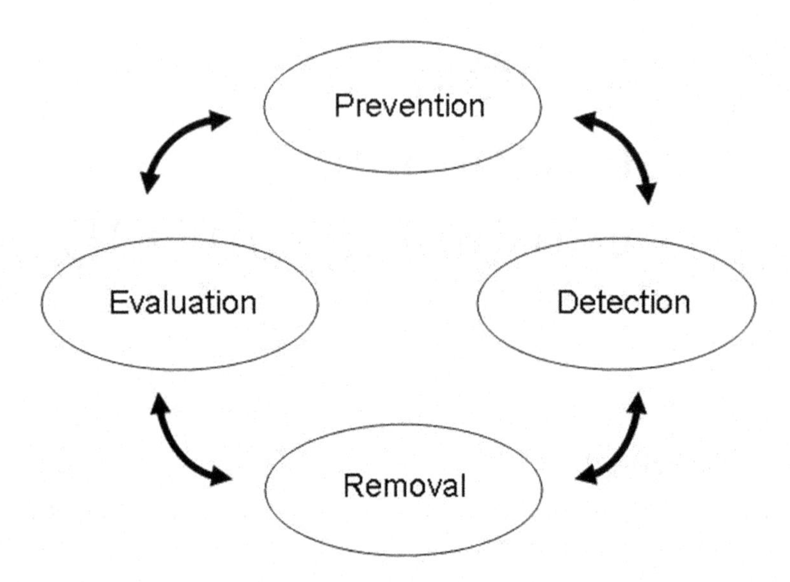

FIGURE 8.1 Relationship between the four main areas in a FOD program [3].

Courtesy: DOT-FAA.

removal equipment through the Airport Improvement Program (AIP) or the Passenger Facility Charge (PFC) Program.

A detailed description of the FAA program is given in [3, 4]. It includes the following four main components of a FOD management program (Figure 8.1):

1. FOD prevention
   a. Awareness
      i. Program existence and status
      ii. FOD policy and management support
      iii. Safety culture
   b. Training and education
      i. Audience
      ii. Features
      iii. Training objectives
      iv. Training documentation.
   c. Maintenance programs
2. FOD detection
   a. FOD risk assessment
   b. FOD detection operations
      i. Inspection areas
      ii. Methods and techniques
   c. FOD detection equipment
3. FOD removal

   a. Equipment characteristics
      i. Mechanical systems
      ii. Non-mechanical systems
      iii. Storage systems
   b. Performance
      i. Operational standards
      ii. Testing/validation
   c. Removal operations
4. FOD evaluation
   a. Data collection and analysis
      i. Documentation
      ii. Reporting
      iii. Investigation
      iv. Database
   b. Continuous program improvement.

8.2.2 The FOD Control Corporation

Several other proposals for FOD management are also available in various literature and circulars of aviation companies. Dave Larrigan [5] suggested the following ten elements for FOD management:

1. Management's strong, visible commitment
2. Local FOD committees
3. Housekeeping performance standards
4. Training and awareness
5. Selection and maintenance of ground support equipment and airfield maintenance equipment
6. Spare parts and tools
7. Airport construction projects
8. Motivating construction crews to understand FOD threats
9. Monitoring and inspection
10. Seasonal considerations.

8.2.3 Transportation Research Board; US National Research Council

The finding of an Airport Cooperative Research Program, conducted by the Transportation Research Board with the approval of the Governing Board of the National Research Council, is published in [9] focusing on the following five main areas that carefully structure a FOD management program:

1. Inspection
   a. Inspection areas
   b. Inspection techniques
   c. Prevention techniques
   d. Current equipment and technology available for inspection

   e. Current airport inspection and practices
2. Detection
   a. Current Equipment and Technology
      i. Manual
      ii. Supplemental
      iii. Automated (radar, electro-optical sensors, hybrid)
   b. Foreign object debris risk assessment
3. Removal
   a. Current equipment
      i. Nonmechanized FOD removal
      ii. Mechanized foreign object debris removal
      iii. Foreign object debris storage
   b. Current airport removal practices
      i. Common removal methods
      ii. Common foreign object debris removal equipment
4. Documentation and analysis of data
   a. Documented items
   b. Database
   c. Assessing performance
   d. Improving foreign object debris management
   e. Current equipment and technology available for documentation
   f. Current airport documentation practices
5. Training and promotion
   a. Human factors
   b. Training
   c. Promotion
   d. Visibility
   e. Awareness
   f. Current airport training, promotion, and management practices
      i. Programs and practices in use
      ii. Level of importance
      iii. Participation
      iv. Additional practices (FOD manager, training program, quality assurance, adaptations during low visibility and nighttime, additional resources, liability).

8.2.4 Boeing Company

Boeing suggested the following program to control airport FOD which addresses the four main areas [7]:

1. Training
2. Inspection by the airline, airport, and airplane handling agency personnel
3. Maintenance
4. Coordination

8.3 OTHER TECHNIQUES FOR FOD PREVENTION

The operational target in any FOD prevention program should always be "zero." However, it is impossible to achieve such a goal.

8.3.1 Proactive-Reactive Measures

Kraus and Watson [8] suggested the following two interconnected loops for FOD prevention: namely, proactive (preventative) measures and reactive (corrective) measures.

Proactive measures can be used to eliminate and prevent foreign object damage in the aviation maintenance processes. Moreover, its guidelines are not standards, but each organization can modify to fit its particular function.

8.3.1.1 Proactive Measures

They cover the following areas:

a. Management support
   i. Create a FOD prevention program
   ii. Establish a FOD point of contact (FOD POC)
   iii. Establish and maintain a FOD prevention culture
   iv. Actively participate in NAFPI
b. FOD awareness
c. FOD education and training
   i. Technical FOD training
   ii. Initial and recurrent training
d. Housekeeping
   i. General
   ii. Individual
e. Maintenance activities
   i. Maintenance guidelines
   ii. Material handling
   iii. Tool control and accountability
   iv. Use of borescope, X-ray, and other state-of-the-art equipment for FOD inspection and retrieval
f. FOD inspection and audits.

8.3.1.2 Reactive Measures

They cover the following items:

a. Determine the level of effort
b. Conduct accident/incident investigation
   i. Human factors investigative models and tools
   ii. Standardized FOD investigation reporting form
   iii. Data entry into database
c. Analysis of data – root cause analysis

d. Develop and implement corrective actions
   i. Corrective action plan content
   ii. Safety aspects of error
e. Implementation and evaluation of corrective actions
f. Concluding comments on structured reactive measures.

8.3.2 Twelve Elements for FOD Prevention Program

Ringger [9] suggested the following 12 basic elements of a FOD prevention program:

1. FOD prevention training.
2. Early design consideration for FOD prevention, resistance to damage, foreign object entrapment, etc.
3. Assembly sequencing and maintenance/manufacturing techniques that include proper care and use of assembly/maintenance equipment and parts protective devices
4. Handling of material
5. Housekeeping
6. Control of tools and personal items
7. Control of hardware/consumables
8. Measuring techniques for analysis, trending, and feedback
9. Incident investigation/reporting, "Lessons learned"
10. Control of hazardous material
11. Access controls
12. Focal point

8.3.3 Seven Areas for FOD Prevention

Simmons and Stephan [10] suggested the following components of a FOD prevention program:

1. Organization
2. Policies and procedures
3. Vision
4. Measurement tools
5. Investigations of incidents and accidents
6. Feedback procedure
7. Establishing goals.

8.4 FAA FOD PROGRAM DURING OPERATION

As outlined in Section 8.2.1 and illustrated in Figure 8.1, the four disciplines of the FOD program proposed by FAA are prevention, detection, removal, and evaluation. I will follow this program in this book.

8.4.1 FOD Prevention

A breakdown for the prevention measures suggested by FAA are awareness, teaching and education, and maintenance programs, as illustrated in Figure 8.2.

8.4.1.1 Awareness

Awareness is further subdivided into the following three categories [3]:

- Program existence and status
- FOD policy and management support
- Safety culture

Figure 8.3 shows its three components.

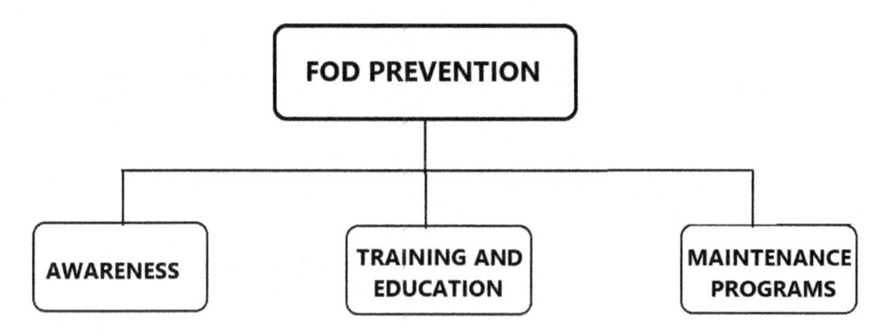

FIGURE 8.2 Elements of FAA FOD prevention methods [3].

Courtesy: DOT-FAA.

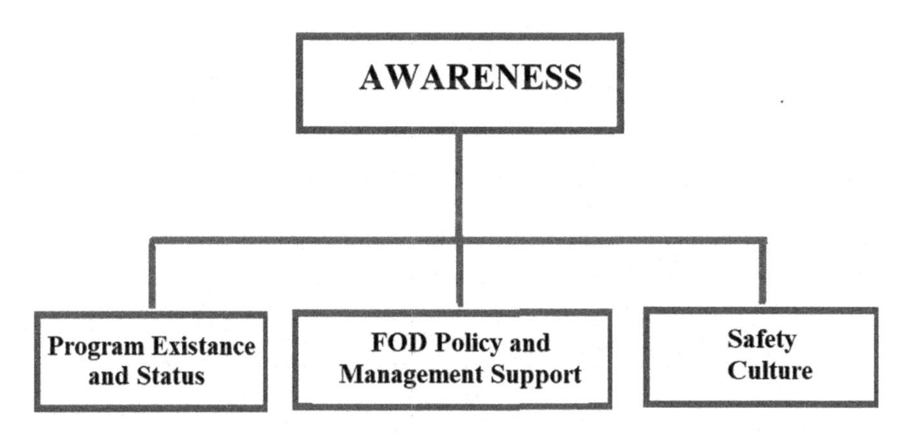

FIGURE 8.3 Three modules of awareness section of FAA FOD management program [3].

Courtesy: DOT-FAA.

a. Program existence and status

Applicable personnel are aware of the program's existence [3]

- FOD letters, notices, and bulletins (Figure 8.4)
- FOD boosters
- FOD lessons learned
- Safety reporting drop boxes and electronic reporting through websites or email
- Exchange safety-related information with other airport operators through regional offices or professional organizations
- FOD discussion at employee staff meetings
- T-shirts with FOD stickers
- Disposal cans with FOD stickers

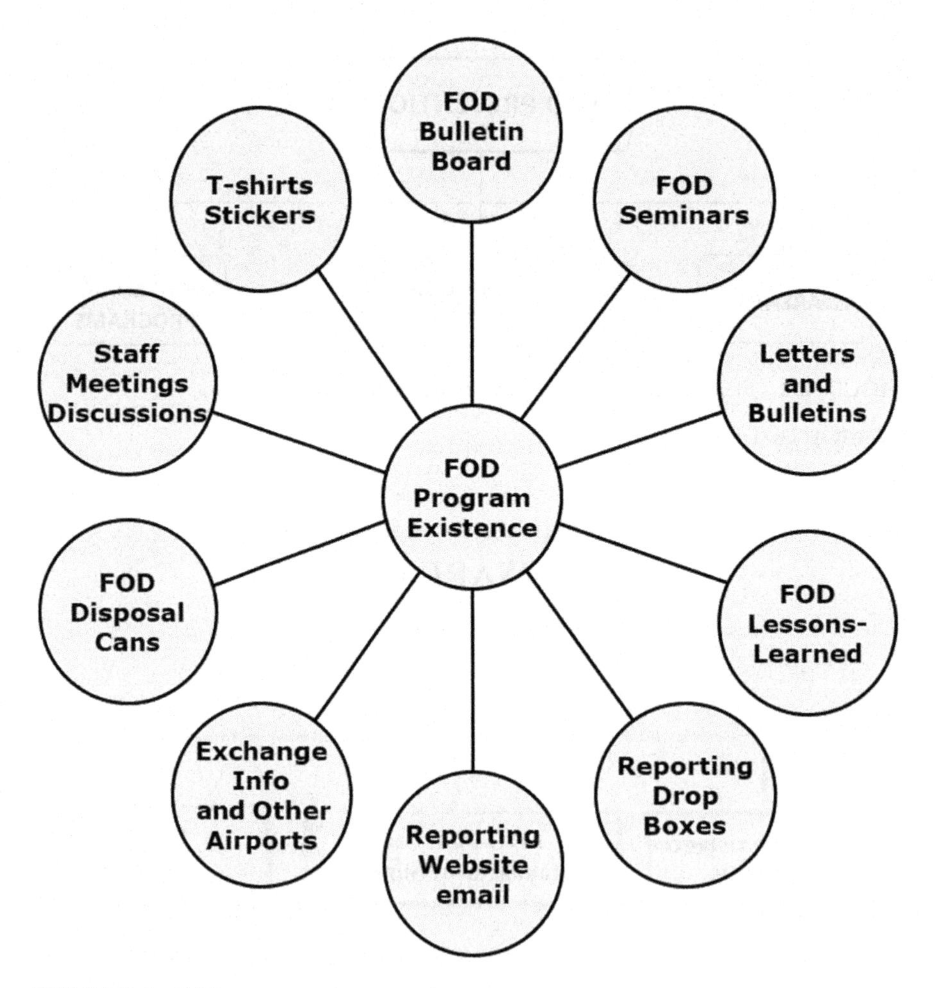

FIGURE 8.4 FOD program existence and status.

b. FOD policy and management support

It includes [3]:

- The FOD program
- The FOD program manager
- The FOD committee

b.1. The FOD program

Must be displayed in clear locations and identifying:

- Methods and processes to achieve safety
- Personnel responsibilities and accountabilities

b.2. The FOD program manager

a. Review and assess the airport's FOD management program and make necessary revisions
b. Conduct scheduled and unscheduled evaluations/inspections of work areas to assess the effectiveness of the FOD management program
c. Assure implementation of corrective actions for FOD prevention
d. Assure that FOD incidents are thoroughly investigated, and that incident reports are accomplished.
e. Assure that causes of FOD incidents are thoroughly analyzed to identify corrective measures
f. Notify affected contractor/tenant organizations and personnel of unique FOD prevention requirements
g. Develop techniques and assign responsibilities for publication of special FOD prevention instructions
h. Review results of the FOD incident investigations and evaluate the adequacy of corrective actions
i. Evaluate the amount and kind of foreign objects found and how they were found (e.g., during daily inspections, by pilots, airport operations staff, etc.)
j. Review and approve FOD prevention training curricula, designate training personnel, and assure that airport/contractor/tenant personnel receive required training
k. Assure that written procedures provide for adequate records attesting to the current status and adequacy of the FOD management program
l. Manage any additional program activities, including the scheduling of the FOD committee meetings, as required.

b.3. The FOD committee

It includes [3]:

a. A number of airports of varying sizes and complexities have found it helpful to establish a FOD committee, set by the airport's executive management.
b. Typical FOD committee members include those stakeholders with a direct relationship to FOD (such as those in a position to produce or remove FOD),

including tenant representatives, air carriers, airport operations and public safety staff, and contractor representatives, etc.

The FOD manager would typically chair the committee.

c. Important functions of the FOD committee are:
   i. Serve as a resource for the FOD manager
   ii. The determination of potentially hazardous FOD situations
   iii. Performing an evaluation of collected FOD data.

C. Safety culture

- An effective FOD management program requires more than the rules and procedures but management have to establish the attitude, decisions, for the priority to safety
- In effective safety cultures, there are clear reporting lines, clearly defined duties, and well-understood procedures.
- Personnel fully understand their responsibilities and know what to report, to whom, and when. Though it is an intangible aspect of a safety program, proper personal attitudes and corporate commitment enable or facilitate the elimination of unsafe acts and conditions that are the precursors to accidents and incidents.

8.4.1.2 Training and Educations

a. Audience
b. Features
c. Training objectives
d. Training documentation

a. Audience

- Everyone with access to the AOA
- Airport operations, construction, aircraft maintenance and permanent/seasonal servicing staff (e.g., catering, fuel, cabin cleaning, baggage and cargo handling, waste disposal, etc.), and any other contractors.
- New employees with safety, security, communications, and vehicle operations
- Supplement to the general FOD awareness for the driver training curriculum (or apron walking privileges) already in place at many airports

b. Features

1. A documented process to identify training requirements
2. A validation process that measures the effectiveness of training
3. Recurrent training and education (to help maintain awareness)
4. Human (and organizational) factors

c. Training objectives

1. Overview of the FOD management program in place at the airport
2. Safety of personnel and air carrier passengers

3. Causes and principal contributing factors of FOD
4. The consequences of ignoring FOD and/or the incentives of preventing FOD
5. Practicing clean-as-you-go work habits, and the general cleanliness and inspection standards of work areas (including the apron and AOA)
6. Proper care, use, and stowage of material and component or equipment items used around aircraft while in maintenance or on airport surfaces
7. Control of debris in the performance of work assignments (e.g., loose items associated with luggage, ramp equipment, and construction materials)
8. Control over personal items and equipment
9. Proper control/accountability and care of tools and hardware
10. Requirements and procedures for regular inspection and cleaning of aircraft and apron areas
11. How to report FOD incidents or potential incidents
12. Continual vigilance for potential sources of hazardous foreign objects
13. FOD detection procedures, including the proper use of detection technologies (if applicable)
14. FOD removal procedures

d. Training documentation

- Training requirements and activities should be documented for each area of activity within the organization
- A training file should be developed for each employee, including management, to assist in identifying and tracking employee training, training requirements, and verifying that the personnel have received the planned training
- Any training program should be adapted to fit the needs and complexity of the airport in question
- At certificated airports, training is required by 14 CFR Part 139, Certification of Airports.

8.4.1.3 Maintenance Programs

1. Aircraft servicing
2. Aircraft maintenance
3. Air cargo
4. Construction
5. Airfield maintenance operations
6. Pavements

8.4.2 FOD Detection

First, let me state that in the present book, FOD detection step will combine the two steps or stages: inspection and detection adopted in other FOD management literatures; see for example [6]. Detection of debris is achieved by FOD inspection of the AOA. If upon inspection there is no debris, then no FOD will be detected, whereas if debris exists it is detected and next removed.

A continuous monitoring at any time in the day is an essential step in detecting FOD. Time may be the most crucial aspect of the detection phase, especially in busy airports. Sometimes it is required to inspect a long wide runway (two miles long and ten lanes across) in a short time (could be a few minutes) prior to the landing/departure of an aircraft. [11].

The different detection methods belong to two categories:

- Manual
- Automation

Manual methods are performed by airport personnel and users, who resemble the primary "sensor" to detect FOD on airport surfaces. These manual detections are intermittent and scheduled through regular frequencies. It is either performed via FOD walks or using road vehicles.

Automation provides a continuous detection of FOD on runways and other aircraft movement areas. It supplements the capabilities of airport personnel through devices including radar and/or electro-optical sensors.

Drones are a new addition to automation methods. Within the last couple of years, drones have proven to be helpful devices rather than harmful ones as usually considered. It is used for inspection purposes in airports in several countries.

Another classification for detection (or inspection) methods could be:

- Fixed
- Mobile

8.4.2.1 Manual/Visual Detection

8.4.2.1.1 Introduction

Manual inspection/detection is also called a "visual inspection." It refers to both FOD walks and mobile vehicles. FOD squad walks are performed by airport employees who walk along the runway or taxiways in search of the FODs. They detect and remove any debris in the same time. This method is more prone to man-made mistakes, and it requires quite an amount of time.

The other manual inspection method is carried out using some airport vehicles.

Airport vehicle is driven by a trained employee who performs multiple passes of runways and taxiways to detect any existing debris, as illustrated in Figure 8.5.

During driving, he completes a paper-based inspection or FOD checklist. To improve the efficiency of the human observations, instead of completing a paper-based checklist, two improvements evolved:

1. Implementing an electronic checklist, typically loaded onto a personal digital assistance tablet or notebook computer. Thus, an electronic daily self-inspection checklist or a FOD reporting form will be completed, and airport's inspection database records are easily maintained
2. Using a GPS/GIS-based inspection and database application, the inspector can pinpoint the exact coordinates of FOD using GPS coordinates. Next, the

FIGURE 8.5 Ground vehicle inspecting airports for FOD [12].

Courtesy: DOT, FAA.

maintenance personnel can quickly locate and remove that debris rather than relying on a written description of where a debris is located [6].

Inspection personnel should drive toward the direction of landing aircraft on the runway with a high intensity flashing beacon and headlights on day and night. However, it is recommended that a runway inspection be done in both directions. For a more effective inspection and due to the width of a runway and the runway strip, the vehicle operator drives along both sides of the runway. Also, inspectors should pass in each direction on all taxiways (stub and parallel). The minimum speed of the vehicle should be 20 mph (30 km/h) to allow the inspector to conduct an effective inspection, taking into consideration how busy the aerodrome is [13].

Effective manual detection of FOD relies on training airport personnel who regularly monitor their surroundings and detect the presence of any debris. Training is based on the inspection criteria of ICAO [14] and complies with the 14 Code of Federal Regulation (CFR) Part 139 requirements of FAA [15].

The persons are normally working on the terminal or ramps, attending gates, handling baggage, fueling aircraft, piloting aircraft, controlling air traffic, monitoring construction activities as well as anyone who works on the AOA.

In addition, some airports, like Vancouver International Airport, also rely on Aircraft Rescue and Fire Fighting personnel and police personnel [6].

Based on the regulations set by ICAO Annex 14 [14], it is required to conduct inspections four times daily, or about every 6 h. This standard was set during the ICAO Chicago conference in 1944 and is still maintained in Europe. But, in the United States, most airports are so busy that they decided to conduct a single FOD check of the runways in the morning before operations start [11]. Based on the FAA standards, this "single daylight inspection" seemed acceptable.

For military airfields (including runways and taxiways), FOD checks at least once during the shift [17].

8.4.2.1.2 Statistics of Manual/Visual Detection

Statistics for visual inspection reflects the following features:

1. Comparing FOD walks to mobile vehicles, it is noticed that more FOD are found during walks than when driving [11]
2. Regardless of the costs of the training for airport personnel, the manual detection of FOD is the least expensive detection method compared to other options
3. Visual/manual inspection is subject to multiple environmental factors such as low visibility, fog, or rain which inhibit positive FOD identification
4. Sometimes the information on the location of FOD is received in a faulty time which results in unnecessary delays [16]
5. The outcome of FOD visual inspection at Hartsfield-Jackson Atlanta International (ATL) airport carried for 486 days 2008–2009 outlined that 886 pieces of FOD were collected. It implies an average of 1.8 collections per day, or 10.6 pieces of FOD per 10K commercial aircraft movements, equivalent to one piece of FOD found for every thousand flights in or out [11].

Table 8.1 summarizes the percentage and type of FOD found by visual inspection at Hartsfield-Jackson Atlanta International Airport over 486 days during 2008–2009.

6. Moreover, considering the locations where those FOD were found at Atlanta airport, it had the following breakdown:
   - 83% of debris was on the ramp, or in the "throat" area leading from the ramp to the taxiway
   - 14% of debris was found on the taxiway
   - 2% of the debris was on the runway itself
7. Visual FOD inspection in several large European airports had similar results, where 50–60 pieces of debris were detected at the airport per month (or nearly

TABLE 8.1
FOD Found at Atlanta (ATL) for 486 Days [11]

FOD Type	Aircraft Parts	Ground Vehicles and Tools	Luggage and Passenger Equipment	Concrete, Bitumen, etc.	Other
Percentage %	1.0	18.0	46.0	21.0	14.0

two pieces per day), and one debris every four to nine weeks on the runway (1.7%–3.3 %).

8. At Singapore Changi International airport, they found 11 pieces of FOD on their runways per year, roughly one piece every five weeks using visual inspection (2.8%).

8.4.2.2 Automated FOD Detection

8.4.2.2.1 Introduction

The increase of concern over aviation safety worldwide has driven commercial airports to look for more efficient and effective FOD detection and surveillance systems. Eventually, such systems will replace the traditional manual sweeps of runways, taxiways, and even areas where aircraft are parked.

Automated FOD detection systems are nonintrusive ways of using technology to automatically detect debris on runways. It applies a radar-based scanning technology as well as video technology and image processing. They are more expensive than manual methods but also more effective. Vancouver international airport was the first airport in the world to adopt automated detection technology in 2005 [11].

At Vancouver airport, for example, visual inspections detect one debris on the runways every two months [11], while with the automated scanning system that frequency jumped to one piece of FOD every two days. Thus, automated methods detect 30 times more FOD on the runway than the visual/manual inspections. Alternatively, it may be said that airports relying on visual/manual inspections are missing as much as 97% of the debris.

With such findings, it is expected that due to the importance of aviation safety, automatic detection equipment will be mandatory at airports soon.

Other advantages of automated FOD detection systems over manual methods are:

- Continuous round-the-clock operations
- Not affected by distractions, boredom, and fatigue
- Objective, consistent, and repeatable results

The direct benefits of automated FOD detection [11] are:

- Detect, find, and identify FOD on AOA
- Improve operational safety
- Reduce operating costs of both airlines and airports
- Provide uniform risk exposure for all aircraft movements

Indirect benefits [11]:

- Control the risk profile of airport
- Reduce the aircraft delays which in turn reduces carbon emissions
- Reduce total manpower as systems roll out
- Flexible decision-making by tower and airport

- Preserve existing airport runway capacity
- Low visibility operations
- Reduce runway closure times
- Major improvements in safety data recording and risk management.

8.4.2.2.2 FOD Automated Detection Systems

Four different FOD automated detection types are available:

- Stationary radar
- Stationary electro-optical
- Stationary hybrid radar and electro-optical
- Mobile radar

These types can be further described as follows.

8.4.2.2.2.1 Stationary Radar It is a radar detection system which contains a sensor that uses radio detection. It transmits and receives radio signals, as the primary means to detect objects. It can detect a metallic cylindrical target measuring 1.2 inch (3.0 cm) high and 1.5 inch (3.8 cm) in diameter at ranges of up to 0.6 mile (1 km). Sensors are located 165 ft (50.0 m) or more from the runway center line. Generally, two or three sensors are required per runway, depending on airport requirements [18].

8.4.2.2.2.2 Electro-Optical Detection System It is a system having a sensor that uses visual light wavelength as the primary means to detect objects. It can detect a 0.80 inch (2.0 cm) object target at ranges of up to 985 ft (300 m) using ambient lighting only. Sensors are located 490 ft (150 m) or more from the runway center line. Generally, five to eight sensors are required per runway, depending on airport requirements [18].

8.4.2.2.2.3 Hybrid Detection System Stationary hybrid detection system uses both an electro-optical and radar sensor in a unit collocated with the runway edge lights. The system can detect a 0.8 inch (2 cm) target on the runway. Generally, sensors are located on every, or every other, edge light, depending on airport requirements [18].

8.4.2.2.2.4 Mobile Radar It is a radar system mounted on top of a vehicle that scans the surface in front of the vehicle when moving. The radar scans an area (600 ft × 600 ft or 183 m × 183 m) to detect FOD items of height 1.2 inches (3.0 cm) and diameter of 1.5 inches (3.8 cm). The system can operate at speeds of up to 30 mph (50 km/h), supplementing human/visual inspections [18].

A layout for these different detection methods is illustrated in Figure 8.6.

8.4.2.2.3 FAA Automated Detection Systems

The FAA initiated an evaluation for the following four automated detection systems in 2007 in different test locations and approved them all:

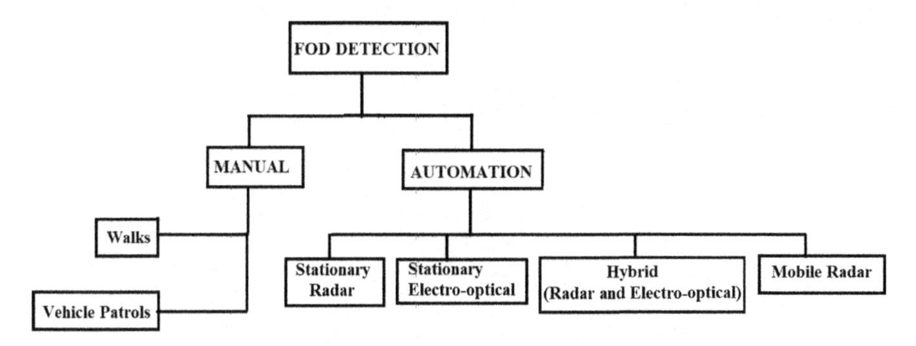

FIGURE 8.6 Breakdown of different detection methods.

- Stationary radar: QinetiQ – Tarsier Radar
- Electro-optical: Stratech – iFerret
- Hybrid radar: X-Sight Systems FODetect
- Mobile radar: Trex Enterprises

Details of each system will be given hereafter.

The FAA initiated an evaluation for the above four systems in 2007 in different test locations and approved them all.

8.4.2.2.3.1 Stationary Radar

a. *QinetiQ Tarsier Radar* QinetiQ Tarsier is a stationary millimeter wave radar mounted on rigid towers (Figure 8.7). Later versions incorporate high-resolution day and night cameras to allow operators to confirm suspect objects.

Tarsier FOD radar detection system sweeps the runway around the clock and compares a baseline image of a clear runway against the seen image. It is programmed to ignore aircraft and ground vehicles. When FOD has been located, the high-zoom camera pans in and sends a live image for visual confirmation, together with the object's GPS coordinates, enabling accurate retrieval of the object within minutes.

Tarsier has more than 100,000 h of operation at several international commercial airports across the globe, including Providence Rhode Island, Dubai, Doha, Vancouver to Heathrow, and at MOD Boscombe Down RAF airbase, and others.

Advantages and features of QinetiQ Tarsier detection system [19–21]:

1. QinetiQ Tarsier system has a high-resolution radar ideally suited for detecting debris day or night in clear or raining conditions
2. The most precise, reliable, and lowest cost detection system in the world
3. Tarsier's radar performance is not affected by dust or heat waves
4. It pinpoints debris location in precise range and bearing
5. Tarsier continues to provide timely and accurate debris detection in adverse weather, including snow, sandstorms and dense fog up to zero visibility to ensure runways are safe, efficient and fully operational every day of the year

FIGURE 8.7 QinetiQ Tarsier detection system [57].

Courtesy: FAA.

6. The system has been proven to detect metal, plastic, rubber, glass, and organic matter. Status information is relayed to airport operators through a single intuitive graphical display
7. A high-resolution night camera combined with a near-infrared illuminator tuned to the lens system far exceeds any competing night vision system
8. Tarsier system detects FOD, birds, and wildlife with a high level of sensitivity
9. Tarsier's specific millimeter wave radar/software combination can continuously monitor pavement and infrastructure
10. Tarsier can determine if there is a pavement crack or change in surface height and alert airport operations before it becomes a hazard.

b. *Nordic Radar* Nordic Radar Solutions is developing an independent long-range system for multiple object detection and real-time threat alerts within the airport and its surroundings [22].

The system is based upon one or more radar sensors with high-resolution cameras and is marketed under the brand name Foreign Object Debris, Drone and Bird-Aircraft Strike Avoidance (FODBASA).

The FODBASA system is an integrated sensor system solution consisting of a FODA subsystem and a BASA subsystem. Each subsystem, in turn, consists of one, two, or multiple individual radars.

For optimal performance, it is usually recommended to use two or more radars. Likewise, it is recommended to use two or more high-resolution cameras.

The FODA subsystem is based on Ka-band technology as a compromise between high-resolution capability and reduced susceptibility to clutter and attenuation from precipitation but also price. The coverage range will typically be 3 km.

The BASA subsystem is based on X-band technology that provides enough resolution capability and at the same time all-weather performance. An extended coverage range up to approximately 10 km is provided. Two separate subsystems are responsible for coverage of a flat 360° horizontal sector and a vertical (elevation) corridor centered along the runways, respectively.

Moreover, the FODA subsystem and a BASA subsystem can transfer complementary information in a way to warrant dual-band operation covering the critical runway surface zone.

c. *Stratech iFerret* The iFerret is the world's first *intelligent* vision-based system, which is certified by FAA [23]. iFerret is a collaboration project between CAAS and Stratech. Under this collaboration, Stratech provided the vision-based technology while CAAS provided the domain knowledge to customize the technology for use on the runway [24–26].

It is capable of real-time automatic detection of FOD, identification of FOD location, sending alerts to user and recording and post-event analysis. Five to eight sensors are required per runway. The iFerret is the only system with HD and color panoramic view of the entire runway. It delivers HD resolution color images that enables night visibility and superior visual clarity for FOD identification. Post-incident analysis is made possible with its video recording capability.

The FAA-certified iFerret™ system has been adopted by the world's major commercial airline hubs, including Singapore's Changi Airport, Dubai International Airport (United Arab Emirates), Hong Kong International Airport, Miami International Airport (MIA), Düsseldorf International (DUS), Chicago O'Hare (ORD), as well as other airports and air force bases.
Advantages and features of Stratech iFerret:

- It is a FOD Detection and Airfield Surveillance System
- It was one of four FOD detection systems approved by the FAA
- It uses high-resolution camera mounted on rigid towers
- Five to eight sensors are required per runway (Figure 8.8)

FIGURE 8.8 iFerret sensors mounted along the runway [57].

Courtesy: FAA.

- It is an advanced vision-based detection, which automatically detects, pinpoint FOD location, classifies and records FOD in the airport/airbase environment
- It is the only system with HD and color panoramic view of the entire runway.
- It has all-weather capabilities with superior night vision capabilities without assisted lighting
- Post-event analysis is made possible with its video recording capability
- It can track the increasingly high volume of aircraft and ground vehicle traffic in non-movement areas around the terminals and ground ramps.

Exhaustive tests of iFerret system were conducted on the runways during day and night for clear and different rainfall intensity ranging from 4 mm/h up to 22 mm/h. A summary of these results is given in Table 8.2. The test objects were the determination of the distance range that radar can detect FODs having different sizes (ranging from 1 to 4 cm).

Finally, a comparison between the previously described three fixed automated detection method and the manual methods are given in Table 8.3.

d. *X-Sight Systems FODetect* Xsight is called FODetect® (Foreign Object Debris Detect). It is an automated FOD detection system attached to existing runway and taxiway lighting structure and utilizes power from their hosts. It is a hybrid radar-optical sensing technology that uses a charge coupled device (CCD) camera with zoom capability and 77 Ghz frequency modulated continuous wave (FMCW) radar [30, 31].

Xsight (FO Detect) is operational at Logan airport in Boston, Seattle-Tacoma airport, Beijing Daxing International Airport, and Bangkok Suvarnabhumi International Airport and others.

Xsight has the following features and advantages:

1. 77GHz FMCW radar
2. Powerful processor for signal and image processing

TABLE 8.2
Detection Range for Different FOD Sizes [24]

FOD Size (cm)	Rainfall Intensity mm/h	Distance: Up to (m)	
		Day	Night
4	Clear	1,100 m	600 m
	16	890	-----
	22	-----	520
2	Clear	780	390
	16	590	------
	4	-------	364
1	Clear	780	310
	4	460	364

TABLE 8.3
Manual Versus Automated Airport Inspection [25]

Manual	Automated
Tedious manual inspection	Fully automated FOD detection
Scheduled infrequent inspection	24 × 7 continuous operation
Subject to human error, fatigue, or complacency	Learning system with minimal false alarm
Difficult to inspect/search for FOD after dark	Night vision capabilities without assisted lighting
No visibility from ATC/Ground Operations Control Centre	Increased awareness through visual verification/ confirmation
Rough estimation of FOD location	Pinpoint location for rapid recovery
No clue of FOD occurrence and no means of identifying cause/source	Record playback for post-event analysis/ investigation
Inefficient compliance with ICAO standards or FAA regulations	Full compliance without compromising airfield/ runway operation
FOD incidents endanger airport/ airbase personnel	Safety for pilots, flight crews, ground staff, and passengers

3. Supports wireless (WiFi), copper-based LAN (10/100 Ethernet), and fiber-optic based LAN communications
4. Mechanical protection door for vision sensor protection during extreme weather conditions
5. Automatic scan and manual motion control capabilities
6. Data recording capability
7. It provides a continuous monitoring of airport runways (during day and night) for rapid removal of FOD and wildlife in all weather conditions
8. The system automatically separates birds from other FOD types, and then automatically alerts the airport's wildlife team in real time
9. It can scan the whole runway (no blind spots) continuously without interfering takeoffs and landings in less than 60 sec
10. It guides personnel all the way to the location of the detected FOD by a laser beam activated from the system sensor
11. It improves runway safety, operational efficiency, and increase runway capacity
12. Operators validate any FOD using both radar and the sensor camera before dispatching personnel to remove it, thus eliminating unnecessary runway closures.

8.4.2.2.3.2 Mobile radar Trex Enterprises FOD Finder™ XM is a ground-breaking technology and the only mobile debris detecting and clearing equipment in the world designed and manufactured by TREX Aviation Systems, Inc [27–29].

FIGURE 8.9 Mobile debris detecting and clearing equipment [26].

Courtesy: FAA.

Like Tarsier, the system incorporates a Millimeter Wave Radar and Infrared Cameras mounted on the roof of airport vehicle (Figure 8.9). It can detect any metallic cylindrical target 1.2 inches (3.0 cm) high.

It was tested and evaluated by the FAA and approved by the Federal Communications Commission for use at airport areas.

The FOD Finder™ XM-M is a fully integrated system that includes:

- A heavy-duty pick-up truck
- A Trex FOD detection radar system
- A Trex FOD vacuum retrieval system
- An in-vehicle display with an intuitive user interface
- Data acquisition and storage.

It is operating in the following airfields:

- Dyess Air Force Base in Texas
- Marine Corps Air Station Yuma in Arizona
- Honolulu International Airport
- Chicago O'Hare International Airport
- Dallas/Fort Worth International Airport
- Tianjin Binhai International Airport in China

FOD Finder™ XM has the following features and advantages:

- It does not interfere with airport and aircraft communication, navigation, and surveillance systems.
- It can be used in all operational areas to include aircraft hangars, taxiways, and aprons.
- It reduces runway closure time and increases airport efficiency.
- It reduces the repair cost for FOD damages.
- It improves the safety conditions at airports and airfields.

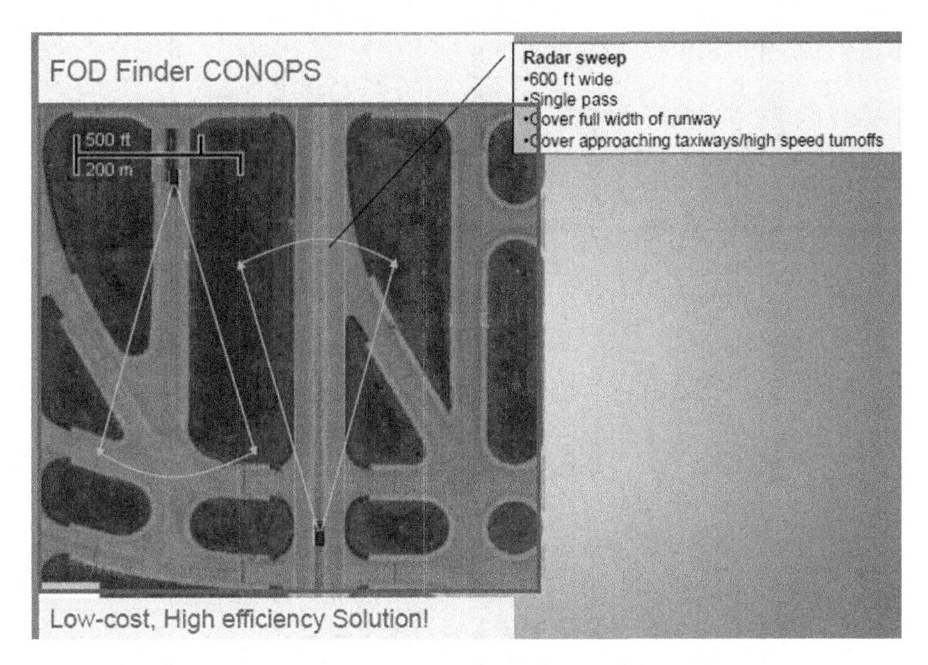

FIGURE 8.10 The movement of the FOD vehicle and the sweeping radar arcs [26].

Courtesy: FAA.

- It improves the war-fighting readiness for military operations.
- It has 100% detection capability at night.
- It is an all-weather detection system and can operate in rain, mist, fog, haze, snow, and sandstorms.
- It can operate on a variety of airport surfaces including grooved or smooth asphalt and concrete.
- It has a limited system downtime with little to no maintenance required.
- It has a touch-screen control panel inside of the vehicle cab like a standard GPS device mounted in an automobile.
- It provides an immediate automated reporting of airfield conditions to operations management and ground control which expedites corrective actions and facilitating FAA required Part 139 reporting.

However, the major disadvantage is the requirement for air operations to stop while the unit is moved through the area. The movement of the FOD vehicle and the sweeping radar arcs are illustrated in Figure 8.10.

A summary and a comparison between the four radar systems used for FOD detection is given in Table 8.4 [23].

8.4.2.3 Drones

An unmanned aerial system (also known as drones) equipped with a high-resolution digital camera has recently been added to the inspection techniques. Several airports worldwide adopted it now. Though for many years, drones were a severe nuisance to

TABLE 8.4
Comparison of Different Automated Inspection Methods

Name	Mobility	Type	Object Size	Object Range	Sensor Location	Sensor Number
QinetiQ Tarsier	Stationary	Radar	Metallic cylindrical target 1.2 in (3.0 cm) High and 1.5 in (3.8 cm) diameter	Up to 0.6 mile (1.0 km)	165 ft (50.0 m) or more from runway centerline	2 or 3 per runway
Startech – iFerret@ SIN	Stationary	Electro-optical	0.8 in (2.0 cm) object	Up to 985 ft (300.0 m) Using only ambient lighting	490 ft (150.0 m) or more from runway centerline	5–8 per runway
Xsight	Stationary	Hybrid radar and Electro-optical	0.8 in (2.0 cm) object	Up to 985 ft (300.0 m) Using only ambient lighting	Unit collocated with the runway edge light	Every or every other edge light
Trex Enterprises	Mobile	Radar	Metallic cylindrical target 1.2 in (3.0 cm) High and 1.5 in (3.8 cm) diameter	Scans area 600 × 600 ft (183 × 183m) In front of vehicle	Mobile	One sensor The system can operate at speed 30 mph (50 km/h), supplementing human/visual inspections

airports resulting in delayed flights and safety hazards, the situation is now twisted. With the introduction of FAA drone regulations 14 CFR Part 107 (and its Canadian similar TP312) and the FAA new low-altitude authorization and notification capability, a new horizon for drones became visible. A new peaceful culture is created between unmanned aircraft and manned aircraft. Over the last five years, the use of drones has exploded. For example, in 2017, some three million drones were sold worldwide, and more than one million drones were registered with the FAA. Also, the global commercial drone market is expected to grow by 26% each year from 2016 to reach a value of $10.738 trillion by 2022 [32].

Cameras installed on drones are its most useful feature. Most commercial drones flying today are used for photography as both still and video. The quality of the photography images has increased with the development of 4K video resolution cameras.

Drones have been used in several airports worldwide, including Luxembourg (Lux) Airport in Luxembourg [33], Erik Nielsen Whitehorse International Airport (YXY), Whitehorse, Yukon, Canada [34], Edmonton International Airport (EIA), Alberta, Canada [35], Paris-Orly, France, Milan-Linate, Italy, Geneva International Airport, Switzerland [32], Saline County Regional Airport, Bryant, Arkansas, AR, United States [36], Golden Triangle Regional Airport in Columbus, MS, United States [36].

- The inspection activities covered by drones are:
  - FOD
  - Preventative maintenance on runways, taxiways, and aircraft handling aprons, enhancing safety
  - Pavement, building, and marking conditions
  - Rooftops, parking lots, perimeter fencing, windsocks, safety areas, signage, and taxiway and runway markings

Finally, the cost of the drones may vary as follows:

- A commercial-grade drone may cost between $1,500 and $2,500 (not including spare equipment as batteries, propellers, transport cases, etc.)
- A medium-grade aircraft package may cost around $3,500–$4,000.
- Sophisticated equipment (such as miniature infrared cameras or LIDAR units) may cost $15,000–$20,000.

8.4.2.4 Inspection

8.4.2.4.1 Disciplines

As described in [1, 38], the following areas and operations are typically prone to having FOD and should be carefully inspected. Inspection here is concerned with foreign object debris:

1. Movement areas (runways and taxiways)
   a. Runway: the portion used by aircraft to take off
   b. Pavement should be free of holes, cracks, mud, dirt, sand, loose aggregate, debris, foreign objects, rubber deposits, and other contaminants

   c. Service roads that cross taxiways, debris falling from construction vehicles
   d. Shoulders: unpaved areas adjacent to pavement
   e. Pavement joints
   f. Turf areas: grass and ditches collect and hold light debris (paper, cardboard, plastic, etc.)
   g. Fence-lines: fences can collect trash on windy days
   h. Pedestrian and ground vehicle operations
2. Aircraft servicing operations
   Refueling, catering, cabin cleaning, baggage, and cargo
   Baggage pieces, including bag tags and wheels
   Areas at both ends of the conveyor, and that between the baggage cart and the conveyor belt
3. Construction operations
   - Regular and thorough cleaning of the construction site
   - Particular attention to construction vehicle routes that cross or are adjacent to active pavements
   - Briefing the contractors about the risks that FOD creates will curb construction debris and regularly clean up the construction site.
   - Routine checks for construction debris are to be arranged for each construction project on AOA.

For additional guidance on airport actions during construction activities, refer to the FAA instruction [39].

4. Aircraft maintenance activities
   - Variety of small objects, like rivets, safety wire, and bolts, which become FOD when they are inadvertently left behind.
   - All tools should be accounted for using checklists. For more information refer to NAS 412, Tool Accountability [37].
   - All vehicles should be driven on clean, paved surfaces when possible. However, if a vehicle must be driven on unpaved surfaces, the operator should check the vehicle tires for foreign objects immediately after returning to the pavement.
5. Airport apron
   - Anywhere on the apron where ground vehicles operate
6. Air cargo operations
   - Blowing debris such as plastic cargo wrappers
   - Regular cleaning for fences employed to contain debris.

The above-listed inspection areas are influenced by several parameters [11], including airport age, aircraft types and loading practices, cargo types, mix of passenger versus cargo traffic. Other important parameters include airline ramp practices, maintenance practices, mix of ground vehicles in use, weather, and the number of winter operations. Finally, the inspection procedure depends on the following airport characteristics: distance to repair hangars, the nearness of buildings, construction activity, pavement types, and surface cleaning/sweeping regime.

8.4.2.4.2 Inspection Frequency and Techniques

As identified in [14] and [15], for airports certificated under 14 CFR Part 139, the self-inspection program is a key component of an airport operator's airport certification program and is required under §139.327. The primary attention in inspection processes should be given to the operational items such as pavement areas, safety areas, FOD, fueling operations, ground vehicles, wildlife hazard management, construction, snow and ice control, markings, signs, lighting, aircraft rescue and firefighting, navigational aids, and obstructions.

There are four types of inspections with different frequencies as follows.

8.4.2.4.3 Regularly Scheduled Inspection

Operational areas must be inspected at least once each day, during times when aircraft activity is minimal in order to create the least impact on airport operations. For airports that serve air carriers after dark, a part of this inspection should be done during the hours of darkness. During night inspections, personnel and vehicles should be equipped with additional lights/lighting systems for better FOD detection. Pavement should be free from cracks, scaling, spalling, bumps, low spots, vegetation along runway and taxiways edges, as well as any debris that could cause foreign object damage to aircraft. The inspector should be sure that no foreign objects are left on the pavement from snow removal operations, and runway conditions comply with the FAA circular [40] for Airport Winter Safety and Operations. Inspectors should also ensure that debris and foreign objects are continuously being picked up around construction areas [39].

8.4.2.4.4 Continuous Surveillance Inspection

The continuous surveillance inspection may be at any time since hazardous conditions can occur at any time and in a short period of time. It consists of observation of activities and any apparent abnormalities with physical facilities in the air operations area. The continuous surveillance inspection should assure that the snow and ice will not affect the safety of aircraft operations. Moreover, the inspector should check for, and remove, any FOD on haul roads adjacent to movement areas that can be tracked onto taxiways, aprons, ramp, as well as aircraft parking areas. Continuous surveillance inspection also includes checks of birds or mammals on or adjacent to the runways, taxiways, aprons, and ramps to determine if there is a potential wildlife hazard problem.

8.4.2.4.5 Periodic Condition Inspection

Periodic condition inspection of activities and physical facilities can be conducted on a regularly scheduled basis but less frequently than daily. The time interval could be weekly, monthly, or quarterly, depending on the activity or facility.

8.4.2.4.6 Special Inspection

Special inspections of activities and facilities should be conducted after receipt of a complaint or as triggered by an unusual condition like a significant meteorological event or an accident or incident. Special inspections also are adopted during snow and ice conditions.

Special inspections must be conducted at the end of construction activity to ensure that there are no unsafe conditions present. A special inspection should be conducted prior to construction personnel leaving the airport if corrective actions are necessary.

Special inspections should be documented on the appropriate portions of the regularly scheduled inspection checklist.

8.4.2.5 Comparison between FOD Detection Systems

Table 8.5 illustrates a summary of the features of the detection systems used to detect FOD on runways and AOA surfaces [18].

8.4.2.6 Runway Closure

Finally, let us discuss the cases for runway closure after detection of FOD. As identified by the FAA CertAlert on March 17, 2009, not all types of FOD will necessitate an immediate runway closure. However, a quick and decisive action should be taken, in all cases, to assess the threat posed by reported FOD. FAA [1] discusses the following two cases:

1. If the location or characteristics of the FOD present, no immediate safety hazard the object should be removed as soon as the operational schedule permits.

TABLE 8.5
FOD Detection Systems [18]

System	Detection Principles	Capability
Human/Visual	Fundamental baseline for the performance of FOD detection systems. Human observation provides detection and human judgment provide the hazard assessment capability to assure safety.	Supports regularly scheduled, periodic condition, and special inspections
Radar	Uses radio transmission data as the primary means to detect FOD on runways and AOA surfaces.	Fixed systems support continuous surveillance; mobile systems supplement human/visual inspections
Electro-optical	Uses video technology and image processing data as the primary means to detect FOD on runways and AOA surfaces	Support continuous surveillance
Hybrid	Uses a combination of radar and electro-optical data as the primary means to detect FOD on runways and AOA surfaces.	Support continuous surveillance
Drones	Use high-resolution digital camera for photography as both still and video to detect FOD on runways and AOA surfaces	Supports regularly scheduled, periodic condition, and special inspections

2. If the location or characteristics of the FOD present an immediate safety hazard, the FOD manager will inform the airport supervisor that a hazard exists, who will in turn act and temporarily cease operations. If the FOD source is an aircraft or airport equipment, the FOD manager will notify the equipment operator

In large European airports, the runways were closed due to FOD and wildlife for an average of 200–240 min per month. The total cost of this problem is 10,000 aircraft movements and US$ 1 million for large airports [41].

8.4.3 FOD Removal

8.4.3.1 Introduction

FOD removal from the AOA is a critical stage of any successful FOD management program. It may start with a pilot reporting FOD to the tower. For United States and EU airports, the tower next contacts airport operations who dispatch someone to pick the debris [11]. This process takes 5 to 10 min for someone just to get to the runway and start looking for the debris. The search itself can take another 5 to 10 min. Thus, in total, the entire process from pilot report to debris removal might last anywhere from 10 to 20 min. So, the runway must be closed for this precious time.

Concerning FOD removal, two cases may be encountered:

- An isolated FOD object is detected on a runway/taxiway. In this case, a manual removal will be the most efficient.
- Multiple FOD objects are detected, say in maintenance areas or near construction areas. Removal equipment is the most appropriate.

The complexity of FOD removal is another issue. Two possibilities are visible [6]:

- Fairly simple and straightforward removal, for example, if a luggage tag is found on the apron surface, a line crew staff can easily remove it.
- Very complex and dangerous case, say if a fastener or a tool is found on the runway.

In developing a FOD removal plan, airport managers may wish to consider:

- Whom, when, how, and with what method FOD piece will be removed?
- Whether the FOD removal process will contradict the Safety Management System plan?
- How to expedite the FOD removal without interference with any airport operations?
- How to maintain the highest levels of the safety of the personnel during debris removal?

Though FOD removal is the responsibility of the airport operator, each airline or tenant can be asked to keep their area free from FOD. To accomplish this, airports may adopt manual as well as equipment removal.

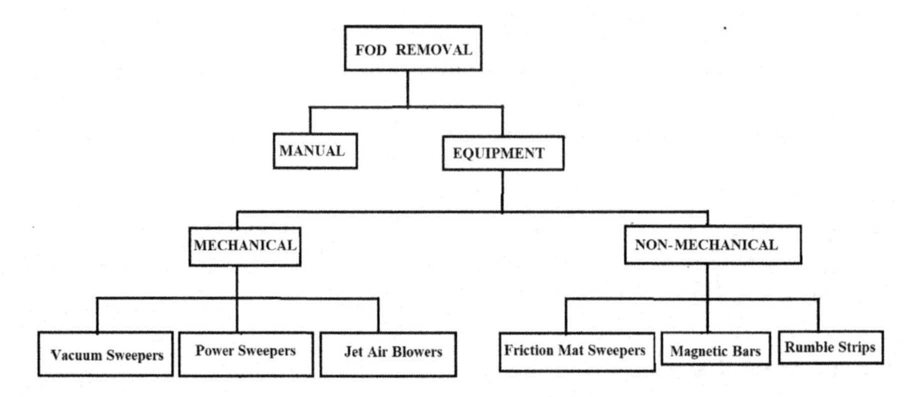

FIGURE 8.11 FOD removal methods.

Federal Aviation Authority FAA in its circular AC 150/5210-24: Airport Foreign Object Debris (FOD) Management [1] outlined that the use of that circular is not mandatory, but its use is mandatory for the acquisition of FOD removal equipment through the AIP or the PFC Program.

FOD removal equipment is either mechanical or nonmechanical. Mechanical systems are categorized into power sweepers, vacuum systems, and jet air blowers. Nonmechanical is either magnetic bars, friction mats, or rumple strips [6]. Figure 8.11 illustrates the different types of FOD removal methods.

FOD removal operations are not meant to occur when a given area is contaminated with snow or ice. In such winter conditions, the procedures listed in AC 150/5200-30, Airport Winter Safety and Operations, are used to clear the AOA surfaces.

8.4.3.2 Manual Removal

As defined in [16], manual labor is the first and most reliable method to remove FOD. Manual removal may be carried out by individuals or by a group, which is known as FOD squad walks. At a major airport, there are service teams who go around in their cars picking up FOD. For runways, the teams are allowed a short period where they can walk and scan the runway for FOD and pick anything up that looks dangerous.

At every gate, a fast and easy cleanup can be made using both push brooms and shovels [42, 43].

Manual sweeping is employed to retrieve any discarded debris such as paper, plastics, cans, cargo wrappings, and the like. The operational areas are divided into sections with an employee in charge of his or her area.

In the US Air Force (USAF), based on the FOD regulations (AF Instruction 21-101, paragraph 14.19.2.12), FOD walks are mandatory to remove FO Debris from ramps, uncontrolled runways, and access roads. In such FOD walks, the flightline personnel line up shoulder-to-shoulder and walk down uncontrolled movement areas and taxiways picking up debris. This is usually the first task of the day before flight operations begin, and all air operations personnel participate [44].

Annual FOD walk [44] is a large-scale planned activity, where many USAF staff are assembled based on commander instructions at a predetermined time and location and walk the entire base or flightline to retrieve trash, rocks, loose metal objects, or any other item that may become a FOD hazard. Such activity increases awareness and cooperation.

8.4.3.3 Mechanical Systems

Most airports utilize mechanized systems (such as power sweepers or vacuum systems) to remove the debris. Some airports rely on jet air blowers to displace FOD [4]. Mechanized FOD removal is more costly for an airport; however, such extra expense is justified by the enhanced efficiency provided.

Proper maintenance is necessary to ensure successful operation with a minimal breakdown of equipment. Equipment in this category varies in size and is found in sizes from small push units to large area systems that are truck mounted.

Types of mechanical removal systems include:

8.4.3.3.1 Power Sweepers

Power sweepers are employed in all airports, including private, small regional, world's leading international airports, heliports, and military airbases (Figure 8.12). They are employed for routinely removing any debris from the aircraft maneuvering areas, aprons, runways, gates, and the areas adjacent to them [45–47].

Power sweepers are available in different configurations: namely, a self-propelled walk-behind sweeper, sweeper attached to a tractor, and a sweeper truck.

FIGURE 8.12 Sweeper truck (Ravo street sweeper) [58].

For sweeper trucks, it is recommended to have versatile applications and facilities including:

- Easy maneuverable
- High-speed airport runway sweeping
- Suitability for removing fine dust from the narrowest spaces up to big space grounds
- Collection of liquids, such as surface water or de-icing agents (glycol) [47–50]
- Efficient performance in both summer and winter
- The ability of broom both to work in snow and carry out normal cleaning
- Collection of magnetic metals

The FAA circular AC150/5210-24 [1] recommends that:

> *Since bristles can detach from brooms and become a source of FOD, brushes made with metal bristles or spines should not be used for FOD removal purposes.*
>
> *Instead, Plastic or combination plastic/metal bristles may be appropriate. However, the user should consult the equipment manufacturer for specific recommendations. Regardless of the equipment used, a thorough check of the pavement should be conducted after the sweeping procedure.*

Another power sweeper called FODBUSTER® is used in several airports and air bases [51], as illustrated in Figures 8.13 and 8.14.

FIGURE 8.13 FODBUSTER® power sweeper [53].

With permission.

FODBUSTER® is a simple, reliable, low cost, low maintenance, compact, and efficient rotary-brush airport runway sweeper. It is a nonmotorized and all-weather operation. It can clean the entire runway, taxiway, or ramp in just a few passes [59].

Only one person can operate it. He/she attaches such a light trailer to a motorized vehicle and continue sweeping while performing other duties, such as patrol or delivery. Neither extensive training nor certification is required to operate or service it.

FODBUSTER® can quickly and efficiently remove sand, rocks, metal parts, chipped concrete, asphalt, safety wire, various types of plastic, paper, baggage parts, cargo wrappings, and other ferrous and nonferrous objects from FOD-sensitive areas.

FODBUSTER® has the following characteristics:

- Sweeps at optimal speeds of 10 to 15 mph (16 to 24 km/h)
- Picks up larger objects, such as pavement chunks and maintenance debris
- Collects 40 to 60 pounds (18 to 27 kg) of FOD on each side-mounted, removable debris hopper
- Vehicles towing FODBUSTER® between sweeping locations can be driven at speeds of up to 35 mph (58 km/h)
- Little routine maintenance is needed, except for periodic lubrication of the traction system and normal tire maintenance
- Nylon brushes generally last 2–3 years (or 6,000+ miles) and are easy and inexpensive to replace.

8.4.3.3.2 Vacuum Systems

Vacuum systems rely on airflow as the primary means of removing FOD. Airports often utilize a unit that combines a vacuum system with a mechanical broom and

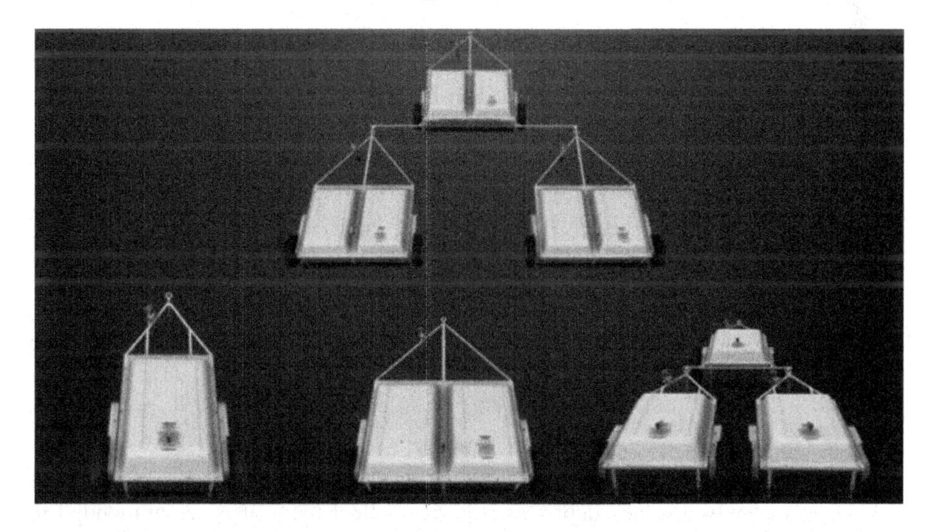

FIGURE 8.14 Different FODBUSTER® configuration [53].

With permission.

TABLE 8.6
Different FODBUSTER® Configuration

Number in Figure	Name	Sweeping Path
1	FBR-4	48 inch (122 cm)
2	FBR-8	96 inch (244 cm)
3	FBR-11	11 ft 5 inch (348 cm)
4	FBR-22	22 ft (670 cm)

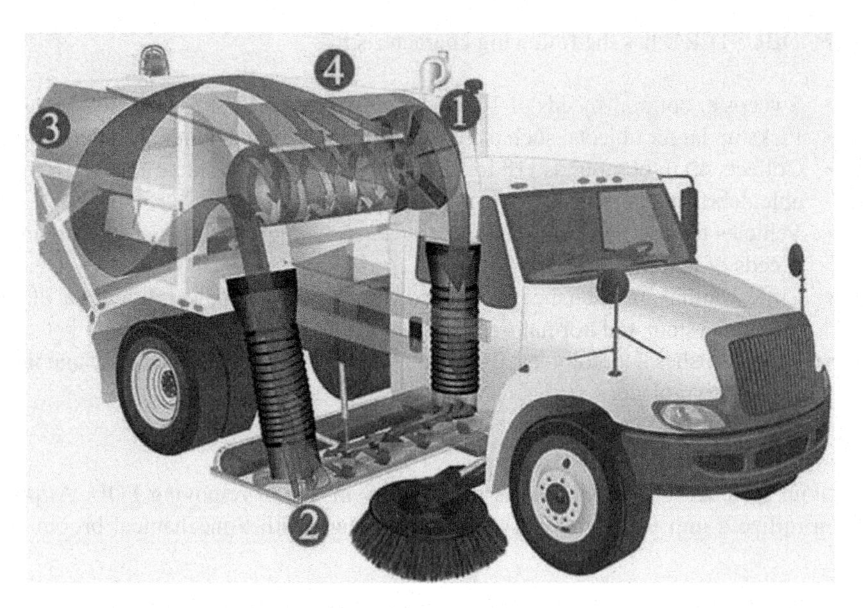

FIGURE 8.15 Regenerative air sweeper [52].

With permission.

a regenerative or recirculating air feature (Figure 8.15). By utilizing a constantly moving windrow broom to transfer debris over to a suction nozzle at one side of the sweeper, debris is removed through a suction tube.

A regenerative air sweeper uses a controlled blast of air to dislodge debris from the surface [52]. All debris picked up by the pick-up head is directed up the large diameter heavy-duty suction hose into the hopper. Truck removes trash, dirt, and fines from the entire area beneath the full-width pick-up head.

8.4.3.3.3 Jet Air Blowers

Jet air blowers direct a stream of high-velocity air toward the pavement surface. Since these systems only displace debris and do not collect them, it is recommended to acquire jet air blowers that only incorporate a debris collection mechanism to avoid relocating FOD to other areas in the AOA [1].

Some jet air blowers are used in airports to remove ice, dirt, snow, leaves, and other debris.

8.4.3.4 Nonmechanical Systems

Nonmechanical systems are simply attached to or towed behind a vehicle. These nonmechanized units are versatile and can be attached to a tug or any motorized vehicle. The vehicle then tows it over areas needing to be cleared. Moreover, they are less costly to operate and rarely out of service as a result of mechanical issues [6]. Such nonmechanized FOD removal systems include three types: namely, tow-behind friction mat sweepers, magnetic bars and rumble strips.

8.4.3.4.1 Tow-behind friction mat sweepers

A rectangular assembly towed behind a motorized vehicle employs a series of bristle brushes and friction to sweep FOD into sets of capture scoops, which are covered by a retaining mesh to hold collected debris [1].

FOD*BOSS is considered the ultimate friction mat sweeper used in airports ranging from small country and regional ones with limited budgets and operational restraints to medium and large airports as well as military bases across the globe. For 20 years, the FOD*BOSS has been trusted and used to control the costly and very dangerous problem of FOD. It is a patented sweeping system designed for collecting FOD from tarmac and aircraft movement surfaces (Figure 8.16).

FIGURE 8.16 FOD*BOSS [53].

With permission.

Each FOD*BOSS incorporates a series of brushes that activate the FOD, which is collected by rows of specially formulated Aerothane™ blades and entrapped by a

FIGURE 8.17 Single FOD*BOSS [53].

With permission.

FIGURE 8.18 Duplex FOD*BOSS tow-behind friction mat [53].

With permission.

FIGURE 8.19 Triplex FOD*BOSS tow-behind friction mat [53].

With permission.

retaining mesh cover. It has three configurations: namely, single, duplex, and triplex, as illustrated in Figures 8.16–8.19 [53–55].

Its sweep widths of 8 ft, 16 ft, or 22 ft, respectively. It can be used by one person and towed behind any vehicle. The FOD*BOSS works at speeds up to 25 mph. The complete unit is portable and folds away into a roller/storage bag in minutes.

FOD*BOSS has the following advantages:

1. Can be used both in civilian airports and military bases
2. Easy to deploy
3. Can remove and collect nearly all FOD types on a site, including small rocks, broken pavement, paperclips, employee ID badges, earplugs, screws, nuts, bolts, and wires
4. High speed and efficiency
5. Has no motor and few moving parts, making it virtually maintenance-free
6. Constructed for multiple uses over the long haul
7. Enable military installations to achieve wartime goals and protect its invaluable assets from preventable harm

8. Works on concrete, asphalt, interlocking pavement, fine-textured seal, smooth or grooved surfaces
9. Can be used in both wet and dry conditions and still maintain a 99%+ pick-up rate
10. The debris remains in the collection area behind our patented barrier if the driven vehicle is slowing speeds, stopping and starting, or cornering.

8.4.3.4.2 Magnetic Bars

These magnetic bars can be suspended beneath tugs and trucks to pick up metallic material. Figure 8.20 shows a magnetic bar attached to the front of a pick-up truck. Also, installing a magnetic bar on baggage tugs is quite effective at removing metal from ramp areas and sweep the metal FOD falling off passenger luggage.

Common magnetic materials include ceramic, rare earth, and alnico metals. With the majority of FOD collected at airports being metal, these bars offer a simple solution to that specific FOD source. But these magnetic bars will not be able to pick up the following types of common FOD materials: titanium and aluminum alloys, some stainless steel, and plastics. Magnetic bars should be inspected and cleaned regularly

FIGURE 8.20 Truck with magnetic bar [56].

With permission.

to remove all the collected metallic debris. Otherwise, the collected debris may fall off the vehicle and become FOD yet again [56].

8.4.3.4.3 Rumble Strips

Rumble strips or "FOD Shakers" are long devices, which are 10 to 15 ft (3 to 4.60 m) long, that are positioned on the pavement to dislodge FOD from vehicles that drive over them. They are used at transitions from the landside to the airside or adjacent to airside construction areas.

While these devices were used in the past, they are no longer widely used now as their effectiveness at removing debris from tires or vehicle undercarriages is negligible. The current best practice for removing FOD from tires is to use a hand tool to remove detected debris manually.

8.4.3.5 Foreign Object Debris Storage

FOD are collected in containers or bins, placed at all gates, hangers, maintenance areas, and in the ramp areas. Such containers or bins should be easily visible and marked. They must be emptied frequently to prevent them from overflowing and becoming a source of FOD themselves. FAA suggests that airport employees wear waist pouches to collect debris. FAA also recommends that FOD containers have covers or lids to prevent wind, jet, or prop-wash from stirring up or shifting debris inside the container, thus creating more FOD [1].

The following instructions should be followed in collecting FODs:

- It should not be used for aircraft rubbish bags, oil cans, or other non-FOD rubbish.
- Environmental Protection Agency requires that empty oil cans have a separate barrel.
- It should have lids to prevent wind from redistributing FOD back onto the ramp areas and to preclude water fouling the bottom of bins [1].
- It should be emptied on a scheduled basis or at the end of the employee shift assignments.
- For hazardous materials, specialized containers, in accordance with appropriate regulations, must be used.

Finally, evaluating the debris collected in containers and bins can reveal its sources and indicate where personnel and equipment should be deployed for more effective control.

8.4.4 FOD Evaluation

8.4.4.1 Data Collection and Analysis

The following points should be reviewed:

- How the FOD object was detected
- Date and time of FOD detection and retrieval
- Description of FOD retrieved (category, size, color) and/or image (if available)
- Location of FOD object (coordinates and reference to the AOA location)

- Possible source
- Name of personnel detecting/investigating FOD item
- Airport operations and weather data during the FOD detection event

1. Reporting
   - FOD manager may designate and train certain personnel to collect, tag, store, and report on the collected FOD for future data analysis efforts.
2. Investigation
   - Major FOD incidents are investigated by the FOD manager/appropriate airport personnel to determine the source of FOD and damage caused.

3. Database
   - Records for accidents or serious incidents help to identify any trends, repeats, unusual conditions, etc., in order for corrective action and risk assessment.

8.4.4.2 Continuous Program Improvement

A survey of FOD activities in 56 airport operators, including domestic, international, and military bases, was performed [6].

A summary of the findings from this synthesis, representing 50 airports throughout the United States and internationally, revealed that:

- Almost two-thirds of airports have a FOD management program.
- Most airports conduct inspections for FOD daily, relying on human/visual means.
- Many airports also conduct FOD walks, typically either weekly, monthly, or annually.
- Most airports detect FOD visually, with only some using fixed or mobile systems supporting either continuous or periodic surveillance.
- In addition to manually removing FOD by hand, most airports also utilize mechanized systems (such as power sweepers or vacuum systems) to remove the debris, with some airports relying on jet air blowers to displace FOD. Of the nonmechanized systems, only magnetic bars are used by most airports.
- Although most airports document FOD when it is removed, just over one-quarter of airports use an electronic database for documenting FOD.
- Just over half of airports use FOD letters, notices, and/or bulletins to maintain FOD awareness among airport employees.
- Less than one-fifth of airports employ a FOD manager. At half of the airports, responsibility for the FOD management program is carried out as part of an employee's existing job duties.
- To ensure the quality of their FOD management programs, most airports implement initial and recurrent training, as well as management oversight.
- During reduced visibility and nighttime conditions, only one-third of airports perform more frequent inspections to ensure effective FOD detection and removal. Almost half of the airports have not implemented any additional measures during these conditions.
- If resources were available to enhance an existing FOD management program, almost three-quarters of airports would acquire equipment or technology for detection and/or removal.

REFERENCES

[1] Improving Aviation Safety: Questions and Answers, USDA, www.aphis.usda.gov/wildlife_damage/airline_safety/pdfs/Aviation%20Program%20FAQs.pdf
[2] Aerospace Industries Association: Flight Test Operations Group, Foreign Object Damage/Foreign Object Debris (FOD) Prevention, National Aerospace Standard (NAS 412), Washington, DC: Aerospace Industries Association of America, Inc. (1997).
[3] Foreign Object Debris (FOD) Management, FAA AC No. 150/5210-24, September 30, 2010.
[4] Chaplin, G. FOD Prevention–It's Up to You! The FOD Control Corporation.
[5] Larrigan, D. Designing an Effective Aviation FOD program, Chapter Five, *Make it FOD Free!* pp. 70–77, 2004.
[6] Prather, C.D. Current Airport Inspection Practices Regarding FOD (Foreign Object Debris/Damage), Airport Cooperative Research Program (ACRP-26), 2011.
[7] Safety, Foreign Object Debris and Damage Prevention, Aero 1998, QTR_1 www.boeing.com/commercial/aeromagazine/aero_01/textonly/s01
[8] Kraus, D.C. and Watson, J. Guidelines for the Prevention and Elimination of Foreign Object Damage/Debris (FOD) in the Aviation Maintenance Environment Through Improved Human Performance, December 21, 2001 www.faa.gov/about/initiatives/maintenance_hf/library/documents/media/human_factors_maintenance/guidelines_for_the_prevention_and_elimination_of_fod.pdf
[9] Ringger, G. FOD Awareness, ASA Annual Conference, Phoenix, AZ, 2015 www.aviationsuppliers.org/ASA/files/ccLibraryFiles/Filename/000000001318/Workshop%20K%20-%20Ringger.pdf
[10] Simmons, D. and Stephan, J. Operational Considerations for Airline Management. In *Make it FOD Free: The Ultimate FOD Prevention Manual,* G. Chaplin, Ed., The FOD Control Corporation, Tucson, AZ (pp. 95–97), 2004.
[11] McCreary, I. Runway Safety: FOD, Birds, and the Case for Automated Scanning. *Insight*, SRI, Washington, DC, 2010.
[12] FAA Guide to Ground Vehicle Operations: A Comprehensive Guide to Safe Driving on the Airport Surface, www.faa.gov/airports/runway_safety/media/ground_vehicle_guide_proof_final.pdf
[13] The Self Inspection Process, Presentation to: Caribbean Aviation Professionals, ICAO/FAA Comprehensive Aerodrome Certification Inspector Workshop, FAA, October 2015.
[14] Annex 14, Volume I – Aerodromes Design and Operations, 8th edition, July 2018.
[15] FAA Part 139, Certification of Airports, Electronic Code of Federal Regulation, e-CFR data, March 24, 2020.
[16] Procaccio, F.A. Effectiveness of FOD Control Measures, Master of Aeronautical Science, Embry-Riddle Aeronautical University, Worldwide Campus, October 2008.
[17] Airfield and Flight Operations Procedures, Department of the Army, FM 3-04.300, August 2008.
[18] Airport Foreign Object Debris (FOD) Detection Equipment, FAA Advisory Circular AC 150/5220-24, September 30, 2009.
[19] QINETIQ: Automatic Foreign Object Debris Detection, www.qinetiq.com/sectors/aviation-and-aerospace/automatic-foreign-object-debris-detection
[20] TARSIER|FOD Detection System www.moog.com/content/dam/moog/literature/Aircraft/tarsier/Moog-TarsierAutomaticRunwayFODDetectionSystem-Brochure.pdf
[21] TARSIER: Automatic Runway FOD Detection System, www.tarsierfod.com/

[22] Nordic Radar Solutions, Airport safety, https://nordicradarsolutions.com/airport-safety/
[23] Stratech Systems Ltd. Re-rating Catalyst: Global Acceptance for iFerret at a Tipping Point, November 12, 2014, http://internetfileserver.phillip.com.sg/POEMS/Stocks/Research/ResearchCoverage/SG/Stratech20141112.pdf
[24] Chew, D. and Gwee, J. iFerret™ intelligent Airfield/Runway Surveillance & Foreign Object & Debris (FOD) Detection System, https://aci.aero/Media/aci/file/2008%20Events/Safety%20Seminar%202008/Speakers/day%202/Jeffrey%20Gwee%20Dr%20David%20Chew%20FOD%20Detection%20System_12Nov08.pdf
[25] Gwee, J. and Karuppiah, G. The Future of FOD Detection at Airports, Journal of Aviation Management, 2007, © Civil Aviation Authority of Singapore 2007, pp. 37–50, https://saa.caas.gov.sg/documents/21216/30126/SAA_Journal_2007.pdf/2c419d02-5d2e-46d2-958f-2ee52e7f4271
[26] Weller, J.R. FOD Detection Administration System: Evaluation, Performance Assessment and Regulatory Guidance, Wildlife and Foreign Object Debris (FOD) Workshop, Cairo, Egypt, March 24–26, 2014, www.icao.int/MID/Documents/2014/Wildlife%20and%20FOD%20Workshop/Assessing%20Risk%20FAA.pdf
[27] FOD Finder™ XM Is the Only Mobile Debris Detecting and Clearing Equipment in the World Designed, Developed and Manufactured by Trex Aviation Systems, Inc., Airport Improvement, October 12, 2016 https://airportimprovement.com/news/us-government-invests-fod-finder-xm-m-state-art-technology-clear-debris-airfields-and-runways
[28] FOD Finder™ XM-C, Trex Aviation Systems, www.aerospecialties.com/app/uploads/2016/12/Trex_FOD-Finder-Systems.pdf
[29] FOD Finder™, www.trexenterprises.com/fodfinderSite/pages/fodfinder.html
[30] Automated FODetect Enhances Runway Safety: ODetect from Xsight Systems Is an Automated FOD Solution, September 25, 2014, www.aviationpros.com/directory/product/11701486/xsight-systems-inc-automated-fodetect-enhances-runway-safety
[31] FOD Detection, Be Ready for the Unexpected, © 2018 Xsight Systems, www.xsightsys.com/index.php/fodetect/
[32] Marcellin, F. Good Drones: UAVs Changing Airport Operations for the Better, February 20, 2020, www.airport-technology.com/features/positive-uses-of-drones-in-aviation/
[33] Fürbas, S. Drones at Airports – New Opportunities Through New Technology, Airport Business, November 21, 2018, www.airport-business.com/2018/11/drones-airports-new-opportunities-new-technology/
[34] Schwanz, M. Whitehorse Int'l Embraces Drone Technology for Airfield Surveys, Airport Improvement Magazine, March 4, 2020, pp. 58–61. Airportimprovement.com
[35] Airport Inspections Conducted by Drones, October 24, 2019, www.novuslight.com/index.html
[36] Guillot, B. and Dowell, M. Airport Benefits of Drone Technology, AviationBros, April 15, 2019, www.aviationpros.com/aircraft/unmanned/article/12436848/airport-benefits-of-drone-technology
[37] Airport Foreign Object Debris (FOD) Management, FAA Advisory Circular AC 150/5210-24, September 30, 2009.
[38] Airport Safety Self-Inspection, FAA Advisory Circular AC 150/5200-18C, March 24, 2004.
[39] Operational Safety on Airports During Construction, FAA Advisory Circular AC 150/5370-2.

[40] Airport Field Condition Assessments and Winter Operations Safety, FAA Advisory Circular AC 150/5200-30D, July 3, 2017.
[41] Hussin, R., Ismail, N. and Mustapa, S. A Study of Foreign Object Damage (FOD) and Prevention Method at the Airport and Aircraft Maintenance Area, IOP Conf. Series: Materials Science and Engineering 152 012038 (2016).
[42] Tom Brothers and Simmons, D. Setting up a FOD Program, 2019, www.fodcontrol.com/setting-up-a-fod-program/
[43] ACI World Operational Safety Sub-Committee, Airside Safety Handbook, 4th edition, 2010, ACI World, Geneva, Switzerland, www.skybrary.aero/bookshelf/books/3171.pdf
[44] Messenger, P.J. Annual FOD Walk, www.fodcontrol.com/annual-fod-walk/
[45] www.alibaba.com/product-detail/Airport-runway-sweeper-HS600E-3-in_60741639081.html
[46] Ravo street sweeper.JPG, https://commons.wikimedia.org/wiki/File:Ravo_street_sweeper.JPG
[47] https://commons.wikimedia.org/wiki/File:Ravo_street_sweeper.jpg
[48] TYMCO – Airport Sweeping: High-Speed Airport Runway Sweeping, www.tymco.com/sweepers/airport-sweeping/
[49] Airport Sweeper: AS 990 www.aebi-schmidt.com/en/products/airport-sweepers/schmidt-as-990
[50] Airport Sweeping, http://schwarze.com/en/airport-sweeping/
[51] The FODBUSTER® FOD Sweeper, www.fodcontrol.com/fodbuster/
[52] Regenerative Sweepers, www.tymco.com/sweepers/regenerative-air-system/
[53] The FOD*BOSS Helps Airfields Focus on Safety, May 1, 2018, www.aerosweep.com/the-fodboss-helps-airfields-focus-on-safety/
[54] Protecting Military Aircraft from FOD, September 13, 2017, www.aerosweep.com/protecting-military-aircraft-fod/
[55] FOD*BOSS. The Ultimate FOD Sweeper www.aerosweep.com/fbaw/?gclid=CjwKCAjw3-bzBRBhEiwAgnnLCksnIa6tpKqiUSmGEDHEINbIvGuUCTo5yxjqfe5GL8GDu1fn4fju1xoCSc4QAvD_BwE
[56] Commercial Truck Magnets, www.bluestreakequipment.com/magnetic-sweepers/commercial-truck-magnet/
[57] Performance Assessment of a Radar-based Foreign Object Debris Detection System, DOT/FAA/AR-10/33, February 2011.
[58] https://upload.wikimedia.org/wikipedia/commons/4/4b/Ravo_street_sweeper.jpg
[59] www.fodcontrol.com/fodbuster/

9 Management of Animate (Biological) FOD

9.1 INTRODUCTION

Animate FOD may be categorized into ground or airborne groups. Ground animates include plants, ground insects, mammals, reptiles, and human. Airborne animates include flying insects, birds, and bats.

Birds are not the only wildlife causing serious problems for departing and landing aircraft. Deer, coyotes, and even alligators wandering onto runways are also hazardous.

As discussed in the Federal Aviation Administration (FAA) studies, the total number of wildlife-aircraft strike reports has increased, but the damaging strikes in the airport environment have decreased [1]:

- The reported strikes were 1,850 in 1990, 6,000 in 2000, 14,500 in 2017, and 17,228 in 2019. This increase is due to the increased number of aircraft movements. In addition, in some countries such as the UK and Australia since 2006, wildlife strike reporting changed from voluntary to mandatory [2].
- Damage due to wildlife strikes was 14% of reported strikes in 2000, while in 2017, the rate was only 4%.
- In 2019, birds were involved in 94% of the reported strikes, bats in 3.2%, terrestrial mammals in 2.3%, and reptiles in 0.5% [3].
- However, until now tens of thousands of animals die each year as a direct result of collisions with aircraft.

Although the economic costs of wildlife strikes are extreme, human lives lost when aircraft crash due to wildlife strikes is the most important. Such human fatalities and injuries pushed airlines and airports to adopt FOD programs. Such FOD programs will help reduce employee injuries from animate debris as well as save money. Moreover, numerous court adjudications stated that wildlife FOD management is responsible for operating a safe, wildlife-free facility and to warn flight crews of wildlife activity [4, 5].

DOI: 10.1201/9781003133087-9

9.2 WILDLIFE NATURAL AND ARTIFICIAL ATTRACTANTS

Most wildlife is attracted to the airport environment because it has something they want, generally food, water, or shelter, habitat, and a secure environment [6]. Hazardous wildlife finds food in different sources like grass, seed, rodents, agricultural operations, waste disposal, and landfills. Municipal solid waste landfills (municipal landfills) are known to attract large numbers of hazardous wildlife, particularly birds [7].

Hazardous wildlife is also attracted to water sources for drinking, feeding, bathing, and roosting. Many airports have permanent bodies of drinking and wastewater near or between runways for landscaping or flood control. At the same time, temporary water pools may be found in rainy periods.

The grass, trees, brush, shrubs, holes, patches of weeds, culverts, and airport structures resemble wildlife that needed cover for resting, roosting, and nesting.

Figure 9.1 depicts separation distances recommended by the FAA between the closest point of the aircraft operations area, with special attention for approach and departure airspace, and the hazardous wildlife attractant. It illustrates the three perimeters A, B, and C. Perimeter A is for airports serving piston-powered aircraft

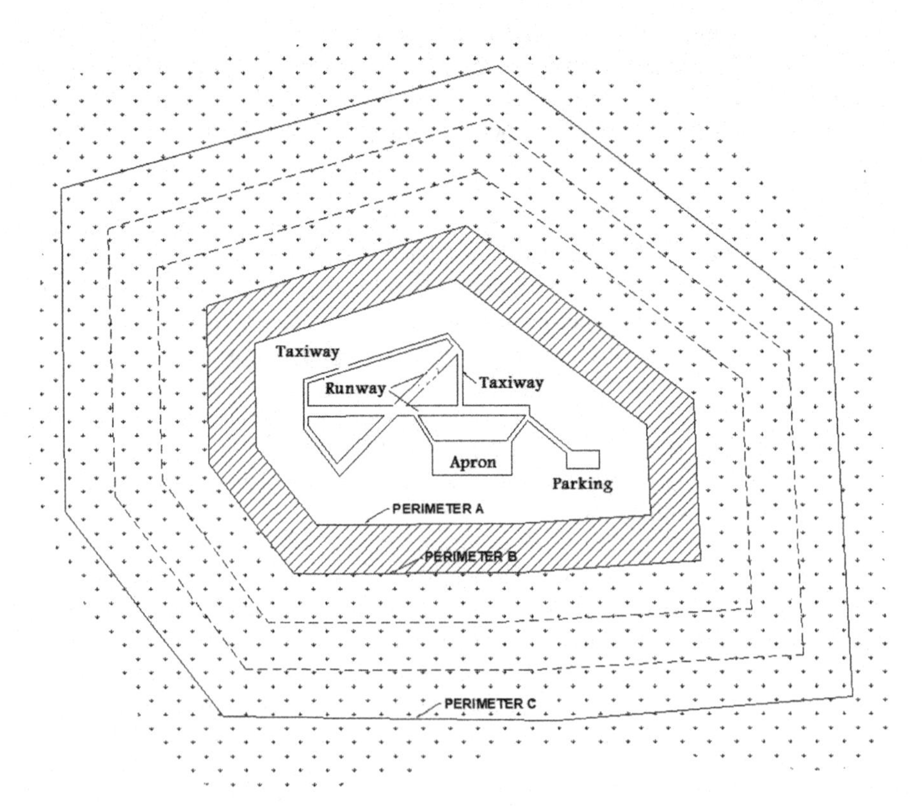

FIGURE 9.1 FAA recommended separation distances within which hazardous wildlife attractants should be avoided, eliminated, or mitigated [7].

Courtesy: US Department of Transportation.

which recommends hazardous wildlife attractants be 5,000 ft from the nearest aircraft operations area. Perimeter B is for airports serving turbine-powered aircraft and recommends hazardous wildlife attractants be 10,000 ft from the nearest aircraft operations area. Finally, perimeter C is for all airports and recommends a 5-mile range to protect approach, departure, and circling airspace.

Finally, the most cost-effective way of managing wildlife at airports is to eliminate or modify the attractions to ensure that wildlife avoids the airport. A famous saying in airports: wildlife should be treated like unwanted relatives at your home (never let them get comfortable!).

9.3 AIRPORT WILDLIFE HAZARD WORKING GROUP (WHWG)

Wildlife hazard management on an airport needs communication, cooperation, and coordination between various groups on the airport and with various local, state, and federal agencies and private entities [8]. This necessitates the establishment of a Wildlife Hazards Working Group (WHWG) to facilitate these communication, cooperation, and coordination.

The members of this WHWG are:

- A representative from each of the key groups and agencies involved in wildlife issues
- A representative from each of the security, maintenance, operations, and air traffic control departments
- Representatives from the state wildlife agency, the US Fish and Wildlife Service and USDA/WS
- Representatives from any facility near the airport that significantly attracts wildlife (such as a landfill or wildlife refuge).

Generally, WHWG should not exceed ten members. Typically, someone from airport management chairs the WHWG. The chair can be rotated among various airport departments. At least WHWG should hold one annual meeting.

If there is no regular meeting scheduled for the near future, special meetings of the entire WHWG or a subgroup might be scheduled after significant strike events or other developments affecting wildlife hazards.

9.4 WILDLIFE HAZARD PROGRAMS

All airport operators must be ready to take immediate action to deal with hazardous wildlife on or near the airport operation area (AOA), loading ramps, or parking areas in accordance with the FAA 14 CFR 139.337(a).

9.4.1 FAA Title 14 CFR 139.337 (a)

In accordance with its Airport Certification Manual and the requirement of this section, each certificate holder shall take immediate action to alleviate wildlife hazards whenever they are detected.

9.4.2 FAA Title 14 CFR 139.337 (b)

It requires that each airport certificate holder must be sure that if any of the following four events occurred on or near the airport, a Wildlife Hazard Assessment (WHA) is conducted [8]:

1. An air carrier aircraft experiences multiple wildlife strikes
2. An air carrier aircraft experiences substantial damage from striking wildlife. Substantial damage means damage or structural failure that affects the structural strength, performance, or aircraft flight characteristics and requires major repair or replacement of the affected component
3. An air carrier aircraft experiences an engine ingestion of wildlife
4. Wildlife of a size, or in numbers, capable of causing an event described in the above three items is observed to have access to any airport flight pattern or aircraft movement area.

Where an air carrier aircraft is categorized as: *either a large air carrier aircraft if designed for at least 31 passenger seats or a small air carrier aircraft if designed for more than 9 passenger seats but less than 31 passenger seats, as determined by the aircraft type certificate issued by a competent civil aviation authority (14 CFR 139.5) [8].*

The WHA shall be conducted by a biologist manager having training or experience in wildlife hazard management in airports.

9.4.3 FAA Title 14 CFR 139.337 (c)

It states that WHA at least should contain the following [8]:

1. An analysis of the events or circumstances that prompted the assessment
2. Identification of the wildlife species observed and their numbers, locations, local movements, and daily and seasonal occurrences. It requires a 12-month assessment for a proper documentation of the seasonal patterns of wildlife using the airport and surrounding area during an annual cycle
3. Identification and location of features on and near the airport that attract wildlife.
4. A description of wildlife hazards to air carrier operations
5. Recommended actions for reducing identified wildlife hazards to air carrier operations.

The certificate holder may look to WS – Wildlife Services (USDA) – or to private consultants to conduct the required WHA.

9.4.4 Wildlife Hazard Management Plan

The goal of an airport's Wildlife Hazard Management Plan (WHMP) is to minimize any risks to aviation safety, airport structures or equipment, or human health arising from elevated populations of hazardous wildlife on and around the airport.

Upon completion of the WHA, it is submitted to the FAA for evaluation and determination whether a Wildlife Hazard Management is needed or not. If no management plan is needed, then the airport manager develops and implements a plan to deal with any hazardous wildlife attractants or situations identified in the WHA.

9.4.4.1 FAA Title 14 CFR 139.337 (e) (1–3)

If the FAA determines that a WHMP is needed, then the airport operator must formulate and implement a WHMP, based on Title 14 CFR 139.337 (e) (1–3). The FAA regional coordinator will contact the local US Fish and Wildlife Service and Ecological Services Field Office to identify and inform the airport operator about any federally listed threatened species or critical habitat on or near the airport.

The airport operator must submit his final draft plan, together with a copy of the Biological Assessment, to the FAA for approval.

Once the FAA approves the draft WHMP, the plan is returned to the airport sponsor for inclusion in the airport's Airport Certification Manual and is enforceable.

The WHMP must identify the following:

- Responsible personal for implementing each phase of the plan
- Information on hazardous wildlife attractants on or near the airport
- Appropriate wildlife management techniques to minimize the wildlife hazard
- Appropriate management measures
- Necessary equipment and supplies
- Training requirements for the airport personnel to implement the WHMP
- When and how the plan will be reviewed and updated.

Airport manager must appoint a WHWG for a periodical review of the airport's WHMP and future refinements or modifications.

9.4.4.2 FAA Title 14 CFR 139.337 (f) (1–7)

The WHMP is described in detail in 14 CFR 139.337 (f) which includes:

1. The responsible person may identify a list for individual who have the authority and responsibility for implementing different phases of the plan. It may include the airport director and representatives from airport departments like operations, maintenance, security, planning, finance, and wildlife hazard working group. It may also include local law enforcement authorities like US Fish and Wildlife Services, State Wildlife Agency, City Police, and County Sheriff.
2. Provide a prioritized list of problem wildlife populations and wildlife attractants. It will include a list of completed wildlife population management projects and habitat modification projects to reduce wildlife strike in the AOA, airport structure, and within 2 miles and 5 miles from AOA designed. It addresses:
    - Species-specific population management like deer, gulls, geese, coyotes (will be discussed later), which includes resource protection, repelling/exclusion, and removal
    - Habitat modification and land-use changes (will be discussed later), which includes:
        - Food/prey management
        - Vegetation management
        - Water management
        - Airport buildings

3. Protected wildlife by federal, state, and local laws.

Some wildlife might be protected by laws:

- Federal – 50 CFR 1 to 199
- Stata – fish and game (or equivalent)
- City, county – ordinances
- If pesticides are used, a review for using federal and state regulations should be reviewed

4. Identification of resources that the certificate holder will provide to implement the plan in terms of Personal, Time, Equipment (e.g., radios, vehicles, guns, traps, propane cannons, etc.), Supplies (e.g., pyrotechnics), Pesticides (restricted/nonrestricted use)
5. Procedures to be followed include:
   - Designation of personnel responsible for implementing the procedures
   - Physical inspections of the aircraft movement areas before air carrier operations begin, which include runway, taxiway sweeps, AOA monitoring, and other areas attractive to wildlife
   - Methods to be used repel, capture, and kill
   - Communication methods between personnel conducting or observing wildlife hazard and air traffic control tower: training and equipment
6. Meeting to review and evaluate to Wildlife Hazard Management Plan annually or following an event described in Section 9.4.1; FAA Title 14 CFR 139.337 (b)
7. A training program conducted by a qualified wildlife management biologist for wildlife control personnel (WCP), other airport personnel, and pesticide user and certification.

9.5 WILDLIFE CONTROL STRATEGIES

Two general methods are used in wildlife control: namely, passive control and active control. Each has two basic stratifies.

Passive control deals with:

1. Aircraft flight schedule modification
2. Habitat modification and exclusion

Active control employs:

1. Repellent and harassment techniques
2. Wildlife removal

9.6 WILDLIFE PASSIVE CONTROL

9.6.1 Aircraft Flight Schedule Modification

Whenever possible, adjust flight schedules and altitudes of some aircraft to minimize the chance of a strike with a wildlife species that has a predictable pattern of movement. Examples:

1. Pilots are advised not to depart during a 20-minute period at sunrise or sunset during winter when large flocks of blackbirds cross an airport going to and from an off-airport roosting site
2. During parts of the year, scheduling nighttime arrivals and departures to avoid strikes with birds that do not fly at night like albatrosses and other seabirds
3. Temporarily closure of a runway with unusually high bird activity or a large mammal (e.g., deer) incursion until WCP can disperse the animals
4. During bird migration seasons, adopt a policy of keeping flight routes away from bird migration roots
5. Flying civil airliners at altitudes higher than those achieved by birds. Since about 95% of bird strikes occur below 6,000 ft, and about 60% below 2,000 ft, keep the long-distance airliners cruise at above 30,000 ft
6. Since 95% of military bird strikes happen below 2,000 ft, and about 70% below 500 ft, keep military aircraft flying at higher altitudes.

9.6.2 Habitat Modification and Exclusion

This procedure for passive control aims at changing the environment to be less attractive or accessible to wildlife. Take any action at airport to reduce, eliminate, or exclude one or more of the wildlife attractants to wildlife (food, water, and cover).

9.6.2.1 Food

Food attractants include:

1. Urban food sources near airports like handouts from people in taxi stands or parks, grain elevators, feed mills, food waste near grocery stores, restaurants, catering services, and sewer treatment plants
2. Rural food sources like sanitary landfills, feedlots, some agricultural crops, especially cereal grains or sunflower, and spilled grain along road and rail rights-of-way

The following recommendations must be followed:

- Do not use airport property for agricultural production, including hay crops, within the separation zones identified in Figure 9.1 as stated by FAA [9].
- Do not use trees and other landscaping plants that produce fruits or seeds attractive to birds for the street side of airports.
- More research is needed to define grass height or vegetation type for airside areas as there is no consensus on the utility of tall-grass management for airports. Tall grass, by interfering with visibility and ground movements, is thought to discourage many species of birds from loafing and feeding. Also, it does not provide birds the space to achieve the wingbeat needed for takeoff. However, some birds as Canada geese do not appear to be discouraged by tall grass.
- The US Air Force now requires airfield grass to be maintained at the height of 17 to 35 cm. As per the studies in England since the 1960s and the

recommendation of USDA, civil airports should maintain grass height of 6–10 inches (15–25 cm). On the contrary, Vancouver International Airport has undertaken an experimental program using Reed Canary Grass maintained in some areas at the height of over 75 cm.
- Eliminate all human food waste.
- Prohibits the construction or establishment of new municipal landfills within 6 miles of certain public-use airports
- Manage existing landfills and other waste disposal sites that attract large numbers of birds by netting or relocation.

9.6.2.2 Water

Water acts as a magnet for birds. It is recommended that:

- Eliminate all standing water on an airport to the greatest extent possible.
- Fill or modify to allow rapid drainage of depressions.
- Do not establish retention ponds, open drainage ditches, outdoor fountains, and other wetland sites on or adjacent to airports.
- When it is not possible to drain a large detention pond completely, use physical barriers, such as bird balls, wires grids, pillows, or netting, to deter birds and other hazardous wildlife. Evaluate the use of physical barriers and ensure they will not adversely affect water rescue.

9.6.2.3 Cover

All wildlife needs cover for resting, roosting, escape, and reproduction. As examples for covers:

- Corporate lawns, golf courses, and even building roofs nearby ponds are used by non-migratory Canada geese in urban areas.
- Building ledges, abandoned buildings, open girders and bridge work, and dense vegetation are used by pigeons, house sparrows, and European starlings.
- March vegetation are used by blackbirds.

For airport spacing and landscaping, avoid:

- Plants that produce fruits and seeds desired by birds
- The creation of areas of dense stands of trees as it provides excellent cover for deer, coyotes, nesting geese, raptors, European starlings, blackbirds, rodents, and other wildlife for roosting

Other methods that reduce wildlife attraction to airport ground cover are:

- Thinning the canopy of trees, or selectively removing trees to increase their spacing to eliminate bird roosts that form in trees of airport, employ
- Eliminate from airport areas all piles of construction debris, discarded equipment, unmowed fence rows, and other unmanaged areas which provide

excellent cover for commensal rodents like rats or mice, and den sites for woodchucks, feral dogs, and coyotes

- The use of undesirable vegetation or mildly toxic to wildlife as:
  - Fescue grass that contains fungal endophytes which are unpalatable to grazing birds, such as geese, as well as to rodents, deer, and fewer insect numbers
  - Wedelia or Bermuda grass, which is appropriate for subtropical airfields
  - Artificial (synthetic) turf provides a more sterile environment for wildlife at airports.

9.6.2.4 Exclusion Techniques

Exclusion techniques are employed when food, water, and cover cannot eliminate wildlife hazards. Examples are the use of physical barriers to deny wildlife access to a particular area or installing a covered drainage ditch instead of an open ditch.

9.6.2.4.1 Exclusion for Birds

Several methods may be used as follows:

- Use of tubular steel beams rather than the I-beams as they are much less attractive as perching sites for starlings and pigeons.
- Netting may be used for old structures, having rafter and girded areas in hangers, warehouses, and under bridges.
- Netting can be installed over small ponds and similar areas.
- Curtains made of heavy-duty plastic sheeting, cut into 12-inch strips, and hung in the doorways of warehouse or hanger, will prevent birds from entering these openings.
- Anti-perching devices, such as spikes, can be installed on ledges, roof peaks, rafters, signs, posts, and other roosting and perching areas to keep certain birds from using them (Figure 9.2).
- Changing the angle of building ledges to 45° or more will deter birds.
- Over-head wire systems for retention ponds and drainage ditches can reduce gulls and waterfowl hazard.
- Plastic, 3-inch diameter "bird balls" or floating mats over ponds will completely exclude birds and yet allow evaporation of water.
- Designing ponds with steep slopes to discourage wading birds such as herons.

9.6.2.4.2 Exclusion for Mammals

Proper fencing is the best method for the exclusion of mammals. The fence line should be free of excess vegetation. The fence line should be patrolled at least daily, and fix any breaks, or other holes in the fence as soon as they are discovered. The mostly used fences are:

- A 10–12-foot chain link fence with 3-strand barbed wire outriggers recommended by FAA with a 4-ft skirt of chain-link attached to the bottom of the fence and buried at a 45° angle on the outside of the fence to prevent animals from digging under the fence and reduce the chance of washouts

FIGURE 9.2 A well-maintained fence at least 10-feet high [8]

Courtesy: US Department of Transportation, FAA.

- Electric fences for excluding deer [10] are not as costly as permanent fencing but have drawbacks in safety and maintenance.

9.7 WILDLIFE ACTIVE CONTROL

Wildlife active control methods are categorized into two main groups: namely, repellent and harassment methods and wildlife removal.

9.7.1 Repellent and Harassment Techniques

The objectives of repellent and harassment techniques are to make the airport area unattractive to wildlife and make them uncomfortable or fearful. These techniques include chemical, auditory, or visual means. However, the serious problem arises when birds and mammals become accustomed to these methods. Such a habituation can be minimized by:

- Using each technique sparingly and appropriately when the target wildlife is present
- Using an integration of appropriate repellent techniques
- Reinforcing repellents with occasional lethal control when there is abundant species such as gulls or geese.

9.7.1.1 Chemical Repellents for Birds

Chemical repellents and toxicants are listed in [10]. They include polybutene, anthraquinone, methyl anthranilate, aminopyridine, and aluminum ammonium sulfate. In the United States, all chemical repellents must be registered in each state as well as either the US Environmental Protection Agency or the Food and Drug Administration (FDA), before they can be used. In several other countries, including the Netherlands, chemical repellents are not used due to toxic features [11].

Chemical products can be applied to perching structures, vegetation, turf grass, and standing waters in the airport. Sometimes, they are applied to temporary pools of standing water on airports (Figure 9.3).

The detailed description of the chemical repellents is given hereafter.

9.7.1.1.1 Polybutene (for Perching Structure)

Polybutene is found in either liquid or paste formulations. Due to its sticky nature, it makes the birds feel uncomfortable. The effective lifetime of polybutene is 6–12 months under normal conditions but reduces substantially in dusty environments. Polybutene is effective for controlling pigeons and preventing raptors from perching on antennas, but less effective for smaller birds such as sparrows.

9.7.1.1.2 Anthraquinone (for Turf Feeding)

Anthraquinone is usually available in liquid formulations and applied by sprayer to the vegetation [13] (Figure 9.3). It is registered as a chemical used for repelling

FIGURE 9.3 Chemical repellents applied to temporary pools of standing water [12].

Courtesy: USDA.

geese from turf since it acts as an aversion repellent to birds. Anthraquinone is used only in high-risk areas of runways as well as bird grazing areas, and is not allowed in the UK.

9.7.1.1.3 Methyl Anthranilate (for Turf Feeding and Water)

Methyl anthranilate is registered by Homestead Air Reserve Station in Florida as a feeding repellent for birds (geese, gulls, waterfowl, and starlings) on turf and golf courses [8]. It is not registered in the UK. It is a nontoxic active compound in ReJeX-iT. Anthranilate products are also liquid formulations that are applied by sprayer to the vegetation. Its effectiveness in repelling geese varies depending on its growing conditions, rainfall, mowing, and the presence of alternate feeding areas.

9.7.1.1.4 Naphthalene

Naphthalene is a repellent that works on the sense of smell. It was tested at airfields in the UK. It was applied to the field as "mothballs" [14].

9.7.1.1.5 Avitrol (Frightening Agent)

Avitrol is a toxic repellent. It is registered for repelling several bird species, including pigeons, house sparrows, blackbirds, and gulls. When birds eat baits (normally grain) that are strongly treated with Avitrol, then they will have distress behavior. Consequently, it frightens other nearby birds [15].

9.7.1.2 Chemical Repellents for Mammals

There are some taste and odor repellents in the market to repel deer, rabbits, and other mammals from browsing on vegetation [10]. Odor repellent includes some products to be applied directly to the vegetation and general area (like predator urine). Some of these products might be appropriate for short-term protection of valuable landscaping plants and fruit trees. However, they have negligible effects on repelling deer or other mammals from the airport operating area.

9.7.1.3 Audio Repellant for Birds

It includes propane cannons, distress-call and electronic noise-generating systems, shell crackers and other pyrotechnics, and ultrasonic devices.

9.7.1.3.1 Propane Cannons

Gas or propane cannons (or "exploders" as they are known in the United States) are mechanical devices that produce loud banging noises by igniting either propane or acetylene gas [16]. Gas cannons have the following features:

- Produce 130 dB, which is high enough for bird harassment
- Fired at fixed intervals from a control room with the aid of closed-circuit television cameras
- Fired in dusk and dawn, which are the times of feed and roost of birds

FIGURE 9.4 A propane cannon used at Baltimore-Washington airport [17].

Courtesy: USDA.

- Should not have a permanent site but relocated frequently to avoid bird habituation
- Cannot alone disperse all bird species; thus, it should be used together with other depredation techniques to be effective for gulls, blackbirds, waterfowl, pheasants, and other game birds
- Its price ranges from US$ 450 to US$ 1400 worldwide, while in the UK it ranges from GBP 165 to GBP 475
- Costs vary depending on whether it is a single, double, or multi-bang cannon.

Figure 9.4 illustrates a propane cannon used at Baltimore-Washington airport [17].

9.7.1.3.2 Distress-Call and Electronic Noise-generating Systems

Recorded distress calls are an artificial reproduction of bird calls mostly found in airports like gulls, crows, and starlings or a bird annoyance sound. It produces noise levels up to 110 dB and a frequency response between 12,000 and 14,000 Hz. Such calls are either broadcasted from stationary or movable equipment. Static systems are used for smaller areas that sometimes need a louder volume. The calls are broadcasted for about 90 sec from a stationary vehicle approximately 100 m from the target flock. However, birds habituate rapidly to such sound sounds from stationary speakers.

The movable vehicle must be driven close to the birds, with 100–200 m as maximum distance. The distress call is to be played for 15–20 sec and played again if the birds do not respond within 20 sec. If birds have not moved by the third attempt, it means that distress calls are not successful and other methods have to be used [18].

9.7.1.3.3 Shell Crackers and other Pyrotechnics

Projectiles that can be fired from breech-loaded shotguns or from specialized launchers will provide an auditory blast or scream, in addition to smoke and flashing light, to frighten birds. Recent cartridges have ranges of up to 300 yards. If these pyrotechnics are used together with other harassment techniques and limited lethal ones like shooting, it will be very useful in driving birds away from airports.

9.7.1.3.4 Ultrasonic Devices

Ultrasonic devices have not proven any success as bird repellents. For example, pigeons showed no response when exposed within 10 ft to a fully functional, high-frequency sound generating device. It is not recommended to deploy these devices in hangers or other airport settings to deter birds [8].

9.7.1.4 Audio Repellant for Mammals

Propane cannons are the most used audio scaring device for deer. However, deer rapidly habituate to propane cannons. Consequently, it is recommended to use them only for short term (i.e., several days), emergency situations until a permanent technique like fencing or deer removal can be achieved. Pyrotechnic and other electronic noise-generating devices are also ineffective in repelling deer or other mammals for more than a few days.

9.7.1.5 Visual Repellant for Birds

9.7.1.5.1 Scarecrow

Scarecrows are the traditional method first used to scare birds. Some of them mimic the appearance of a predator that frighten birds and cause them to fly away to avoid potential predation. Many scarecrows are motionless human-shaped effigies constructed from inexpensive materials, which are either ineffective or provide only short-term protection. The recent versions of scarecrow include inflatable human effigies (like the Scary Man®) and the flashing Hawkeye [19].

9.7.1.5.2 Laser

A laser is a nonlethal harmless bird repelling technique, which is effective and accurate over long distances (up to 1,000 m). It is one of the most effective repellents. It is silent and can be safely used to disperse birds around runways, structures, or water ponds. Its advantages are:

- Birds will not get used to its threat, unlike some deterrent devices
- Under low-light conditions, like sunrise and sunset, and in overcast, rainy, or foggy weather conditions, lasers can be effective in dispersing several bird species like geese, cormorants, and crows
- Lack of noise

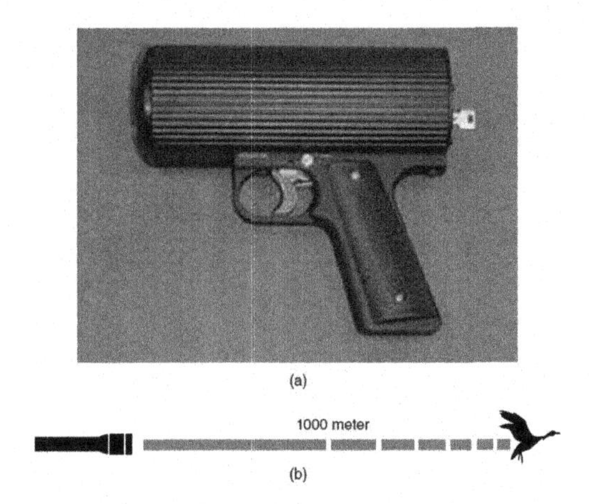

FIGURE 9.5 Laser devices: (a) hand-held laser device; (b) effective distance.

Courtesy: USDA-APHIS [8].

However, its effectiveness decreases or even completely vanishes in daylight conditions and its usage needs special caution in an airport environment [20]. Figure 9.5 illustrates a laser pistol and the effective distance for a laser beam. It projects a 1-inch diameter red or green beam.

9.7.1.5.3 Pulsating Lights

The Pulselite System is an effective system that substantially reduces bird strikes by some 30%–66% and significantly increases the visibility of aircraft. It was tested by several airlines, including Alaska, Horizon, and Qantas Airways. These reductions result in increased profitability, enhanced safety, improved customer service, and higher customer satisfaction ratings [21].

9.7.1.5.4 Scaring Aircraft

Frightening paints may be added to either civil aircraft or helicopter to scare birds [22, 23]. Flying on such frighteningly painted aircraft at Vancouver Airport successfully deterred gulls, ducks, and geese.

9.7.1.5.5 The Robotic Birds

Robotic birds are used to repel birds from airfields. Remote-controlled falcons and eagles (sometimes called "Robirds") are designed to fly over airfield areas, scaring off intruders [24]. The Robirds were developed by Clear Flight Solutions in the Netherlands and Canada (Figure 9.6). In the Netherlands, there are two remote-controlled Robirds: namely, the falcon and eagle. The falcon robot has a body length of 23 inches, a wingspan of 47 inches, and a flight speed of 50 mph. The eagle robot is 46 inches long and 86-inch wingspan. Manufacturers hope to make them autonomous in the future. Both birds are 3D printed using glass fiber and nylon composite material and painted to look as real as possible. In Canada, the Robirds are robotic birds of prey that employ a flapping wing motion during flight, the same as real birds.

FIGURE 9.6 Flying Robirds.

Courtesy: USDA-APHIS [8].

Another robotic bird is identified as Robop and designed by Robop. It has the shape, behavior, and sound of a hunting peregrine falcon to scare birds such as seagulls and pigeons. Robop can move its head and wings and emits four peregrine calls to deter birds on airport runways. It can withstand harsh weather conditions for many years with minimal maintenance.

9.7.1.5.6 Corpses (Dead Birds)

Dead bodies are a successful visual deterrent. Experiments and field demonstrations have confirmed that a dead turkey vulture (freeze-dried taxidermy mount with wings spread) hung by its feet will cause vultures to abandon the site. Previous tests using dead gulls and ravens suspended from poles have also shown promising results in dispersing these species from feeding and resting sites. The dead bird must be hung in a "death pose" to be effective. Dead birds lying supine on the ground are generally ignored or even attract other birds.

9.7.1.6 Visual Repellant for Mammals

Unfortunately, visual repellents such as flags, effigies, and laser have proven ineffective for repelling mammals [8].

9.7.1.7 Falconry (Trained Falcons) to Repel Birds

Falconry is generally defined as fighting birds with birds. Trained falcons and other birds of prey have been used intermittently to disperse birds since the late 1940s in several airports in Europe (Scotland and Spain) and North America (United States and Canada) (Figure 9.7). For good results, daily operations on a year-round basis are required in most cases. The advantage of falconry is that the birds on the airport are exposed to a natural predator for which they have an innate fear. However, falconry programs have the following disadvantages [25–28]:

- Expensive, as they require several birds that must be maintained and cared for by a crew of at least two full-time trained, highly motivated personnel.
- Their effectiveness in reducing strikes has been difficult to evaluate.
- It is effective only when used in combination with other frightening techniques.

FIGURE 9.7 Falconry.

Courtesy: USDA, IBSC/WBA [8, 25].

- Most birds of prey used in falconry programs are unsuccessful in dispersing large birds such as waterfowl (particularly geese) as well as other birds of prey.

9.7.1.8 Trained Dogs to Repel Birds

The use of trained dogs, especially border collies, keep unwanted geese waders and wildfowl from airports (Figure 9.8). As with falcons, the advantage is exposure to a natural predator. A single border collie together with its handler can keep an area nearly $50\,km^2$ free of larger birds. The first commercial airport in the world to employ a border collie was Southwest Florida International Airport in 1999.

Likewise, the disadvantage is that the dog must be always under the control of a trained person, and the dog must be cared for and exercised 365 days a year. A dog will have little influence on birds that are flying over the airport, such as raptors and swallows and especially gulls.

Generally, all types of dogs will not hurt or even touch the birds as they cannot catch them.

9.7.1.9 Radio-Controlled Model Aircraft to Repel Birds

Radio-controlled (RC) model aircraft have been used to disperse birds since the early 1980s in both civilian and military airports. The RC airplane resembles audio and visual repellent method. It can direct the flying birds away from the airfield [8]. An example of a RC aircraft is the Robo-Falcon™ shown in Figure 9.9a, which may be used together with other depredation techniques for dispersing birds from airports. At Whiteman Air Force Base, Missouri, balsa wood RC aircraft are used to keep the airfield clear of raptors and other large birds. They also proved effective at dispersing the base's redwing blackbird roost [29].

The main advantages of RC aircraft are:

- Under the control of a person and can be directed precisely to disperse the birds away from the airport runway
- It can be deployed on an "as needed" basis and need only a little between flights.

FIGURE 9.8 Trained dogs [8].

Courtesy: USDA.

The disadvantage of RC aircraft is that a trained person is required to operate the RC aircraft in an airport environment.

The Korean Atomic Energy Group and LIG Nex1 designed the world's first bird strike robot. It is a six-wheeled unmanned ground vehicle that uses both acoustics and visual techniques (laser) to scare birds (Figure 9.9b).

9.7.1.10 Nonlethal Projectiles to Repel Birds

Paintballs and rubber or plastic projectiles, fired from paint-ball guns and 12-gauge shotguns, can be used to repel several bird species like Canada geese and roosting vultures [8, 30]. Paintballs are spherical gelatin capsules containing primarily polyethylene glycol and powered by carbon dioxide or nitrogen. When fired, these balls usually break once they hit the target. There are several types of rubber or plastic projectiles (slugs, buck shot, pellets, beads) for use in a shotgun.

Paintballs are fired by a high-quality paint-ball gun from 20 to 100 ft from a bird. The proper distance from the bird for firing varies by projectile and species of bird. The objective of shooting from a sufficient distance is not to kill it right on the spot but

FIGURE 9.9 Radio controlled vehicles: (top) aircraft; (bottom) unmanned ground vehicle.
Courtesy: USDA-APHIS [8].

may only cause the bird to have fractured wings, legs, or internal injuries. However, shooting a small bird particularly in sensitive areas like its head may immediately kill it [30]. Also, a squirrel, if shot from a short distance, might die right away. Paintball guns will not kill a rabbit, a cat, a dog, or a bear, but induce temporary pain [30].

9.7.2 Wildlife Removal

In several cases, hazardous wildlife must be removed from an airport by capturing and relocating or by killing the target animals. A state permit is mostly required, while a federal Migratory Bird Depredation Permit is rarely required before any migratory or protected birds may be taken (captured or killed). A state permit is also necessary before any state-protected mammals may be taken. Any capturing or killing must be done humanely and only by trained [8].

9.7.2.1 Capturing Birds and Mammals

Live-captured birds and mammals and their relocation depend on the legal, political, and social facts for each species. State wildlife agencies are increasingly concerned about the relocation of captured wild animals due to disease possibilities and the creation of additional wildlife problems at release sites. It is recommended that:

- Euthanize unprotected birds (like pigeons, house sparrows, and European starlings), using procedures recommended by the American Association of Wildlife Veterinarians (AAWV)
- Dispose of mammals captured on airports (like raccoons, woodchucks, and coyotes) according to state regulations
- Euthanize and donate resident Canada geese captured during molt or by nets to soup kitchens or food banks, provided that all federal and state regulations are maintained.

9.7.2.1.1 Chemical Capture of Birds

Alpha chloralose (A-C or AC) is registered with the FDA as an immobilizing agent and can only be used by certified people or by people working under the authority of personnel with the USDA/WS. It can be used for capturing waterfowl, coots, and pigeons. A-C, incorporated into bread baits, is ideal for capturing ducks, geese, and coots, while corn baits are recommended for pigeons or groups of waterfowl. Birds eating a clinical dose of A-C can be captured in 30–90 min. Complete recovery normally occurs within 8 h but can take up to 24 h [31].

9.7.2.1.2 Chemical Capture of Mammals

Large mammals, such as deer, can be captured with tranquilizer guns (nonlethal air guns that reduce agitation in animals via anesthetic drugs). However, it is not recommended or permitted in most states. Only if the use of firearms is not safe or practical, the use of tranquilizer guns might be appropriate. The use of tranquilizer guns requires personnel with a high degree of skill and experience. If used in an airport environment, safeguards must be in place to ensure partially tranquilized deer do not enter runway areas.

9.7.2.1.3 Live-Trapping Birds

Live trapping is generally used against federally and state-protected bird species. Birds will be next relocated away from the airport (at least 50 miles) [32]. Live trapping with walk-in traps on roofs or isolated sites may be used to remove pigeons at airports. However, if relocated, pigeons can fly long distances to return to the site of capture. Therefore, it is recommended to euthanize captured pigeons following AAWV guidelines.

Live trapping may use mist nets, cage traps, cannon nets, or large funnel-shaped lead-in traps. Trapping should be done by professionals having the skills and proper tools to remove the birds without injuring them.

They may be used to capture pigeons, house sparrows, and raptors (hawks and owls) in the AOA (Figure 9.10).

The major advantage of live trapping is selectivity; thus, any nontarget birds can be released unharmed. However, the major disadvantages are they often labor intensive, time consuming, and relatively costly.

9.7.2.1.4 Live-Trapping Mammals

Trapping wild animals requires a high degree of knowledge and skill and should be carried out by skilled professionals. Trappers must be knowledgeable in procedures for handling and euthanizing mammals. Detailed description of various trap designs is given in [10]. State and local regulations might restrict the use of some types of traps. Specialized drop-door traps, drop nets, or rocket net setups can be used to capture deer, while smaller box-type or basket traps can be used to capture medium-sized mammals,

FIGURE 9.10 Pigeon trap [32].

(Source: JFK)

like raccoons, woodchucks, beavers, and feral dogs. Leg-hold traps and snares can be used to capture coyotes, feral dogs, and raccoons. Traps must be checked frequently (at least once every 24 h and more frequently in both hot and cold weather).

9.7.2.1.5 Killing Birds and Mammals

Killing of wildlife in an airport is the last option for repelling them if all other methods (habitat modification, exclusion techniques, and repellent actions) were not successful. However, killing a particular animal or reducing the numbers of a wildlife species by lethal means may be a temporary option for wildlife management until a long-term, nonlethal solution is implemented (e.g., erection of deer-proof fence, relocation of nearby gull nesting colony).

A justification for using lethal methods is achieved via the following information:

- A proof that the wildlife species is a threat to economy, safety, or health
- A justification for the failure of nonlethal methods to solve the problem
- An assessment of that killing will have no impact on local and regional populations of the species
- Assurance that the killing procedure for a specific species is appropriate (i.e., safe, effective, and humane)
- A proof that the killing program will help in reducing the bird strike problems.

9.7.2.1.6 Destroying Eggs and Nests

Egg addling (puncturing, shaking, or oiling) is very effective in controlling birds by preventing hatching (Figure 9.11). Engine oiling involves coating the eggshells with oil such as liquid paraffin. Though pricking of eggs with a needle allows bacteria to enter the egg and desiccate its contents, some pricked eggs may still hatch. It is preferred to break their eggs.

FIGURE 9.11 Egg oiling at SLC [32].

(*Source*: SLC)

However, egg addling encourages the nesting birds and any nonbreeding birds to stay at the airport.

For this reason, do not allow Canada geese, mute swans, and gulls to nest on airport property and remove their nests if available. Also, destroy the nests of pigeons, starlings, house sparrows, and any other birds hazardous to aviation. Next, install physical barriers to prevent renesting.

At the time of nest destruction, harass the adult birds from the airport or shoot them. Check the nesting area weekly for renesting until the end of the nesting season (end of June).

Since the migratory bird nests are protected by federal law, they may need a Depredation Permit before destruction.

9.7.2.1.7 Shooting Birds

Shooting birds is restricted and used only after all other wildlife control methods have failed to repel the birds [33, 34]. It is used in cases where the immediate removal of problematic birds is necessary. FAA and USDA-APHIS support live-ammunition shooting as for bird population reduction. Most airports worldwide have adopted shooting birds to decrease the bird population. All shooters must have permission from the USDA-APHIS or PANYNJ.

Figure 9.12 illustrates typical shooters using a pistol and a 12-gauge shotgun for shooting birds. Along the southern boundaries of JFK, two to five shooters direct their guns away from the airport but toward flying gulls that came within range (about 40 m). Generally, shooting is done during daylight in the open so that other birds can witness the action. Shooting a shotgun has the following advantages:

- It reinforces other audio or visual repelling techniques.
- The loud noise, coupled with the death of one or more of the flock members, can frighten the rest of the flock away.
- The target birds are permanently removed.

FIGURE 9.12 Bird shooters using pistol and gun.

Courtesy: IBSC/WBA [33].

Four prime rules apply when using bird shooting at airports [8]:

1. Use only trained personnel who have an excellent knowledge of wildlife identification
2. Use the proper gun and ammunition for the situation
3. Have necessary federal, state, and municipal wildlife kill permits and record killed bird species and date
4. Notify airport security, air traffic control, and the local law enforcement authority (if needed) before instituting a shooting program.

The number of killed birds by shooting in a few airports will be reviewed hereafter.

The number of killed birds by shooting in 2012 were 1,487 in Sacramento, 1,125 in Oakland International Airport, and 410 in San Francisco International Airport.

In 2011, the number of killed birds in the same airports were 237, 1,250, and 554, respectively [35]. Nearly 70,000 gulls, starling, geese, and other birds have been slaughtered in the New York City area, mostly by shooting and trapping, in the period 2009–2017 [36].

9.7.2.1.8 Shooting Mammals

Shooting is the best procedure for removing deer in either of the following cases: fencing is inadequate to keep deer off an airport or if it has gotten inside the airport's fence. Shooting must be done by professional sharpshooters, using non-ricocheting bullets in rifles equipped with night-vision scopes and noise suppressers. Stationary or mobile elevated shooting stands can be erected on the ground or on a truck bed to direct shots toward the ground. Shooting of deer at airports must be coordinated through the state wildlife agency [37, 38].

9.8 CONTROL METHODS FOR MOST HAZARDOUS BIRDS

9.8.1 Introduction

Table 9.1 is extracted from FAA survey for ranking hazardous wildlife species [39]. It is a short list for only the most ten hazardous wildlife species.

1. Aircraft incurred at least some damage (destroyed, substantial, minor, or unknown) from strike
2. Aircraft incurred damage or structural failure, which adversely affected the structure strength, performance, or flight characteristics, and which would normally require major repair or replacement of the affected component, or the damage sustained made it inadvisable to restore aircraft to airworthy condition
3. Aborted takeoff, engine shutdown, precautionary landing, or other negative effect on flight
4. Based on the mean value for percent of strikes with damage, major damage (substantial damage or destroyed), and negative effect-on-flight

TABLE 9.1
Composite Ranking of Hazardous Wildlife Species

	% of Strike with:					
Wildlife Species	**Damage[1]**	**Major Damage[2]**	**Effect on Flight[3]**	**Mean Hazard Level[4]**	**Composite Ranking**	**Relative Hazard Score[5]**
White Tailed Deer	84	36	46	55	1	100
Snow Goose	77	41	39	53	2	95
Turkey Vulture	51	19	35	35	3	63
Canada Goose	50	17	28	31	4	57
Sandhill Crane	41	13	27	27	5	48
Bald Eagle	41	12	28	27	6	48
Double Crested Cormorant	34	15	24	24	7	44
Mallard	23	9	13	15	8	27
Osprey	22	7	15	15	9	26
Great Blue Heron	21	6	16	15	10	26

[1] Aircraft incurred at least some damage (destroyed, substantial, minor, or unknown) from strike.

[2] Aircraft incurred damage or structural failure, which adversely affected the structure strength, performance, or flight characteristics, and which would normally require major repair or replacement of the affected component, or the damage sustained made it inadvisable to restore aircraft to airworthy condition.

[3] Aborted takeoff, engine shutdown, precautionary landing, or other negative effect on flight.

[4] Based on the mean value for percent of strikes with damage, major damage (substantial damage or destroyed), and negative effect-on-flight.

[5] Mean hazard level (see footnote 5) was scaled down from 100, with 100 as the score for the species with the maximum mean hazard level and thus the greatest potential hazard to aircraft.

5. Mean hazard level (see footnote 5) was scaled down from 100, with 100 as the score for the species with the maximum mean hazard level and thus the greatest potential hazard to aircraft.

9.8.2 Methods for Controlling Some Hazardous Birds (ACRP 32)

9.8.2.1 Goose

Canada goose population is increasing drastically. Figure 9.13 illustrates its population from 1970 to 2012.

Management methods include [40]:

- Auditory frightening devices including gas-operated cannons and pyrotechnics
- Visual frightening devices including lasers, avian systems corporation (ASC) 7500 Rotating Laser System, and Coyote Predator Models
- Auditory-visual frightening devices including radio-controlled model aircraft, helicopters and boats, and dogs
- Nest removal and treatment
- Trapping and relocation
- Shooting
- Habitat modification
- Exclusion using fences, nets, wire-grid systems, floating membrane covers, and plastic balls or spheres.

9.8.2.2 Vultures

Black and turkey vultures cause problems in several ways [41]. The most common problems associated with vultures are structural damage, depredation to livestock

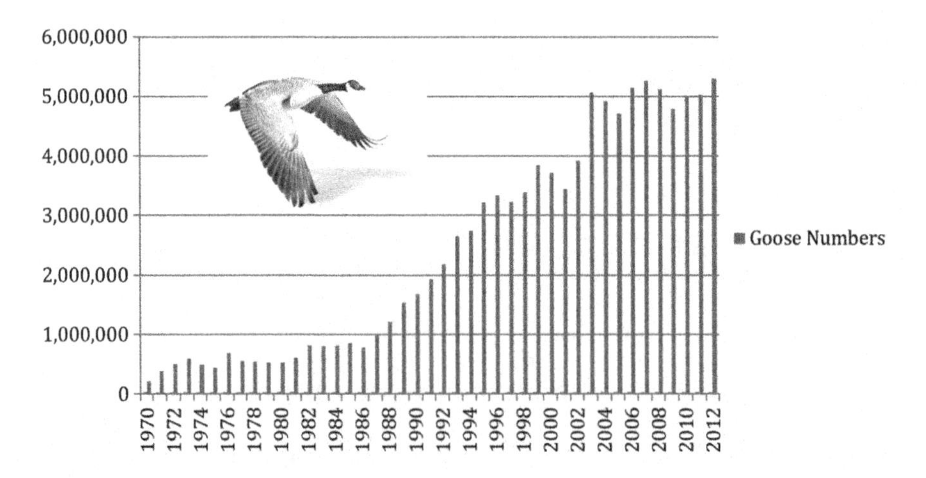

FIGURE 9.13 Canada goose population estimates from 1970 to 2012 [40].

Courtesy: USDA.

and pets, and air traffic safety. The following techniques are employed for their management:

- Habitat modification

Altering the vegetation structure of a given roost via thinning branches on trees or removing some trees to open the roost site. Disposal of dead livestock and removal of other human-made foods may reduce vulture use in some areas.

- Exclusion

Spikes that are short, sharp, tightly spaced, and resistant to bending are normally effective.

- Scare devices including effigies, lasers, and pyrotechnic
- Shooting
- Translocation

9.8.2.3 Sandhill and Whooping Cranes

Sandhill crane populations continue to grow in the United States [42]. The following management methods are employed:

- Habitat modification

Two forms of habitat modification are used to manage cranes: namely, structural modification and supplemental feeding. Extensive supplemental feeding programs have been used worldwide to prevent migrating and wintering cranes from damaging nearby agricultural fields.

- Translocation/relocation

The capture and translocation/relocation of cranes is not recommended as they will continue to cause conflicts, if released.

- Frightening devices including auditory (propane cannons and pyrotechnics), visual (powerline diverters and reflective streamers), and dogs
- Chemical repellents as Avipel®, 9,10 anthraquinone (AQ) which leads to an aversion to the food
- Trapping
- Shooting

9.8.2.4 Double Crested Cormorant

Following a low point in the 1970s, populations of cormorants expanded in North America since 1970s [43]. The following management methods are employed:

- Habitat modification

Several methods are used, including:

- Removal or destroying of nest trees
- Fisheries management
- Translocation/relocation
- Frightening devices including auditory (shell crackers, screamers, whistling or exploding projectiles, bird bangers, propane cannons), visual (scarecrows, human effigies, and balloons)
- Nest removal and treatment
- Chemical repellents as Avipel®, 9,10 anthraquinone (AQ) which leads to an aversion to the food
- Trapping using net and foot-hold traps
- Shooting

9.8.2.5 Herons

The following management methods are employed [44]:

- Habitat modification

Since diseased fish are highly vulnerable to heron and egret depredation, it is recommended to maintain quality fish health standards to minimize bird depredation:

- Exclusion using netting or wire mesh
- Frightening devices including auditory (shell crackers, screamers, whistling or exploding projectiles, bird bangers, propane cannons), visual (scarecrows, human effigies, and balloons).

9.9 CONTROL METHOD FOR MOST HAZARDOUS MAMMALS

9.9.1 Deer

The total US deer population in 2017 was about 33.5 million. Because of increasing urbanization and rapidly expanding deer populations, deer are adapting to living around airports, where they often find food and shelter [8]. From 1990 to 2004, over 650 deer-aircraft collisions were reported to the FAA. Of these reports, over 500 indicated the aircraft was damaged because of the collision.

Management and control of deer may be accomplished via:

- Use tall fences: 10- to 12-ft chain-link with 3-strand barbed wire outriggers and a 4-ft skirt of chain-link fence material, attached to the bottom of the fence and buried at a 45° angle on the outside of the fence to prevent animals from digging under the fence
- At least a daily patrol for the fence, and any washouts, breaks or other holes in the fence should be repaired as soon as they are discovered

- To reduce the costs of the abovementioned fences, the USDA, National Wildlife Research Center confirmed that some 4 to 6-ft, 5 to 9-strand electric fences designs can be 99% effective at stopping deer. Installation of such electric fences neither needs specialized equipment nor training and can be accomplished by airport personnel [8]
- Adopting hunting programs to eliminate deer within the airfield area (standing and light mobile shooting) if safe and legal [31, 45]
- An aggressive pyrotechnic and depredation program should be employed to chase deer away from airports
- Pre-flight and pre-landing runway sweeps during nighttime and low-visibility operations
- Routine clearing of tree, brush, and forest to enable effective harassment

Deer are protected in all US states and many countries worldwide. Permission must be obtained from natural resources management agency whenever deer are shot outside of the normal hunting season. Depredation permits are generally issued by the state natural resources management agency.

9.9.2 Bats

Many bat species are federally protected, and special permits may be required. Control of bat colonies is accomplished by modifying the location where they roost, drink, or feed as follows:

- Locating the daytime roosts and eliminating or modifying the structures
- Removal of open water near the airfield can also decrease bat usage near the active runway
- Reducing or changing the type of lighting near the active airfield may reduce bat usage the airfield

9.9.3 Coyotes

These animals can be controlled by one or more of the following methods:

- Install 8- to 10-ft chain link fencing with a 4-ft skirt and 3-strand barbed wire outriggers
- Net wire and electric fencing
- Rodent control may reduce the numbers of these animals
- Shooting
- Use leg-hold traps (Nos. 3 or 4) or snares
- Pyrotechnics, gas cartridges, radios and other noise makers, scarecrows and moveable, human effigies, strobe lights may be used to scare them
- Mow airside vegetation short to eliminate rabbit and field mouse habitat

Check the laws regulating coyote control in the county, state, or country to determine the coyote's status and legal take methods.

9.9.4 Buffalos and Cows

Airport perimeter fences should be used and frequently checked.

9.9.5 Feral Pigs

They can be controlled using:

- Trapping with euthanization is the most effective method with proper disposal of all remains
- Fence made of metal with a buried bottom fence edge (at least 2 ft) to prevent "dig-unders"
- Harassment noise such as bangers and screamers may have a short-term effect.

9.9.6 Foxes

It is controlled via the following methods:

- Use fences as with coyotes and deer.
- Use toxicants; the M-44™ is registered for control of red and gray foxes in the United States.
- Fumigate fox dens (only in North Dakota, South Dakota, and Nebraska).
- Trap foxes using nos. 1½, 1¾, and 2 double coil spring traps and nos. 2 and 3 double long spring traps.
- Shoot foxes.
- Eliminate trees, brush, and other cover within the AOA.
- Control rodent prey based on airport property.

9.9.7 Raccoons

It is controlled via the following methods:

- Secure trash cans inside buildings or wire lids down
- Use dumpsters with lids that lock down

9.9.8 Rabbits and Hares

- The only effective control program for them is to initiate a poison bait station program
- All carcasses should be removed from the airfield environment and properly disposed

9.9.9 Rodents (Squirrels, Woodchucks, and Rats)

Control of these animals is best performed through:

- The installation's Pest Management Program.
- Poison and traps

9.9.10 Reptiles (Snakes, Alligators, and Turtles)

Reptiles should be looked for during airfield sweeps and, if found, returned to their habitats. However, care should be taken since some snakes are venomous.

9.10 WILDLIFE HAZARD MANAGEMENT TRAINING FOR AIRPORT PERSONNEL

The following areas of training and levels of skill are suggested for WCP to implement control activities in airports under the WHMP [1].

9.10.1 Bird Identification

Many years of training and practice are needed for any person to become an expert in field identification of all bird species at a location because:

- Hundreds of bird species are available (over 600 species in the United States only).
- Adults and subadults may have different plumage patterns and bill colors.
- Some birds (gulls, European starlings, and black-bellied plovers) may have different plumage patterns and bill colors in summer and winter.
- Some species are present in an area all year, others only in winter or in summer, and others only in migration (spring, fall).

WCP require basic training to identify, in all plumages, most hazardous birds, and other rarer species but considered hazardous when present. Table 9.1 lists the relative hazard of various species groups based on the percent of reported strikes that cause damage or an effect-on-flight.

There are bird identification guides available in hard or soft copies books that provide useful life history information and vocalizations.

9.10.2 Mammals Identification

There are few mammal species of importance on an airport. Training WCP to identify mammals includes:

- Identification by sight all common large and mid-sized mammals (e.g., deer, raccoons, woodchucks, coyotes) that live around the airport
- Identification by signs (e.g., trails in grass, burrows, and fecal material) which will indicate the population eruption of field rodents, such as voles, deer mice, or rats
- A survey by a biologist using snap traps will identify the species and relative abundance of rodents in various airport habitats

9.10.3 Basic Life Histories and Behavior of Common Species

WCP should have some understanding of the biology and behavior of these species. For example, training shall identify:

- Whether birds are present year-round or only in summer, in winter, or during migration
- For locally breeding bird species when they nest and when young are fledged from nests
- The daily movement patterns between roosting, feeding, and loafing areas for birds in the airport
- Preferred food for birds
- Influence of weather on the presence and behavior of various wildlife species on the airport
- Reaction of wildlife to approaching aircraft and to various repellent devices

Information on the geographic range, feeding habits, and habitat preferences for each wildlife species are given in [46–48]. They provide concise summaries of life history information (nesting, feeding, habitat preferences) for most birds in North America.

9.10.4 Wildlife and Environmental Laws and Regulations

All WCP should be familiar with the following laws, permits, and records [1]:

- Federal and state laws protecting wildlife and regulating the issuance of permits to take (capture or kill) individuals causing problems
- Environmental laws and regulations regarding pesticide applications, drainage of wetlands, and endangered species must be considered in implementing WHMPs
- Federal Migratory Bird Treaty Act (MBTA), whereby almost all native migratory birds are protected regardless of their abundance
- Federal and often state permits must be issued before protected species can be taken on an airport
- State regulations regarding permits for activities involving removal (killing or trapping/relocating) of mammals
- State protection for non-native birds, such as pigeons, house sparrows, and starlings, and gallinaceous game birds, such as turkeys, grouse, and pheasants
- A clear understanding of which species have no legal protection, and, for all others, the species and numbers allowed to be taken under permits issued
- The methods of removal allowed and acceptable procedures for disposing of removed wildlife. Detailed records must be maintained of wildlife taken under permit.

9.10.5 Wildlife Control Techniques

WCP will need training to deploy the following techniques safely and effectively:

- Firearms
- Pyrotechnic
- Pesticide application
- Distress call tapes, propane cannons, and miscellaneous techniques

9.10.6 Record Keeping and Strike Reporting

WHMP develops a system to

1. Document the daily activities of WCP
2. Log information about wildlife numbers and behavior on the airport
3. Record all wildlife strikes with aircraft

This information records the effort made by the airport in reducing wildlife hazards. It is also extremely useful during periodic evaluations of the WHMP and when revisions to the plan are proposed. Figure 9.14 presents an example for the daily log of wildlife control activities.

Figure 9.15 illustrates a compilation of monthly statistics and a summary for wildlife control activities in the airport.

Figure 9.16 provides the annual summary of wildlife control activities derived from monthly reports from monthly activities listed in Figure 9.15.

9.10.7 Sources of Training

It includes:

- Books, manuals, and videos *handling* Field Guides – Birds, Field Guides – Mammals and Life Histories [1]
- Workshops on airport wildlife control offered by the USDA/WS or other entities
- World Bird strike Association (WBA)
- Hunter safety and firearms courses

Example of a daily log of wildlife control activities.

Airport					Month	Year	
		Location (Grid)	Wildlife		Control method	Results/comments	Initials
Date	Time		Species	No.			

FIGURE 9.14 Example of a daily log of wildlife control activities [1].

Courtesy: USDA.

Example of a form to provide monthly summary of wildlife control activities.

Airport			Month	Year
Control activity (modify list as appropriate)	This month	Same month last year	Comments (list wildlife dispersed or removed by species and method)	
No. of pyrotechnics fired				
No. of times distress calls deployed				
No. of runway sweeps to clear birds				
No. of wildlife removed				
Miles driven by wildlife patrol				
No. of reported strikes				
No. of reported strikes with damage				
No. of carcasses found (no strike reported)				

Summary paragraph of other wildlife control activities:

FIGURE 9.15 Example for monthly report [1].

Courtesy: USDA.

9.11 BIRD AVOIDANCE

Bird avoidance represents one of the expensive but very fruitful in wildlife hazard management. It includes two methods:

- Avian radars: which can complement other management practices (e.g., habitat modification) to reduce the risk of bird collisions with aircraft [14]
- Optical methods: which provide not only target detection but also support target identification [49]

Airport				Year					
Number of:									
Month	Pyro-technics fired	Times distress calls deployed	Runway sweeps to clear birds	Wildlife dispersed	Wildlife removed[1]	Miles driven by wildlife patrol	Reported strikes[2]	Reported strikes with damage	Carcasses found (no strike reported)[2]
Jan									
Feb									
Mar									
Apr									
May									
Jun									
Jul									
Aug									
Sep									
Oct									
Nov									
Dec									
Total									

[1]Provide separate list by species and method.
[2]Provide separate list by species.

FIGURE 9.16 Example for annually report [1].

Courtesy: USDA.

9.11.1 Avian Radars

In the late 1990s, radar companies introduced commercially avian radar systems to the market. They identify any sudden bird hazards that are not detected by traditional bird hazard management methods.

However, avian radars are only available in several military bases as well as some hub airports like JFK (in New York City), Chicago O'Hare, Logan/Boston in the United States, Vancouver International Airport in Canada, Schiphol in the Netherlands, Haneda in Japan, Taoyuan in Taiwan, as well as some airports in other countries.

The main function of avian radar is to detect birds at or close to airports [50]. Radar information is transferred to both the aircraft cockpit and the airport operation center to avoid collisions with a large flock of birds.

FIGURE 9.17 Avian radar operation.

Courtesy: US Department of Transportation, FAA [51].

Any avian radar system is composed of four modules [51]: namely, the radar unit, the scanning unit/antenna, the digital radar signal processor, and the visual display. The avian radar system generates either an electromagnetic energy wave (pulse) or a radio signal. This radio will be transmitted through an antenna. A part of this signal is reflected from surrounding objects and returns to the system [51], as illustrated in Figure 9.17.

The radio signal generated by the radar must produce an echo different from background noise which is characterized by what is called the radar cross section (RCS). Figure 9.18 illustrates the measured RCS for a crow.

The antenna is rotated either mechanically or electrically, in either a vertical or a horizontal direction, as needed. There are two general types of antennas: namely, the slotted array antenna and the parabolic dish antenna. Data from the radar system are transferred to several airports authorities and departments.

Avian radar systems provide the following capabilities:

- Day and night monitoring of any bird movements
- Permanent automatic recording of all targets detected and tracked
- Developing hourly, daily, weekly, monthly, and seasonal summaries of bird activities
- Extending surveillance areas (both in distance and altitude).

Avian radar systems yield the following benefits to airports:

- They transmit radar images to airport authorities and thus help managers to have quick responses to bird threats
- They add any unusual bird activities to the update of information in airport ATIS announcements or NOTAMs

FIGURE 9.18 The measured RCS for a crow.

Courtesy: US Department of Transportation, FAA [51].

- Developing an archive for information to be used in the airport or shared with other airports (based on 37 AC 150/5220-25 11/23/10)
- Identifying any activities of birds (time, movement origin, routes, and destination)

Avian radar may be fixed or small mobile devices (Figure 9.19). The famous types of radar are Robin, Merlin, and Accipiter [52].

9.11.2 Optical Systems

Optical systems include video, image intensification, and thermal systems. Video cameras provide target detection and support target identification. A single camera may be mounted on rigid towers [49, 53] (Figures 8.8 and 8.12), while multiple cameras may also be fixed along runways. Single camera is installed in several

FIGURE 9.19 A mobile avian radar.

Courtesy: FAA [49, 54].

airports, including Chicago O'Hare (United States), Heathrow (UK), and Vancouver (Canada) (Figure 9.19).

Multiple cameras are installed in Chicago O'Hare (Figure 8.13). Magnification can reveal the detail needed for identification [49, 54] (Figure 9.20). An infrared camera uses an illuminator to capture images. A combination of radar and a set of cameras is used to monitor bird movements close to [55] and far away.

9.12 COVID-19: AIRPORT WILDLIFE HAZARD MANAGEMENT IN A TIME OF REDUCED OPERATIONS

During the COVID-19 pandemic (since March 2019 until now), many airports are partially closed, and most airlines suspended their flights and grounded 30%–85% of their fleet [56]. Air traffic is greatly avoided due to the crowds of passengers in a closed environment of airports. Some airliners take off with few passengers. This calmness in airports and staff activities may provide the best environment for bird life and breeding, where shelter, food, and water are available. Thus, wildlife approaches the area, makes nests, and creates permanent homes on the airport runway. In this case, a serious threat to aviation safety is encountered and should be avoided by continuous dispersal activities. The question arising is: What are the implications for an airport WHMP [57]? The situation varies from "no change at all" with the Wildlife Risk Management Plan

FIGURE 9.20 Bird magnification.

Courtesy: FAA [49, 54].

continuing as before through to a "full stop" of all actions for a period of time [58]. The optimum performance is to continue WHMP. Thus, the movement area is continuously checked for birds, even if there is closed runways. It is important not to provide birds with a chance to settle on and around the aircraft. Also, bird controllers should work day and night, and do not reduce their capacity during the COVID-19 outbreak. This was exactly followed in Schiphol airport. However, in many airports, due to financial constraints, airports reduced staff and curtail wildlife management activities such as wildlife patrols, mowing, equipment purchases, and so on.

As suggested in [56], not all safety aspects of a WHMP (14 CFR Part 139.337) may be applied. A priority would be FAA 14 CFR Part 139.337(a), "all certificated airports MUST take immediate action to alleviate wildlife hazards whenever they are detected." This would seem a higher priority than ensuring all bird surveys are conducted in a timely manner.

9.13 METHODS TO IDENTIFY BIRD STRIKE REMAINS

9.13.1 Introduction

It is an easy mission to identify large mammals such as deer and coyotes striking an aircraft. But, it needs careful examination to identify which bird has struck an aircraft from tiny bits of "snarge" (wildlife remains) containing, for example, bat hair and fragments.

Specialized people from the following institutions or research labs can provide identification of wildlife:

- Biological Survey Unit of the USGS Patuxent Wildlife Research Center
- Smithsonian Institution's Feather Identification Lab
- Feather Identification Lab and the Air Force Safety Automated System (AFSAS)

They did for many years by providing identification of fragmentary hair, bones, and claws from mammals involved in wildlife strikes. A comparison between the remains with the material in museum reference collections or by examining microscopic hair or feather fragments, the identity of the wildlife becomes clear.

9.13.2 Collecting Bird Strike Material

Following a collision between a bird and an aircraft, small feather fragments and blood stains are usually the only evidence that a bird was involved in the incident. The remains of strike may be feathers, and tissue/blood snarge, either as dry or fresh material.

9.13.2.1 Feathers

Three cases are encountered:

- A whole bird is available, then pluck a variety of feathers (breast, back, wing, tail)
- A partial bird is found, collect a variety of feathers that are colored or patterned
- Feathers are only available, send all available material

The following instructions must be followed [59]:

- Do not cut feathers from the bird, as the downy part at the base of the feathers is needed
- Do not use any sticky substance, so do not use tape or glue
- Include a copy of the report (AFSAS, WESS, or FAA 5200-7)
- Secure all remains in a resealable plastic bag.

9.13.2.2 Tissue/Blood ("Snarge")

The word "snarge" means the dry or fresh residue smeared on an aircraft after a bird/aircraft collision. For dry material, use one of the following methods [59]:

- Scrape or wipe off the material into a clean reclosable bag
- Wipe the area with a prepackaged alcohol wipe
- Spray the area with alcohol to loosen material, then wipe with clean cloth/gauze

Note: do not use water, bleach, or other cleansers as they destroy DNA.
For fresh material, use either of the following methods:

- Wipe the area with an alcohol wipe and clean cloth/gauze
- Apply fresh tissue/blood to an FTA® DNA collecting card

9.13.3 Reporting and Shipping

When reporting and shipping bird strike remains, there are several guidelines for collecting and submitting feather or other bird/wildlife remains for species identification [9]. The guidelines listed below will help to maintain species identification accuracy, reduce turnaround time, and maintain a comprehensive database:

1. Collect and submit remains as soon as possible
2. Provide a completed report of the FAA 5200-7
3. Collect as much material as possible
4. Place the material in clean plastic/zip-lock bags
5. Mail the report and material to the Smithsonian Institution.

Feather identification is free of charge to all US aircraft owners/operators at all US airport operators, and to any foreign air carrier if the strike happened at a US airport.

The turnaround time for species identification is usually 24 h from receipt. The reports and species identification information are then sent to the FAA, Office of Airport Safety and Standards, and USDA, NWRC, in Sandusky, Ohio.

9.13.4 Methods Used to Identify Bird Strike Remains

The method used to identify bird strike remains depends on what kind of material is available.

9.13.4.1 Examination by Eye

Eye examination of bird is a macroscopic method that depends on the skill of the examiner and the condition of the feather remains. Eye examination was employed in Europe before 1978 [60]. Experienced ornithologists in Smithsonian Institution, Feather Identification Lab, examine feathers by eye to determine the species or group involved and compare it with specimens in a museum collection.

Figure 9.21 illustrates the procedure for identifying the feather remains of a bird strike by comparing tail feathers with a museum specimen of a killdeer (*Charadrius vociferus*) [61].

9.13.4.2 Microscopic Examination

Identification of samples of small feather fragments, blood, and tissue involves examination under a microscope [61]. This technique will identify the family or genus of bird involved, but usually does not provide a species identification. Usually, preparation of the sample for microscopic analysis is first done before sandwiching between the microscope slide and a coverslip. Special preparation is followed for very dirty or greasy samples.

The downy part of a feather is unique for different groups of birds (e.g., duck, raptor, or passerine). Figure 9.22 illustrates the microscopic structure of a meadowlark feather, which is a type of many passerines.

Bloodstains can be used in microscopic identification [62]. Though avian blood/tissue from aircraft is of paramount importance, simple laboratory procedures are full of discrepancies. Other methods, like chemical examination of blood by the benzidine/orthotoludine test under simulated air crash situations, are more useful [62].

FIGURE 9.21 Matching an unknown tail feather with a museum specimen of a killdeer.

Courtesy: US Department of Transportation, FAA [63].

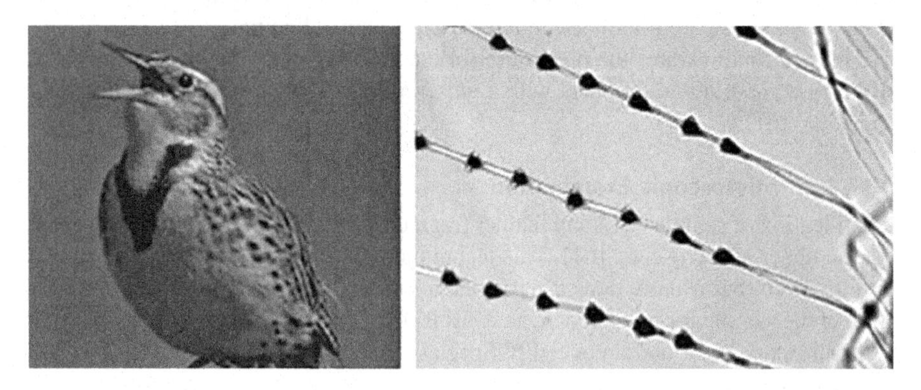

FIGURE 9.22 The microscopic structure of meadowlark feather.

Courtesy: US Department of Transportation, FAA [63].

9.13.4.3 Keratin Electrophoresis

Identification of feathers by visual means is much elementary, particularly at the lower taxonomic levels. Optical microscopy and scanning electron microscopy improve results; however, nearly 25% of the samples cannot be identified below the family level.

Electrophoresis of proteins extracted from feather keratin can provide reliable identification of feather remnants from any source. However, the sample must be at least 10 mg for identifications to the species level [62]. In keratin electrophoresis, feather proteins from an unknown sample are compared with samples from known specimens stored in databases. Their results are more processible in identification and exceeds the results obtained by other means for samples that cannot be identified visually.

9.13.4.4 DNA Analysis

Small amounts of blood or tissue may be used for DNA analysis, providing very accurate results (97%–99% efficiency). However, DNA analysis may cost US$ 15,000 for each sample.

DNA barcoding is the newest tool available for bird strikes. It involves extracting the mtDNA for the cytochrome c oxidase subunit 1, COI or COX1 gene, known as the "barcode gene." This is extracted from bird strike samples that consist of blood and tissue and then the unknown sequence is matched to a DNA library available on the Barcode of Life Database (BoLD) [56, 63].

REFERENCES

[1] Improving Aviation Safety: Questions and Answers, USDA, www.aphis.usda.gov/wildlife_damage/airline_safety/pdfs/Aviation%20Program%20FAQs.pdf

[2] McKee, J. and Phillip, S. Arie Dekker, and Kylie Patrick, Approaches to Wildlife Management in Aviation, Chapter 22: Problematic Wildlife, a Cross-Disciplinary Approach, Francesco M. Angelici (Editor), Springer, 2016.

[3] Dolbeer, R.A., Begier, M.J., Miller, P.R., Weller, J.R., and Anderson, A.L. Wildlife strike to Civil Aircraft in United States 1990-2019, FAA, National Wildlife Strike database, Report no. 26, Feb 2021.

[4] Managing Wildlife for Airport Operators, The FOD Control Corporation, www.fodcontrol.com/managing-wildlife-airport-operators/

[5] Aerospace Industries Association: Flight Test Operations Group, Foreign Object Damage/Foreign Object Debris (FOD) Prevention, National Aerospace Standard (NAS 412), Washington, DC: Aerospace Industries Association of America, Inc. (1997).

[6] DeVault, T.L., Blackwell, B.F., Belant, J.L., and Begier, M.J. Wildlife at Airports: Wildlife Damage Management. Technical Series, US Department of Agriculture Animal & Plant Health Inspection Service Wildlife Services (2017).

[7] Hazardous Wildlife Attractants on or near Airports, FAA AC 150/5200-33C, February 21, 2020.

[8] Cleary, E.C and Dolbeer, R.A. Wildlife Hazard Management at Airports: A Manual for Airport Personnel, FAA, US Department of Agriculture, Second Edition, July 2005.

[9] FAA AC 150/5300-13, Airport Design, Appendix 19.
[10] Hygnstrom, S.C., Timm, R.M., and Larson, G.E. editors. *Prevention and control of wildlife damage*. University of Nebraska Cooperative Extension Division, Lincoln, Nebraska. (2-volume manual) (1994).
[11] Desoky, A.E.A.S.S. A review of bird control methods at airports. Global Journal of Science Frontier Research: E Interdisciplinary 14(2), 41–50, 2014.
[12] Dolbeer, R.A., Begier, M.J., Miller, P.R., Weller, J. R., and Anderson, A.L. Wildlife strikes to civil aircraft in the United States, 1990–2019. U.S. Department of Transportation, Federal Aviation Administration, Office of Airport Safety and Standards, Serial Report Number 26, 2021.
[13] Dolbeer, R.A., Seamans, T.W., Blackwell, B.F., and Belant, J.L. Anthraquinone formulation (Flight Control™) shows promise as avian feeding repellent. The Journal of Wildlife Management, 62(4), 1558–1564, 1988.
[14] Blokpoel, H. *Bird Hazards to Aircraft*, 236. 1974, Ottawa, Ontario, Canada: Canadian Wildlife Service. Ministry of Supply and Services.
[15] Bird/Wildlife Aircraft Strike Hazard (BASH) Management Techniques, Air Force Pamphlet 91–212. February 1, 2004.
[16] Bishop, J., McKay, H., Parrott, D., and Allan, J. (2003). Review of international research literature regarding the effectiveness of auditory bird scaring techniques and potential alternatives. www.researchgate.net/publication/242454383_Review_of_international_research_literature_regarding_the_effectiveness_of_auditory_bird_scaring_techniques_and_potential_alternatives (accessed May 9, 2021).
[17] DeVault, T.L., Blackwell, B.F., Belant, J.L., and Begier, M.J. Wildlife at Airports: Wildlife Damage Management. Technical Series, US Department of Agriculture Animal & Plant Health Inspection Service Wildlife Services (2017).
[18] Bird/Wildlife Aircraft Strike Hazard (BASH) Management Techniques, Air Force Pamphlet 91–212. February 1, 2004.
[19] Wind Powered Flashing Hawkeye Bird Scarer. www.scaringbirds.com/windpowered-scarers/flashing-hawkeye-ground-mounted (accessed May 10, 2021).
[20] Blackwell, B.F., Bernhardt, G.E., and Dolbeer, R.A. Lasers as non-lethal avian repellents. Journal of Wildlife Management, 66(1), 250–258, 2002.
[21] Precise Flight. www.preciseflight.com/commercial (accessed May 9, 2021).
[22] Harris, R.E. and Davis, R.A. Evaluation of the efficacy of products and techniques for Airport Bird Control. LGL Report TA2193. LGL Limited, Environmental Research Associates (1998).
[23] Eagle Helicopter. www.pinterest.ie/pin/290200769713828754/ (accessed May 9, 2021).
[24] O'Callaghan J. (2014). Is it a bird? Is it a plane? No, it's Robird: Robotic falcons and eagles mimic predators to keep pests away from airports and farms. www.dailymail. co.uk/sciencetech/article-2743272/Is-bird-Is-plane-No-s-ROBIRD-Robotic-falconseagles-mimic-real-predators-pests-away-airports-farms.html#ixzz4X4HvwaLs (accessed May 9, 2021).
[25] International Birdstrike Committee (2016). Recommended Practices No. 1: Standards For Aerodrome Bird/Wildlife Control. www.int-birdstrike.org/Standards_for_Aerodrome_bird_wildlife%20control.pdf (accessed May 10, 2021).
[26] Battistoni, V., Montemaggiori, A., and Iori, P. Beyond falconry between tradition and modernity: a new device for bird strike hazard prevention at airports. In: *Proceedings of International Bird Strike Committee, IBSC Meeting, and Seminario Internacional Perigo Aviario e Fauna*, 1–13. Brasilia (2008).

[27] Watermann, U. Experimental falconry program to reduce the gull strike hazard to aircraft at John F. Kennedy International Airport, New York. Report prepared for Port Authority of New York and New Jersey by Bird Control International Inc., Georgetown, Ontario, Canada (1997).

[28] Blokpoel, H. Bird hazards to aircraft: problems and prevention of bird/aircraft collisions. Clarke, Irwin, Ottawa, Ontario, Canada (1976).

[29] Bird/Wildlife Aircraft Strike Hazard (BASH) Management Techniques, Air Force Pamphlet 91–212. February 1, 2004.

[30] Can a Paintball Gun Kill a Rat, Bird, Cat, Squirrel or Rabbit – Complete Guide https://blasterexpert.com/can-a-paintball-gun-kill/#:~:text=with%20one%20shot.-,Can%20Paintball%20Gun%20Kill%20a%20Bird%3F,wings%2C%20legs%20or%20internal%20injuries.

[31] Cleary, E.C. and Dickey, A. Guidebook for Addressing Aircraft/Wildlife Hazards at General Aviation Airports, ACRP 32, 2010.

[32] DeFusco, R.P. and Unangst, E.T. Airport Wildlife Population Management: A Synthesis of Airport Practice. Airport Cooperative Research Program, ACRP Synthesis Volume 39 (2013).

[33] Bird Detection System. NEC Corporation, Air Transportation Solutions Division, Tokyo, Japan. www.nec.com/en/global/solutions/bird/common/pdf/bird_detection_system.pdf (accessed May 10, 2021).

[34] Alfred J.G. "BIRDS AT AIRPORTS" (1994). The Handbook: Prevention and Control of Wildlife Damage. 56. https://digitalcommons.unl.edu/icwdmhandbook/56

[35] Stock, S., Villlarreal, M., and Nious, K. Birds Shot Daily by Bay Area Airport Workers: Over 3,000 birds shot in the last two years, NBC, May 21, 2013, Updated September 4, 2015, www.nbcbayarea.com/news/local/birds-shot-by-bay-area-airports/1942485/

[36] Nearly 70,000 birds killed in New York in attempt to clear safer path for planes, the Guardian, January 14, 2017 www.theguardian.com/world/2017/jan/14/new-york-birds-killed-airport-miracle-on-hudson-sully

[37] Bird Harassment, Repellent, and Deterrent Techniques for Use on and Near Airports, ACRP 23, 2011.

[38] International Birdstrike Committee: Recommended Practices No. 1 Standards For Aerodrome Bird/Wildlife Control, Issue 1 – October 2006.

[39] Protocol for the Conduct and Review of Wildlife Hazard Site Visits, Wildlife Hazard Assessments, and Wildlife Hazard Management Plans, FAA AC150/5200-38, 8/20/2018.

[40] Cummings, J. Geese, Ducks and Coots, Wildlife Damage Management Technical Series, USDA, www.aphis.usda.gov/wildlife_damage/reports/Wildlife%20Damage%20Management%20Technical%20Series/GeeseDucksCoots-WDM-Technical-Series.pdf (accessed May 16, 2021).

[41] Avery, M.L. and Lowney, M. Vultures, Wildlife Damage Management Technical Series, USDA, www.aphis.usda.gov/wildlife_damage/reports/Wildlife%20Damage%20Management%20Technical%20Series/Vultures.pdf (accessed May 16, 2021)

[42] Jeb Barzen Founder and Ken Ballinger, Sandhill and Whooping Cranes, Wildlife Damage Management Technical Series, USDA, www.aphis.usda.gov/wildlife_damage/reports/Wildlife%20Damage%20Management%20Technical%20Series/Cranes-WDM-Technical-Series.pdf (accessed May 16, 2021)

[43] Dorr, B.S., Sullivan, K.L., Curtis, P.D., Chipman, R.B., and McCullough, R.D. Double-crested Cormorants, Wildlife Damage Management Technical Series, USDA, www.aphis.usda.gov/wildlife_damage/reports/Wildlife%20Damage%20Management%20Technical%20Series/Cormorants-WDM-Technical-Series.pdf (accessed May 16, 2021).

[44] Hoy, M.D. Herons and Egrets, Wildlife Damage Management Technical Series, USDA, www.aphis.usda.gov/wildlife_damage/reports/Wildlife%20Damage%20Management%20Technical%20Series/Herons-and-Egrets-WDM-Technical-Series.pdf (accessed May 16, 2021).

[45] Luna, M. "Joint Base McGuire-Dix-Lakehurst deer management for aviation safety": Managing white-tailed deer populations on military installations to minimize the risk of deer strikes. Proceedings of the North American Bird Strike Conference 17:139–156. Halifax, Nova Scotia, 2019, https://canadianbirdstrike.ca/wp-content/uploads/2020/03/Luna_2019.pdf

[46] Alsop, F.J., III. *Birds of North America, Eastern Region, Western Region* (752 pages). DK Publishing, Inc., New York (2001).

[47] Ehrlich, P.R., Dobkin, D.S., and Wheye, D. *The birder's handbook: a field guide to the natural history of North American birds, including all species that regularly breed north of Mexico.* Simon and Schuster, New York (1988).

[48] Sibley, D.A. *The Sibley guide to bird life and behavior.* Alfred A. Knopf, New York (New York).

[49] Weller, J.R. (2014). FOD Detection System: Evaluation, Performance Assessment and Regulatory Guidance. Wildlife and Foreign Object Debris (FOD) Workshop, Cairo, Egypt, March 24–26, 2014. www.icao.int/MID/Documents/2014/Wildlife%20and%20FOD%20Workshop/Assessing%20Risk%20FAA.pdf (accessed May 10, 2021).

[50] Air Line Pilots Association, International (2009). Wildlife Hazard Mitigation Strategies for Pilots. www.alpa.org/-/media/ALPA/Files/pdfs/news-events/white-papers/wildlife-hazard.pdf?la=en (accessed May 10, 2021).

[51] US Department of Transportation (2010). Airport Avian Radar Systems, Federal Aviation Administration, Advisory Circular, AC No: 150/5220-25. www.faa.gov/documentLibrary/media/Advisory_Circular/AC_150_5220-25.pdf (accessed May 10, 2021).

[52] Avian Radar: Does it Work? United States Department of Agriculture, Animal and Plant Health Inspection Service. www.aphis.usda.gov/aphis/ourfocus/wildlifedamage/programs/nwrc/sa_spotlight/avian+radar+does+it+work (accessed May 10, 2021).

[53] Wildlife Strikes to Civil Aircraft in the United States 1990–2014. FAA National Wildlife Strike Database: Serial Report Number 21. www.fwspubs.org/doi/suppl/10.3996/022017-JFWM-019/suppl_file/10.3996022017-jfwm-019.s8.pdf (accessed May 12, 2021).

[54] Tarsier. www.tarsier.qinetiq.com/solution/Pages/tarsier.aspx#!prettyPhoto/4 (accessed May 13, 2021).

[55] Bird Detection System. NEC Corporation, Air Transportation Solutions Division, Tokyo, Japan. www.nec.com/en/global/solutions/bird/common/pdf/bird detection_system.pdf (accessed May 14, 2021).

[56] Gaikwad, S., Munot, H., and Shouche, Y. Utility of DNA barcoding for identification of bird-strike samples from India. *Current Science*, 110(1), 10, 2016 www.researchgate.net/publication/309634391_Utility_of_DNA_barcoding_for_identification_of_bird-strike_samples_from_India (accessed January 15, 2019).

[57] Brom, T.G. Microscopic identification of feathers in order to improve birdstrike statistics. Proc. Conf. Wildlife Hazards to Aircraft (Charleston, SC), report no. DOT/FAA/AAS/84-1: pp. 107–120, 1984.

[58] Franklin, J. Wildlife Hazard Management, April 14, 2021, www.easa.europa.eu/community/topics/wildlife-hazard-management (accessed May 17, 2021).

[59] Dove, C.J., Dahlan, N.F., and Heacker, M. Forensic bird-strike identification techniques used in an accident investigation at Wiley Post Airport, Oklahoma, 2008. *Human-Wildlife Interactions* 3(2): article 6, 2009. https://digitalcommons.usu.edu/hwi/vol3/iss2/6/ (accessed May 14, 2021).

[60] Brom, T.G. Microscopic identification of feathers in order to improve birdstrike statistics. Proc. Conf. Wildlife Hazards to Aircraft (Charleston, SC), report no. DOT/FAA/AAS/84-1: pp. 107–120, 1984.

[61] Smithsonian Institution, Feather Identification Lab, www.faa.gov/airports/airport_safety/wildlife/smithsonian/ (accessed May 17, 2021).

[62] Ouellet, H. and van Zyll de Jong, S.A. (1990). Feather Identification by Means of Keratin Protein Electrophoresis. Bird Strike Committee Europe, Working Papers, 20th Meeting, Helsinki. www.int-birdstrike.org/Helsinki_Papers/IBSC20%20WP8.pdf (accessed 15 May 2021).

[63] Dove, C.J., Heacker, M., and Rotzel, N. (2007). The Birdstrike Identification Program at the Smithsonian Institution and New Recommendations for DNA Sampling. 2007 Bird Strike Committee USA/Canada, 9th Annual Meeting, Kingston, Ontario. www.semanticscholar.org/paper/the-birdstrike-identification-program-at-the-and-dove-Heacker/4f3e6163331a6020518f18bb6af3cf678f3d4b5f?p2df b (accessed May 12, 2021).

10 Numerical Studies for the Interaction of Solid and Liquid Debris with Aircraft Modules

10.1 INTRODUCTION

Fixed-wing, as well as rotary-wing aircraft, may encounter particulate-laden flows during ground and airborne operation. These include solid particles, liquid particles, runway gravel, rain, hailstones, ice crystals, and chunks. Thus, the airframe and engines of aircraft and helicopters are susceptible to many physical problems such as erosion, corrosion, fouling, built-up dirt, foreign object damage, and icing. These problems will influence the aerodynamics (lift and drag), weight of aircraft, increase the fuel consumption, may decrease aircraft thrust force or power, which will influence aircraft safety especially during takeoff and climb.

Moreover, it may have drastic impacts on aircraft engines like changes in air/gas path boundaries of its components. Such changes will result in deterioration of the engine performance parameters, including thrust force, specific fuel consumption, and overall efficiency. It may also cause engine surging, excessive heating, or both. Due to the airframe/engine compatibility, the engine deterioration may lead to the aircraft failure to satisfy the required operating conditions at different phases of their flight envelop. Condition monitoring techniques are applied for airframe and engine diagnosis, and hence suitable maintenance corrective actions are carried out. Both solid and liquid particulate flows resemble typical two-phase flow phenomena. Extensive numerical analyses are devoted to these problems. There are two numerical methods applied: namely, Eulerian and Lagrangian ones. Commercial codes including FENSAP-ICE, CANICE, CIRAAMIL FLUENT, and CFX are used.

The first part of this chapter is devoted to solid particles or debris including sand, dust, and volcanic ash, which are often referred to as calcia–magnesia–alumina–silica. Such solid particles may cause erosion and fouling of aeroengines. Erosion mostly affects the cold section of aeroengines (propeller, intake, fan, and compressors, as well as the main and tail rotors of helicopters), while fouling mostly influences the hot section of aeroengines (combustion chamber and turbines). However, in some cases, turbines may be eroded, and compressors may be fouled.

DOI: 10.1201/9781003133087-10

Erosion analysis is carried out in three steps: namely, calculation of air/gas flow, next tracing of solid particle trajectories inside or around the aircraft/engine module, and finally estimation of erosion rates and the encountered wear limits of different modules. Reduced lifetime of such modules may be identified based on the structural repair manual of the manufacturer and the calculated rate of wear in such studies. Air/gas flow analysis around or within aircraft module is calculated by solving Navier–Stokes (N-S) or Reynolds-averaged Navier–Stokes (RANS) equations with one or two-equation modeling turbulence. Grid is generated at first, and the initial and boundary conditions for air/gas are stated. Newton's second law with appropriate forces is solved using numerical integration techniques like Runge–Kutta. The cloud particles are identified first based on material and diameter distribution. Particles' trajectories are then traced and when collide with the internal surface, empirical formulae are used for defining postimpact velocities based on the pre-impact values. Erosion (the rate of target material removal at particle's impact location) is then calculated using another set of empirical relations based on particle properties and impacted target material. The wear rates at each element of the surface of the impacted target are summed up, and the critical target areas are defined. Finally, the global effects of erosion on the performance of these modules and its lifetime are estimated.

The above procedure is applied for the intake [1–4], and fan of high bypass turbofan engines [5] like GE CF-6, the two types of compressors (axial [6] and centrifugal [7–10]), the axial [11, 12] and radial turbines [13–19, 20, 21], the propeller of turboprop engines like that installed on C-130 aircraft [21–25]. The erosion of piston engines powering small-fixed and rotary-wing aircraft is investigated for the first time including their pistons, valves, and cylinder walls [26, 27]. The role of secondary flow in modifying the erosion pattern particularly close to the hub sections is also investigated [28–33]. The erosion of the rotor of helicopters is also reviewed in detail [34–37].

Volcanic ash is a critical issue for aviation safety. It may cause erosion or fouling of the compressors or turbines of aeroengines. It influences their performance, and engine reliability, service life, and even may result in in-flight failure. Fouling of compressors is a reversible process treated by water washing either online or off-line methods. Turbine fouling is an irreversible process since particles crossing the combustion chambers may melt, or at least soften, and if they strike solid surfaces of combustor or turbine blades, they will adhere to them on impact. Such deposited particles may increase the blade surface roughness, change their shape, and even block the film-cooling holes [38–40]. Numerical modeling of fouling of both compressors and turbines is described, and several case studies are presented.

The second part of the chapter will focus on liquid particles including rain and ice. The world's global warming has its impact on ice, frost, heavy rain, fog, typhoon, tornado, thunderstorm, lighting, etc. [41]. Rainfall and rainstorm have been a concern for aircraft and missiles since World War II. Experimental investigation of rain effect on aircraft flight has started in 1941 [42], which continued in many institutions. Numerical simulation of the rain phenomenon is treated in [43–46]. Ice accretion is encountered when an aircraft flies through clouds below 26,000 ft at subsonic speeds. It may occur on most aircraft components like radome, wings, empennages, engine intake, propellers, main and tail rotors of helicopters. Ice accretion results from small

(5 to 50μm) supercooled droplets (droplets cooled below freezing) which can freeze upon impact with the aircraft surface. Icing protection systems are designed to prevent ice from forming on critical surfaces listed before. If the ice protection system fails to function or if a human error occurs during operation, a hazardous flight condition may result [47].

10.2 SOLID PARTICLE EROSION (SPE)

Solid particle erosion (SPE) is very common for both fixed- and rotary-wing aircraft [1]. Computational fluid dynamics (CFD) is used for erosion prediction. Solid particulate flow in aviation incorporates a continuous carrier phase (air or gas) and the dispersed phase, the solid particles. Fixed-wing aircraft are vulnerable to erosion during both ground and airborne flight phases. On ground operation from the paved or unpaved runway, solid particles may strike the different modules of aircraft and or be ingested into its engines or other openings. During the engine's high-power setting at takeoff, a ground vortex (GV) is developed which has the capability of picking up particles from the ground and ingesting them into the engine [48–55]. Thrust reversal can also blow solid particles into the engine intake during landing [56]. During airborne operations, the aircraft can be impacted by solid particles, including dust, sandstorms, and volcanic ash.

For helicopters hovering close to the ground, its main rotor will generate severe dust clouds around them due to its downwash. This cloud of particles will enter the engine. Moreover, helicopters normally fly at low altitudes and thus will encounter dust and small diameter fly ashes. Most helicopters are fitted with particle separators that can remove large particles of the damaging fly particles. However, significant amounts of small particles ranging in size between 1 and 20 *microns* still pass through and enter the engine. These particles do erode the compressor blades' surfaces and result in an unacceptably short operating life.

The three steps for estimating the erosion damage of any components or module of aircraft are air/gas flow analysis, tracing particle trajectories, and calculating the erosion rates at critical areas frequently impacted by particles. These will be described hereafter in detail.

10.2.1 Classification of Particulate Flow (Dense and Dilute)

The order of coupling between the dispersed and continuous phase is determined by the volume fraction α_p of the solid material,

$$\alpha_p = \frac{V_p}{V} = \frac{V_p}{V_p + V_g} \tag{10.1}$$

Three cases for the volume fraction [57] are considered as defined hereafter:

- Highly diluted flow $(\alpha_p \le 10^{-6})$ is a one-way coupling case, where the particles are influenced by the suspending air/gas, but it does not affect the suspending media

- For a volume fraction in the range ($10^{-6} \leq \alpha_p \leq 10^{-3}$), a two-way coupling case is assumed where particles are influenced by the suspending fluid and at the same time affect the flow
- As the volume fraction exceeds 10^{-3} (which is the case of dense flow), particle–particle interactions must be considered, and this case is referred to as four-way coupling.

10.2.2 Diluted Flow

Such multiphase flows are described by the Lagrangian–Eulerian approach, where the N-S equations are solved for the continuous carrier phase, and Lagrangian approach is used for the dispersed particle phase. Particles are idealized as "point-particles" (PP). The air/gas flow is solved using Reynolds-averaged N-S (RANS) equations adopting large eddy simulation (LES) or direct numerical simulations (DNS). Solution of particle-laden flow is denoted either PP-LES or PP-DNS [58].

Several commercial codes solvers including ANSYS FLUENT, ANSYS CFX, STAR CCM, and OpenFOAM are employed to solve these particulate flows [59, 60].

10.2.3 Governing Equations

10.2.3.1 Air/Gas Flow

The equations governing the air/gas flow are the conservation of mass, conservation of momentum, conservation of energy, and some other auxiliary equations.

The air/gas flow is assumed a compressible, turbulent Newtonian fluid which will be described by the RANS equations as follows.

10.2.3.1.1 Conservation of Mass

The mass conservation equation applied to a fluid passing through an infinitesimal, fixed control volume for steady flow in tensor notation as:

$$\frac{\partial \rho}{\partial t} + \frac{\partial}{\partial x_i}\left(\rho \overline{V_i} \right) = 0 \tag{10 .2}$$

Where:

$\overline{V_i}$: is the Reynolds-averaged flow velocity in the ith direction.
x_i: is the coordinate in the ith direction.
ρ: is the air density.

10.2.3.1.2 Conservation of Momentum

The conservation of momentum equation for steady and laminar flow can be written as:

$$\frac{\partial\left(\rho \overline{V_i}\right)}{\partial t} + \frac{\partial}{\partial x_j}\left(\rho \overline{V_i}\,\overline{V_j}\right) = -\frac{\partial \overline{p}}{\partial x_i} + \frac{\partial \tau_{ij}}{\partial x_j} + \frac{\partial}{\partial x_j}\left(-\rho \overline{V_i' V_j'}\right) + \overline{f_D} \tag{10.3}$$

Where $\bar{p}$ is the Reynolds-averaged flow pressure, and τ_{ij} is the Reynolds stress tensor given by:

$$\tau_{ij} = \mu\left[\frac{\partial \overline{V_i}}{\partial x_j} + \frac{\partial \overline{V_j}}{\partial x_i} - \frac{2}{3}\delta_{ij}\frac{\partial \overline{V_k}}{\partial x_k}\right] \tag{10.4}$$

$$\text{Where}: \delta_{ij} = \begin{cases} 1 \; if \; i = j \\ 0 \; if \; i \neq j \end{cases}$$

$\left(-\rho\overline{V_i'V_j'}\right)$ is the Reynolds stresses and $\overline{f_D}$ is the additional body force.

For highly diluted suspension, that is, one-way coupling, the effect of the dispersed phase on the fluid flow is negligible, thus $\overline{f_D} = 0$.

If the volume fraction of the dispersed phase increases, the influence of the dispersed phase on the fluid flow must be considered. A two-way coupling is assumed and thus $\overline{f_D} \neq 0$, which is identified here as a source term. It is the force of the particle that will be transferred to the continuous phase momentum in the time step Δt, and depends on the particle mass flow rate $\dot{m}_p$.

Furthermore, at high volume loadings [58], a four-way coupling must be enabled. Thus, the interactions between particles must be considered, at which point particle collisions are resolved within the Lagrange solver.

Several turbulence models are used, namely the standard k − ε model [61], Renormalization Group (RNG) k − ε model [62], the k − ω-SST model [63], the Spalart–Allmaras turbulence model is known to be one of the best turbulence models for the prediction of turbomachinery aerodynamics (high subsonic and transonic speeds) [64].

Where:

k is the turbulent kinetic energy $[J/kg]$
ε is the turbulent dissipation turbulent kinetic energy dissipation rate (m^2s^{-3})
ω is the specific turbulence dissipation rate (s^{-1})

10.2.3.1.3 Conservation of Energy

The steady-state conservation of energy equation is given by the following equation

$$\frac{\partial}{\partial x_i}\left[\overline{V_i}\left(\rho E + \bar{p}\right)\right] = \frac{\partial}{\partial x_i}\left[K_{eff}\frac{\partial \bar{T}}{\partial x_i} + \overline{V_j}\left(\tau_{ij}\right)_{eff}\right] \tag{10.5}$$

$$\text{Where} \quad K_{eff} = K + K_t \text{ and } K_t = \frac{\mu_t C_p}{P_{rt}} \tag{10.6}$$

E: is the total energy of the air
K_{eff} is the effective thermal conductivity

K: is the air thermal conductivity.
K_t: is the turbulent thermal conductivity
P_{rt}:is the turbulent Prandtl number
$\left(\tau_{ij}\right)_{eff}$: is the viscous heating

10.2.3.1.4 Auxiliary Equations

- Equation of State

$$p = \rho R T \tag{10.7}$$

The density of an ideal gas is computed through the equation of state, and other auxiliary fundamental equations are the isentropic relations describing the relation between the total and static conditions.

Where R is the gas constant.

Sutherland viscosity law

$$\mu = \mu_0 \left(\frac{T}{T_0}\right)^{3/2} \frac{T_0 + S}{T + S} \tag{10.8}$$

The air viscosity is computed according to the Sutherland viscosity law. Sutherland's viscosity law resulted from a kinetic theory by Sutherland using an idealized intermolecular-force potential. The formula is specified for two or three coefficients. Sutherland's with three coefficients is used in the present work, and it is expressed as:

Where μ_0 is a reference viscosity corresponding to a reference temperature T_0, and S is an effective temperature called Sutherland constant which is a characteristic of the gas. For air at moderate temperatures and pressures, $\mu_0 = 1.7894 \times 10^{-5}$ *Pa.s*, $T_0 = 273.11K$, $S = 110.56K$.

10.2.3.2 Particle Trajectories

The particles are first ingested into the flow domain. Due to their high inertia, they do not follow the carrier phase streamlines and tend to impact the surfaces. For low particle concentration, the particles trajectory codes employ Lagrangian methods of solution as the particle equations of motion can be decoupled from the flow field.

10.2.3.3 Lagrangian Approach

When the particle's volume fraction is low, the Lagrangian modeling of the particulate phase will be a reasonable approach [65]. A set of three-dimensional differential equations is solved to calculate the particle location and velocity when moving in a compressible airflow around or inside aircraft modules.

The equations that govern the solid particle motion can be formulated using Newton's second law as follows:

$$m_p \frac{d\vec{V}}{dt} = F_D + F_p + F_b + F_B + F_g + F_{TD} + F_S + F_M \tag{10.9}$$

Where F_D: Drag force, F_p: Pressure gradient force, F_b: Buoyancy force, F_B: Basset force, F_g: Gravity force, F_{TD}: Drift force due to the turbulence diffusion, F_S: Saffman force, and F_M: Magnus force.

In most aircraft applications, only the drag, gravitational, and buoyancy forces are the most dominant forces. The governing equation is reduced to:

$$m_p \frac{d\vec{V}}{dt} = C_D \frac{\rho_a}{2} \left|\vec{U} - \vec{V}\right| \left(\vec{U} - \vec{V}\right) A + m_p \vec{g} + \vec{B} \tag{10.10}$$

Where $m_p, D_p, \vec{V}$ mass, diameter, and speed of the spherical particle

$m_p = \rho_p \pi D_p^3/6$ and $A = \pi D_p^2/4$ is the projected area of a spherical particle

C_D is the drag coefficient, which is dependent on the Reynolds number ($C_D = f(R_e)$) which is expressed as:

$$R_e = \frac{\rho_a \left|\vec{U} - \vec{V}\right| D_p}{\mu} \tag{10.11}$$

$\vec{U}, \rho_a$ Speed and density of the airflow

$$\text{Buoyancy Force } \vec{B} \mid \left|\vec{B}\right| = \rho_a\, g \pi D_p^3/6 \tag{10.12}$$

When aircraft is airborne, the flow within engine modules has a high speed, and any suspended particles will be governed only by the drag force. For small Reynolds numbers $Re < 0.1$, the viscous effect is dominating, and this is referred as the Stokes regime CD = 24/*Re*. Above the value of $R_e = 2.6 \times 10^5$, the drag coefficient is constant CD = 0.4.

When drag force is the only dominant one, then the equation of motion is expressed by:

$$m_p \frac{d\vec{V}}{dt} = \frac{\pi}{8} C_D \rho_a d_p^2 \left|\vec{U} - \vec{V}\right| \left(\vec{U} - \vec{V}\right) \tag{10.13}$$

10.2.3.4 The Mechanism of Particle Impact and Rebound

The impact dynamics of particles is governed by restitution coefficients. The restitution coefficients are a measure of the kinetic energy exchange upon impact. They define the velocity change of particles due to impact. This change in velocity includes

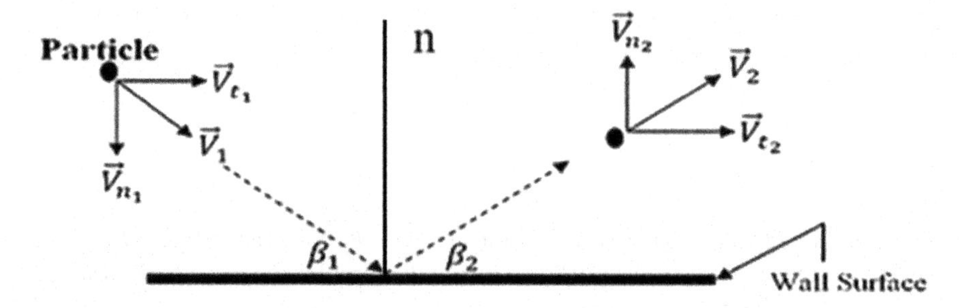

FIGURE 10.1 Impact dynamics [66].

both the magnitude and direction (angle), and therefore it may be expressed as normal $(\eta_N)$and tangential (η_T) restitution coefficients as illustrated in Figure 10.1. Particle impact and rebound from a surface

$$\eta_N = \frac{V_{n_2}}{V_{n_1}} \tag{10.14}$$

$$\eta_T = \frac{V_{t_2}}{V_{t_1}} \tag{10.15}$$

The rebound characteristics depend on the particle size, shape, speed of rotation, orientation at the point of impact, material properties of particle and wall, and the impact angle and velocity [66].

10.2.3.5 The Mechanism of Erosion

The erosion phenomenon depends on material properties (brittle or ductile), particle impinging characteristics (velocity and angle), and particle properties (size and shape) [67]. In one of the earliest analyses of ductile erosion [68], the particles were assumed as a cutting tool that moves into the surface causing plastic deformation of the material and removal of the debris, as shown in Figure 10.2. For brittle material, the erosion leads to surface and subsurface cracking and spalling of the target. The maximum erosion for the ductile material occurs at some angle between 20° and 30°, while for brittle material occurs at 90°.

An empirical formula for the erosion per unit mass of impacting particles for different particle and ductile target material combinations [69] is expressed by Equation (10.16):

$$\varepsilon = k_1 f(\beta) V^2 \cos^2 \beta \left(1 - R_T^2\right) + k_3 \left(V \sin \beta\right)^4 \tag{10.16}$$

Where ε is the erosion parameter (mg/gm), V is the impact velocity (m/s), and β is the impact angle (degree). The values for the constants k_1, R_T, k_3 depend on the particle and target materials.

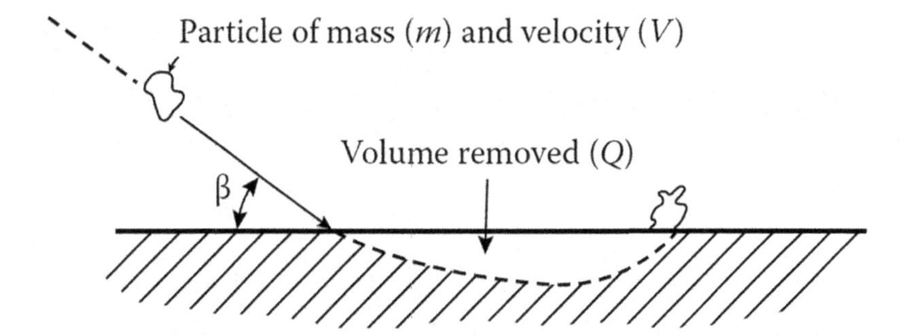

FIGURE 10.2 Ductile erosion simulation [66].

The erosion rate depends on the particle concentration (α), the number of impacts, and the erosion parameter (ε).

Any cloud of particles is composed of particles having different materials and different sizes. The actual distribution of particle sizes is to be replaced by a piecewise continuous species having average ranges. The number of species is equal to the number of materials multiplied by the number of average diameters. Figure 10.3 illustrates the terrain particle size distribution in different locations or cities in the United States) [68].

The mass flow rate of any individual species of the particle $(\dot{m}_{pi})$ will be defined as:

$$\dot{m}_{pi} = \dot{m}_p r_i \tag{10.17}$$

Where (r_i) is the species mass ratio. The number of particles in each particle streamlines $(\dot{n}_{pi})$ for a specified species (i) is given by

$$\dot{n}_{pi} = \frac{\dot{m}_{pi}}{M_{pi} N_s} \tag{10.18}$$

Where (N_s) is the number of particle's streamlines for each species (i), and (M_{pi}) is the mass of a particle in species (i). The rate of volume removed by the impact of one particle of an individual species is given by:

$$Q_i = \frac{M_{pi}\epsilon}{\rho_s} \tag{10.19}$$

Where ϵ is given by Equation (10.16), and ρ_s is the eroded target density. Thus, if the area of the eroded region on the target surface is denoted by "A," then the average depth of the material removed over the complete surface by each impact of a particle is given by:

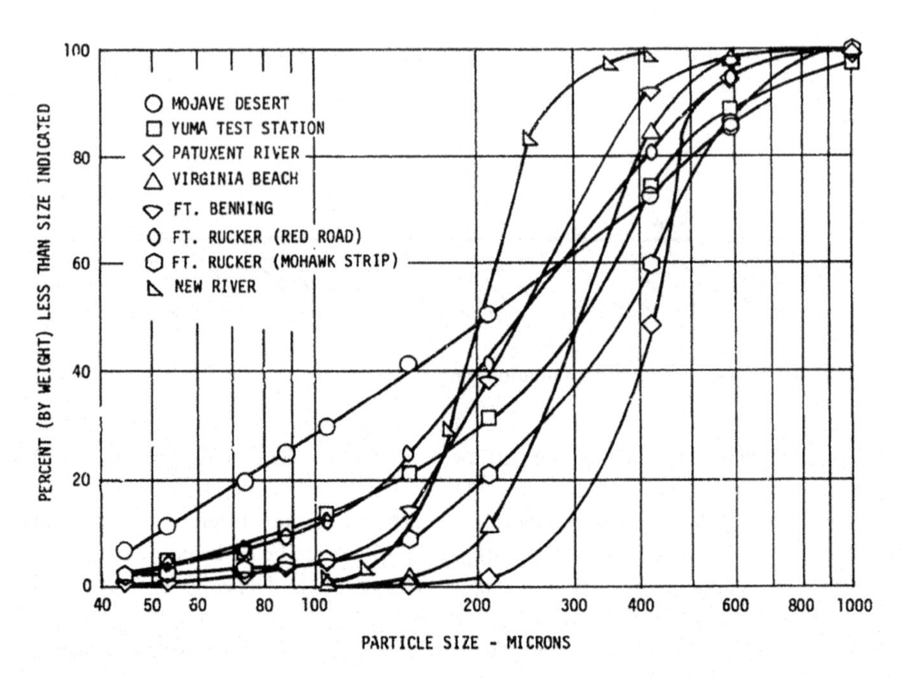

FIGURE 10.3 Terrain particle size distribution (United States) [68].

Courtesy: US Air Force.

$$\Delta h_i = \frac{Q_i}{A} \tag{10.20}$$

The rate at which the surface contour changes for all particles species will be the cumulative or the sum of all:

$$\left(\frac{\Delta h_i}{\Delta t}\right)_t = \sum_{j=1}^{j=\dot{N}_{imp}} \left.\frac{\Delta h_i}{\Delta t}\right|_j \times P_i \tag{10.21}$$

Where $\dot{N}_{imp}$ is number of particles impacts at a certain area of the surface per unit time, and P_i represents the probability that an individual species of particle will strike the surface.

10.2.3.6 Summary for the Numerical Steps for Calculating Erosion Rates

Several case studies outlined hereafter use commercial codes like ANSYS FLUENT [22]. The numerical schemes can be summarized as follows:

A. Gas flow
    1. The main program reads the geometry of the case study, operating conditions and generates the appropriate mesh (mostly finite volume)

2. It initializes the values for velocities, the pressure of the air/gas at each node
3. It calculates new values for air/gas properties at all grid nodes by solving RANS equations
4. Iteration is employed as necessary to reach correct values of air/gas velocities at each node that satisfies the continuity equation
5. Energy and turbulence equations are solved using the previously updated values of the other variables and a convergence criterion is set like the residuals of all variables to be less than 10^{-6}

B. Particle trajectories
1. Initialize the conditions of the particles (injection locations and velocity) and properties (density, diameter distribution, and mass flow rate)
2. Solve Lagrange equations to calculate the particle trajectories
3. If the particle impacts a wall, the main program calls user-defined functions to compute the rebound velocities and continue tracing particle trajectory until it exits the specified domain
4. Another particle is traced in the same manner until all particle trajectories are calculated

C. Erosion estimate
1. As particle impacts the surface, use the erosion data obtained on the boundary wall elements as input to another developed Fortran program to localize the precise erosion location using inverse interpolation calculations
2. Call user-defined functions to calculate the erosion rate, penetration rate, and the impact frequency
3. Use the erosion data to define the new geometry of the eroded part after a certain operating period and compare it to the structural acceptable wear limit set by the manufacturer, and thus the life time of this component is defined
4. Calculate the performance deterioration (drop of efficiency and pressure ratio of compressor, or increase in turbine or exhaust gas temperature [EGT])

10.2.4 Particulate Flow in the Intakes of High Bypass Ratio (HBPR) Turbofan Engine

During the ground run of an aircraft engine, a ground vortex (GV) will extend from the ground to the engine fan face of as illustrated in Figure 10.4 into the intake.

The requirement for vortex formation is the existence of a stagnation point on the ground plane. Such a stagnation point acts as a focal point for vorticity upstream. It is concentrated at first and next stretched into the intake. It depends on the two parameters:

- Height-to-diameter ratio (H/D_i) of the intake
- The velocity ratio $\left(U^* = \dfrac{U_i}{U_\infty}\right)$

FIGURE 10.4 Intake ground vortex [56].

Published with author permission.

Where H is the distance from the engine centerline to the ground, D_i is the engine inlet diameter, U_i is the inlet air velocity, and U_∞ is the far upstream air velocity. The vortex map of an intake of an HBPR turbofan engine like the GE-CF6-50 engine is evaluated in [2, 3, 55] and plotted in Figure 10.5.

Ingestion of debris into intakes is dominated by the effect of the suction force from the reduced static pressure at the GV core acting on the upper surface of the particle. The maximum size of foreign object debris that can be lifted is dependent on the GV, the strength of the wind blowing on to the intake, the intake configuration, particle material density, and ground surface roughness [70, 71]. A case study for simulating the motion of sand particles into the engine of 737 aircraft using the solver ANSYS CFX 17.2 and one-way coupling is illustrated in Figure 10.6 [71, 72]. The air had inlet total pressure and temperature of 101 kPa, 298 K, mass flow rate per unit area of $250\,kg/m^2 sec$, and crosswind velocity of 6 *m/s*. Sand particles were assumed smooth spherical having radii $R_p = 0.00034$ m and density ρ_p=1,650 kg/m^3.

10.2.5 Erosion of the Fan of High Bypass Ratio (HBPR) Turbofan Engine

This case study handled here is the intake and fan of the General Electric CF6-50 HBPR turbofan. The axial length of the intake was 1.5555 m. The fan rotor has 38 tapered and highly twisted blades. The fan blade rotor has a diameter of 2.181 m, a

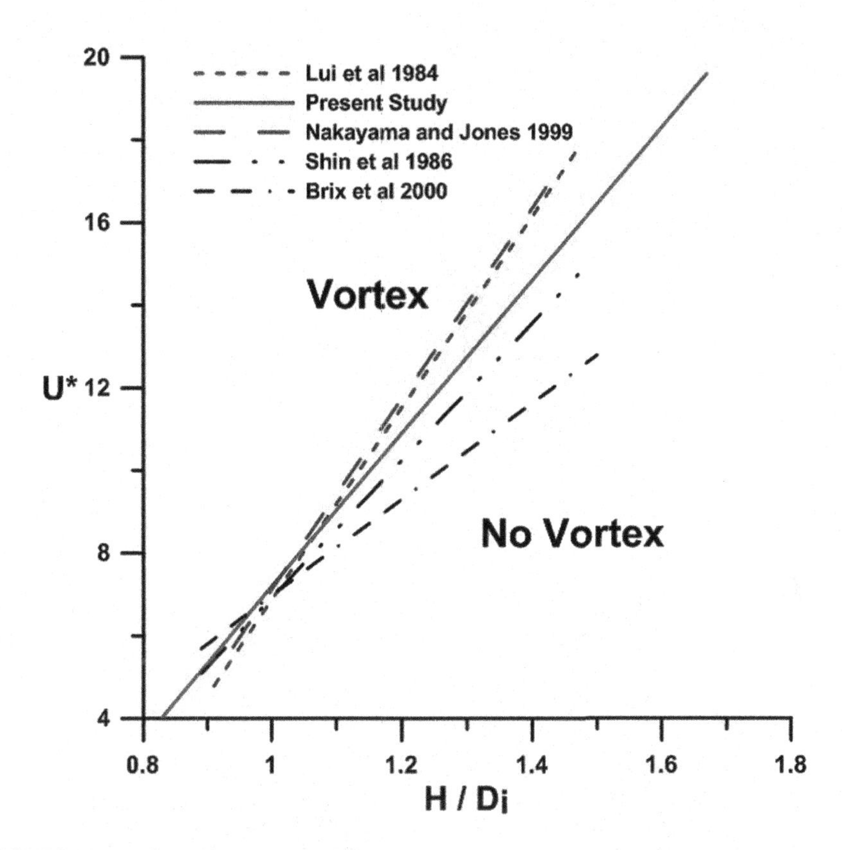

FIGURE 10.5 Ground-vortex map [2].

low hub to tip ratio of 0.4033, a span of 0.6507 m, and its nose is simplified as a hemisphere of radius 0.4057 m [5, 73].

The computational domain includes both fan and the intake zones. Due to the point symmetry of them, a periodic sector of an angle (360/38) was employed for both the fan and the intake zones. The commercial code Fluent 6.1 package was used. The two zones were merged to form the required computational domain, and a GAMBIT preprocessor (a software in Fluent 6.1) is used. Two case studies were examined at sea level and 11 km cruise altitude, where the fan rotational speed was $\omega = 404$ and $385\,rad/s$, respectively. Sand particles were assumed for takeoff having a density of $2650\,\frac{kg}{m^3}$, concentration of $0.176\frac{gm}{m^3}$, mass flow rate of $97.5\frac{gm}{s}$, size distribution varied from 50 to 300 microns, and initial velocity of $128.6\frac{m}{s}$. The absolute Mach number for takeoff is illustrated in Figure 10.7.

The trajectories of sand particle having diameters 250 mm during takeoff are shown in Figure 10.8. Most of these particles have two successive impacts. The first impact is observed on the suction side, while the second is with the blade pressure side. The particle impacts the pressure side with a higher velocity than the suction side.

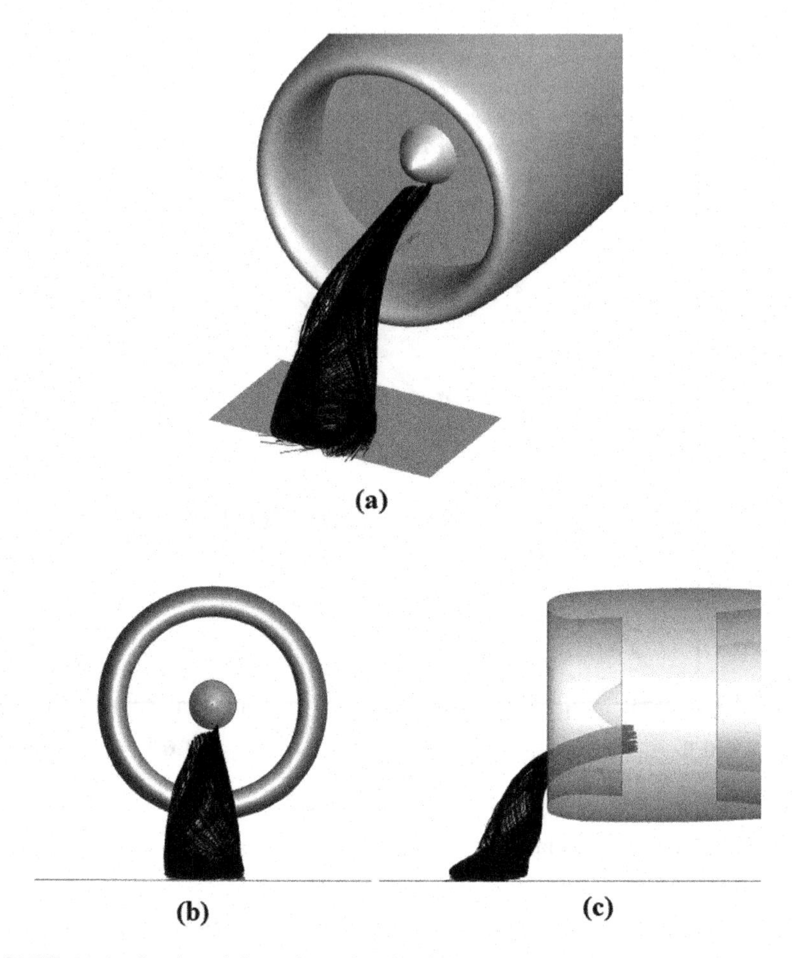

FIGURE 10.6 Sand particles trajectories simulation.

Reproduced from [72] with author permission.

Figure 10.9 shows the impact frequency and erosion rate contours on both pressure and suction sides of the fan blade, at takeoff condition and for particle size of 250 mm. Based on Rosin-Rammler particle diameter distribution [74], a particle sizes distribution in the Middle East having diameters 50, 100, 150, 200, and 250 micron is assumed in the estimation of erosion rate due to sand particles during takeoff. The total erosion rates and the penetration rates were calculated. Figure 10.10 illustrates the blade chord reduction variation with the span. This variation is plotted for 2,000, 4,000, and 6,000 operating cycles. Based on the blend limits listed in the engine manual of CF6 turbofan engine [75] and Figure 10.10, one can predict that the blade lifetime is nearly 5,000 operating cycles. Variation in the blade chord reduction along the blade span for different operating cycles and the blend limits is given by the GE CF6 manual [73].

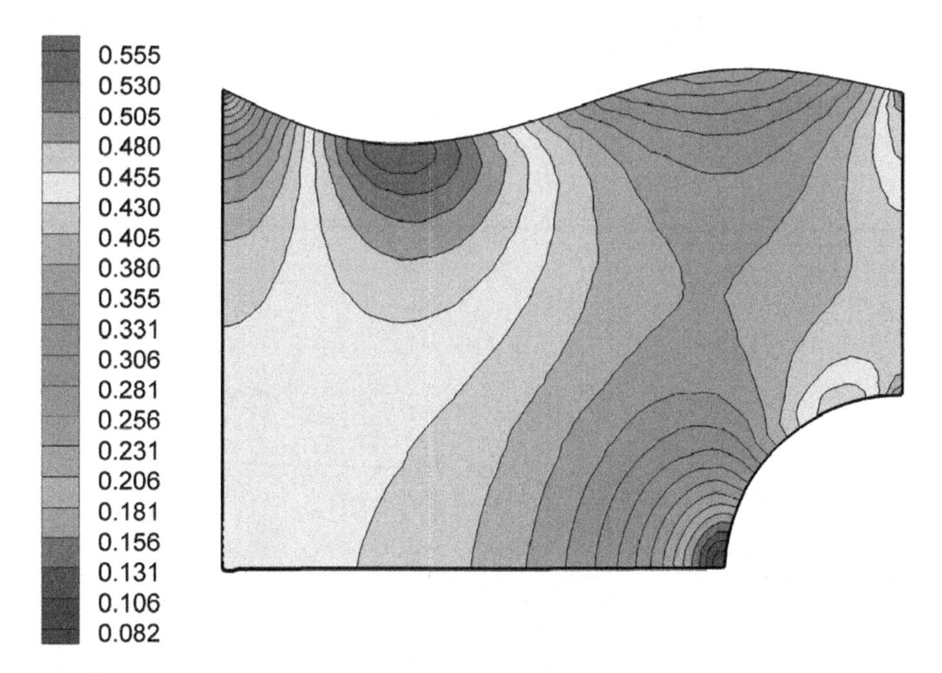

FIGURE 10.7 The absolute Mach number in the intake (takeoff) [73].

A similar analysis for particle trajectories and erosion estimate for cruise conditions are performed [73]. Erosion results in 1.1% and 1.96% loss of the engine overall efficiency after 4,000 and 8,000 cycles, respectively, during cruise operation. Other penalties in engine operation due to fan erosion are given in Table 10.1.

10.2.6 Erosion of Centrifugal Compressor

The author presented a theoretical estimate for the erosion of a centrifugal compressor having a radial impeller in [7–9]. Both silicon dioxide and aluminum oxide particles having radii varying from 3 to 60 μm were traced. These trajectories defined four internal areas subjected to successive impact and consequently high erosion rates: casing surface close to the inlet, hub surface near the outlet of impeller, and two areas on the blade pressure surface. Compressor was made of aluminum or stainless-steel alloy. The first compressor encountered high erosion rates and penetrating depth. After 100 operating hours, the penetration was maximum at the hub, having values of 5.6 and 3.87 mm for the aluminum and stainless-steel alloys, respectively. The least penetration was for the casing with values of 0.068 and 0.04 mm for the aluminum and stainless steel, respectively.

Another configuration of a radial compressor having two different size splitters was examined in [76]. The critical areas were defined.

A third case of a backswept compressor was examined by the author and his group [77, 78]. This type of centrifugal compressor is more stable and efficient than radial impellers since its blade loading is improved. It is used mostly in small

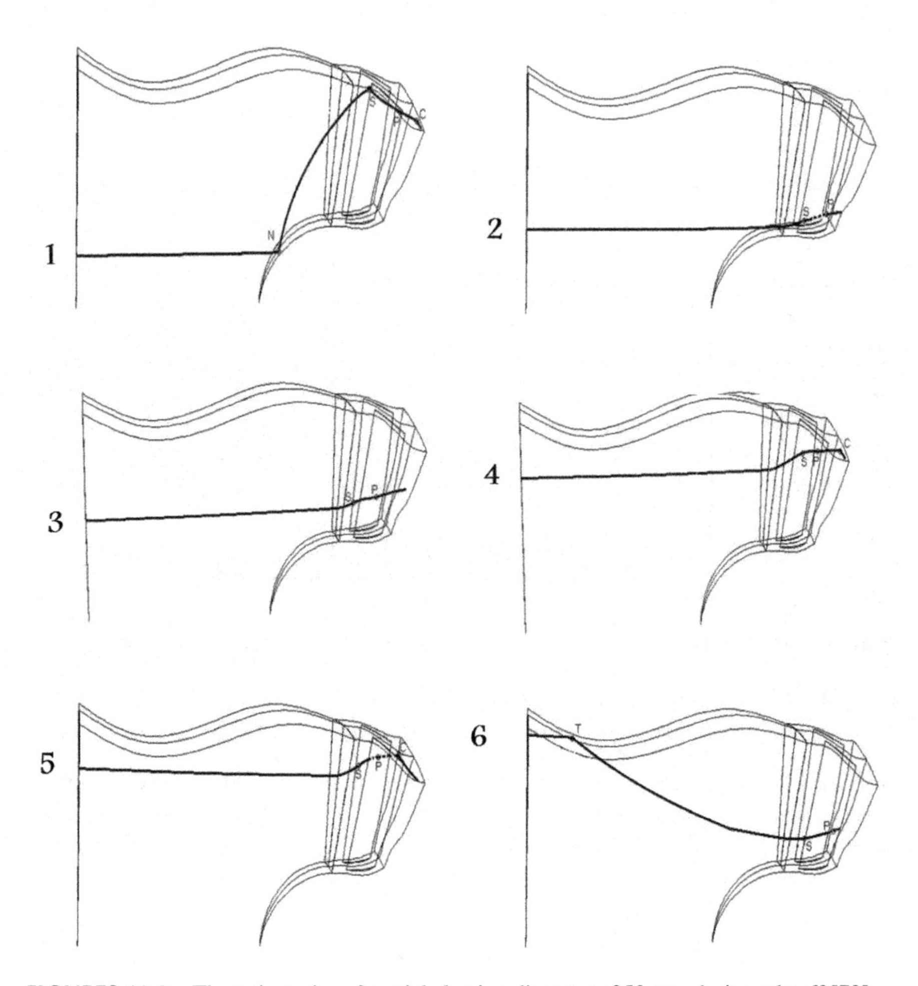

FIGURES 10.8 The trajectories of particle having diameters 250 mm during takeoff [73].

turboprops, turbo shafts engines, and auxiliary power units (APUs). It is also used in turbo charging of internal combustion engines, and air compression in gas turbines used in aircraft and marine propulsion. The back sweep angle was 60^0. Number of impeller blades was 20, rotational speed was 14,000 rpm, tip speed was 393.2 m/s, impeller total pressure ratio was 1.72, and mass flow rate was 4.12 kg/s. A periodic sector of an angle 360°/20 was employed. The commercial code FLUENT V6.0.12 was used. The grid network of the whole computational domain was composed of 156,837 cells, 320,971 faces and 32,665 nodes. FLUENT was also used for tracing the particle trajectories and their impact on impeller walls (surfaces). Silicon dioxide particles had a material density of 2,650 kg/m^3, diameters ranging from 5 to 150 microns, nine particle initial positions, and initial velocity referred to gas flow was ($V_p / V_g = 50\%, 100\%$). Figure 10.11 illustrates the trajectories of particles having a diameter of 100 μm at different locations and velocity ratio ($\boldsymbol{Vp / Vg} = 100\%$). The location of the particle impacts with the internal surface of compressor are denoted

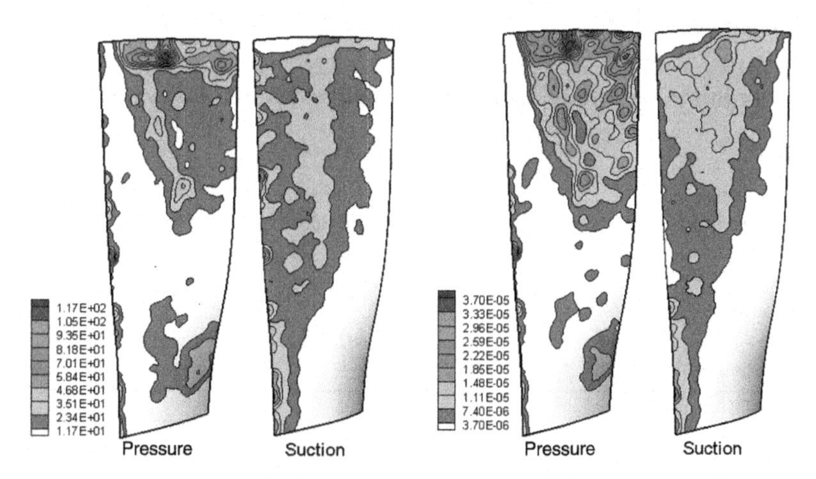

FIGURE 10.9 The impact frequency (left) and erosion rate contours (right) on pressure and suction sides of the fan blade due to sand particles of size of 250 mm [73].

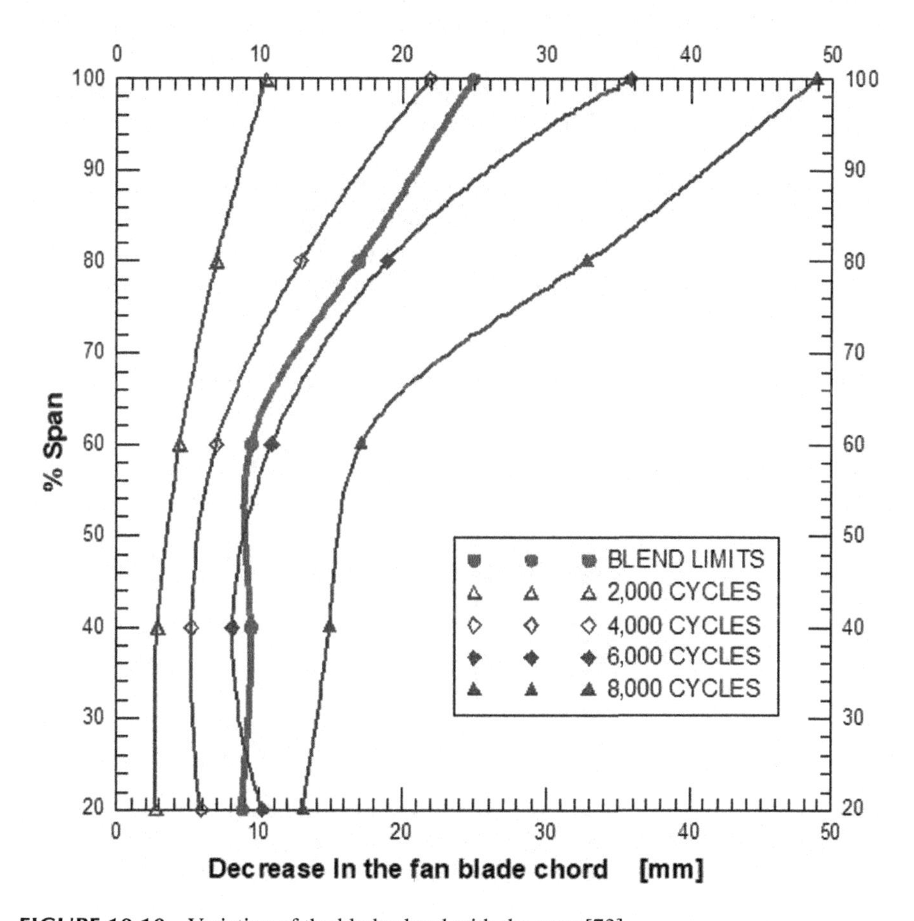

FIGURE 10.10 Variation of the blade chord with the span [73].

TABLE 10.1
Deterioration in the Fan and Engine Parameters After 8,000 cycles [73]

Parameter	Takeoff	Cruise
Surge margin %	1.496	2.51415
Fan pressure ratio %	−2.333	−2.356
Air mass flow rate %	−2.6425	−2.422
Fan efficiency %	−3.5518	−3.554
Engine thrust %	−4.9292	−5.367
SFC %	+0.9942	+1.9758
Engine overall efficiency %	----	−1.9579

as: (C) for the casing, (P) for blade pressure surface, (S) for blade suction surface, and (H) for the hub of impeller.

To evaluate erosion damage to the compressor, sand particles were assumed having particle diameter distribution equivalent to those in the Middle East. The Rosin-Rammler particle diameter distribution is adopted for erosion prediction. The particle diameters range from 5 to 150 μm with the mean value at 75 μm. This distribution simulates the particle sizes in the Middle East [68]. Only 622 sand particles were injected into the impeller inlet. However, more particles are needed for a better estimate of erosion rates.

The erosion rate on the pressure surface was greater than the suction surface. The maximum erosion rate on the pressure and suction surfaces were 2.156×10^{-2} and 4.047×10^{-3} mg/gm, respectively. Maximum erosion on the pressure surface appears in a region near the hub of the impeller at the trailing edge which was exposed to the highest particle velocity impacts.

The penetration rate contours on the impeller blade surfaces are shown in Figure 10.12. Its distribution is the same as the erosion rate with the maximum value of $8.704 \times 10^{-9}\, m/s$ and $4.794 \times 10^{-9}\, m/s$ of the pressure and suction surfaces, respectively.

The lifetime prediction may be obtained by calculating the total exposure time after which the full penetration will occur. For a known wall thickness, the time required to get full penetration of the wall is given in hours by:

$$t_p = \frac{T_w}{3600 * \left(\frac{\Delta h}{\Delta t}\right)} \tag{10.22}$$

Where (T_w) is the wall thickness, and $\left(\frac{\Delta h}{\Delta t}\right)$ is the maximum value of total depth of penetration over the wall. Here the maximum impeller wall thickness was 4 mm and the maximum total depth of penetration was $8.7 * 10^{-9}$ (m/s), then:

Time until full penetration $= 0.004/\left(8.7 * 10^{-9}\right) = 4597{,}770$ (s)/3600.
$= 127.71$ h (continuous exposure).

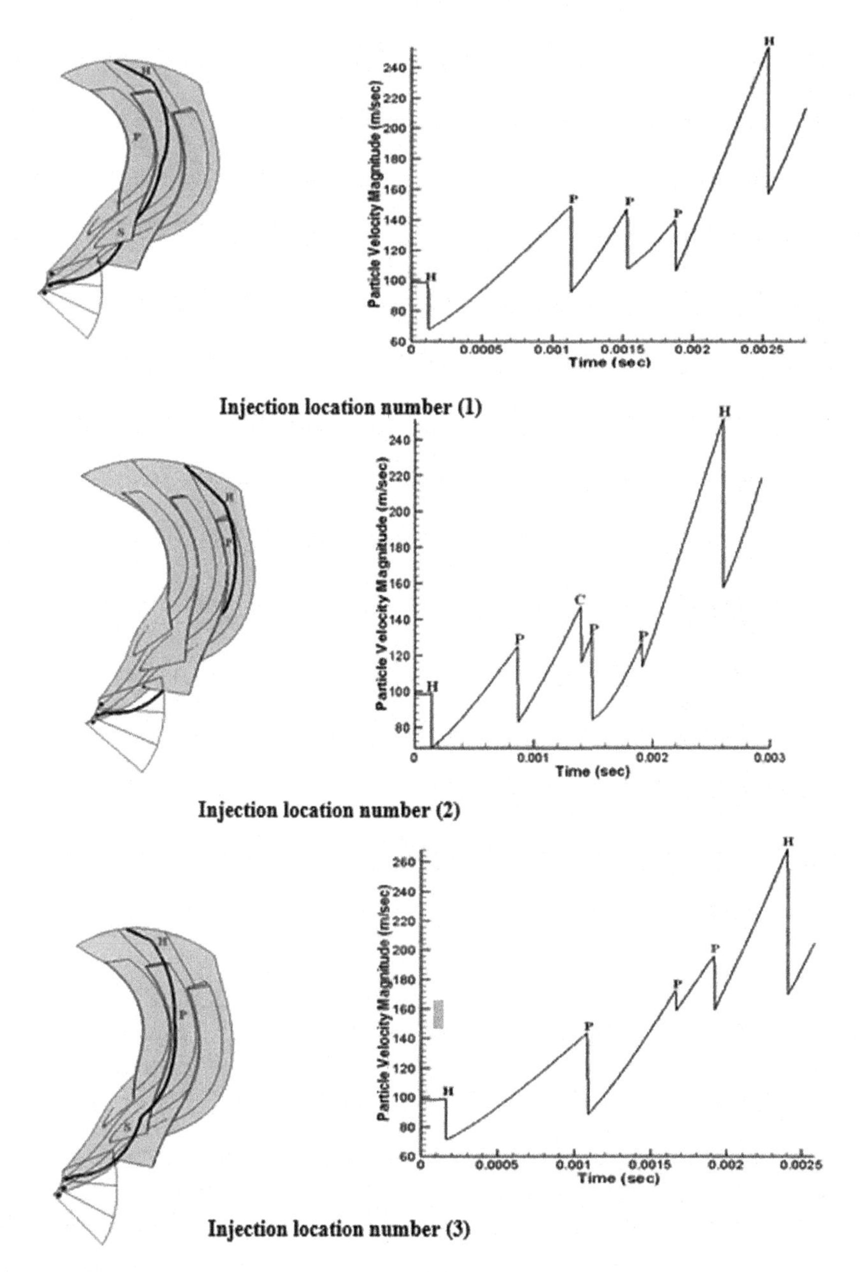

FIGURE 10.11 Trajectory of particle having diameter of 50 micron injected at locations (1), (2), (3) and $V_p/V_g = 100\%$ [78].

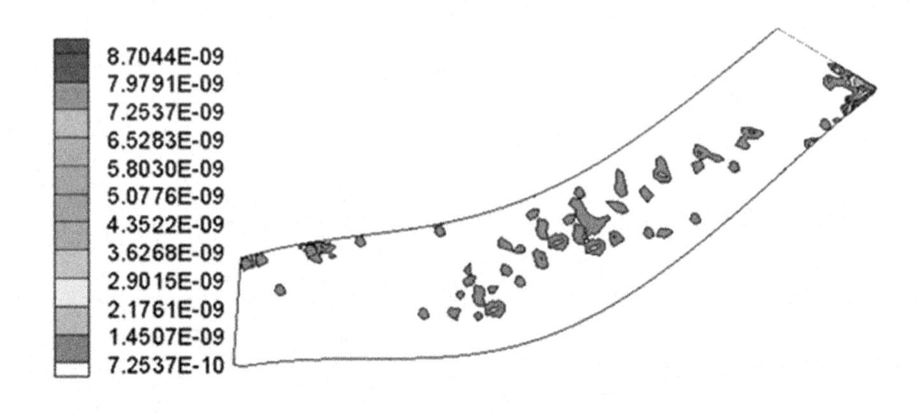

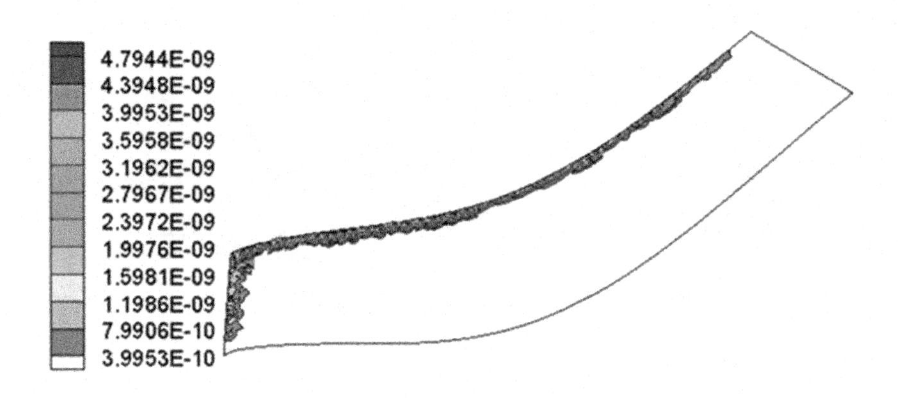

FIGURE 10.12 Contours of penetration rate distribution (m/sec) on pressure and suction surface due to cloud of particles in the Middle East [78].

Such erosion damage will increase the tip clearance, which in turn will decrease the compressor efficiency, and the engine lifetime. Moreover, it will increase the engine thrust specific fuel consumption TSFC and the cost of repair and maintenance.

10.2.7 Erosion of Axial Compressor

The ingestion of solid particulate flow into the axial compressor will lead to aerodynamic penalties, structural wear, and reduction in a working lifetime. It may affect both engine reliability and safety. Aerodynamic penalties for compressors and the whole engine are emphasized in the degradation of the compressor pressure ratio, isentropic efficiency, flow capacity, an overall drop in power output, thermal efficiency, and increase in fuel consumption and EGT [79, 80]. Structural wear led to the following imperfections [81–83]:

1. Leading-edge blunting, and variation in the incidence angle and in the leading-edge radius
2. Thinning of the blade trailing edge
3. Variation of the blade trailing edge angle
4. Increase in surface roughness
5. Variation of camber
6. Increase of blade tip clearance
7. Chord shortening
8. Spalling of the casing.

Based on GE data for its CF6 engine, a survey of field engines at three different customer locations, led to the following conclusions regarding blade erosion [84]:

- Leading-edge chord wears at a rate of 0.25 mm per 1,000 flight cycles (equivalent to about 2,400 h)
- Trailing edge thinning occurs on the concave side of the airfoil tip at a rate of 0.06 mm per 1,000 flight cycles.

SPE can be reduced by using surface protection methods among which is the erosion resistant coatings. Examples are metal nitrides (e.g., TiN), multi-layered structures of metal/ceramic, and coatings of nanocomposite material [85, 86].

Protection coatings are deposited thin layer/s on the surface of the material to improve their erosion resistance.

The benefits of coatings may include [87]:

1. Up to 1% ageing recovery of aeroengines
2. Minimization of erosion, cracking, spalling, and wear
3. Surface roughness reduction

Secondary flow is normally generated by the turning of the fluid within the blade row. This secondary flow has a reasonable contribution in modifying particle paths [29–32]. Secondary flow may lead to a higher concentration of particles in the hub wall boundary layer compared to the streamwise or primary flow alone. This may lead to severe erosion at the roots of the blades in the next row.

An experimental technique using laser-Doppler anemometry (LDA) was adopted to define the three-dimensional viscous flow field in an axial stationary cascade [30]. The experimental apparatus used in this study was described in [32]. The LDA was operated in the back-scatter mode. A four-bladed cascade test section was used to model the flow through the blade passage of a typical axial stator. The central passage of the cascade was divided into 14 cross sections, representing the measuring planes [31]. The cross sections were normal to the tangents of the curve defining the passage centerline. Each cross section was divided by an 11 × 11 square grid, and at each node five components of the velocity vector were measured. These components were resolved to give the primary velocity component (normal to cross section) and the secondary velocity components (in the plane of the cross section).

The trajectories of several particles having diameters ranging from 1 to 5 microns were traced twice, one time with consideration of secondary flow and the second when the secondary flow is ignored. The initial velocity of the particle was assumed to be one-half of the corresponding gas velocity at the same inlet location. Secondary flow drastically changed the particle trajectories. Particles moved from one blade surface toward the other surface. Particles were also concentrated in the corner made by the blade surface and lower endwall (hub).

The total material loss was calculated by integrating the erosion damage curve over the entire length of the blade. The secondary flow decreased the local erosion rate on the suction surface; however, it increased the overall damage because impacts occur over a sixfold greater length. On the pressure surface, secondary flow increased the local erosion rates, and since impacts with and without secondary flow occurred on the same area, thus, secondary flow increased the overall (integrated) erosion [31].

10.2.8 Erosion of Radial and Axial Turbines

Radial inflow turbines are found in APUs and turbochargers. The ingestion of solid particle (sand or fly ash produced during combustion) erodes its internal surfaces. It will in turn influence the reliability, performance, and structural integrity of these radial inflow turbines. Extensive studies were performed to analyze the erosion damage of radial inflow turbines [13–18]. A case study was considered having the following characteristics: An equivalent flow rate and rotational speed of 0.220 kg/sec and 30,800 rpm, a total-to-total efficiency of 88%. The number of blades of nozzle and rotor were 29 and 12 blades, respectively. The turbine had the following radii in centimeters: scroll mean inlet of 17.11, nozzle inlet 9.8, rotor inlet of 7.52 cm, and rotor mean exit of 3.52. The rotor inlet width was 0.805 cm. A cloud of particles had a concentration $\alpha = 0.0143\%$, and composed of silicon dioxide (70%) and aluminum oxide (30%), diameters ranged from 15 to 300 microns. The critical areas of the turbine elements were the pressure side of the stator and the tip region of the rotor. Lifetime was assessed for a stator made of titanium alloy Ti-6AL-4V and rotor manufactured from stainless steel or titanium alloy.

The author examined the erosion of a two-stage axial turbine (stators and rotors) [11, 12]. The three steps used for estimating erosion rates are used. The geometry of the two stage (stators and rotors) changes due to erosion (removal of material). The change of geometry affects all the gas flows, particle trajectories, and erosion rates. The lifetime of this turbine stages was assumed to be 12,000 hours and it was divided into four segments of durations: 2,000, 3,000, 3,000, 4,000 hours. The pressure and suction surfaces of each blade were covered by a square grid network of 10 × 10 elements. The location of every impact through this coarse grid was denoted by an element number. The erosion was calculated for the first 2,000 h based on the original blade geometries and, therefore, constant gas flows and particle trajectories. After the initial 2,000-h engine running time, new blade geometries were estimated from the

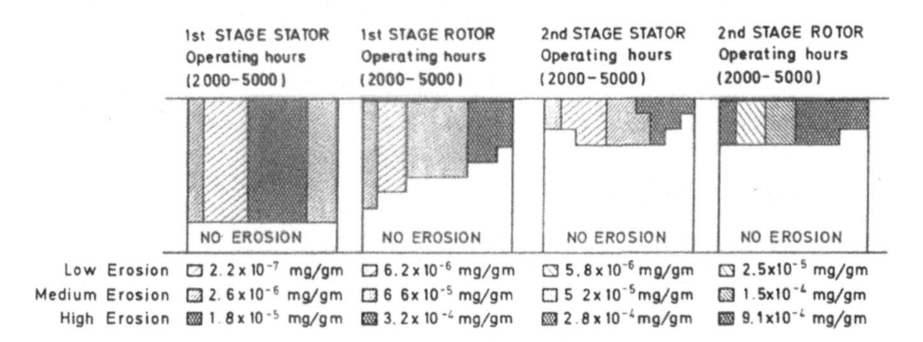

FIGURE 10.13 Blade erosion values after 2,000 operating hours [12].

calculated erosion. Then, the gas flows and particle trajectories were redetermined. Erosion rates were reestimated for the next 3,000 h of engine running time, and the new blade geometries were calculated after 5,000 operating hours. The quasi-dynamic procedure was repeated several times, so that blade geometries were determined for 2,000, 5,000, 8,000, and 12,000 h of engine running time.

The erosion patterns in the periods (2,000–5,000) operating hours are shown in Figures 10.13. Erosion is concentrated on the pressure surfaces of all blades. This was due to the usage of inviscid gas flow calculations. Adopting N-S or RANS methods will provide a more accurate 3D gas flow solution that accounts for boundary layer and secondary flow. Such accurate gas flow will provide more accurate particle trajectories and erosion estimates.

The maximum erosion rate for the period (5,000–8,000) in the first stator and rotor were 2.4×10^{-5}, 3.7×10^{-4}. Maximum erosion for the same period in the second stator and rotor were 3.84×10^{-4}, 9.6×10^{-4}.

10.2.9 Erosion of Aircraft Propellers

Propellers are installed to both turboprop and piston engines. Aircraft propellers are subject to wear, fatigue, corrosion, foreign object damage, erosion, and icing when operating from unpaved runways or during unfavored weather conditions [88, 89]. All of which can lead to propeller failure which will not only result in a loss of thrust or power but also may lead to dramatic accidents. If even a small part of the propeller blade is lost, an imbalance of blades will follow, and may the engine be torn from the aircraft making the aircraft uncontrollable. Erosion of the propeller will lead to an increase of blade surface roughness and changes in propeller profile. This will in turn cause deterioration in engine performance parameters like equivalent power, thrust, specific fuel consumption, and overall efficiency.

Erosion of propellers is also analyzed by the same three steps listed before. A case study of subsonic three-bladed tapered and highly twisted straight propeller blades installed to the turboprop engine Pratt & Whitney Canada PT6-25A powering Pilatus PC-7 Turbo Trainer was examined by the author and his group [22–25]. The tip and root radii are 1,115 and 80 mm.

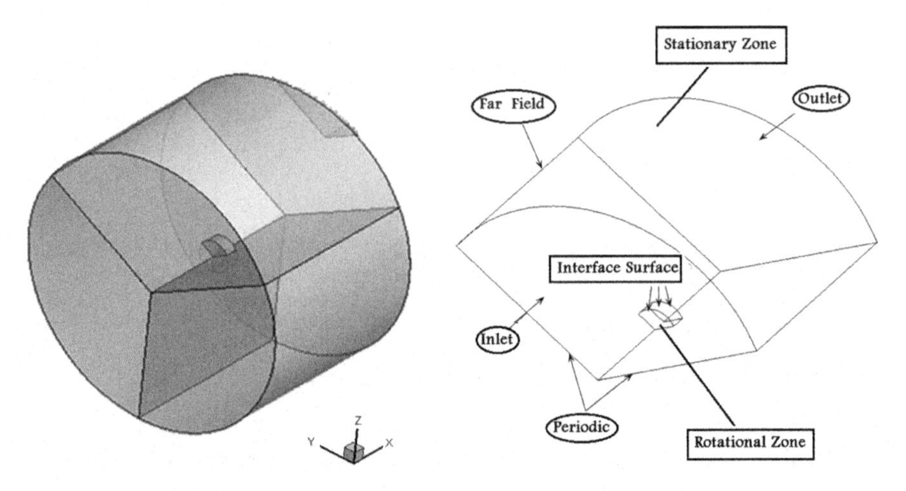

FIGURE 10.14 The entire solution domain [23].

The air 3-D governing Equations (10.2–10.8) are solved using the commercial code FLUENT V6.3.26 software. Like the fan of a HBPR Turbofan engine, the computational domain was divided into two zones: stationary and rotational (here the propeller), which was next merged together to form the required computational domain [22]. The stationary or external zone had a cylindrical section (120°) and was composed of three domains: an upstream, downstream, and top of blade region. The upstream zone extends 14,000 mm (approximately 13.5 blade lengths) in front of the origin (chosen as propeller nose or spinner) and the downstream zone also extends 14,000 mm behind the origin. Due to the rotating reference frame, a top region having a constant radius of 21,500 mm (approximately 21.5 blade lengths) from the origin was adopted. The internal or rotational zone was selected to have 1,000 mm (approximately 1 blade length) in front of the origin, 1000 mm behind the origin and a 120° cylindrical sector having a constant radius of 2,500 mm (approximately 2.5 blade lengths) from the origin (Figure 10.14). The GAMBIT preprocessor (a module in the FLUENT package) is used to generate the suitable mesh for both the external and internal zones using an unstructured tetrahedral grid. Next, the two meshes were merged using the TMERGE subroutine. During takeoff, the inlet Mach number and the rotational speed were 0.2 and 1,500 rpm.

After solving the airflow equations, the particles' governing Equations (10.9–10.13) were solved based on the one-way coupling Lagrangian approach also using FLUENT V6.3.26. Upon striking the propeller surface, the particle rebounds according to the prescribed restitution coefficients described by Equations (10.14) and (10.15). FLUENT code calls a function named "IMPACT" to calculate the rebound velocity. Also, when any particle reaches a wall, it calls the functions named "ER_RATE," "PEN_RATE," and "IMP_FREQ" to compute the erosion rate, penetration rate, and the impact frequency, respectively. These values are returned to the element face that the particle impacts.

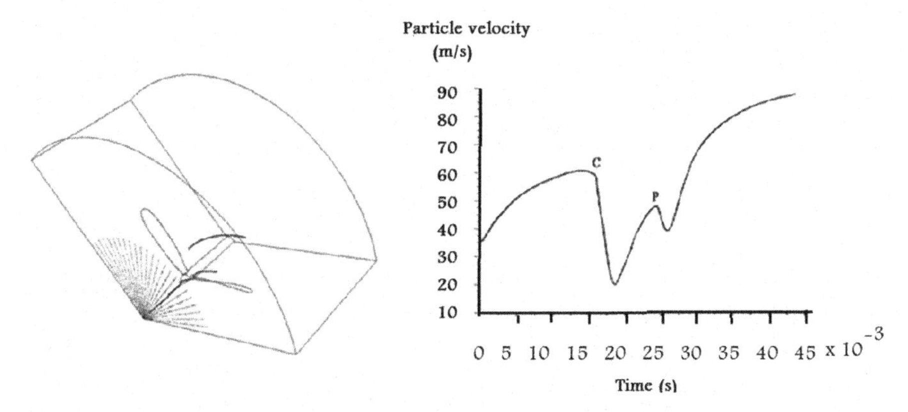

FIGURE 10.15 Trajectory of particles with diameter dp = 50 μm (takeoff) (V_p/V_a = 50%) [23].

Sand particles having a density of 2650 kg/m^3, concentration of $0.176 * 10^{-3}$, mass flow rate of 94.55 gm/sec, initial velocity ratios $\frac{V_p}{V_a} = 50\%$ and 100%,and diameters of 50, 100, 150, and 200 μm were considered. For uniform distribution of particle inlet locations, each 120° sector is divided into 23 injection lines 5° spaced apart and the number of injection points were 37, 36, and 35 per line for center line, lines with even angles and lines with odd angles, respectively.

Figure 10.15 shows the particle trajectories for a 50 μm particle diameter of and velocity ratio of $\frac{V_p}{V_a} = 50\%$, injected toward the propeller nose location. It was slightly accelerated in the axial direction then impacting the propeller nose (point C), velocity decreased at first then after entering the propeller domain was accelerated due to centrifugal force and impacted the blade pressure side near the trailing edge (point P) at about 5% the blade span with a velocity about 45 m/s. It was accelerated again and left the domain after speeding for nearly 0.045 sec. The trajectories for several other particles are found in [22–25].

The Rosin-Rammler particle diameter distribution was assumed with a mean diameter of 162.5 μm to simulate desert particles. For takeoff flight phase and assuming $\mathbf{V_p} / \mathbf{V_a} = 100\ \%$, the calculated impact frequency and erosion rate are shown in Figure 10.16. The impact frequency on the left side of the figure clarifies that the pressure side is exposed to a great number of particles impacts than the suction side. The maximum impact frequency appears at the hub portion of the pressure side near the leading edge. This is due to the rebound of large diameter particles from the propeller nose (spinner). The particle impacts on the suction side are concentrated at the blade leading edge, especially from hub to 40% of the blade span.

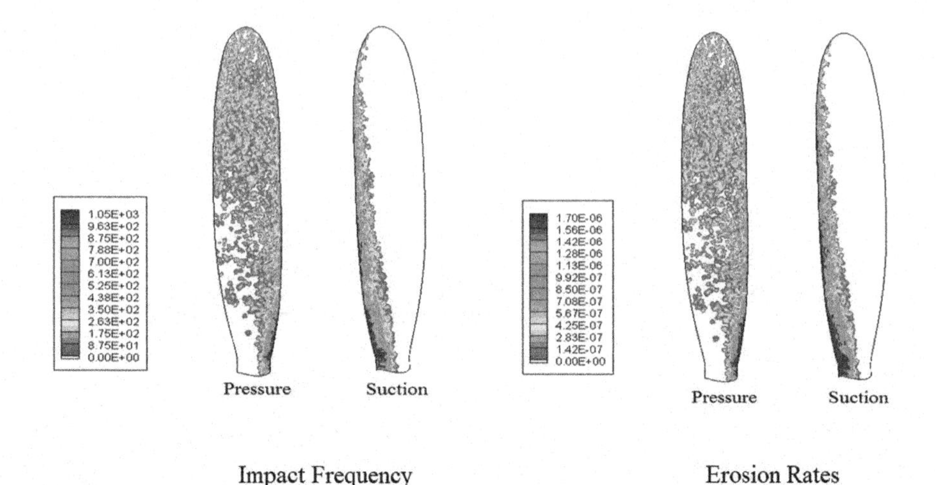

FIGURE 10.16 Left: impact frequency (*imp/cm².gm*), right: erosion rate (mg/gm), Takeoff flight phase and ($V_p/V_a = 100\%$) [24].

10.2.10 Piston Engines

Piston engines power many small fixed- and rotary-wing aircraft. Few research works handled the erosion of these engines. The author and his group examined their erosion [26, 27]. A three-dimensional mathematical model has been developed, and a modified version of KIVA-II code was used for its solution. Two approaches for the particles rebound behavior, including both the deterministic and the stochastic approaches, have been investigated.

A case study of a spark-ignition engine (SIE) having a bore and stroke of 100, 120 mm, ratio between lengths of connecting rod and crank of 2.7, compression ratio: 9, rated speed: 1,500 rpm, cooling water temperature: 400 K. Spark timing 20^0 before Bottom Dead Center (BTDC), intake valve opens at 5^0 BTDC and closes at 30^0 after Bottom Dead Center (ABDC), exhaust valve opens 30^0 before BDC (BBDC) and closes 5^0 after Top Deas Center (ATDC). Particles have mean diameter of 20 microns and volume fraction ratio of 35 $\frac{mg}{m^3}$.

The gas flow characteristics in the four strokes are calculated. The gas flow in the suction stroke is shown in Figure 10.17.

Figure 10.18 illustrates the particle motion inside the cylinder during a complete cycle. During the suction stroke, particles being influenced by the airflow are accelerated to the right (below the exhaust valve) impact the inlet valve face, then rebound to impact the valve seat in the cylinder head. The particles follow the air tumble motion during the suction stroke. During the pressure stroke, particles impact the piston face and the left side of the cylinder.

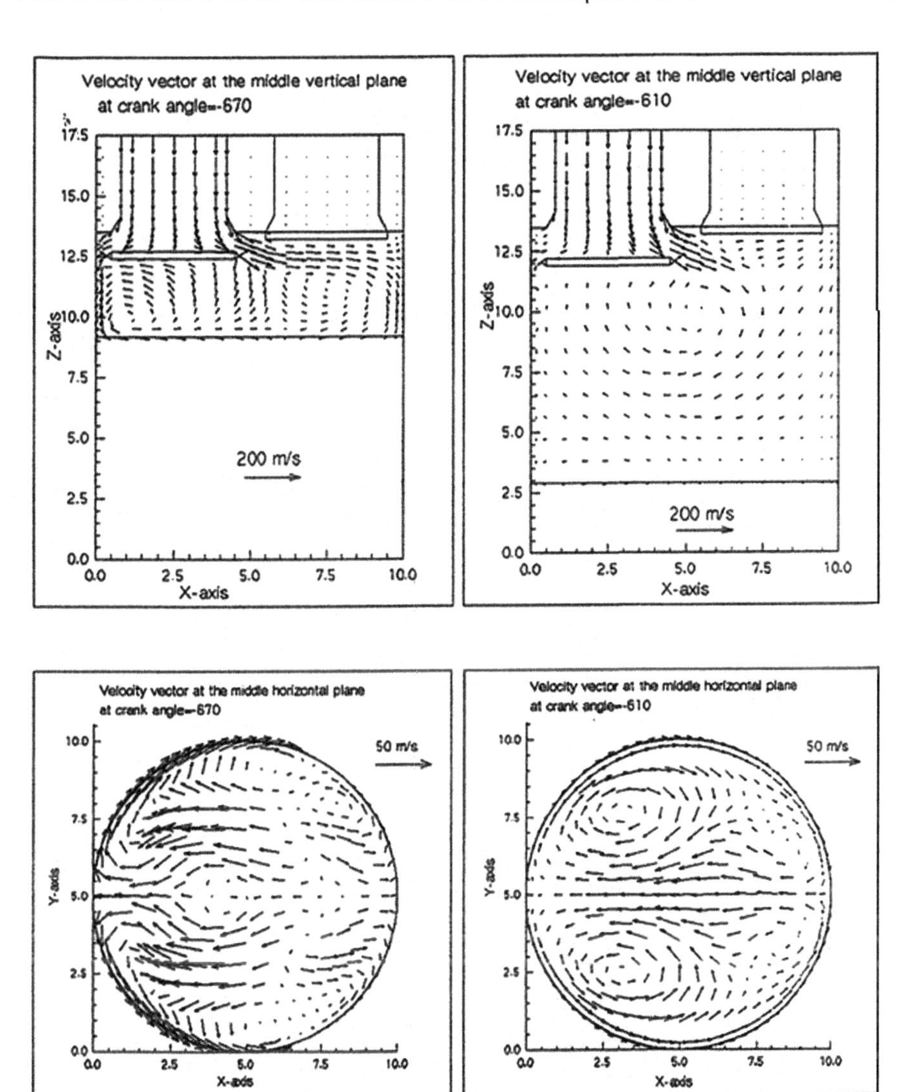

FIGURE 10.17 Airflow into the cylinder in suction stroke [26].

10.2.11 Volcanic Ash

When an aircraft flies through a volcanic eruption plume, it will suffer from both volcanic ash and gas aerosols. Plumes of ash spewed from volcanic eruptions may pose a significant flight safety hazard for aviation [90] as they decrease visibility [91, 92] and cause dangerous damage to airframes and especially to aeroengines [93].

Volcanic ash has the following impact on turbine-based engines: abrasion of engine inlet, electrostatic discharge (St. Elmo's fire) on engine cowls, erosion of compressor and turbine blades (thus reducing their efficiency), deposition on fuel nozzles and high-pressure turbine nozzles, clogging the cooling holes of turbine blades (causing

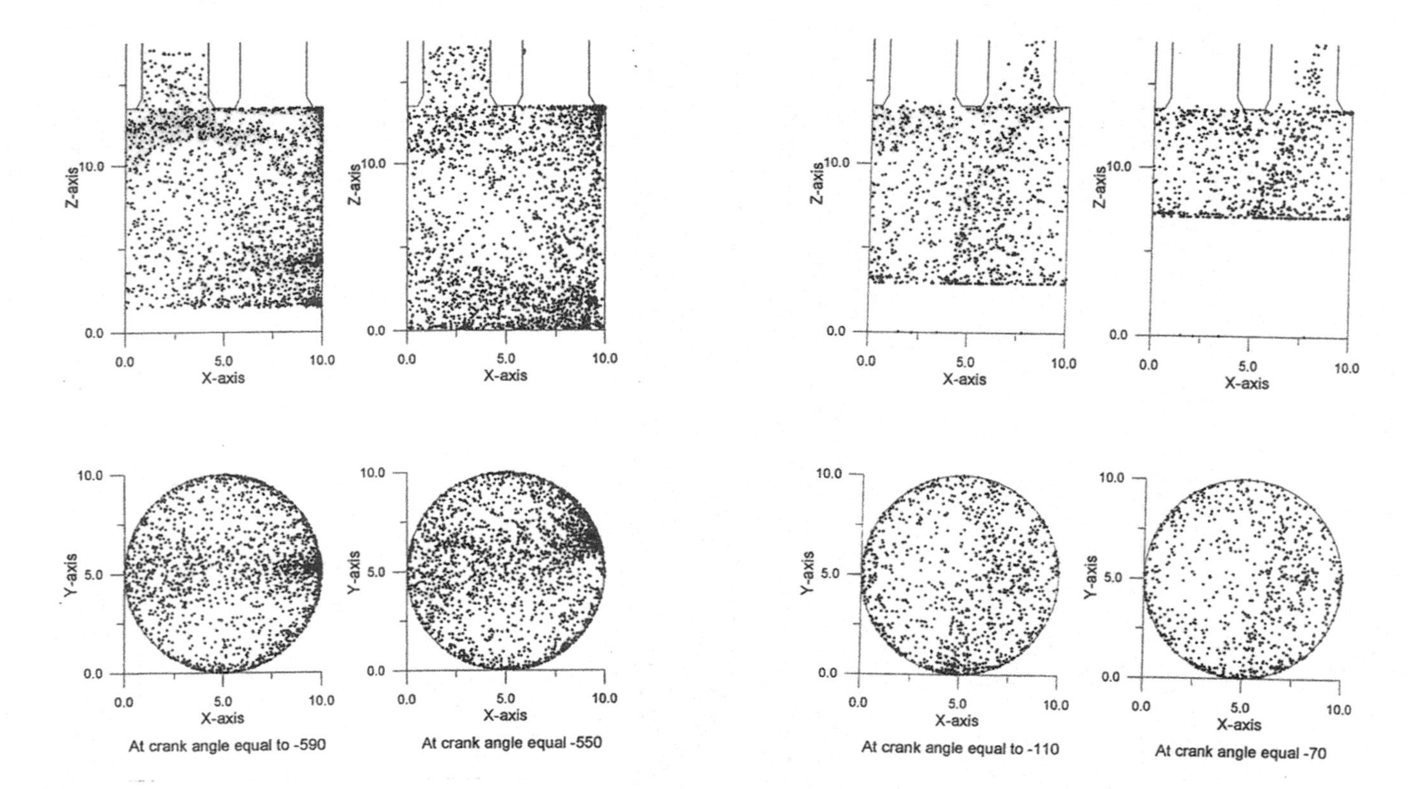

FIGURE 10.18 Particle distribution for the deterministic rebound case with dry walls at different crank angles [26].

the turbine to overheat and failure), torching from the tailpipe, and flameouts, vibration or surging of engine(s), engine failure or other damage leading to crash, and fluctuations in EGT with a return to normal values.

For shaft-based engines, it may cause clogging of engine oil filter with large diameter particles (greater than 10 microns), erosion of propeller of turboprop engine, and erosion of rotor of helicopter attached to turboshaft engine.

For APU, it may lead to abrasion of its inlet duct, erosion of compressor blades and other internal parts, and deposition of hot section parts

For aircraft windshield and passenger windows, it leads to fogging and abrasion and reduces visibility for pilots (partially or completely opaque), abrasion of passenger's windows, pitting, frosting, or breaking, and electrostatic discharge (St. Elmo's fire).

The numerical fluid dynamics particle tracking model (CD Adapco Star-CCM+) was employed for solving the ingestion of volcanic ash particles into the engine core section [94, 95] and the interaction with the fan surface. Two steps were followed:

1. Calculation of the airflow characteristics through realistic dimensions of an axial high bypass turbofan, engine intake, and other aerodynamically relevant parts. The 3-D steady-state fluid flow equations for the conservation of mass, momentum (Reynolds-averaged N-S equations of motion), and energy (Equations (10.2), (10.3), and (10.5)) were solved simultaneously using an implicit coupled flow and energy algorithm for compressible high-speed flows including a two-equation turbulence model (k–ε).
2. A stochastic Lagrangian particle-tracking model was used for the calculation of volcanic ash particle trajectories. This model is based on the translational force balance (Equations (10.9) and (10.10)) and rotational momentum balance of volcanic ash particles and considering the turbulent velocity fluctuations of the particles within the engine. The normal and tangential restitution coefficients for particle rebound after colliding with the internal surfaces of the turbofan were modified as follows:

$$\eta_T = \eta_N = 1.0 \tag{10.23}$$

The trajectories of volcanic ash inside the HBPR turbofan engine were traced in the following six cases [94–96]; two different flight altitudes (10,000 ft and 35,000 ft having different atmospheric pressure and temperature) and three fan power conditions: namely, low, intermediate, and high power, representing 35%, 75%, and 90% of the fan power maximal rotational speeds of the fan.

The volcanic ash particles were assumed to be spherical, rigid, and internally homogeneous solid particles with a mass density of $\rho_p = 2700\, kg/m^3$ which resembles most frequent volcanic ash. The particles were seeded from 400 injector points located 6 m in front of the engine intake and arranged in a 20 × 20 grid. The particle had the same velocity as the air at the same inlet point, 1,000 particles were injected at each of the 400 injection points. Simulations were carried out for 16 sizes ranging from $1\, to\, 100\ \mu m$. Thus, the total number of particles of each size was 4×10^5, and the total number of particles was 6.4×10^6.

The following findings of the study were:

1. The probability of particle collision (P_I) with the fan section of the turbofan engine was considered as follows:
    - At all fan speeds, the fine particles having diameters $D_p = 1-2\ \mu m$ do not often interact with any of the fan surfaces (P_I <5%)
    - For particles having diameters $D_p = 2-5\ \mu m$, ($P_I = 20-25\%$) at 35,000 ft and from ($P_I = 20-36\%$) at 10,000 ft
    - For particles having diameters $D_p \leq 20\ \mu m$, ($P_I = 40-60\%$) at different speeds and altitudes
    - For particles having diameters $D_p = 50-100\,\mu m$, ($P_I > 90\%$) at different speeds and altitudes
2. The ratio of the number of particles ingested into the engine core to the total number of particles seeded into the engine.
    - For fine particles having diameters $1-2\,\mu m$, at both altitudes and all engine operation conditions, no particle–fan interaction, or in other words all the ingested particles into the engine inlet, passed into the engine core
    - For large diameter particles (specifically $D_p = 100\,\mu m$), and at altitude 35,000 ft, core reduction factor = 2.53 for the high-power condition and = 2.02 for the low power condition. While for an altitude of 10,000 ft, core reduction factor = 2.32 for the high-power condition and =1.94 for the low power condition.

10.3 FOULING OF AIRBREATHING ENGINES

Operation of fixed-wing aircraft in a harsh environment characterized by a very poor air quality (or highly contaminated airborne) like that in the Far East, Middle East, and Africa may ingest many solid particles into airbreathing engines powering airliners [97]. Such *contaminants are a mix of organic and inorganic fine particles including sand, dust, volcanic ash, oil, soot, and salt.* They build up to form a coating on the internal surfaces is referred to as fouling.

Fouling of aeroengine modules has the following definitions:

- Accumulation of deposits on target surfaces causing an increase in surface roughness
- Adherence of particles to airfoils and annulus surfaces due to the presence of oil and water mists

It may then melt or soften during the combustion process and combine with sulfur and oxygen. When they strike the solid surfaces of the hot section some of them will adhere to the turbine guide blade, while others might rebound.

These particles may cause erosion or deposition of the internal modules of aeroengines depending on their sizes.

Large diameter particles ($\geq 20\,\mu m$) cause erosion of the fan, propeller, and compressors of turbine-based or shaft-based engines. However, these large particles may be broken-up by the compressor and arrive in the hot-section with sizes in the range 5–10 μm.

Moreover, in pursuit of higher efficiency, the peak temperatures of aeroengines must be increased, which is only possible via turbine cooling. Unfortunately, the deposited particles may clog the cooling holes adopted in turbines. Therefore, particle deposition has become an important issue in the 21st century and extensively researched both numerically and experimentally.

10.3.1 Axial Compressor

Small particles having sizes (0.1–10 μm) may deposit on the modules of cold section of engine or pass into the hot section. They may be fine dirt, oil vapors, soot, exhaust fumes, dust, pollen, insects, water vapor, salt, sticky industrial chemicals, and unburnt hydrocarbons. Fouling of compressor reduces their isentropic efficiency, tip clearance. However, it increases stalling potential and surface roughness. It also degrades the whole engine performance, which is manifested in the increase in fuel consumption, EGT, corrosion, flameout, as well as the decrease in the thermal efficiency, mass flow rate, thrust (or power) and time between overhauls (TBOs) [98, 99], which increases the overhaul costs and even lead to premature component failure. Turboshaft engines powering helicopters are strongly vulnerable to damage due to the deposition of sand and dust during brownout [100]. The rate of fouling in a compressor depends on the atmospheric condition (rain, humidity, or fog) and the engine installation and elevation of the intake. The compressor fouling can be controlled through inlet air filtration, or particle separators, and a suitable mix of online and offline compressor washing.

10.3.1.1 Compressor Washing

The deterioration of compressors is usually reversible, as the particles can be removed through water washing. Chemicals used in compressor wash break down the organic bonds of the contaminants and dissolve the inorganic material. This allows the contamination to be flushed out of the engine [101]. There are two broad categories of cleaners: solvent-based and detergents. Solvents work well to dissolve organic contamination, but inefficient with inorganic contaminations like salt. There are multiple types of detergent material, but it is recommended to use those with biodegradable, nontoxic, and nonflammable properties. These are safer to use, easier to store, and disposed of at a lower cost.

The frequency of compressor wash depends on the rate of decay in engine performance and the cost of the engine wash. Helicopters with shorter trips and low elevations or near the coast will accelerate the fouling rate. The operator can perform an economic analysis to define the break-even point for optimum wash frequency which may be daily for the saline environment, while helicopters infrequently used could be washed every six months.

There are two types of washing: online ("HOT" Washing) and offline ("COLD" Washing). For online washing, the fluid is commonly heated as listed in the engine manufacturers' maintenance manual. It is performed with the aeroengine operating and under load and recommended every week or so. *Offline washing* is performed with the engine shutdown. It is more thorough than online cleaning due to both the greater volume of cleaning solution that is passed through the engine and the much greater dwell (soak) time for the cleaner to work. It is generally recommended three to four times per year [102].

10.3.1.2 Numerical Modeling of Fouled Compressor

Stage-stacking method is used for calculating the performance of fouled multi-stage axial flow compressors [103, 104]. Since the deposited mass of particles is not uniform in different stages, the three compressor parameters flow coefficient ϕ, the *loading* (pressure rise) coefficient Ψ, and stage efficiency η will have different values at the same time for different stages. A fouling severity coefficient (FS) is introduced. It takes values between 0 (new and clean) and 10 for the most severe case. Fouling severity also depends on the aeroengine layout (single, double, or triple shaft) and expressed by the parameter (c_g)[99]. The following case study involved the abovementioned stacking method for examining the deterioration of fouled compressor. An eight-stage axial compressor have mass flow rate: 10.84 kg/s, inlet total conditions: 101.4 KPa and 288 K, the total pressure ratio: 11.53, and polytropic efficiency: 88.8%, and uniform fouling level of the first four stages. Figure 10.19 shows the relation between pressure ratio and mass flow rate in design and fouled condition after 2,000 h. As the compressor rotational speed increases, the drop in pressure ratio

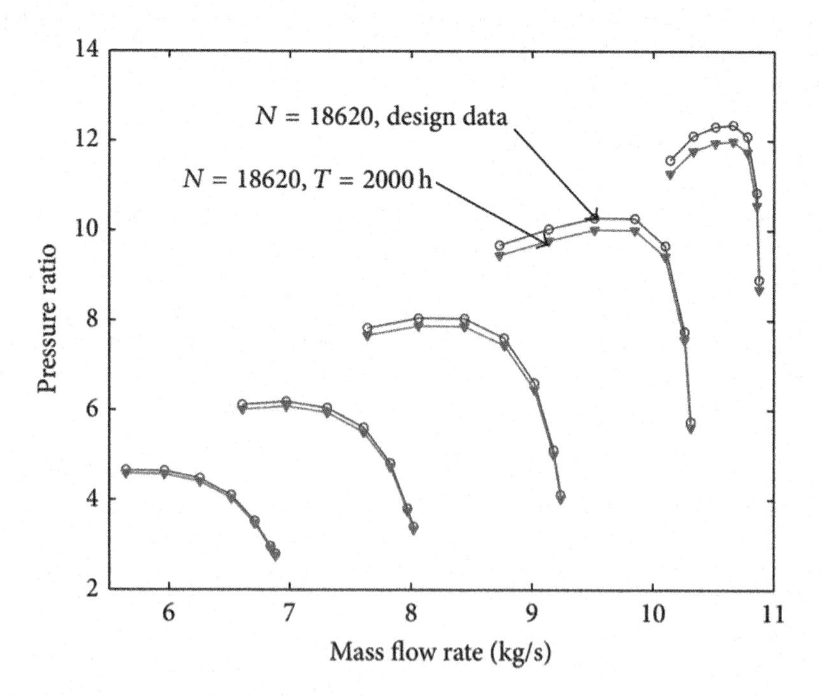

FIGURE 10.19 The pressure ratio curve for design and fouled compressor with different relative fouling severity coefficient after 2,000 h [103].

of the fouled compressor increases for 2,000 operating hours. Same trend is noticed for compressor efficiency.

10.3.2 Axial Turbines

The high-pressure turbine of aeroengine (especially the first stage stator) is subjected to the interaction with gas flowing from the combustion chamber. This gas is very hot and usually laden with ashes and unburned particles produced during fuel combustion. Such particles may deposit on the blade surface, changing the blade geometry, clogging the cooling holes which affect the heat transfer coefficient on the blade surface. Fouling caused the fatal crash of Bell-Boeing MV-22 powered by Rolls-Royce AE1107C turboshaft engine in Hawaii on May 17, 2015. This led to the stall of the compressor on the left Rolls-Royce AE1107C turboshaft engine and material buildup on the turbine blades [105]. Extensive numerical and experimental investigations for fouling were performed in the last 50 years [106–109].

Here an integrated approach based on an Eulerian (flow)–Lagrangian (particle) scheme is used for particle-laden flow in turbines. The three-dimensional aerodynamics of fouled turbines is analyzed following three steps for erosion analysis: namely, gas flow analysis, particle motion trajectories, and particle-wall interaction. The first two steps are described earlier in detail. The particle-wall interaction is investigated based on the bounce-stick model described in [107]. Particles will stick (or deposit) on the blade surface if the particle normal impact velocity is smaller than the particle capture velocity. Another model for particle sticking is proposed in [110]. The sticking propensity of a particle depends on the particle and target surface properties and temperature. Since the gas temperature is very high in gas turbines, the particles start to melt and become very sticky. The "sticking probability' approach is used to model the particle's stick/bounce based on the particle viscosity [111].

To examine the effect of film cooling on particle deposition, a high-pressure turbine, the AGTB-B1 turbine, was used. Its cascade had chord length, height, and pitch of 250, 300, and 178.5 mm, respectively. The stagger angle, inlet flow angle, and exit flow angles were $\beta_s = 73^0, \beta_1 = 133^0, \beta_2 = 28.3^0$, respectively. The turbine total inlet temperature TIT was (T_{t1} =1453 K). The inlet and exit Mach numbers were 0.37 and 0.95. The cooling hole diameter was 3 mm and located on the suction and pressure surfaces at a percentage of the chord length of 0.02 and 0.03. The particles had mass flow rate of $5.71\times10^{-6}\frac{kg}{s}$, density of $990\,kg/m^3$, and specific heat of 984 J/ (kg. K). Cooling was performed with a plenum and two rows of film-cooling holes, and each had 20 holes at the leading edge. One row was on the pressure surface (PS) inclined at angle 70^0 to the x-axis, and the second row was on the suction side (SS) inclined at angle 38^0 to the x-axis [106]. ANSYS FLUENT version 16.2 was applied for a structured multiblock grid with cell refinement near the blade surface having 0.1 million cells. Figure 10.20 illustrates the effect of particle size on its trajectories for the case of blowing ratio M = 0.93 and β_1 =133°. Small diameter particles ($dp = 1$ $and\, 5\,\mu m$) follow the gas streamlines and few of them impacted the blade surface. More collisions with the blade surface (especially its pressure surface PS) are

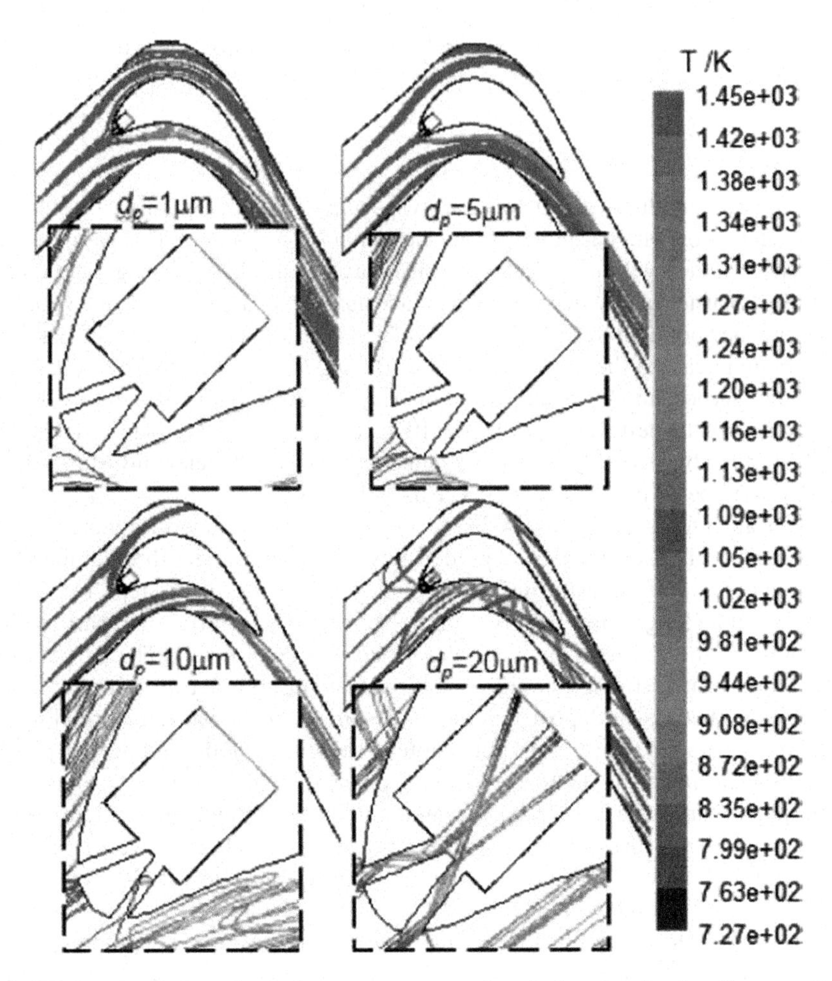

FIGURE 10.20 Trajectories and gas temperatures of particulate flow having diameters (1, 5, 10, 20 μm) and blowing ratio M = 0.93.

With Permission Taylor & Francis [106].

experienced by larger size particles. Few particles having a diameter of $dp = 10\ \mu m$ invade the film cooling holes. More particles had diameters of $dp = 20\ \mu m$ flow into the injection cavity of the coolant air. Coolant reduced the temperature of blades surface close to the two cooling holes with some 200°.

10.4 RAINFALL

When an aircraft flies in a rain condition, particularly during the stages of takeoff and landing, its flight safety is threatened by massive detrimental effects of rain. Rain effects an aircraft are as follows [41–45]: reduction of visibility, impart a downward and backward momentum, increase of the aircraft mass due to the thin water film, increase of the surface roughness and stresses producing aerodynamic lift and drag penalties, and affects the accuracy of measuring instruments on an aircraft like aircraft stall warning system, pitot tubes, and angle of attack (AOA) sensor. Severity of rainfall depends on its accumulation expressed in millimeters per hour. A 2,000 mm/h rain is characterized as incredible, a 500 mm/h rain is considered severe, and 100 mm/h is considered a heavy rate.

Aircraft materials subjected to rain erosion are normally either bulk type or coated. High-velocity rain droplet impact can generate pressures of many tons per square inch on aircraft surface. This pressure causes high stress. For ductile materials, it may create deformation. For brittle material like glass or thermosetting plastic domes and fiber-reinforced, it may cause cracking. In the case of laminates, the interface of the layers can be subjected to shear stresses, causing delamination [112, 113].

Erosion of aircraft modules [114] is normally caused by rain drops of size 1–5 mm. For civil aircraft, the impact speed is normally 250 m/s [115]. For fans of its aeroengines the impact speeds are normally 200–400 m/s [116]. While for military aircraft, the impact velocity occurs at Mach numbers greater than one [117]. The major factors influencing the erosion rate due to water drop impact are [118]: collision (impact) speed, water droplet size and shape, time, and design. Some liquid erosion theories are given by Thiruvengadam [119] and Springer [120], and others summarized in [121].

10.4.1 Numerical Simulation of Rain Effects on Aircraft

Air-rain problem is a typical gas-fluid two-phase flow problem, which can be treated either as Eulerian two-phase flow or Lagrangian particle tracking method [122, 123]. Eulerian method (two fluids) is employed when the concentration of the dispersed phase (here water droplets) is large. It treats the droplets distributed in the airflow field as quasi fluid permeating the gas phase and solves the conservation equations (mass, momentum, and energy) for each phase. Interphase exchanges of mass, momentum, and energy between the two phases are included as source terms in the appropriate conservation equations. A turbulence model like k-ε (RNG) or the Spalart–Allmaras turbulence models are employed. This model is better and easy to implement for particles of uniform size. The second method is adopted when the particle's concentration is small enough; thus, the particles' (raindrops) equations of motion based on Newton's second law are solved to track the trajectories of the droplets in the airflow field [123]. Starting from the particles' injection points, the equations of motion of particles are integrated. When the particle strikes the surface, it forms a thin liquid film. The main assumptions for such water film model are as follows [124, 125]: the layer is thin, less than 500 microns in thickness, the temperature in the film particles changes slowly and always below the liquid boiling temperature, film particles are in direct contact with the wall surface and only conduction heat transfer between the liquid film and the wall is considered. Particle's post-impact behavior depends on the Weber number defined as:

$$W_e = \frac{\rho_p D_p \bar{u}_r^2}{\sigma_p} \tag{10.24}$$

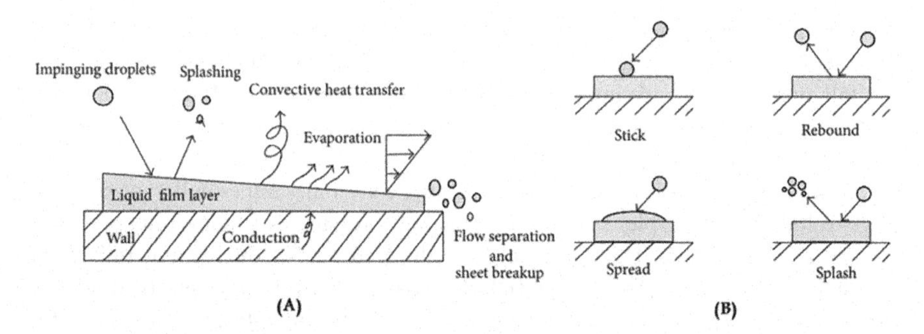

FIGURE 10.21 Water droplet impingement with aircraft surface. (A) Mechanisms of mass, momentum, and energy transfers for the wall-film model; (B) the four impingement regimes for drop impact with the wall.

$$\bar{u}_r = \bar{u}_p - \bar{u}_{wall} \tag{10.25}$$

Where $\rho_p, D_p, \sigma_p, \bar{u}_r$ are the density, diameter, surface tension of the rain drop particle and the relative velocity with respect to the wall, respectively.

Four regimes are encountered as shown in Figure 10.21: namely, stick, rebound, spread, and splash. The (stick) regime ($W_e < 5$) occurs when an impinging drop has extremely low impact energy and adheres to the film surface [46], the (rebound) regime ($5 < W_e < 10$) occurs when the impinging drop has low impact energy and bounces off the film, the (spread) regime ($10 < W_e < 3672$) occurs when the particle has high energy and merges with the liquid film upon impact; the fourth regime (splash); ($W_e > 3672$) occurs when the particle has effectively high energy, and thus splashes and breaks up into a cloud of secondary droplets.

The following case study is discussed in detail in [126] treating the NACA64-210 airfoil having a chord length of 0.762 m, and the air free-stream velocity is 48.4632 m/s and the corresponding Reynolds number was 2.6×10^6. The flow is turbulent and both the $k-\varepsilon$ and Spalart–Allmaras models are used. The operating conditions are air-free stream total pressure and total temperature of $101 kPa$ and 273 K, liquid water content (LWC) of 25 g/m^3 The computational domain consists of an extruded O-type grid around the NACA 64-210 airfoil, and AOA up to 20°. The main findings of the study were: a shift of stall angle from 15° in the dry condition to 13° in the wet condition, decrease of the maximum C_L by 13.2%, increase in C_D by 47.6%. Figure 10.22 shows the raindrop traces by at AOA = 0°, 4°, 8°, 12°, 16°, and 20° at time of $t = 0.125$ s.

At AOA = 0°, raindrops concentrate on both the top and the bottom surfaces of the airfoil. At AOA = 4°, there have been few raindrops on the top surface close to the trailing edge and many others on the bottom surface were flowing from the stagnation point to downstream. At AOA = 8°, there were no raindrops on the bottom surface (a full separation), and the discrete phase began to separate on the top surface. At AOA = 20°, a maximum degree of coupled separation is reached, and large-scale vortices appear leading to the runback of rivulets on the upper surface.

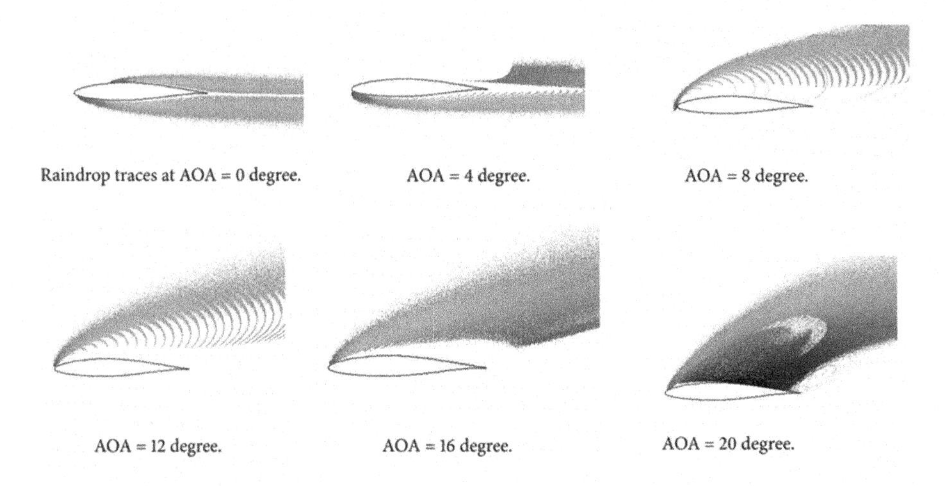

FIGURE 10.22 Raindrop traces at AOA = 0°, 4°, 8°, 12°, 16°, and 20° [125]

10.5 ICING

10.5.1 Introduction

Ice accretion is encountered when an aircraft flies through clouds below 26,000 ft at subsonic speeds. The shape, location, and type of the accreted ice depend on both meteorological and aircraft-specific factors [127]. Meteorological factors include cloud types, droplet diameter, LWC, outside air temperature, pressure altitude, terrain factors, variations with the season. A cloud is a mass of water drops or ice crystals suspended in the atmosphere. There are many different types of clouds [128–130].

Clouds are categorized primarily by two major factors – location and shape (Figure 10.23) [131]. One way categorizes clouds based on its height in the sky (high, middle, and low). Another way clouds are named by their shape. Cirrus clouds are high clouds. They look like feathers. Cumulus clouds are middle clouds. These clouds look like giant cotton balls in the sky. Stratus clouds are low clouds. They cover the sky like bed sheets.

Icing simulation can be performed through the following three methods [132, 133]:

- Flight testing
- Experimental testing
- Computational simulation

10.5.2 Computational Methods

Computational methods are composed of three steps. The first two are same as those used in rainfalls [134]: namely, the Eulerian two-phase flow method and the Lagrangian particle tracking method. In the Eulerian approach, the following commercial codes

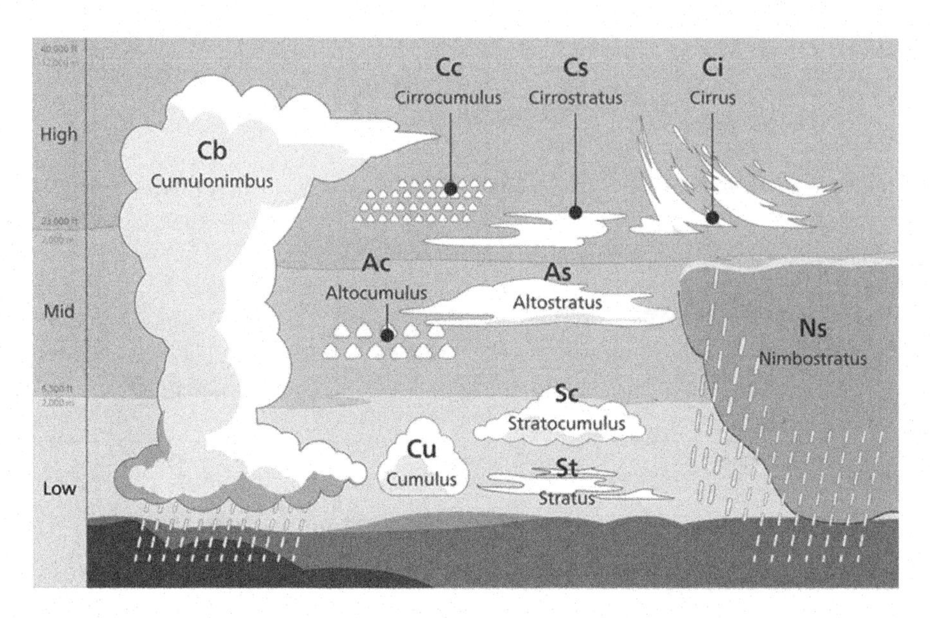

FIGURE 10.23 Tropospheric cloud classification by altitude of occurrence [131].

are used: FENSAP-ICE [135], CANICE [136], CIRAAMIL [137], FLUENT, and CFX. The codes using Lagrangian approach to compute droplet trajectories through the air include: LEWICE 2D/3D [138–141], ONICE 2D/3D [142, 143], ONERA 2D [144, 145], and TRAJICE2 [146, 147].

The airflow is first calculated using the Eulerian method as described earlier; next, the water droplet trajectories are traced using the Lagrangian method. The third step is the calculation of local and total impingement (or collection) efficiencies [148–149] for multi-disperse droplet distributions.

The collection efficiency is a measure of how efficient a surface at collecting super-cooled liquid water droplets. There are two collection efficiencies: local and total [97, 148, 149]. Figure 10.24 illustrates a 2-D and 3-D icings. For a 2-D case, (S) is assumed to be zero at the leading edge (or stagnation point) and is assumed to be positive on the upper surface and negative on the lower surface. Now consider a small segment of (Yo) or (δYo), and divide this by the corresponding small segment of the airfoil arc length, (δS), that droplets in (δYo) would impinge. Denote (S_u) and (S_l) as the upper and lower limits of droplet impingement on the airfoil. Now the local collection efficiency, β, is defined as the limit of (δYo) as (δS) approaches zero, or:

$$\beta = lim_{\delta S \to 0}\left(\frac{\delta Y_0}{\delta S}\right) = \left(\frac{DY_0}{DS}\right) \tag{10.26}$$

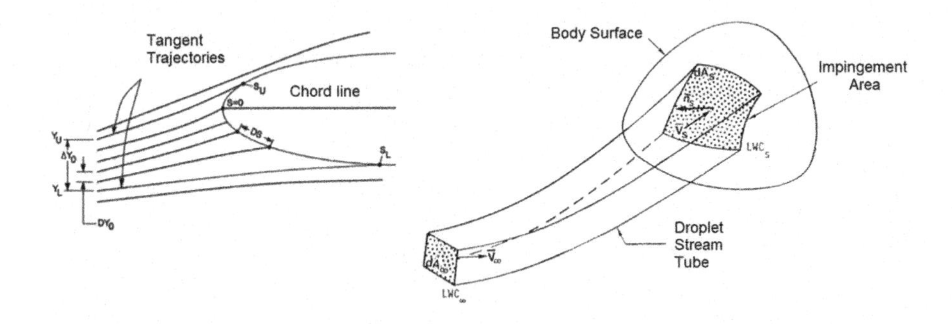

FIGURE 10.24 Definition of three-dimensional local and total impingement efficiency [97, 149].

Courtesy: NASA and Department of Transportation.

The total and local collection efficiencies will then be related by the following expression:

$$\Delta Y_0 = \int_{S_L}^{S_u} \beta DS \tag{10.27}$$

For a three-dimensional case, the local impingement efficiency values may be calculated from the ratio of the area delimited by some trajectories (usually four trajectories) away from the body to the corresponding impinged region formed by the impact trajectories intersect the face and is expressed by Equation (10.28):

$$\beta = \frac{dA_\infty}{dA_s} \tag{10.28}$$

Where dA_s is the area of the element on the body surface and dA_∞ is the far upstream tube formed by the position from droplets injection points which will impact the panel surface.

The total collection efficiency for multi-sized water droplets is given by:

$$\beta = \sum_{bin=1}^{N_{bin}} F_{bin} \beta_{bin} \tag{10.29a}$$

Where each bin corresponds to a specific droplet diameter, and the sum of F_{bin} for all droplet-bins must be equal to one, or

$$\sum_{bin=1}^{N_{bin}} F_{bin} = 1.0 \tag{10.29b}$$

Based on the local and total collection efficiencies, the ice accumulated on the surface (of airfoil or engine intake, compressor, etc.) within a specified time is calculated, and the new surface shape is defined after a specified time. The effect of ice accretion on the aerodynamic performance (drag, lift, stall angle, and speed) is next evaluated by incorporating an interactive boundary layer code that calculates lift and drag changes [150, 151]. For a second time interval, airflow is recalculated using the Eulerian approach and droplet trajectories are recalculated using the Lagrangian method. The new shape of the surface after a second-time interval is recalculated. The effect of ice accretion on the aerodynamic performance (drag, lift, stall angle, and speed) is also evaluated. The previous steps are repeated up to the total time planned where the final shape of the module surface is obtained. The above-described steps were applied for different case studies, as shown hereafter.

10.5.3 Icing of Wing Section NACA 23012

The NACA 23012 airfoil is representative of general aviation and commuter aircraft wing sections [152]. The simulated ice shapes for airfoil were determined using the NASA Glenn LEWICE 2.2 computer code. The icing conditions are: $V\infty = 175\,mph$,

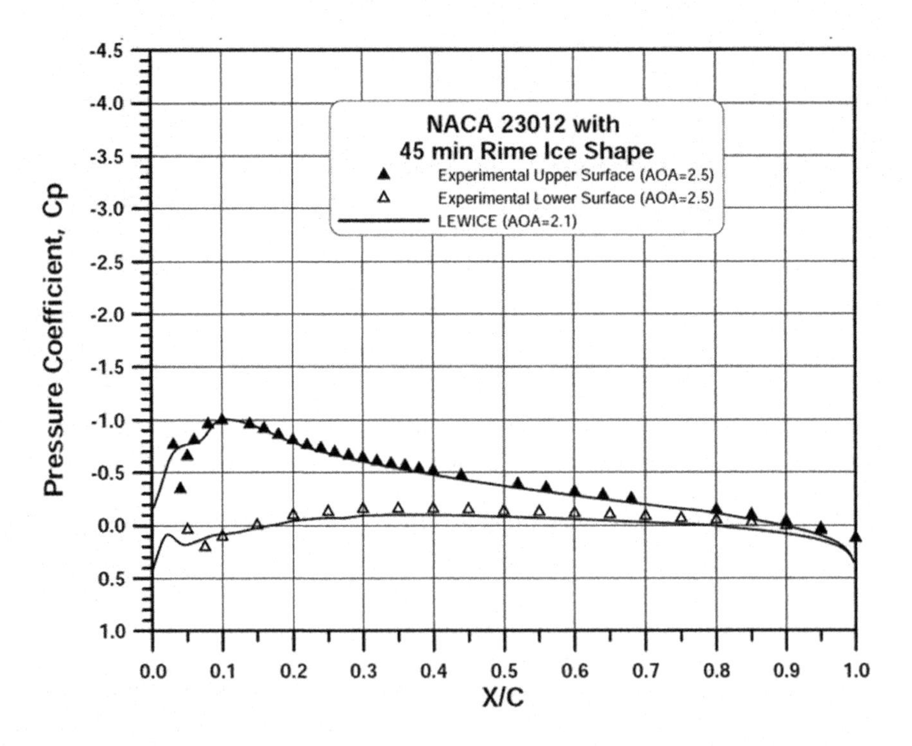

FIGURE 10.25 Comparison of pressure distribution for NACA 23012 with 45-min rime ice shape at $\alpha = 2.5$ [152].

Courtesy: NASA.

$AOA = 2.5^0$, $MVD = 20\mu m$, $LWC = 0.5 g/m^3$ and the pressure altitude is 1,800 ft, where the static pressure is approximately 94,806 Pa. Figure 10.25 illustrates a comparison of the experimental and computational pressure distribution for rime ice pattern after 45 min and AOA $\alpha = 2.5$.

The trajectories of droplets were computed using the LEWICE 1.7 code for a 45-min glaze ice shape for different mean volume diameters (MVD) as presented in Figure 10.26. The trajectories of the smallest droplet having diameter of 16.3 μm experienced considerable deflection in between the horns. The remaining nine droplet sizes in the distribution ranging from 63.7 to 1046.8 μm experienced small deflection of the trajectories until nearly 508.5 μm. For droplets larger than 508.5 μm, the trajectories were practically straight. Since the LEWICE code does not simulate large droplet splashing, then the trajectory simulations for the large MVD cases do not include droplet splashing effects.

Another study for icing of NACA 23012 is performed by the author and his group [153, 154]. The three cases of icing (glaze, mixed, and rime) were handled based on the ice shape accredited on NACA 23012 airfoil after different times. The same operating conditions stated above were followed. The total temperatures were 268 K for glaze icing, 264 K for the mixed case, and 252.3 K for rime icing. FLUENT 6.3.26 code with multi-block grid technique was used. Aerodynamics were calculated using N-S flow solver and Spalart–Allarams turbulence model. The computational domain has a total number of grids of 9,720, the volume of the smallest grid $6.782134e(-006)\ m^3$. Figure 10.27 illustrates the velocity profile after 45 min for clean, glaze, mixed, and rime cases.

10.5.4 Icing of Intake of Turbofan Engines

The icing of intake of an HBPR turbofan engine reduces its air mass flow rate, the engine propulsive characteristics, and the aircraft aerodynamic characteristics. The intake studied here by the author and his group [155–157] is the subsonic "Pitot-type" circular intake of an HBPR turbofan engine like GE-CF6-50, GEnx, GE90, P&W1500, RR Trent 1000. Figure 10.28 shows the main dimensions of the examined intake.

The computational domain is shown in Figure 10.29, including a sector having an angle of 15° (as the fan contains 18 blades).

Numerous case studies were examined for flight Mach numbers varying from 0.02 during takeoff and 0.85 for the cruise flight phase. Figure 10.30 illustrates progressive ice accretion for fan spinner and intake lips after 5, 10, 15, and 20 min for flight at altitude 7,000 m, Mach number 0.6, 50-micron droplet diameter, and LWC of 10 gm/m^3.

The local collection efficiency reached 0.95 at the leading edge of intake lips and extended deeply in its interior.

10.5.5 Icing of Turbofan Engine

Icing of commercial high bypass turbofan engine may take place at high altitudes due to ice crystal ingestion into the intake and compression system components [158–159].

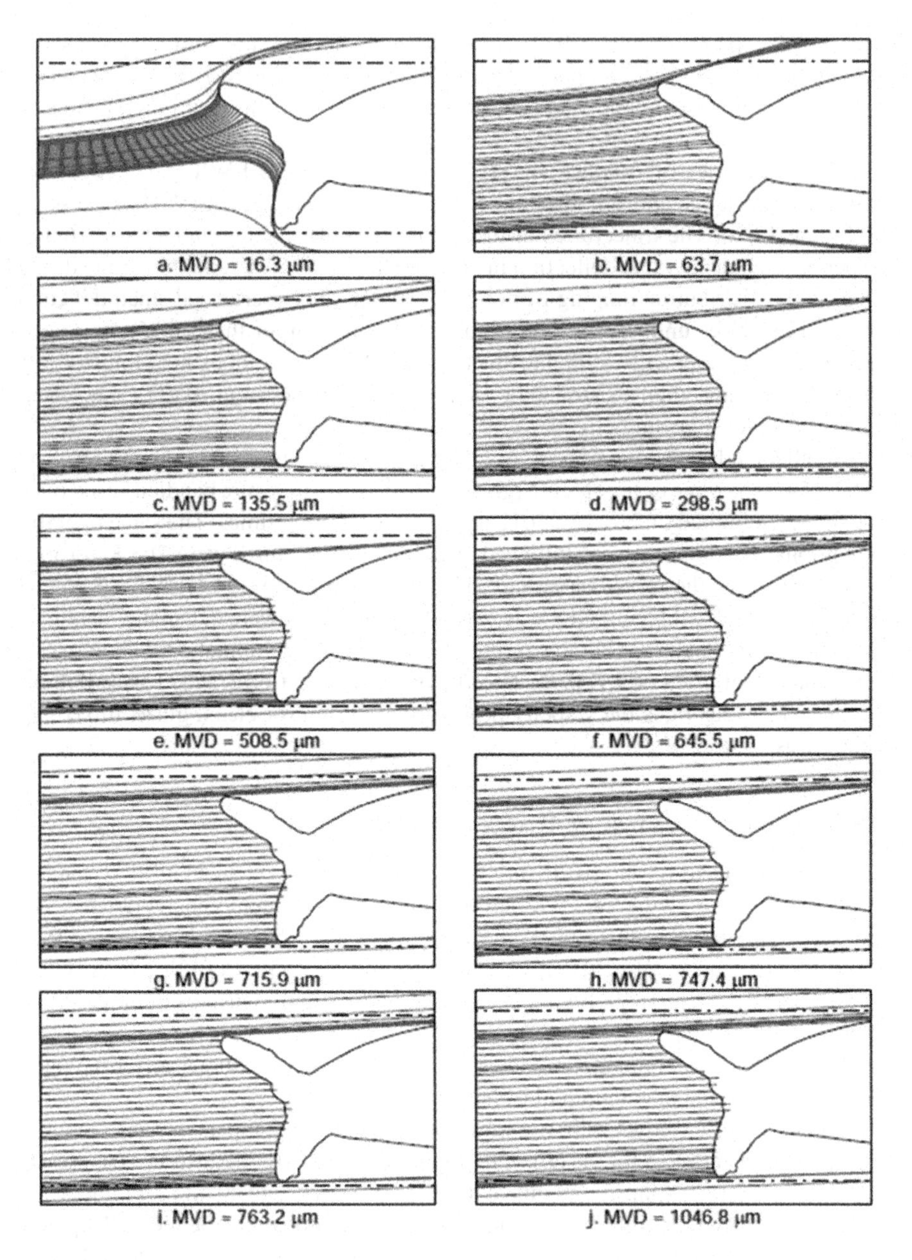

FIGURE 10.26 Computed trajectories for 45-min glaze icing of NACA 23012 using LEWICE code [152].

Courtesy: NASA.

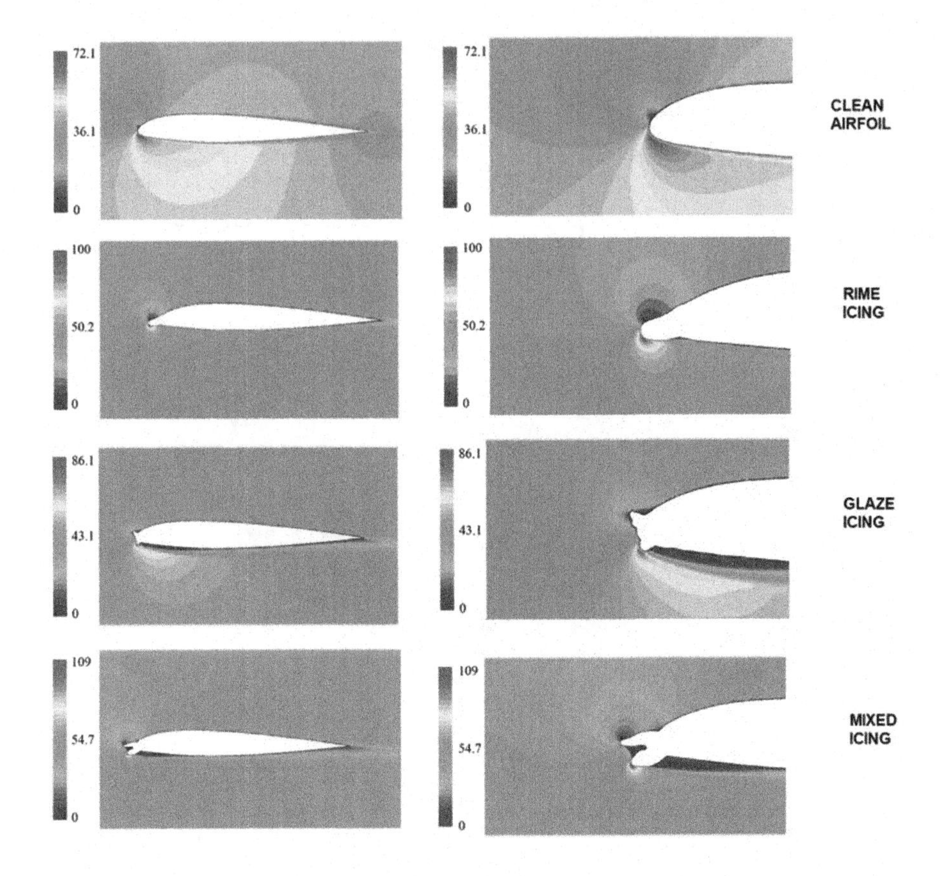

FIGURE 10.27 The velocity profile after 45 min for clean, glaze, mixed, and rime cases [153].

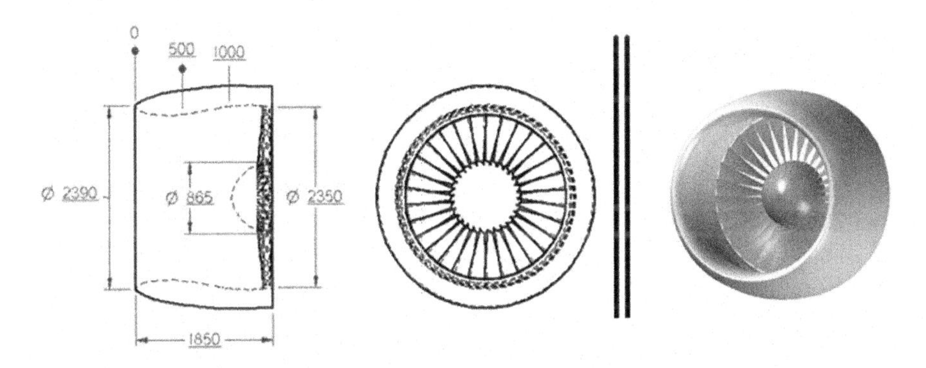

FIGURE 10.28 The main dimensions of the intake of high bypass ratio turbofan General Electric CF6-50 engine [155].

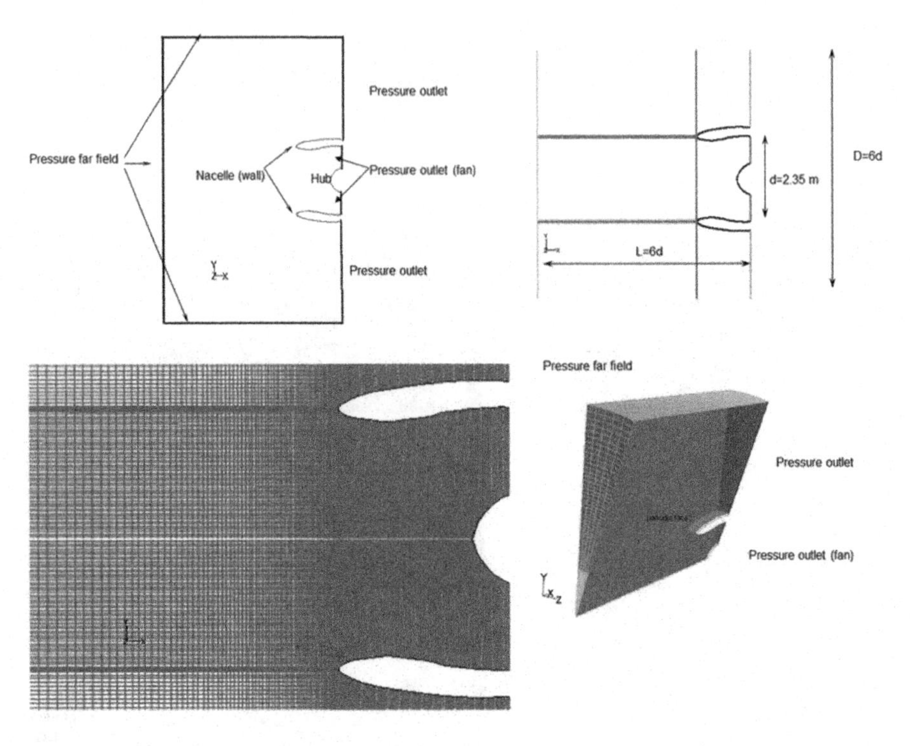

FIGURE 10.29 Computational domain [156].

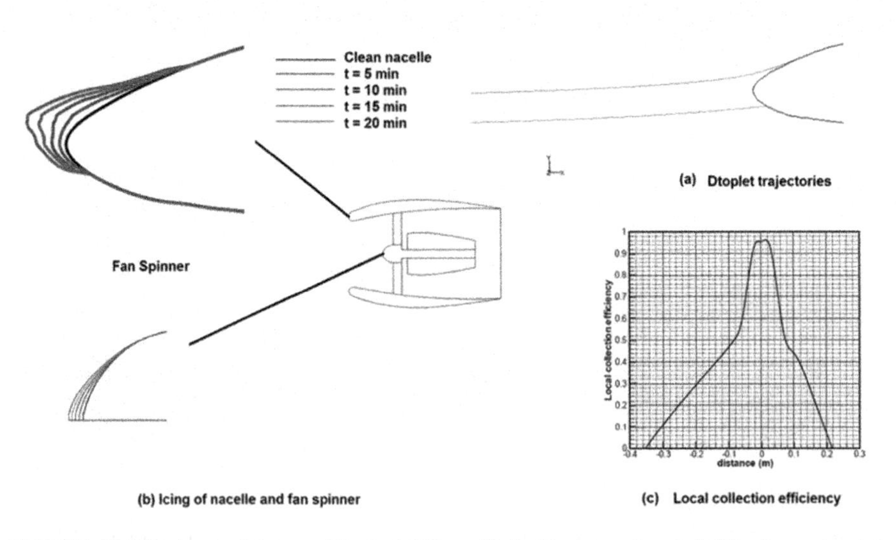

FIGURE 10.30 Intake icing at altitude 7,000 m, flight Mach number 0.6, 50-micron droplet diameter, and LWC equals 10 gm/m^3 [155, 156].

This may lead to engine performance degradation and one or more of the following: an engine rollback (an un-commanded decrease in thrust accompanied by a decrease in fan speed and an increase in turbine temperature), compressor surge or stall, and flameout of the combustor. Ice shedding may also result in mechanical damage to the engine, which is hazardous during flight.

As ice crystals ingest into the fan and low-pressure compression system, the increase in air temperature causes melting of a portion of them. Then, the ice–water mixture will cover the metal surfaces of the compressor stationary components and will lead to ice accretion through evaporative cooling.

In NASA Glenn Research Center, the following icing codes are used to simulate the flow through a turbofan engine:

- An aero-thermodynamics of engine system cycle code (CD)
- A compressor flow analysis code (COMDES)
- An ice particle melt code for examining the rate of sublimation, melting, and evaporation through the compressor flow path (MELT)
- C-MAPSS40k code for calculating and plotting compressor map.

An analysis of icing effects on the performance of a turbofan engine powering an aircraft during its cruise condition: 25,000 ft altitude, Mach 0.75, ISA+10, and a cruise throttle setting is performed. A comparison of the nominal LPC map and the 20% blocked map is shown in Figure 10.31. It shows that:

1. There is a large decrease in the LPC surge line.
2. As the shaft speed increases, the speed lines get dramatically steeper.
3. A decrease in the core mass flow rate, which will force the engine controller to increase the fuel flow rate into the combustor to generate enough power to produce the desired EPR setpoint.

Another detailed study for ice accretion of The Fan of GE-NASA energy efficient engine (E^3) is described in [160, 161]. Numerical simulation of ice accretion in aircraft engines is performed using a Eulerian–Lagrangian approach for the continuous and discrete phases with a one-way coupling model to simulate momentum and energy exchange between the phases.

Simulations are performed in four distinct interconnected modules:

1. Airflow aerodynamics is obtained by ADPAC flow solver by solving the three-dimensional RANS equations in rotating frame of reference
2. Calculation of droplet trajectory path and impingement location in a Lagrangian approach
3. Impingement statistics calculation and determination of flux-based collection efficiency using TURBODROP code
4. Ice shape predictions using the LEWICE icing physics.

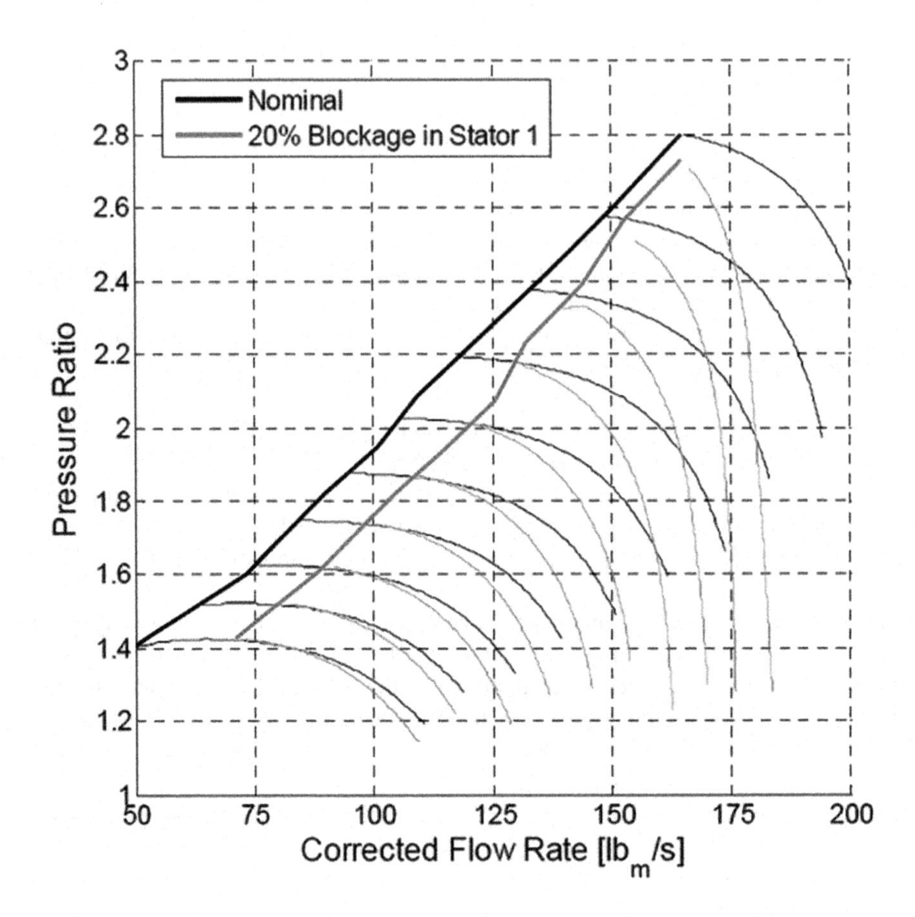

FIGURE 10.31 Constant corrected speed lines and surge lines of the nominal LPC map compared with those of the LPC map with a 20% blockage in Stator 1 [159].

Courtesy: NASA.

The main findings were:

- Collection efficiency over the rotor blade pressure surface is maximum at the leading edge near the tip, increases with reduced rotor speed, and reaches its maximum at 60% design speed
- The maximum ice accretion takes place near the hub and gradually decreases toward the tip at 60% and 70% design speeds.

10.5.6 Icing of Aircraft Propeller

In-flight icing could occur when an aircraft flies in a temperature range between −40 °C and 0 °C. In these conditions, supercooled large droplets can form from small droplets which freeze upon contacting with a propeller. The shape of the formed ice depends on the

LWC, the droplet diameter, the airspeed, and ambient temperature. For low temperatures (−40 to −10 °C), low airspeed, and low LWC, the ice immediately freezes and forms "Rime ice." At higher temperatures (−18 to 0 °C), higher airspeed, and high LWC, the forming of "Glaze ice" has a double horn. These glaze ices are more dangerous than rime ice, and both alter the shape of the airfoil changing the aerodynamic characteristics, adding weight, and possibly bringing the airfoil out of balance [162].

10.5.6.1 Case Study 1

10.5.6.1.1 2-D Case Study

A propeller has four propeller blades, rotational speed of 420 rpm, axial flow speed of 102.89 m/s along the x-axis, a 0° AOA, ambient static temperature, and density of 263.6 K and $1.190\frac{kg}{m^3}$, and flight Mach number $M_\infty = 0.396$, [163]. Other data is: LWC = $0.5\ g/m^3$, MVD = 15 μm, and spray time = 472 s.

First, FLUENT code was used for calculating the airflow field. For a 2D solution of this problem, the propeller was cut into seven slices having radii as (r = 1.35, 1.95, 2.55, 3.15, 1.65, 2.25, 2.85 m). The relative (or apparent) velocity was constructed by combining the free stream and rotational velocity. Their values were (119.24, 134.77, 153.43, 174.21, 126.54, 143.78, 163.60 m/s). Since the geometry of the blade is curled, all sections needed their own mesh. N-S equations were solved. The geometry got thinner on the outer parts of the blade, causing less distortion of the flow. Second, the trajectories of the droplets were traced. Third, the collection efficiency (β) was calculated for different sections.

The main conclusions of this case study were:

1. For the mid-sections of the propeller geometry, the radial velocity component is nearly zero close to the propeller. However, this is not the case for slices near the nacelle and the propeller tip. The assumption for the two-dimensional flow is therefore only valid for a limited section of the propeller geometry
2. Increasing the propeller radius led to more droplets impinging the profile. This could be explained by the significantly higher velocity the section experiences on these outer regions
3. An increase in the slice radius led to more ice. However, after the geometry started to get slimmer near the tip, less droplets were caught, resulting in less ice accretion.
4. An anti-icing analysis conducted proved that to keep the propeller clean of ice, a power peak is necessary near the stagnation point of the geometry irrespective of the section and amount of ice built up on the different sections. All slices needed about the same amount of total energy, which was 15 kW/m.

10.5.6.2 Case Study 2

A 3D CFD methodology is presented in [164] to simulate ice accretion on propeller blades, with automatic blade pitch variation at constant rpm. The study handled one blade (2 m long) of the six-blade propeller installed to a turboprop engine powering a 70-passenger twin-engine aircraft. The operation parameters were flight speed of

190 knots, altitude of 6,000 ft, and rotation speed of 850 rpm. Two icing conditions were simulated:

- Air static temperature of −23 °C, LWC 0.2 g/m^3, and MVD 20 microns resulting in rime ice
- Air static temperature of −16 °C, LWC 0.3 g/m^3, and MVD 20 microns resulting in mixed ice with rime to glaze transition in the radial direction.

The geometry of the clean and iced propeller was automatically meshed at each shot using ANSYS FENSAP-ICE and Fluent Meshing tools. Solutions for the blade were computed in a periodic domain to reduce computational costs.

10.5.7 Icing of Helicopter Rotor

The cruise height of helicopter is normally less than 6,000 m. Since the height of icing weather condition in FAR-29 Appendix C is less than 6,700 m, then it is likely to encounter icing conditions during flight. Moreover, if the flight speed of helicopter is small, then it is difficult to escape from icing clouds in a short time. Consequently, icing may seriously threaten the flight safety of helicopter and even lead to the casualties and property loss [165, 166]. Ice accretion of helicopter will affect the main rotor, tail rotor, windshield, and engine inlet. Since the main rotor is the most critical component of helicopter, then its icing is more serious than other parts. The icing of helicopter rotor is a very complicated process. Both experimental and numerical research are used to study the rotor icing [167–169].

10.5.7.1 Numerical Method

In a forward flight, the airflow field and droplet impingement characteristics of the main rotor blade vary with the different azimuth angles. Thus, the rotation period is divided into several time steps. In each time step, the airflow field, the droplet trajectory, and the ice growth rate of the rotor blade are calculated. Finally, the ice thickness in each time step is calculated [170].

- Airflow field calculation

The unsteady N-S equations together with Spalart–Allmaras turbulence model are solved. A second-order central difference finite volume scheme is employed.

- Droplet trajectory calculation

The Eulerian method is employed where the supercooled droplets in the clouds are considered as a continuous phase. Then, the N-S equations may be used to solve the droplet trajectory equations. The droplet collection efficiency is next calculated.

- Ice accretion calculation

The mass of ice accretion and ice shapes at different sections of the rotor blade are obtained. The effect of centrifugal force on the icing process should be considered in the rotor icing calculation.

10.5.7.2 Case Study

A Caradonna and Tung rotor in non-lifting forward flight is considered. The Mach number at the blade tip is $M_{atip} = 0.8$ and the advance ratio is m = 0.2. A NACA 0012 airfoil section is chosen for the rotor, and the chord length is set to 1 m. The flow velocity is 138.88 m/s, the AOA is 5°, and the droplet mean volume diameter (MVD) is 16 mm. The ice shapes of different blade sections in forward flight are given in Figure 10.32. The icing occurs mainly at the leading edge of the blade. The icing thickness increases as the spanwise distance increases.

The numerical results are in good agreement with the experimental results. The ice shape of a three-dimensional rotor is different from a two-dimensional airfoil because of the rotation effect.

The local velocity of the rotor blade is influenced by the rotation speed and the advance ratio. The change of local velocity will affect the magnitude and distribution of droplet collection efficiency. The values of droplet collection efficiency and ice growth rate are larger in the windward half-circle ($azimuth\,angle = 0^0 - 180^0$), and smaller in the tailwind half-circle ($azimuth\,angle = 180^0 - 360^0$).

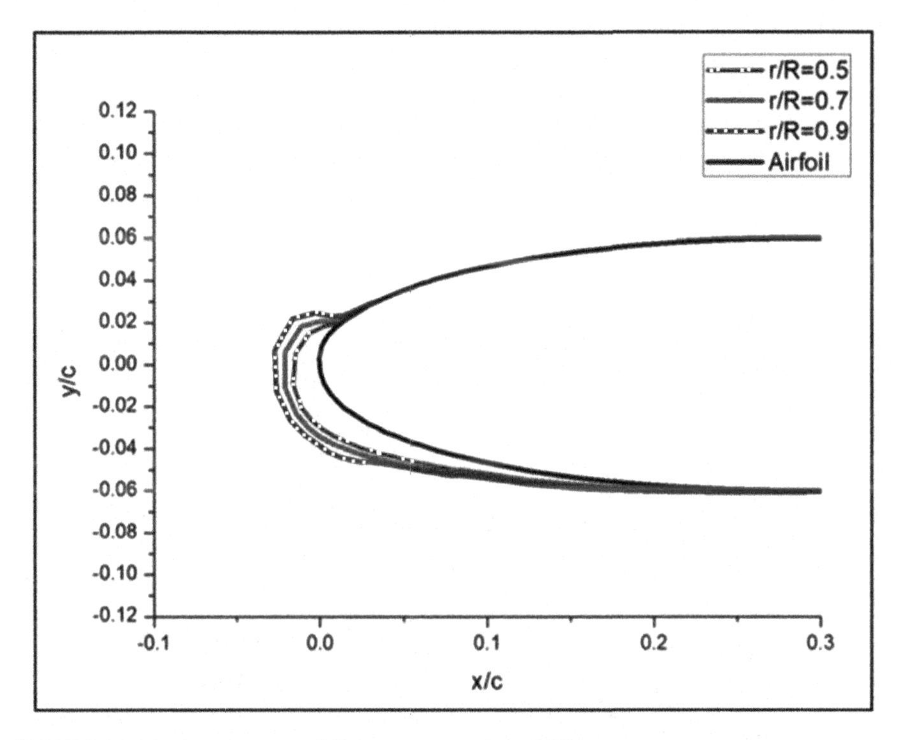

FIGURE 10.32 Ice shapes at different rotor sections. [170].

10.5.8 Anti-icing Systems

10.5.8.1 Introduction

Anti-icing systems are preemptive which are turned on before the flight enters icing conditions. They are applied to wing, tail (horizontal and vertical), and control surfaces such as flaps and ailerons airfoil leading edges as well as engine nacelle lips, propellers, pitot tubes, fuel vents, and windshields. At present, the thermal ice protection method is the most prominent anti-icing method used in large transport aircraft. The thermal method is mainly divided into the engine bleed hot air (used for wing, empennage, radome) and the electro-thermal heater mats (used for propellers, windshields, pitot tubed systems) [171].

Examples for anti-icing systems in aircraft [172]:

- C-130: Engine bleed air for wing and empennage leading edges, radome, engine inlet, and electrical heating for propellers, windshield, and pitot tubes
- Boeing 777: Engine bleed air for engine inlet cowls and three wing leading edge slats
- Boeing 787: Electro-thermal methods for wing leading edges [173].

10.5.8.2 Hot Air Anti-icing System

High temperature, high-pressure air is bled from the engine compressor and channeled through a piccolo tube mounted inside the leading edge of the wing [174, 175]. A series of hot air jets emanate from small holes on the piccolo tube (piccolo holes) and impinge on the internal surface of the leading-edge skin, transferring heat and increasing the skin temperature to prevent ice accumulation (Figure 10.33).

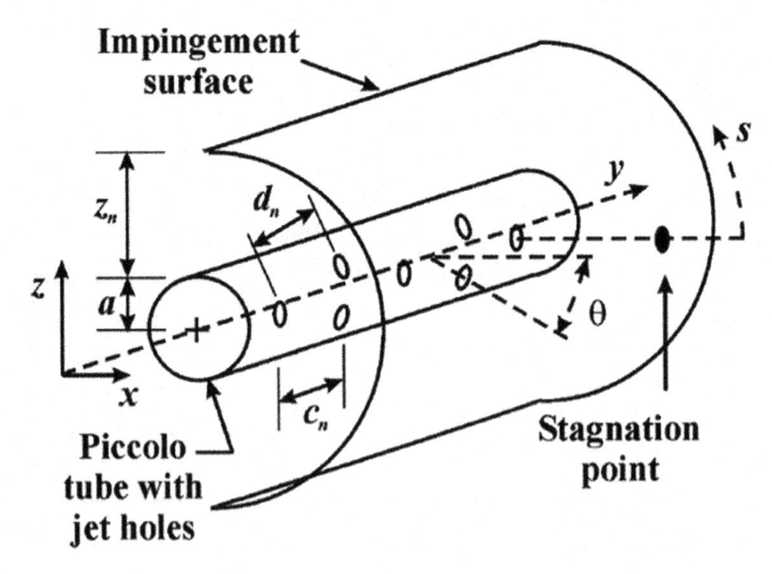

FIGURE 10.33 Geometric parameters for hot air anti-icing system [176].

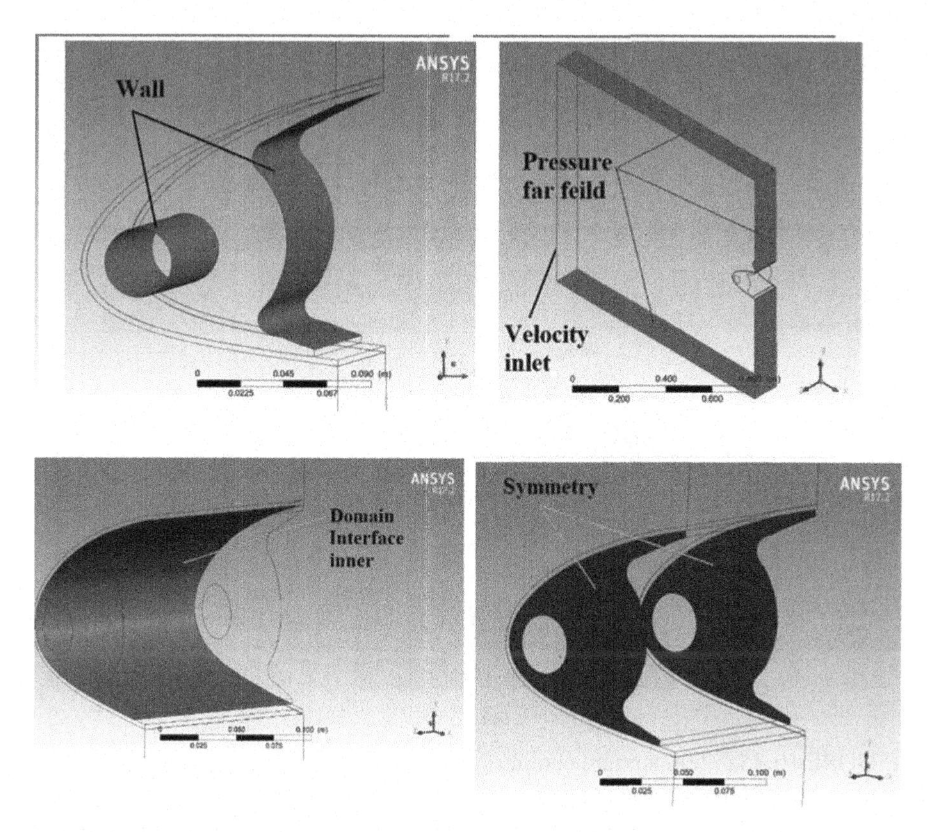

FIGURE 10.34 Numerical model boundary conditions [176].

The parameters governing this technique are the piccolo tube diameter, the distance between piccolo tube and airfoil inner surface, number of jet holes, angle of impingement of hot air ensuing from piccolo tube, and number of jet holes and spacing between two successive holes (c_n, d_n).

An anti-icing analysis using ANSYS CFX 17.2 for NACA 23012 aircraft wing section involving hot air jets from a piccolo tube was investigated numerically. It has the following data: a partial-span model of length 124 mm, three airflow speed values: 59.2, 100, and 200 m/s, two altitudes: 3,000 and 5,000 m, four values of AOA: −6°, 0°, 5°, and 12°, three values of internal bleed temperature: 350, 415, and 449 K, and two values of jet mass flow rate 0.07 and 0.035g/s per hole [175, 176]. Steady-state CFD analysis of the internal (indoor) hot airflow, external (outdoor) dry cold airflow field of the wing, and the thermal conduction in the solid skin were carried out using Commercial ANSYS CFX 17.2 software. Radiation heat transfer was not included. Computational domain was decomposed into three parts (Figure 10.34): namely, solid surface of wing, hot air, and cold air.

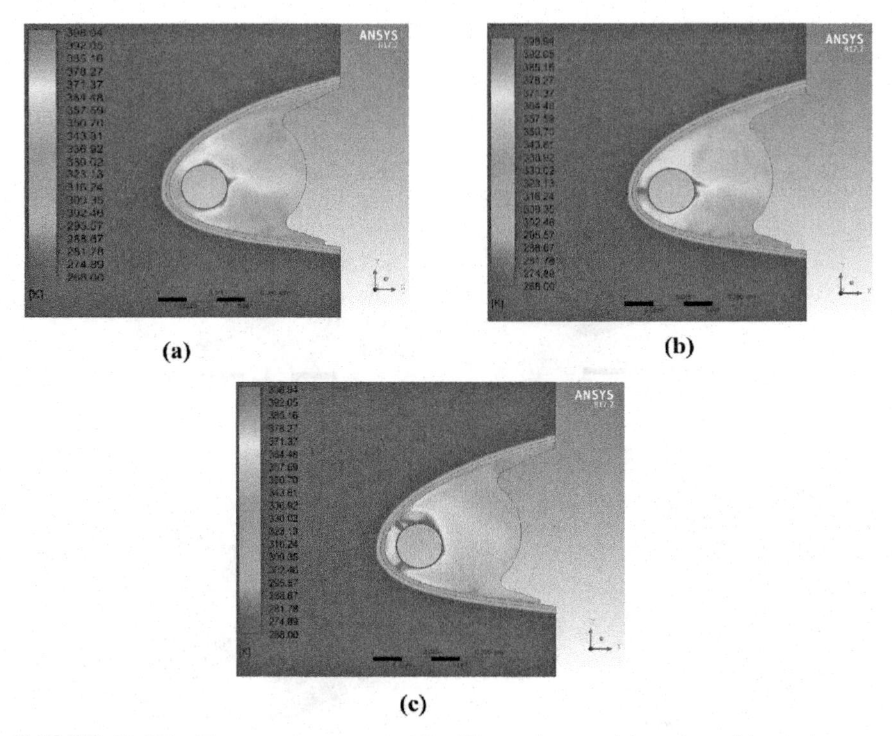

FIGURE 10.35 Temperature contour for 2D sections at (a) no jets, (b) one jet at zero angle, and (c) two jets at angles ±45°) [176].

Figure 10.35 illustrates the temperature contours in 2D sections in the following cases:

No jets: leading to maximum temperature in the inner surface of leading edge of 378 K, one jet at zero angle: leading to maximum temperature in the inner surface of leading edge of 398 K, and two jets at angles ±45° leading to maximum temperature of 398 at the impingement angles (±45°) In all cases, the temperature in the inner surface of leading edge was 364 K. The findings of this study confirmed that:

(a) Performance of anti-icing system was slightly influenced by changing the AOA.

(b) Airflow speed influenced anti-icing efficiency. By increasing the flight speed from 59.2 to 100 m/s, the system efficiency was improved, but decreased for a flight speed of 200 m/s.

(c) When increasing the flight altitude from 3 km (T = 268 K, P= 70.121 kPa) to 5 km (T = 255 K, P = 50.003 kPa), the LE average temperature was reduced by 49 K. So, to ensure an efficient system performance, the bleed air temperature, and mass flow rate should be increased.

(d) Reducing the temperature of bled air from 449 K to 415 K, and 350 K, the air temperatures were insufficient to prevent the formation of ice; thus, the system becomes unsafe.

(f) Reducing bleed air mass flow rate from 0.07g/s per hole to 0.035 g/s per hole reduced the LE average temperature from 318 to 302.7 K, but the system was still safe.

REFERENCES

[1] Ahmed Fayez El-Sayed, A.F. FOD in Intakes – A Case Study for Ice Accretion in the Intake of a High Bypass Turbofan Engine, RTO-EN-AVT-195, September 2011.

[2] El-Sayed, A.F. and Emeara, M.S. Aerodynamics of intakes of high bypass ratio (HBPR) turbofan engines, Internal Report ZU-18-21, Zagazig University, 2021.

[3] El-Sayed, A.F. and Emeara, M.S. Aerodynamics of intakes of high bypass ratio (HBPR) turbofan engines, International Robotics & Automation Journal, 6(2), 88–97, 2020.

[4] Gourgy, J.S. and Bosnyakov, S.M. Sand Particles Motion Simulation in the Vicinity of Turbine Engine Intake, 4th International Conference ERBA-2018, July 3–5, 2018.

[5] Ahmed, H.Z.H. Solid Particulate Flow in an Axial Transonic Fan in a High Bypass Ratio Turbofan Engine, M.Sc. Thesis, Mechanical Power Engineering Department, Zagazig University, 2006.

[6] Balan, C. and Tabakoff, W. 4Axial flow compressor performance deterioration. In Proceedings of the 20th Joint Propulsion Conference, Cincinnati, OH, USA, June 11–13, 1984; p. 1208.

[7] El-Sayed, A.F. and Rashed, M. I. I. Particle Trajectories in Centrifugal Compressors, Proc. of Gas Borne Particles Conf., Inst. of Mechanical Engineers, June 30–July 2, 1981, Paper C61/81, I. Mech. E. 1981, Oxford, UK.

[8] El-Sayed, A.F. and Rashed, M.I.I. Particulate Flow in Centrifugal Compressors Used in Helicopters, Proc. 5th Int. Symposium on Air Breathing Engines (5th ISABE), Bangalore, India, February 16–22, 1981.

[9] Rashed, M.I.I. and El-Sayed, A.F. Erosion in Centrifugal Compressors, Paper No.55, Proc. 5th Int. Conf. on Erosion By liquid and Solid Impact (ELSI V), Newnham College, Cambridge University, UK, September 3–6, 1979.

[10] Elfeki, S. and Tabakoff, W. Particulate Flow Solutions Through Centrifugal Impeller With Two Splitters, 86-GT-130, ASME 1986 International Gas Turbine Conference and Exhibit, June 8–12, 1986, Dusseldorf, West Germany.

[11] Abdel, A.F., El-Sayed, A. and Brown, A. Computer Predictions of Erosion Damage in Gas Turbines, ASME 87-GT-127, 1987.

[12] Abdel Azim El-Sayed, A.F. and Brown, A. Iterative procedure for estimating erosion in aircraft turbomachines, Int. Rep. Royal Military College for Science, RMCS-86-102, 1986.

[13] Clevenger, W.B., Jr. and Tabakoff, W. Erosion in Radial Inflow Turbines – Erosive Particle Trajectory Similarity, vol. 1, NASA CR-134589, 1974.

[14] Clevenger, W.B., Jr. and Tabakoff, W. Erosion in Radial Inflow Turbines –Trajectories of Erosive Particles in Radial Inflow Turbines, vol. 2, NASA CR-134677, 1974.

[15] Clevenger, W.B., Jr. and Tabakoff, W. Erosion in Radial Inflow Turbines – Trajectories of Erosive Particles in Radial Inflow Turbines, vol. 3, NASA CR-134677, 1974.

[16] Clevenger, W.B., Jr. and Tabakoff, W. Erosion in Radial Inflow Turbines – Erosion Rates on Internal Surfaces, vol. 4, NASA CR-134700, 1975.

[17] Clevenger, W.B., Jr. and Tabakoff, W. Erosion in Radial Inflow Turbines – Computer Programs for Tracing Particle Trajectories, vol. 5, NASA CR-134787, 1975.
[18] Eroglu, M. and Tabakoff, W. Effect of Inlet Flow Angle on the Erosion of Radial Turbine Guide Vanes, vol. 112, January 1990 Transactions of the ASME, pp. 64–70.
[19] Ghenaiet, A. Simulations of Particulate Flows Through a Radial Gas Turbine Proceedings of 11th European Conference on Turbomachinery Fluid Dynamics & Thermodynamics ETC11, March 23–27, 2015, Madrid, Spain.
[20] Clevenger, W.B., Jr. and Tabakoff, W. Erosion in Radial Inflow Turbines – Computer Programs for Tracing Particle Trajectories, vol. 5, NASA CR-134787, 1975.
[21] Eroglu, M. and Tabakoff, W. Effect of Inlet Flow Angle on the Erosion of Radial Turbine Guide Vanes, Vol. 112, January 1990 Transactions of the ASME, 64–70.
[22] Farghaly, M.B.S. Erosion of Propeller Blades for Turboprop Engines, M.Sc. Thesis, Aerospace Engineering Department, Cairo University, 2012.
[23] Farghaly, M.B., El-Sayed, A.F., and Salem, G.B. Study the Effect of Erodent Particle Size on the Propeller Blades Erosion for Turboprop Engines, MDPI, J. Fluids, 2018.
[24] Farghaly, M.B., El-Sayed, A.F., and Salem, G.B. Effect of Erodent Particle Initial Velocity on the Erosion of Propeller Blades for Turboprop Engines, ASME 2012 Gas Turbine India Conference, November 2012, 10.1115/GTINDIA2012-9731.
[25] Farghaly, M.B., El-Sayed, A.F., and Salem, G.B. Numerical Simulation of the Aerodynamic Behavior of Propeller Blades at Subsonic Conditions, ICFD11-Eg-4064, Proceedings of ICFD11: Eleventh International Conference of Fluid Dynamics December 19–21, 2013, Alexandria, Egypt.
[26] Abdel Azim El-Sayed, A.F., Salmawy, H.A., Shamloul, M.M., and Mesalhy, O.M. Prediction of Erosion Inside the Engine Cylinder Due to Particulate Flow, SAE 1999-01-0005, Int. Congress and Exposition, March 1–4, 1999, Detroit Michigan, USA.
[27] El-Sayed, A.F.A.A., El-Salmawy, HA., Mesalhy, O.M.S., and Shamloul, M.H. Effect of Automotive Filtration in Cairo City on the Erosion and Oil Contamination of Internal Combustion Engines, Advancing Filtration and Separation Solutions for the Millennium, 12th Annual Technical Conference and Expo, Sponsored by American Filtration & Separation Society, Hynes Connection Center, Boston, MA, April 6–9, 1999, USA.
[28] Ulke, A. An approximate analysis of the effect of secondary flows on the motion of particulates in an axial flow gas turbine, PhD thesis, Carnegie-Mellon University, Pittsburgh, 1975.
[29] Ulke, A. and Rouleau, W. T. The effects of secondary flows on turbine blade erosion. ASME Paper No 76-GT-74.
[30] El-Sayed (Abdel Azim), A.F., Lasser, R., and Rouleau, W.T. Effects of secondary flow on particle motion and erosion in a stationary cascade, International Journal of Heat and Fluid Flow, 7(2), 146–154, June 1986.
[31] Abdel Azim El-Sayed, A.F. and Rouleau, W.T. The effect of different parameters on the trajectories of particulates in a stationary cascade, 3rd multi-phase flow and heat transfer symposium, 1983.
[32] Lasser, R. The effect of secondary flows in an arbitrarily curved passage of converging cross section and in the wake downstream of a cascade of airfoils. PhD thesis, Carnegie-Mellon University, Pittsburgh, 1982.
[33] Abdel Azim El-Sayed, A.F. and Rouleau, W.T. Particulate Flow in stationary cascade, Int. Report CMU-22-1983.
[34] Calvert, M.E. and Wong, T-C. Aerodynamic Impacts of Helicopter Blade Erosion Coatings AIAA-2012-2914.

[35] Rouzee Givens, E. Polyurethane as Erosion Resistant Material for Helicopter Rotor Blades, USAAML Technical Report 65-39, 1965.
[36] Xupeng Bai, Yongming Yao, Zhiwu Han, Junqiu Zhang, and Shuaijun Zhang, Study of Solid Particle Erosion on Helicopter Rotor Blades Surfaces, MDPI, Appl. Sci. 2020, 10, 977; doi:10.3390/app10030977
[37] Gedikli, H. and Özen, İ. Investigation of solid particle erosion behavior on erosion shield of a helicopter rotor blade, Pamukkale University Journal of Engineering Sciences, 26(1), 68–74, January 2020, file:///C:/short%20courses%20KANSAS%20University/EROSION/Helicopter%20erosion%20rotor%20shield.pdf
[38] Osvaldo, M. and Zuñiga, V. Analysis of Gas Turbine Compressor Fouling and Washing Online, Ph.D. Thesis, Cranfield University, School of Engineering, 2007.
[39] Yang, H. and Xu, H. The New Performance Calculation Method of Fouled Axial Flow Compressor, The Scientific World Journal Volume 2014, Article ID 906151, 2014.
[40] Jombo, G., Pecinka, J., Sampath, S., and Mba, D. Influence of Fouling on Compressor Dynamics: Experimental and Modelling Approach, J. Engineering for Gas Turbines and Power, Transaction of the ASME, 140(3), 032603, March 2018, Paper No: GTP-17-1202, https://doi.org/10.1115/1.4037913
[41] Wan, T. and Pan, S-P. Aerodynamic Efficiency Study Under the Influence of Heavy Rain Via Two-Phase Flow Approach, ICAS 2010.
[42] Rhode, R.V. Some effects of rainfall on flight of airplanes and on instrument indications, NASA TN-903, 1941.
[43] Valentine, J.R. Airfoil Performance in Heavy Rain. Transportation Research Record, No. 1428, January 1994, 26–35.
[44] Valentine, J.R. and Decker, R.A. Tracking of Raindrops in flow over an airfoil. Journal of Aircraft, 32(1), 100–105, February 1983.
[45] Valentine, J.R. and Decker, R A. A Lagrangian-Eulerian Scheme for Flow Around an Airfoil in Rain. International Journal of Multiphase Flow, 32(1), 639–648, 1995.
[46] Yihua Cao, Zhenlong Wu, Zhengyu Xu, Effects of rain fall on aircraft aerodynamics, Progress in Aerospace Sciences, 71 (2014) 85–127.
[47] Ranaudo, J. Discoveries on Ice, NASA flight research leads to improved safety for flight in icing conditions, Flight Safety Foundation, February 21, 2017, https://flightsafety.org/asw-article/discoveries-on-ice/
[48] Murphy, J. Intake Ground Vortex Aerodynamics, Ph.D. Thesis, Department of Aerospace Sciences, Cranfield University, Cranfield, UK, 2008.
[49] De Siervi, F., Viguier, H., Greitzer, E., and Tan, C. Mechanisms of inlet vortex formation. Journal of Fluid Mechanics, 124, 173–207, 1982.
[50] Shin, H., Greitzer, E., Cheng, W., Tan, C., and Shippee, C. Circulation measurements and vertical structure in an inlet vortex flow field. Journal of Fluid Mechanics, 162, 463–487, 1986.
[51] Murphy, J., MacManus, D., and Taylor, M. A quantitative study of intake ground vortices, September 2007, ISABE Paper 2007-1209.
[52] Murphy, J. and MacManus, D. Ground vortex aerodynamics under crosswind conditions, Experiments in Fluids, 50, 109–124, January 2011.
[53] Zantopp, S., MacManus, D., and Murphy, J. Computational and experimental study of intake ground vortices. The Aeronautical Journal, 114, 769–784, December 2010.
[54] Mishra, N., MacManus, D., and Murphy, J. Intake ground vortex characteristics, Journal of Aerospace Engineering, 226, 1387–1400, 2012, Proceedings of the Institution of Mechanical Engineers Part G.

[55] Emeara, M. Aerodynamics of Intakes of High Bypass Ratio (HBPR) Turbofan Engines, Ph. D. Thesis, Department of Mechanical Power Engineering, Zagazig University, 2017.
[56] Rossmann, A. Erosion Damage; Aeroengine Safety, https://aeroenginesafety.tugraz.at/doku.php?id=5:53:532:532#prettyPhoto
[57] Elghobashi, S. On predicting particle-laden turbulent flows. Applied Scientific Research, 52(4), 309–329, 1995.
[58] Greifzu, F., Kratzsch, C., Forgber, T., Lindner, F., and Schwarze, R. Assessment of particle-tracking models for dispersed particle-laden flows implemented in OpenFOAM and ANSYS FLUENT, Engineering Applications of Computational Fluid Mechanics, Taylors & Francis, 10(1), 30–43, DOI: 10.1080/19942060.2015.1104266 https://doi.org/10.1080/19942060.2015.1104266
[59] .NSYS FLUENT, 14.5. User's and theory guide. Canonsburg, Pennsylvania, USA: ANSYS, Inc., 2014.
[60] OpenFOAM. User Guide (2nd ed.). OpenCFD Ltd, 2014.
[61] Launder, B. and Spalding, D. The numerical computation of turbulent flows. Computer Methods in Applied Mechanics and Engineering, 3(2), 269–289, 1974.
[62] Batch, H.E. and Haselbacher, H. Numerical Investigation of The Effect of Ash Particle Deposition on the Flow Field Through Turbine Cascades, ASME Paper GT-2002-30600, ASME TURBO EXPO June 3–6, 2002, Amsterdam, The Netherlands.
[63] Menter, F. Two-equation eddy-viscosity turbulence models for engineering applications. AIAA Journal, 32(8), 1598–1605, 1994.
[64] Tschirner, T., Pfitzner, M., and Merz, R. Aerodynamic Optimization of an Aeroengine Bypass Duct OGV-Pylon Configuration, Proceedings of ASME turbo EXPO 2002, June 2002, Amsterdam, The Netherlands.
[65] L´opeza, A. Nichollsa, W., Sticklanda, M.T., and Dempster, W.M. CFD study of Jet Impingement Test erosion using ANSYS Fluent and OpenFOAM, Elsevier B.V. 2015 http://creativecommons.org/licenses/by-nc-nd/4.0/
[66] El-Sayed, A.F. Aerodynamics of Air Carrying Sand – Erosion of Centrifugal compressors, Ph.D. Thesis, Cairo University, 1979.
[67] Smeltzer, C.E., Gulden, M.G., and Compton, W.A. Mechanisms of metal removal by impacting dust particles. Journal of Basic Engineering, 92, 639, 1970.
[68] Duffy, R.J.et al. Integral Engine Inlet Particle Separator, Volume II. Design Guideusaamrdl-TR- 75-31B, 1975.
[69] Grant, G. and Tabakoff, W. Erosion Prediction in Turbomachinery Resulting from Environmental Solid Particles. Journal of Aircraft, 12(5), 471–478, 1975.
[70] Gerhold, M. and Bore, C. Experimental and theoretical investigation into ingestion of debris into air intakes by vortex action, 1984, British Aerospace report, BAe-KGT-R-GEN-01309
[71] Gourgy, J.S. and Bosnyakov, S.M. Sand Particles Motion Simulation in the Vicinity of Turbine Engine Intake, 4th International Conference ERBA-2018, 3–5 July 2018.
[72] Gourgy, J.S. Sand Particles Motion in the Vicinity of Turbine Engine Intake, M. Sc. Thesis, MIPT, Moscow State university, 2018.
[73] EL-Sayed, A.F., Gobran, M.H., and Hassan, H.Z. Erosion of an Axial Transonic Fan due to dust ingestion,. American Journal of Aerospace Engineering, 82(1–1), 47–63, 2015.
[74] Montgomery, J. E. and Clark, J. M., Jr. Dust Erosion Parameters for a Gas Turbine, Soc. of Automotive Engineers Summer Meeting, 1962, Preprint 538A.
[75] CF6-50 turbofan engine repair manual, Egypt Air Company.

[76] Elfeki, S. and Tabakoff, W. Particulate Flow Solutions Through Centrifugal Impeller With Two Splitters, 86-GT-130, ASME 1986 International Gas Turbine Conference and Exhibit, June 8–12, 1986, Dusseldorf, West Germany.

[77] Al-Motawaq, W.I.A. Aerodynamics of Solid Particulate Flow in a Centrifugal Compressor, M.Sc., Cairo University, 2007.

[78] Al-Motawaq, W.I.A., El-Sayed, A.F., and Abdel Rahman, M.M. Erosion of back-swept centrifugal compressor. Ain Shams Journal of Mechanical Engineering (ASJME), 2, 109–120, October 2009.

[79] Kurz, R., Meher-Homji, C., Brun, K., Moore, J.J., and Gonzalez, F. Gas turbine performance and maintenance. In Proceedings of the 42nd Turbomachinery Symposium; Texas A&M University: College Station, TX, USA, 2013.

[80] De Pratti, G.M. Aerodynamical Performance Decay Due to Fouling and Erosion in Axial Compressor for GT Aeroengines, E3S Web of Conferences 197, 11002 (2020) 75° National ATI Congress, https://doi.org/10.1051/e3sconf/202019711002

[81] Natole, R. Gas Turbine Components-Repair or Replace, in IGTI Global Gas Turbine News, May/June 1995, 4–7.

[82] Balan, C. and Tabakoff, W. A Method of Predicting the Performance Deterioration of a Compressor cascade due to Sand Erosion. In Proceedings of the Aerospace Sciences Meeting, Reno, NV, USA, January 10–13, 1983.

[83] Schmücker, J. and Schäffler, A. Performance deterioration of axial compressors due to blade defects. In AGARD.

[84] Fasching, W.A. CF6 Jet Engine Performance Improvement-Program: TASK 1 Feasibility Analysis, NASA CR-159450, March 1979.

[85] Shukla, K., Rane, R., Alphonsa, J., Maity, P., and Mukherjee, S. Structural, mechanical and corrosion resistance properties of Ti/TiN bilayers deposited by magnetron sputtering on AISI 316L. Surface and Coatings Technology, 324, 167–174, 2017.

[86] AG, S. Advanced Compressor Coating for Siemens Gas Turbines, in, Siemens AG. Available online: https://assets.new.siemens.com/siemens/assets/api/uuid:3d6979ec63a_8183e9dc3946e330bf9297534bd/version:1541967370/acc_e50001-g520-a445-x-7600enlrfinal.pdf

[87] Hussein, M.F. The Dynamic Characteristics of Solid Particles in Particulate Flow in Rotating Turbomachinery. Ph.D. Thesis, University of Cincinnati, 1972.

[88] Aircraft Propeller Maintenance, US Department of Transportation, Federal Aviation Administration FAA, Advisory Circular, AC 20-37E Date: September 9, 2005.

[89] Disbrow, M. Propellers are subject to wear, fatigue, corrosion and erosion, all of which can lead to failure if not kept in check, AVIATIONPROS, May 1, 1998, www.aviationpros.com/home/article/10389115/propellers-are-subject-to-wear-fatigue-corrosion-and-erosion-all-of-which-can-lead-to-failure-if-not-kept-in-check

[90] Flight Safety and Volcanic Ash Risk: Management of flight operations with known or forecast volcanic ash contamination, ICAO Doc 9974 AN/487, first Edition-2012.

[91] Weinzierl, B., Sauer, D., Minikin, A., Reitebuch, O., Dahlkötter, F., Mayer, B., Emde, C., Tegen, I., Gasteiger, J., Petzold, A., Veira, A., Kueppers, U., and Schumann, U. On the visibility of airborne volcanic ash and mineral dust from the pilot's perspective in flight. Physics and Chemistry of the Earth. 45–46, 87–102, 2012.

[92] Blake, D.M., Wilson, T.M., and Stewart, C. Visibility in airborne volcanic ash: considerations for surface transportation using a laboratory-based method, Nat Hazards, 2018, https://doi.org/10.1007/s11069-018-3205-3.

[93] Casadevall, T.J. Volcanic ash and aviation safety: Proceedings of the first international symposium on volcanic ash and aviation safety, volume 2047, 1994.

[94] Vogel, A. Volcanic Ash: Properties, Atmospheric Effects, and Impacts on Aero-Engines. Ph. D. Thesis, Department of Geosciences University of Oslo, 2018.
[95] Vogel, A., Durant, A.J., Cassiani, M., Clarkson, R.J., Slaby, M., Diplas, S., and Kruger, K. Simulation of Volcanic Ash Ingestion Into a Large Aero Engine: Particle–Fan Interactions. Journal of Turbomachinery, ASME January 2019, vol. 141/ 011010-1.
[96] Vogel, A., Clarkson, R., Durant, A., Cassiani, M., and Stohl, A. Volcanic ash ingestion by a large gas turbine aeroengine: fan-particle interaction, Geophysical Research Abstracts, Vol. 18, EGU2016-15419, 2016 EGU General Assembly 2016.
[97] Bragg, M.B. Ice Accretion and its Effect on Airfoil Performance, NASA Contractor Report 16559, March 1982.
[98] Bons, R. Prenter, and S. Whitaker, A Simple Physics-Based Model for Particle Rebound and Deposition in Turbomachinery, Journal of Turbomachinery, ASME August 2017, Vol. 139, / 081009-1 to 081009-12.
[99] Jombo, G., Pecinka, J., Sampath, S., and Mba, D. Influence of Fouling on Compressor Dynamics: Experimental and Modelling Approach, Journal of Engineering for Gas Turbines and Power, Transaction of the ASME, March 2018, 140(3), 032603, Paper No: GTP-17-1202, https://doi.org/10.1115/1.4037913
[100] Bojdo, N. and Filippone, A. A Simple Model to Assess the Role of Dust Composition and Size on Deposition in Rotorcraft Engines, Aerospace 2019, 6, 44; doi:10.3390/ aerospace6040044, www.mdpi.com/journal/aerospace
[101] Zuñiga, M.O.V. Analysis of Gas Turbine Compressor Fouling and Washing Online, Ph.D. Thesis, Cranfield University, School Of Engineering, 2007.
[102] Yang, H. and Xu, H. The New Performance Calculation Method of Fouled Axial Flow Compressor, The Scientific World Journal Volume 2014, Article ID 906151, 2014.
[103] Yang, H. and Xu, H. Numerical Simulation of Gas-Solid Two Phase Flow in Fouled Axial Flow Compressor, Conference: ASME Turbo Expo 2014: Turbine Technical Conference and Exposition, June 2014, DOI: 10.1115/GT2014-26365
[104] Tarabrin, A.P., Schurovsky, V.A., Bodrov, A.I., and Stalder, J.-P. Influence of axial compressor fouling on gas turbine unit performance based on different schemes and with different initial parameters, 98-GT-416, Proceedings of the ASME Turbo Expo, edit by ASME, NY (1998).
[105] Drew, J. Fatal MV-22 crash in Hawaii linked to excessive debris ingestion. FlightGlobal, 25 November 2015. www.flightglobal.com/news/articles/fatal-mv-22-crash-in-hawaii-linked-toexcessive-debr-419484/ (accessed on December 6, 2018).
[106] Wang, J., Tian, K., Zhu, H., Zeng, M. and Sundén, B. Numerical investigation of particle deposition in film-cooled blade leading edge, Numerical Heat Transfer, Part A: Applications, Taylor & Francis, https://doi.org/10.1080/10407782.2020.1713692
[107] Ai, W.G and Fletcher, T.H. Computational analysis of conjugate heat transfer and particulate deposition on a high-pressure turbine vane, Journal of Turbomachinery, 134(4), 041020, July 2012. DOI: 10.1115/1.4003716
[108] Singh, K., Premachandran, B., Ravi, M.R., Suresh, B., and Vasudev, S. Prediction of film cooling effectiveness over a flat plate from film heating studies, Numerical Heat Transfer, Part A: Applications, 69(5), 529–544, November 2016. DOI: 10.1080/ 10407782.2015.1090232.
[109] Ardey, S. 3D-Messung des stromungsfeldes um die filmgekuhlte vorderkante einer referenzschaufel, Ph.D. thesis, University of the Armed Forces, Munich, Germany, 1998.
[110] Senior, C.L. and Srinivasachar S. Viscosity of ash particles in combustion systems for prediction of particle sticking. Energy and Fuels, 9, 277–283, 1995.

[111] Seggiani M. Empirical correlations of the ash fusion temperatures and temperature of critical viscosity for coal and biomass ashes. Fuel, 78, 1121–1125, 1999.
[112] Van der Zwaag, S. and Field, J.E. Rain erosion damage in brittle materials. Engineering Fracture Mechanics, 17, 367–379, 1983.
[113] Air Force Research Laboratory (AFRL), Materials and Manufacturing Directorate, Supersonic Rain Erosion (SuRE) Test Apparatus Use Policies, Operating Procedures & Specimen Configurations, University of Dayton Research Institute, September 2020.
[114] Elhadi Ibrahim, M. and Medraj, M. Water Droplet Erosion of Wind Turbine Blades: Mechanics, Testing, Modeling and Future Perspectives, Materials, 13(1), 157, 2020; https://doi.org/10.3390/ma13010157
[115] Gohardani, O. Impact of erosion testing aspects on current and future flight conditions. Progress in Aerospace Sciences, 47, 280–303, 2011.
[116] Burson-Thomas, C.B., Wellman, R., Wood, R., and Harvey, T. Importance of surface curvature in modelling droplet impingement on fan blades. Journal of Engineering for Gas Turbines and Power, 141, 031005, 2019 .
[117] Chapman, B. History of jet fighters. In Global Defense Procurement and the F-35 Joint Strike Fighter; Palgrave Macmillan: Cham, Switzerland, 2019.
[118] Wahl, N.E. Investigation of the Phenomena of Rain at Subsonic and Supersonic Speeds, Technical Report AFML-TR-65-330, October 1965.
[119] Thiruvengadam, A. "Theory of Erosion," Proceedings of Second Meersburg Conference on Rain Erosion and Allied Phenomena, A.A. Fyall, and R. King, Eds., Royal Aircraft Establishment, Farnborough, England, 1967, pp. 605–649.
[120] Springer, G.S. Erosion by Liquidj Ipact, Scripta Publishing Company, John Wiley & Sons Washington, DC, 76, pp. 264.
[121] Adler, W.F. The Mechanics of Liquid Impact. Treaties on Science and Technology, Vol. 16, Academic Press, 1978.
[122] F. Durst D. Miloievic B. Schönung, Eulerian and Lagrangian predictions of particulate two-phase flows: a numerical study, Applied Mathematical Modelling, 8(2), April 1984, 101–115.
[123] Decker, R., and Shafer, C.F. Mixing and demixing processes in multiphase flows with application to propulsion systems. NASA CP-3006; 1989.
[124] Zhenlong, W. and Yihua, C. Numerical simulation of airfoil aerodynamic performance under the coupling effects of heavy rain and ice accretion, Advances in Mechanical Engineering, 8(10), 1–9, 2016.
[125] Haines, P.A. and Lures, J.K. Aerodynamic Penalties of Heavy Rain on a Landing Aircraft, NASA CR 156885, July 1982.
[126] Zhenlong, W., Cao, Y., and Ismail, M. Numerical Simulation of Airfoil Aerodynamic Penalties and Mechanisms in Heavy Rain, International Journal of Aerospace Engineering Volume 2013, Article ID 590924, Hindawi Publishing Corporation, http://dx.doi.org/10.1155/2013/590924
[127] Vukits, T.J. Overview and Risk Assessment of Icing for Transport Category Aircraft and Components, AIAA 2002-0811.
[128] What Are Clouds? December 19, 2017, www.nasa.gov/audience/forstudents/k-4/stories/nasa-knows/what-are-clouds-k4.html
[129] What Are Clouds? April 13, 2011, www.nasa.gov/audience/forstudents/5-8/features/nasa-knows/what-are-clouds-58.html
[130] The Types of Clouds and What They Mean www.jpl.nasa.gov/edu/learn/project/the-types-of-clouds-and-what-they-mean/
[131] Clouds, https://en.wikipedia.org/wiki/Cloud#/media/File:Cloud_types_en.svg

[132] Reinmann, J.J., Shaw, R.J., and Ranaudo, R.J. NASA's Program on Icing Research and Technology. Flight in Adverse Environmental Conditions, AGARD CP-470, 1989 (Also, NASA TM-101989).

[133] Potapczuk, M.G. and Reinmann, J.J. Icing Simulation: A Survey of Computer Models and Experimental Facilities, AGARD-CP-496, Effects of Adverse Weather on Aerodynamics, December 1991.

[134] Yua, C., Kea, P., Yub, G., and Yang, C. Investigation of water impingement on a multi-element high-lift airfoil by Lagrangian and Eulerian approach, Propulsion and Power Research, 4(3), 161–168, September 2015.

[135] Beaugendre, H., Morency, F., and Habashi, W., FENSAP-ICE's Three-Dimensional In-Flight Ice Accretion Module: ICE3D. Journal of Aircraft, 40(2), 2003.

[136] Pueyo, A., Chocron, D., and Kafyeke, F. Improvements to the Ice Accretion Code CANICE, 8th Aerodynamic Symposium of 48th CASI Conference, Toronto, Canada, April 29–May 2, 2001.

[137] Yiqiang, H. Theoretical and Experimental Study of Scaling Methods for Rotor Blade Ice Accretion Testing, M.Sc. Thesis, Penn State Univ., August 2011.

[138] Bidwell, C.S. and Potapczuk, M.G. User's Manual for the NASA Lewis Three-Dimensional Ice Accretion Code (LEWICE3D), Tech. Report, NASA Glenn Research Center, 1993.

[139] Wright, W.B. LEWICE 2.2 Capabilities and Thermal Validation, AIAA Paper 2002-0383, 2002.

[140] Wright, W.B. Validation Results for LEWICE 3.0, 2005, NASA/CR, AIAA-2005-1243

[141] Kim, J.W., Garza, D.P., Sankar, L.N., and Kreeger, R.E. Ice Accretion Modeling using an Eulerian Approach for Droplet Impingement, 51st AIAA Aerospace Sciences Meeting including the New Horizons Forum and Aerospace Exposition 07 – January 10, 2013, Grapevine (Dallas/Ft. Worth Region), Texas, AIAA 2013-0246.

[142] Villedieu, P., Trontin, P., Guffond, D., and Bobo, D. SLD Lagrangian Modeling and Capability Assessment in the Frame of ONERA 3D Icing Suite, AIAA Paper 2012-3132, June 2012, doi: 10.2514/6.2012- 3132.

[143] Trontin, P., Blanchard, G., Kontogiannis, A., and Villedieu, P. Description and Assessment of the New ONERA 2D Icing Suite IGLOO2D," AIAA Paper 2017-3417, June 2017, doi: 10.2514/6.2017-3417.

[144] Henry, R. Development of an electro-thermal deicing/anti-icing model. Chatillon Cedex, Office National d'Etudes et de Recherches Aérospatiales, 1992. Rapport ONERA, TAP-92005.

[145] Guffond, D. and Brunet, L. Validation du programme bidimensionnel de captation. Châtillon Cedex, Office National D'Études et de Recherches Aérospatiales, 1988. ONERA Rapport Technique. RP 20/5146 SY, 52 p.

[146] Gent, R.W., Ford, J.M., Moser, R.J., and Miller, D.R. SLD research in the UK, SAE international No.2003-01-2128, 2003.

[147] Gent, R.W. TRAJICE2—a combined water droplet trajectory and ice accretion prediction program for aerofoils, RAE TR-90054, November 1990.

[148] Heinrich, R., Zumwalt, P., Padmanabhan, Thompsom, and Riley. Aircraft Icing Handbook (Volumes 1–3), FAA Technical Report DOT/FAA/CT–88/8–1, September 1993.

[149] Experimental study of supercooled large droplet impingement effects, DOT/FAA/AR-03/59, September 2003.

[150] Cebeci, T. et al. Prediction of Post-Stall Flows on Airfoils. Fourth Symposium on Numerical and Physical Aspects of Aerodynamic Flows, California State University, 1989.

[151] Shin, J. et al. Prediction of Ice Shapes and Their Effect on Airfoil Performance. AIAA Paper 91-0264, January 1991. (Also, NASA TM-103701.)
[152] Papadakis, M., Rachman, A., Wong, S-C., Yeong, H-W., Hung, K.E., Vu, G.T., and Bidwell, C.S. Water Droplet Impingement on Simulated Glaze, Mixed, and Rime Ice Accretions, NASA/TM—2007-213961, October 2007.
[153] Badry, A.H., El-Sayed, A.F., Salem, G.B., and El-Banna, H.M. Computational investigation (simulation) of flow field and aerodynamic analysis over NACA 23012 airfoil with leading edge glaze ice shape, ICFD11-EG-4061, Proceedings of ICFD11: Eleventh International Conference of Fluid Dynamics, December 19–21, 2013, Alexandria, Egypt.
[154] Badry, A.H. FlowFIeld simulation and aerodynamic performance analysis of iced aerofoils, M.Sc. Thesis, Aerospace Engineering, Faculty of Engineering, Cairo University, Giza, Egypt, 2013.
[155] El-Sayed, A.F. and Hassan El-Hady Hassan Fayed, Icing of the intake of a high bypass ratio turbofan engine, Zagazig University, 2007.
[156] Hassan El-hady Hassan Fayed, Icing Problems in The Aircraft Engine's Intakes, M.Sc. thesis, Zagazig University, 2008.
[157] El-Sayed, A. and Emeara, M.S. Intakes of Aero-Engines: A Case Study, January 2016, Conference: (ICESA) International Conference of Engineering Sciences and Applications, Aswan, Egypt.
[158] Veres, J.P. and Jorgenson, P.C.E. Modeling Commercial Turbofan Engine Icing Risk with Ice Crystal Ingestion, AIAA 2013-2679, 5th AIAA Atmospheric and Space Environments Conference, June 24–27, 2013, San Diego, CA.
[159] May, R.D., Guo, T.H., Veres, J.P., and Jorgenson, P.C.E. Engine Icing Modeling and Simulation (Part 2): Performance Simulation of Engine Rollback Phenomena, NASA/TM—2011-217200
[160] Das, K. Numerical Simulations of Icing in Turbomachinery, Ph.D. Thesis, University of Cincinnati, 2006.
[161] Claus, R.W., Beach, T., Turner, M., Siddappaji, K., and Hendricks, E.S. Geometry and Simulation Results for a Gas Turbine Representative of the Energy Efficient Engine (EEE), NASA/TM—2015-218408, 2015.
[162] Paraschivoiu, I. and Saeed, F. Aircraft icing. Wiley-Interscience publication, 2004.
[163] Laoh, T. Numerical simulation of ice accretion on a propeller blade, University of Twente, https://essay.utwente.nl/69245/1/Report.pdf
[164] Ozcer, I., Baruzzi, G., Desai, M., and Yassin, M., Numerical simulation of aircraft propeller icing with explicit coupling, SAE Technical Paper 2019-01-1954, 2019, https://doi.org/10.4271/2019-01-1954
[165] Korkan, K.D., Dadone, L., and Shaw, R.J. Performance Degradation of Propeller/ Rotor Systems Due to Rime Ice Accretion. AIAA Paper 82-0286, January 1982.
[166] Korkan, K.D., Dadone, L., and Shaw, R.J. Performance Degradation of Helicopter Rotor Systems in Forward Flight Due to Rime Ice Accretion. Journal of Aircraft, 22(8), 713–718, August 1985.
[167] Gent, R. and Cansdale, J. The development of mathematical modelling techniques for helicopter rotor icing, Proceedings of the 23rd aerospace sciences meeting (report no. AIAA-85-0336), Reno, NV, January 14–17, 1985. Reston, VA: American Institute of Aeronautics and Astronautics (AIAA).
[168] Zhao, G.Q., Zhao, Q.J., and Chen, X. New 3-D ice accretion method of hovering rotor including effects of centrifugal force. Aerospace Science and Technology, 48, 122–130, 2016.
[169] Chen, X. and Zhao, Q.J. Numerical simulations for ice accretion on rotors using new three-dimensional icing model. Journal of Aircraft, 54, 1428–1442, 2017.

[170] Zhengzhi Wang, Ning Zhao and Chunling Zhu. Numerical simulation for three-dimensional rotor icing in forward flight, Advances in Mechanical Engineering, 10, 1–12, 2018.
[171] SAE Aerospace. Ice, Rain, Fog, and Frost Protection, AIR1168/4, Society of Automotive Engineers, Inc, PA, 1989.
[172] Roskam, J. Airplane Design. Roskam aviation and engineering corporation Part 4, Ottawa, Kansas, 1985.
[173] "Electro-thermal ice protection system for the B-787," Aircraft Engineering and Aerospace Technology, 79(6), 2007.
[174] Sreedharan, C., Nagpurwala, Q.H., and Subbaramu, S. Effect of Hot Air Jets from a Piccolo Tube in Aircraft Wing Anti-Icing Unit, MSRUAS-SASTech Journal, 13(2), 2011.
[175] Hassaani, A., Elsayed, A.F., and Khalil, E.E. Numerical investigation of thermal anti-icing system of Aircraft wing. International Robotics & Automation Journal, 6(2), 60–66, 2020.
[176] Mohamed, A.H.B. Numerical Investigation of Thermal Anti Icing System Configuration of Aircraft Wing, Ph.D. Thesis, Cairo University, 2020.

Index

G

H

I

J

K

L

M

N

O

P

Q

R

S

T

Printed in the United States
by Baker & Taylor Publisher Services